AF540480

Advances in
Soil Borne Plant Diseases

nipa
Browse Subject
eBooks / eChapters / Articles
eBooks
Explore Now
scan for catalogue
Browse, Search, Read & Buy...
Subject Catalogues
eChapters
Publishers
Print Books
Forthcoming
eBooks
New Titles
NIPA
GENX
ONLINE RESOURCES
Publishing Books and Journals
Ebooks and e articles, Current Affairs
Reasoning, Logic and Aptitude
Competitive Examination Preparation
Language Learning and Development Programme
Document Quality Checker and Improvement Tool
Effective Public Speaking, Presentation and Interpersonal Skills
Online Programmes for Professional Development
Personality Development and Human Values
PAY USING
UPI
PayPal

Advances in Soil Borne Plant Diseases

Editors

M.K. NAIK

Professor & Head, Department of Plant Pathology

G.S. DEVIKA RANI

Department of Plant Pathology

University of Agricultural Sciences, Dharwad

College of Agriculture, Raichur - 584 101, Karnataka, India

NEW INDIA PUBLISHING AGENCY

New Delhi 110 088

NEW INDIA PUBLISHING AGENCY
101, Vikas Surya Plaza, CU Block, LSC Market
Pitam Pura, New Delhi – 110 034, India
Email: info@nipabooks.com
Web: www.nipabooks.com

For customer assistance, please contact
Phone: + 91-11-27 34 17 17 Fax: + 91-11- 27 34 16 16
E-Mail: feedbacks@nipabooks.com

© 2008, Editors

ISBN : 978-81-89422-81-3

All rights reserved. No part of this publication may be reproduced, stored in a retrieval system or transmitted in any form or by any means, including electronic, mechanical, photocopying recording or otherwise without the prior written permission of the publisher or the copyright holder.

This book contains information obtained from authentic and highly reliable sources. Reasonable efforts have been made to publish reliable data and information, but the authors, and publisher cannot assume responsibility for the validity, accuracy or completeness of all materials or information published herein or the consequences of their use. The work is published with the understanding that the publisher and authors are not attempting to render any professional services. The authors, and publisher have attempted to trace and acknowledge the copyright holders of all material reproduced in this publication and apologize to copyright holders if permission and/or acknowledgements to publish in this form have not been taken. If any copyrighted material has not been acknowledged, please write to us and let us know so that we may rectify the error, in subsequent reprints.

Trademark Notice: NIPA, the NIPA logos and their presentations (the way they are written/presented) in this book are the trademarks of the publisher and hence may not be used without written permission, if copied or used without authorization, the infringer will be prosecuted as per law.

NIPA also publishes books in a variety of electronic formats. Some content that appears in print may not be available in electronic books, and vice versa

Composed and Designed by NIPA.

Dedicated to the great teachers and researchers in Plant Pathology

S. D. Garret

R. K. Hegde

S. Nagarajan

Y. L. Nene

Bineeta sen

M. V. Reddy

भारत सरकार
पौधा किस्म और कृषक अधिकार संरक्षण प्राधिकरण
एन.ए.एस.सी. काम्प्लैक्स, (टोड़ापुर के सामने)
डी.पी.एस. मार्ग, नई दिल्ली - 110012 (भारत)
Government of India
Protection of Plant Varieties and Farmer's Rights Authority
N.A.S.C. Complex, (Opp. Village Todapur)
DPS Marg, New Delhi-110012 (India)
दूरभाष/Telphone : 25840777
फैक्स/Fax : 011-25840478
website : www.plantauthority.in

FOREWORD

With great pleasure I browsed through the book entitled "Advances in Soil Borne Plant Diseases edited by Dr. M. K. Naik, Professor and Head and Ms. Devika Rani. G. S. of the Department of Plant Pathology, UAS, College of Agriculture, Raichur, Karnataka. Various aspects of soil borne plant diseases covered in the book on the recent developments in the field of molecular detection, genomics, plant growth promoting rhizobacteria mediated strategy, advances in use of biological control, *Trichoderma,* commercial production, formulation, registration and quality control and use of *Pseudomonas* are extensively presented to enlighten the readers. Several specific and non-specific soil borne pathogens such as *Fusarium, Sclerotium, Verticillium, Rhizoctonia* and *Phytophthora* have been dealt with elegantly. Adequate focus is on wheat, cotton and temperate fruits that badly suffer from soil borne diseases. Management of endoparasitic nematodes such as *Meloidogyne* spp and *Heterodera* spp, the hidden vectors and parasites have been adequately covered along with other group of nematodes. Solarizations, use of mycorrhizae as a biological deterrent of plant pathogens have been duly emphasized.

This "Advances in Soil Borne Plant Diseases" would serve as a comprehensive and critical review on different facets of soil borne plant diseases. The contributors of several chapters and specially the editors deserve the appreciation for brining the information on a single platform.

Hopefully, this book would serve as a reference source and help students, teachers, researchers and extension workers dealing with soil borne plant diseases. I compliment Dr. M. K. Naik and Ms. Devika Rani. G. S. for their efforts and keen desire to provide yet another aid in understanding soil borne plant pathogens.

Dr. S. Nagarajan

Chair person
Protection of Plant Varieties &
Farmer's Rights Authority

भारतीय कृषि अनुसंधान संस्थान, नई दिल्ली – 110 012

INDIAN AGRICULTURAL RESEARCH INSTITUTE
NEW DELHI - 110 012 (INDIA)

डा० एस. ए. पाटिल
निदेशक
Dr. S.A. Patil
Director

Phone : (Off.) 2573 3367, 2584 3378
(Res.) 2573 3973, 2584 6773
Fax : 91-11-2584 6420
E-mail: director@iari.res.in

MESSAGE

The food production has remained stagnant but the challenge before agriculture scientists is to achieve food and nutritional security to the burgeoning population. The plant protection specialists have to play a pivotal role in feeding the increasing number of hungry mouths. Perhaps, the foliar diseases and the above ground symptoms of the plant were the first to get attention of researchers and as a result they are much better understood than many soil borne diseases. The kind of complexicities and intricacies involved in soil along with the living parameters such as pathogen is beyond comprehension compared to foliar diseases. However, the modern agriculture and horticulture practices encourage monoculture in which the same crop is planted repeatedly leading to increased soil sickness by plant pathogenic fungi and bacteria. While there are many books available on plant diseases dealing with foliar and above ground maladies but there are very few books and volumes exclusively on soil borne plant diseases. In this regard, the initiation taken by the editors of this volume "Advances in Soil Borne Plant Diseases" deserves special appreciation in bringing together the leading experts on one platform to review and synthesize the progress made so far and to evolve the future strategies on soil borne diseases.

There are seventeen chapters in the present volume dealing with various facets of soil borne plant diseases. The book begins giving a fine overview of many soil borne plant pathogenic genera highlighting the ecological principles, the importance of disease resistance and rhizosphere manipulation. The genera of soil borne fungi such as *Sclerotium, Rhizoctonia, Phytophthora, Fusarium, Verticillium* and endoparasitic nematodes are extensively dealt with. Advances made in bio-control particularly *Trichoderma* and *Pseudomonas,* their commercial mass production, registration, quality control and the priorities for research are discussed in a befitting way. The technological feasibility of soil solarization its refinement and integration with other techniques, use of mycorrhizae for eco-friendly, sustainable management of diseases are extremely important and relevant topics presented in detail. The philosophy of integrated management of diseases and their long term effects in bringing down the inoculum below target level are given due consideration in the volume. Major soil borne diseases of important crops namely *viz.*, temperate fruit crops, wheat and cotton have been dealt at length.

The book provides a very comprehensive and critical review on different aspects of soil borne plant pathogenic fungi including the molecular detection and characterization. Of late, such molecular tools have an immense potential for detection of soil borne fungi wherever they are not accessible for morphological detection. The genomics of some important bio-control agent such as *Pseudomonas fluorescens* has been emphasized along with gene based detection of antibiotic production which is essential to further explore the bio-control potentiality. We are in an exciting era of biological revolution where genes can be transferred horizontally from one species to another. Inclusion of chapters on genomics of bio-control agents, PGPR mediated resistance, molecular detection *etc.* are very relevant and timely to the current trends of plant disease management applicable to soil borne plant pathogens. I hope the efforts of the editors will certainly enlighten the readers on soil borne plant diseases. I am extremely happy that many authors have consulted the recent literature in presenting various chapters. My heartiest congratulations to the editors for taking all the pain in bringing out this volume on poorly understood group of pathogens in a most presentable form for the readers. The teachers, researchers and students are likely to be benefited by this volume in evolving newer dimensions for research on soil borne plant pathogens. My felicitation to the editors for the wonderful job they have done by presenting this volume.

(S.A. PATIL)

Director
Indian Agricultural Research Institute
New Delhi

PREFACE

Ever growing population, the increased awareness for quality food, dwindling land resource, environmental pollution of mother planet, resistant to chemicals in pathogen population are some of the grave problems faced by the mankind. Need of food for billions in future is a challenge to agricultural professionals including plant pathologists. Further exploitation of hybrid vigor will be a difficult venture in many crops. In this connection, plant protection specialists have to play a vital role in protecting plants and avoiding food losses.

Most obvious diseases such as foliar diseases were studied first unlike their counterparts; soil borne plant diseases which are generally less understood. The soil is much more complex chemically, biologically and physically than the atmosphere above the ground. There exist many intricacies and complexities in rhizosphere, fungistasis, microbial interaction and bio-control processes to a greater extent in soil borne plant pathogens compared to aerial/foliar pathogens. Soil borne plant diseases will continue to be a hazard both in field and horticultural crops, which have been documented through out the history of mankind. Monoculture has become the feature of modern agriculture. Crop plants tend to be produced commercially where they are best adopted and are planted repeatedly in such sites. Crops such as banana, tomato, cotton, tobacco, cucurbits, pigeonpea, chickpea and wheat have suffered the onslaught of soil borne plant pathogens. If a soil borne pathogen is present and favored by that particular environment, the soil infestation increases rapidly. In the absence of proper management much of the best suited land for the crop is lost.

This book makes an attempt to bring together information on soil borne plant diseases which include root rot, crown or collar rot, damping off, wilts, blights in field and horticulture crops. Soil borne plant diseases taken together cause a tremendous loss in world agriculture. Their destructiveness is basically due to their ability to survive for several years in soil as dormant structures like sclerotia, chlamydospores, macroconidia and resting mycelia and they withstand the adverse environmental conditions. Due to substantial losses caused by soil borne plant diseases a thorough understanding of their biology, perpetuation, population dynamic, ecology and molecular aspects of soil borne plant pathogens is essential.

Understanding the nature of soil borne pathogens will also help in determining what opportunities are available for its management. Mode of their survival and dissemination, effect of environment on diseases, problems of host resistance, the role of cultural practices, biological control and the use of novel molecular approaches wherever appropriate are some of the aspects dealt in this volume. The book contains seventeen chapters covering various facets of soil borne plant diseases.

The first chapter gives an overview of soil borne plant pathogens such as *Fusarium, Verticillium, Pythium, Rhizoctonia, Sclerotinia, Sclerotium, Phytophthora and Macrophomina*. The chapter also highlights the ecological and epidemiological tools essential and various approaches for management including rhizosphere manipulation and disease resistance. The book includes three separate chapters on important genera of fungi namely *Rhizoctonia, Sclerotium* and *Phytophthora* (chapters 2nd, 4th and 16th). *Rhizoctonia solani,* a soil inhabiting non-specialized pathogen without any exploitable resistance available among crop species, can only permit some control through cultural practices disadvantaging pathogen by reducing inoculum potential. The chapter on *Sclerotium rolfsii* deals about the morphological and genetic variations, sclerotial biogenesis, mycelial compatability group, recovery of sclerotia and their management. Molecular detection and characterization of soil borne fungal plant pathogens by using chromosome and nucleic acid based technique has been included in the third chapter. Since serological methods are difficult in fungi due to their complex and variable nature of antigens, the techniques like DNA probes can be used in those fungi which are not accessible for morphological detection. Plant growth promoting rhizobacteria mediated disease management strategy has been dealt in chapter five, where as a discussion on genomics of *Pseudomonas fluorescens*, a versatile and most worked out PGPR, along with its biocontrol potential and gene based detection of antibiotic production has been highlighted in sixth chapter. Considering the importance of crops such as wheat, temperate fruits and cotton with respect to soil borne diseases, the editors have included 7th, 8th and 9th chapters respectively on these crops. Soil borne diseases have become a matter of great concern even in crops like wheat due to monoculture and hence major soil borne diseases of wheat such as karnal bunt, blights, flag smut, *Fusarium* head scab, foot and root rot diseases are described in a separate chapter. Cotton has suffered a great loss due to soil borne disease since early days of its cultivation and hence, wilt diseases of cotton have been described in chapter nine. Endoparasitic nematodes particularly *Meloidogyne* spp. and *Heterodera* spp., which are our hidden enemies, have found a place in chapter eleven with emphasis on non-chemical management strategies. Lesion nematode, *Pratylenchus* species, their biology and management has been given adequate attention through chapter fourteen. Use of on-farm wastes such as cruciferous residues combined with sublethal heating to weaken the survival structures of fungi for managing soil borne plant diseases particularly in hot arid region is extensively covered in 12th chapter. The last decade has witnessed a tremendous break-through in research efforts on bio-control especially using *Trichoderma* species. The potential of *Trichoderma* species, various diseases it can control, improvement of *Trichoderma*, the mechanism involved, induced resistance, commercial mass production, formulation, registration, quality control, R and D priorities are reviewed in chapter thirteen. Soil solarization is gaining popularity in more than fifty tropical countries of the world as a viable technology due to growing concern about pesticide residue in food and drinking water. The fifteenth chapter dwells on the principles of solar heating, mechanisms involved, other beneficial effects, limitations and refinement of solarization technology for management of soil borne diseases. Mycorrhizae are equally gaining importance in suppressing soil borne plant pathogens which are essential in sustainable disease management, which has been presented in chapter ten. Lastly, a chapter on integrated management of soil borne diseases has been added emphasizing the philosophy of IDM in an ecologically sound manner with due consideration to economic, social, legislative parameters and usefulness in reducing the quantity and effectiveness of primary inoculum to reach the target of low disease in the crop.

This book is brought out to serve the researchers, extension workers, teachers, post graduate students, scientists and students of general plant pathology, microbiology, microbial ecology, biological control and molecular biology on the current knowledge of soilborne plant diseases. The book chapters with diversified ideas and from different groups are organized systematically to the best of our knowledge.

The publication of this book would not have been possible without the sincere cooperation and the hard work of the contributing authors who are not only specialists in their respective fields and also leaders in their own right from national educational centers, laboratories and universities. Therefore, we have tried to honor their ideas in the original shape. In dealing with such a voluminous task, errors are likely to creep in inspite of our best efforts. However, the onus of the technical contents rests with the contributors. We place on record our reverence and gratitude to all the learned contributors for the cooperation in compiling a very useful information on different aspects of soil borne plant diseases. We are also sure this detailed account on a wide variety of subject concerning soil borne plant diseases would benefit the researchers and teachers for planning future strategies for researching on soil borne plant pathogens.

We are grateful to the Honorable Vice-chancellor, University of Agricultural Sciences, Dharwad and Director of Instruction (Agri.), College of Agriculture, Raichur for providing facilities during the compilation of the book. The senior editor would like to thank his wife Uma, daughter ShreeRaghavi and son ChinmayaKrishna, for their unstinted support, calm inspiration and buoyant temperament in many invisible ways in completion of the task undertaken. The second editor wishes to thank her parents and family members for the kind support and encouragement received during the period. We are extremely thankful to colleagues, staff members and PG students for their help rendered during the preparation of this volume. We gratefully acknowledge the efforts of Mr. G. Earanna for neat typing. We highly appreciate the all-round cooperation and support from the entire staff of New India Publishing Agency, New Delhi for printing and publishing this book with utmost patience and care.

September, 2007
Raichur

M. K. Naik
G. S. Devika Rani

Contents

Foreword *vii*
Message *ix*
Preface *xi*

1. **An Over View of Soil Borne Plant Pathogens** **1-31**
M.K. Naik, G.S. Devika Rani, R.D. Prasad
M.B. Patil and Bineeta Sen

2. **Rhizoctonia solani Kuhn: Soil-Inhabiting Nonspecialized Pathogen in Agroecosystem** **33-53**
Arvinder Kaur and H.S.Chaube

3. **Molecular Detection and Characterization of Soil-Borne Fungal Plant Pathogens** **55-88**
T. R. Sharma and T.K. Jana

4. **Advances in Plant Diseases Caused by Sclerotium rolfsii and Their Management** **89-127**
R. D. Prasad and M. K. Naik

5. **Plant Growth Promoting Rhizobacteria Mediated Strategy for Disease Management** **129-142**
Jayashree Gopalakrishnan

6. **Genomics of Pseudomonas fluorescens and Biological Control of Soil Borne Plant Diseases** **143-164**
Dhananjay Naik, M.S. Rao, Abdul Rahiman Bantwal and M.K. Naik

7. **Soil Borne Diseases of Wheat and their Management** **165-217**
Rashmi Aggarwal, P. Srinivas, K.D. Srivastava and D.V. Singh

8. **Soil Borne Diseases of Temperate Fruits and their Management** **219-231**
J.N. Sharma and R.C. Sharma

9. **Wilt Diseases of Cotton** **233-243**
Kannan, A. and Mayee, C.D

10. **Mycorrhizae as Biological Deterrent of Plant Pathogens** **245-259**
M. N. Sreenivasa

11. **Management of Endoparasitic Phytonematodes Particularly Meloidogyne spp. and Heterodera spp. through Non-chemical Methods** **261-283**
S.K. Patel, N.B. Patel and D.J. Patel

12. **Use of On-farm Wastes in Management of Soil-Borne Diseases in Hot Arid Region** **285-310**
Satish Lodha, Sunil Israel and Ritu Mawar

13. **Biological Control of Soil Borne Plant Pathogens using Trichoderma Species** **311-340**
R. D. Prasad and M. K. Naik

14. **Lesion Nematode, Pratylenchus Species of Pulses, their Biology, Responsible Factors and Management** **341-354**
S.P. Tiwari and M.N. Khare

15. **Soil Solarization by Polyethylene Mulching for the Management of Soil-Borne Diseases** **355-369**
Harender Raj

16. **Integrated Management of Phytophthora Diseases** **371-413**
M. Gunashekaran

17. **Integrated Management of Soil Borne Diseases** **415-427**
V.K. Gupta and R.C. Sharma

Advances in Soil Borne Plant Diseases, 2008
Editors: Naik and Devika Rani, pp. 1-31
New India Publishing Agency, New Delhi

1

An Over-View of Soil Borne Plant Pathogens

[1]M. K. Naik, [1]G.S. Devika Rani, [2]R. D. Prasad
[1]M.B. Patil and [3]Bineeta Sen

INTRODUCTION

Modern agriculture relies more on monoculture, which is highly energy intensive and has a narrow genetic base. This has disturbed the normal ecosystem and sometime results in disease epidemics. The problem becomes much more serious in case of soil borne pathogens, as soil medium is a highly complex living system and is also the habitat of many micro organisms including the pathogens whose activities are influenced by interaction in it. Many serious plant diseases are associated with soil-borne plant pathogens, which cause root rot, crown rot, damping off, blights, fruit decay and wilts in field and horticultural crops. Soil borne plant pathogens, which are most tricky and challenging ones, cause sustainable economic loss. Many important fungal pathogens reside in the soil and infect roots of crops often causing significant yield losses. Soil-borne diseases result from a reduction in biodiversity of soil organisms. Plant diseases occurring in natural environments, rarely run rampant and cause major problems. In contrast,

[1] Department of Plant Pathology, University of Agricultural Sciences, College of Agriculture, Raichur-584 101, Karnataka, India

[2] Directorate of Oilseeds Research, Rajendranagar, Hyderabad-500 030, Andhra Pradesh

[3] Division of Plant Pathology, IARI

the threat of disease epidemics in crop production is constant. The reasons for this are becoming increasingly evident.

Efficient management of soil borne disease is possible only through good knowledge of crop husbandry, physical, chemical and biological properties of soil, ecology and epidemiology of the root disease (Naik, 2003). Therefore this chapter gives an over-view about soilborne plant pathogen, ecological and epidemiological principles and various management options available highlighting the merits and demerits.

SOIL BORNE PLANT PATHOGENS

Fusarium : The greatest number of fungal wilt diseases in higher plants is caused by species of genera *Fusarium*. During the period from 1886 to 1925, *Fusarium* wilt were reported on tomato, crucifers (Smith, 1899), peas, watermelon (Smith, 1899), bananas (Brandes, 1918) etc. Snyder and Hansen (1940) included all species of the genus of *Fusarium* causing vascular wilt. They reduced the entire section into single sp (*Fusarium oxysporium*) based on consistent morphological characteristics and the parasitic members of the species were further categorized into forme and races based on selective pathogenicity. They recognized 25 formae (formae speciales) and 2 races. The number of formae speciales and races of *F. oxysporum* was expanded to include 66 pathogenic Formae speciales and 22 races. Amstrong and Armstrong (1968) proposed a total of 69 formae speciales and 36 races based on studies made for more than 25 years. Booth (1971) mentioned as many as 70 formae speciales. They are found in many areas of tropical and temperate regions. Numerous formae speciales of *Fusarium oxysporum* cause seed rot, damping off and vascular wilts of vegetables, fruits, ornamentals and fiber crops. *Fusarium solani* is non- specific and widely responsible for root rotting, decline in vigour, wilt and mortality in taxonomically diverse groups of plants. Of all soil borne plant pathogens, species of *Fusarium* in particular appears to be the toughest to control as it possess highly resistant chlamydospore, which can survive for many years.

Verticillium : The genus *Verticillium* was established in 1816 by Nees Von Esenbeck based on the characteristic of verticilate conidiophore. As many as 40 species were recognized by Saccardo in 1910. But, the controversy existed concerning the limits of the genus and validity of many species assigned to it. At present, there are five spp that are considered vascular pathogens viz., *V. albo-atrum, V. dahliae, V. nigrescens, V. nubilum* and *V. tricorpus*. Among these, *V. albo-atrum* and *V. dahliae* are considered the most important causing wilt diseases in a wide range of crops and ornamental species. The host range of other three species of *Verticillium* is much more restricted and disease severity is more limited. *Verticillium* wilts are wide spread through out temperate areas of the world. They are especially important in Europe, North America, North Africa, Parts of South America, Russia and Australia. *Verticillum* wilts are not known to occur in many Eastern and South Eastern Asia. *Verticillum* wilt of cotton is an important disease through out cotton growing areas. Potato is also known to be infected by *V. albo-atrum*.

Pythium **:** Species of *Pythium* cause seed rot and damping off of seedlings of all cultivated crops. In India damage is only in summer and rainy seasons, both in glass and open cultivation, in nursery and field. In temperate countries, certain other species of *Pythium* are known to cause loss at low temperatures also. At any temperature high humidity is a predisposing factor.

Rhizoctonia **:** The sclerotial (vegetative) stage of *Rhizoctonia bataticola* (Taub) Butler. which exists in the pycnidial (reproductive) stage as *Macrophamina phaseolina* is an important species infecting 500 plant species. It causes different types of diseases viz., wilt, stalk rot, stem blight, fruit rot, seedling decay and leaf blight in crop plants (Dhingra and Sinclair, 1975). The fungus is particularly prevalent in arid, subtropical and tropical climates. It is recognized as highly valuable pathogen responsible for economically significant damage to a wide variety of crops. *Rhizoctonia solani* is the most destructive plant pathogen occurring globally and causes various Maladies starting from seed decay, damping off, root and stem rots, canker, sheath blight and ear rot in monocots and dicots. Being a pathogen of great diversity with wide host range and lack of sharp differentiation among its specialized strains, *R. solani* poses great difficulty to develop resistance in crop varieties.

Sclerotinia **:** *Sclerotinia sclerotiorium* is an important soil inhabiting pathogen decimating vegetable and ornamental crops at later stages of growth drastically reducing the seed production. It causes a yield loss of 95 per cent in oil seed rape, bean, tomato, sunflower (Conn and Tiwari, 1990). Unlike other fungal genus they are taxonomically well defined by morphological and physiological features. It is hard to kill resting structures oversummering in soil and becoming active at the temperature around 18-20 ^{0}C. It is difficult to reduce the inoculum from sick soil due to wide host range of the pathogen.

Sclerotium **spp :** It is a soil borne plant pathogen which is distributed from temperate to tropical region and cause diseases in a wide variety of plant species belonging to dicotyledonous and a few monocotyledonous species

(Punja, 1988). Isolates of *S. rolfsii, S. coffeicola* and *S. delphinii* show differences in morphology, growth response and enzyme production. These species can be differentiated by colony characteristics such as morphology and size of sclerotia, histochemical staining and composition of sclerotia (Punja and Damiani, 1996). One of the best known members of the genus is *S. rolfsii* which is prevalent in warm temperature and subtropical regions of the world causing diseases in more than 500 plant species.

Phytophthora : It is one of the major genera of soil borne plant pathogenic fungi belonging to oomycetes group. Currently, this genus is included under the phylum oomycota under chromista kingdom. There are about 60 described *Phytophthora* spp. (Erwin and Ribiero, 1996), which are primary invaders of healthy plant tissue with limited saprophytic ability. Some of them have been responsible for economically important diseases such as late blight of potato (*Phytophthora infestans*), quick wilt of pepper (*Phytophthora capsici*) koleroga of arecanut (*Phytophthora palmi* and *Phytophthora meadii*), black shank of tobacco (*P. parasitica* var. *nicotianae*), buck-eye-rot of tomato (*Phytophthora nicotiana* var. *nicotiana*) and *Phytophthora* rot of citrus (*P. cirtophthora*). *Phytophthora* spp are mostly distinguished based on the structure of sporangia and on the basis of whether the taxon is homothallic or heterothallic.

Macrophomina : The genus *Macrophomina* is represented by single species, *phaseolina* which is commonly found on more than 500 plant hosts. It attacks a wide variety of field and vegetable crops besides a range of unrelated plants. It is a warm and dry weather pathogen causing charcoal rot, resulting in 10 - 80 percent loss. The pathogen being difficult to control only cultural practices may to some extent help to avoid the diseases.

ECOLOGY AND EPIDEMIOLOGICAL TOOLS FOR SOIL BORNE PATHOGENS

Quantification of propagules of soil borne plant pathogens

Soil borne fungi survive in the soil in the form of various structures (spores, chlamydospores, sclerotia, hyphal fragments, etc.) that are collectively called 'propagules'. The number of these propagules in any given volume of soil will influence the disease severity or the disease incidence or both. Each of these propagules can act as inoculum and their number in a given volume is sometimes referred to as soil inoculum density. It is important to isolate these pathogens from soil and to quantify them for research into population dynamics of plant pathogens for developing biocontrol strategies, for improving the efficacy of physical and biological methods of management. Studies on population dynamics are also required to understand the suppressiveness of the soils and to isolate specific antagonist from these soils as a biocontrol agent and even also to predict whether it will be worth to plant certain crop in soil.

Studies on rhizosphere population of soil borne plant pathogens in relation to disease incidence and severity are limited. Conventional methods for estimating numbers of microbes in soil such as plating, usually only detect 0.1-10% of the microorganisms in soil. Even if we could detect 100% of the microbes present in soil, we would still be left with tremendous gaps in our understanding of the various processes that take place in the soil, and this stresses the need to examine many factors simultaneously in elucidating those contributing to soilborne diseases. The work on management of soil borne disease should be based on ecological principles namely inoculum density, economic threshold level, quantification, prediction and agronomic manipulation, which is entirely dependent on quantification of innoculum. Events with respect to pathogenesis in soil occur only in the vicinity of nutrient sources (Mitchell, 1979). Thus there is a need to focus attention on microhabitats that are the principal sites of action, but which comprise only a small fraction of the soil mass. Many a times, there is a high concentration of propagules within preferential sites called ecological niches and there are currently no known methods of enumeration of ecological niches. Several methods using light and electron microscopy have been developed for studying microorganisms *in situ* in soil (Foster, 1986). While plating procedures can allow enumeration of pathogen numbers in soil their location can only be determined through the use of microscopy. Calcofluor White M2R New through its non-toxic properties is used to trace the growth of fungal pathogens in soil after first being allowed to take up the compound. Soil has generally proven intractable to electron microscopic studies, using both transmission and scanning methods (Foster, 1986). Such studies using transmission electron microscopy of ultra thin soil sections have shown that there are more soil bacteria in the rhizoplane, all within 5 μm of the root surface. Although the rhizosphere effect can be detected 1000- 3000 mm from the root surface, the majority of studies have been confined to the rhizoplane.

The methods used for estimating inoculum density of soil borne plant pathogens are physical separation of the pathogen from the soil, plating of soil on specific nutrient media and enumeration and introduction of baits into the soil (Buczacki, 1981). However, none of these methods allow the discrimination between parasitic and saprophytic strains of an organism. Plating and baiting procedures are the most widely applicable methods but quantification can be also achieved through the use of the most probable number (MPN) or dilution end point techniques (Menzies, 1963; Buohot, 1979). For MPN, data is required only on the proportion of replicates giving a negative response at each dilution. While dilution end point requires counts of colonies. Caution is needed for interpretation of results, since the mean inoculum density is a crude biological statistic that masks a large amount of stochastic variation (Gilligan, 1983). The major assumptions of MPN are that microorganisms are randomly distributed in the test medium so that number of propagules in aliquots follows a Poisson distribution, and that each propagules is capable of evoking a quick response (growth or infection). Precision can be increased by adjusting the range of dilutions, the dilution ratio and the sample number per dilution. Assessment of inoculum density in naturally infested field soils allows the development of artificial infestation levels for glasshouse disease resistance screening assays which are representative of naturally occurring levels, and also has predictive value for assessing the likelihood of future epidemic development (P fender *et al.,* 1981).

Determination of economic threshold level of inoculum in soil borne fungi

The term "threshold" is variously used and defined in integrated pest management. Literally, threshold indicates the point at which some effect begins (Ferris, 1987). Bruehl (1987) defined that plants tolerate certain population of pathogen with no detectable economic loss. At some threshold density, the losses begin and losses are proportional to increased number of propagules. At some point, the host no longer suffers in proportion to increase in number of pathogen and it is essential to know the threshold level of inoculum as a measure of biological stress imposed by pathogen. Not many soil borne pathogens have been worked out with regards to economic threshold level (ETL) of inoculum. Population of *Fusarium* has a direct impact on severity of wilt (Baker, 1965). Rao and Rao (1963 and 1966) showed that the tempo of disease development on a susceptible variety of cotton against *F. oxysporum* f.sp. *vasinfectum* increased with a rise in the inoculum level from 0.5 per cent to 10 per cent. Douglas (1970) working with *Fusarium* wilt of muskmelon was in favour of inoculum density only over an intermediate range of inoculum concentration, above which the muskmelon plant severely wilted and below which they survived. Netzer (1976) estimated 200-800 propagules of *Fusarium oxysporum* f.sp. *niveum* per q of soil in a field cropped under watermelon and he also noticed that saturation point for infection reached with a population of 300 propagules/g in infested soil kept dry for three months. An attempt was made for the first time to determine the economic threshold level of inoculum of *Fusarium udum* at ICRISAT center, both under field and glasshouse conditions. The ETL was found to be 920 colony forming unit (cfu) per gram soil in vertisol and 830 cfu per gram in alfisol (Naik and Reddy, 1993a). The ETL is an important epidemic tool particularly for soil borne pathogens and is helpful for evaluation of several management measures such as cropping pattern, biocontrol, agronomic manipulation and soil solarization (Naik *et al.*, 1996), etc. Thus, it can be used as decision making tool for management of soilborne pathogens. The ETL has an underlying economic basis and is a function of relationship between cost and stress reduction.

Epidemiological models developed for airborne pathogens are not satisfactory for their soilborne counterparts. Inoculum density, spatial distribution balance between saprophytic and parasitic activities and the infection court were all of greater importance in the soil than in the air. Although many soilborne diseases are monocyclic with no secondary infection within a season, some are demonstrated to be polycyclic. For example, many of the diseases caused by *Phytophthora* spp. are also polycyclic while a few diseases such as caused by cyst nematodes are monocyclic (Gilligan, 1985b). The most common approach to mathematical modeling of disease epidemics has been the application of standard non-linear models such as the monomolecular logistic, Gompertz, Weibull and Richards and Bertslanffy functions (Gilligan, 1985b). It is possible to have biological meaning in these models since amount of disease and the rate of disease can be accounted.

Pre-planting forecasting of soil borne diseases

Disease prediction model is an abstract representation and has been well attempted in foliar diseases. However, quantitative analysis of all significant factors contributing to disease severity in soilborne disease is very challenging and difficult task. Many a times the monocyclic nature of soilborne diseases and complex nature of the soil are the special attributes of soilborne pathogens. An attempt to predict monocyclic epidemic therefore, must account for initial inoculum density. A preliminary attempt was made at ICRISAT by creating a gradient of inoculum in wilt sick plot using differential solarization technique (Naik and Reddy, 1993; Naik *et al.*, 1995). When such gradient of initial inoculum density was used for planting a susceptible cultivar that ended up in a gradient of disease proportional to inoculum. Simple regression of initial inoculum concentration as independent variable on wilt incidence as dependent variable indicated a value of 85 to 94 per cent in pooled analysis (Naik and Reddy 1993b and Naik *et al.*, 1995). Thus, it is possible to give a pre-planting warning of soil borne disease based on the intensity of sickness of the soil for which accurate quantification of inoculum in a given field is most essential.

In all instances of epidemiological modeling, it is essential to collect disease progress data, namely the increase in proportion of plants diseased over time. In generating disease progress data, more meaningful information will be obtained if account can be taken of the spread of lesions within plants (Gilligan, 1985b).

A forecasting system for *Sclerotinia* disease on field grown lettuce based on the response of the pathogen to environmental conditions is currently being developed in a joint project between Warwick HRI and ADAS. The aim of this work is to identify periods of risk so that fungicide sprays can be timed more effectively and unnecessary applications avoided. Germination of *S. sclerotiorum* sclerotia to produce apothecia was found to be governed by soil temperature and moisture while infection by ascospores was controlled by air temperature and humidity. Mathematical equations describing these relationships have now been derived which when incorporated into the Warwick HRI decision support software MORPH, should allow prediction of both these processes using in crop weather data. This information can then be potentially used to determine whether fungicide sprays are required and to optimize their time of application when ascospores are present and environmental conditions are favourable for infection and development of Sclerotinia disease.

APPROACHES TO MANAGEMENT OF SOIL BORNE DISEASES.

Suppressive soils in managing soil borne diseases

A typical teaspoon of native grassland soil would contain between 600 to 800 million individual bacteria that are members of perhaps 10,000 species. There are several miles of fungi, and perhaps 5000 species of fungi per teaspoon of soil. There are 10,000 individual protozoa split into three main groups, i.e., flagellates, amoebae and ciliates, and perhaps 1000 species of protozoa. There are 20 to 30 beneficial nematodes, which are members of as many as 100 species. There are two types of disease suppression, specific and general. Specific suppression results from one organism directly suppressing a known pathogen. These are cases where a biological control agent is introduced into the soil for the specific purpose of reducing disease incidence. The general suppression is the result of a high biodiversity of microbial populations that creates conditions unfavorable for plant disease development.

Soil-borne diseases result from a reduction of biodiversity of soil organisms. Restoring beneficial organisms that attack, repel, or otherwise antagonize disease-causing pathogens will render a soil disease-suppressive. Plants growing in disease-suppressive soil resist diseases much better than in soils low in biological diversity. Beneficial organisms can be added directly, or the soil environment can be made more favourable for them through use of compost and other organic amendments. The compost quality determines its effectiveness at suppressing soil-borne plant diseases. Compost quality can be determined through laboratory testing.

Studies on soil suppressiveness have led to a basic concept of "Soil Receptivity" to diseases (Alabouvette *et al.*, 1982). Every natural soil possesses some ability to suppress the disease. Soils suppressive to some of the most important diseases caused by soil borne plant pathogens are *Pythium* spp., *Phytophthora cinnamomi, Rhizoctonia solani, Thielaviospsis basicola, Gaeumannomyces graminis* var. *tritici* and *Fusarium oxysporum* (Schneider, 1982; Cook and Baker, 1983). Soil suppressive to *Fusarium* wilt limit the incidence of wilt in banana, carnation, cucumber, cotton, flax, muskmelon and tomato. In spite of diversity of crops this suppressiveness is specific to *Fusarium* wilt which is expressed against all formae specials of *Fusarium oxysporum* but ineffective against diseases caused by other soil borne pathogens (Toussoun, 1975; Alabouvette *et al.*, 1980).

The stability of suppressiveness to Fusarium wilt is attributed to soil type, soil properties and associated microbiota (Stover, 1962; Stotzky and Martin, 1963 and Alabouvette *et al.*, 1982). The most important tool needed to study the suppressiveness is to assess the degree of suppressiveness. Measuring the inoculum density of soils after infestation is another method of assessing the suppressiveness. Thus it is possible to rank soil as a function of their suppressive effect and study the influence of various factors on the level of suppressiveness (Rouxel, *et al.*, 1987). More or less, complex microbial interaction between parasitic and saprophytic microflora are responsible for suppressiveness. The fact that the suppressiveness is microbiological in nature does not mean that the soil physico-chemical properties are not involved in the suppression. (Stutz *et al.*, 1989) established a direct relationship between the soil type and suppressiveness to black root rot of tobacco.

A correlation between suppressiveness and physico-chemical properties has been established in *Fusarium* wilt of banana in Central America (Stotzky and Martin, 1963). Amir and Alabouvette (1993) induced suppressiveness in conducive sandy soils of Algerian palm grown by addition of 25 per cent montmorillonite. The pH has been also associated with suppressiveness. Other characteristic of soils such as clay and sand contents, aeration and carbon nitrogen ratio determine the total biomass which in turn has a role in general suppression (Cook and Baker, 1983). The results also suggest that the phenomenon of 'fungistasis' is expressed in high degrees in suppressive soils than in conducive soils (Lockwood, 1981). Similarly, iron which is required for *Fusarium* growth is chelated by siderophore producing fluorescent pseudomonads is responsible for suppression of *Fusarium* in soil (Sneh *et al.*, 1984). Thus, the competition for nutrients is also one of the factors responsible for suppressiveness for *Fusarium* wilt (Lemanceau, 1989). In addition, the evidences for possible role of non-pathogenic *Fusarium* in suppressive mechanism have been also cited. The main difference between the suppressive and conducive soil was the ratio of pathogenic/ non pathogenic *Fusarium* (Alabouvette *et al.*, 1984). The evaluation of various modes of action in natural suppressiveness of soils should produce new progress and vistas in the field of soil suppressiveness.

Disease resistance and Screening techniques against soil borne plant pathogens

Host plant resistance has been an apt choice in all the crop improvement programs and is essential to recommend the cultivars directly for cultivation in endemic area, infested with soil borne fungus. Although the resistant variety is always one of the best ways and will go a long way in reducing loss due to soil borne plant pathogens.

Various techniques were employed by different investigators for inducing wilt symptoms in plants. The techniques included mixing of three weeks old wheat kernel culture at the rate of 50 g to sterilized soil, pouring a heavy spore suspension and mycelial fragment suspension to plants grown in pots containing sterilized soil (Haymaker, 1928) dipping of injured roots of transplants in mycelial suspension (Wellman, 1939), inoculating the soil with fungal culture and transplanting the seedling in wilt sick plot (ICRISAT), pouring suspension around the root zone and rapid-root dip transplanting techniques (Naik *et al.*, 1996) etc. Transmission of *Fusarium oxysporum* through seed and soil has been recorded (Palodhi and Sen, 1983; Naik, 1990). Soil inoculation has been rated as superior technique over seed inoculation in *Fusarium* wilt of muskmelon (Palodhi, 1975), whereas, seed inoculation was found to be the best against wilt of watermelon (Naik, 1990).

Efficient field, greenhouse, and laboratory procedures to evaluate chickpea and pigeonpea lines for resistance to *Fusarium* wilt have been developed and standardized at ICRISAT (Nene *et al.,* 1981; Nene and Kannaiayan, 1982; Nene *et al.,* 1989). Stock isolates of *F. oxysporum* f. sp. *ciceris* should not be maintained by repeated transfers on growth medium because of loss of virulence through mutation (Toussoun and Nelson, 1976). It is recommended that virulent isolates are to be stored as dormant cultures in either sterile soil (Toussoun and Nelson, 1976), silica gel (Windels *et al.,* 1988) or sterile filter paper.

In India, host plant resistance to *Fusarium* wilt of tomato was studied by evaluating a collection of 134 cultivars, breeding lines and alien species of tomato. Of these genotypes, eight possessed high degree of resistance to race-1 of *F. oxysporum* f. sp. *lycopersici* (Kesavan and Chowdhary, 1977). Similarly, Chauhan (1988) studied the reaction of several varieties lines of tomato to *Fusarium* wilt at Hissar. Out of 142 advance lines of chilli, 16 were resistant and 48 were moderately resistant (Ahmed *et al.,* 1994). Among 28 hot pepper cultivars, 12 were resistant and 6 were moderately resistant. Nayeema *et al.,* (1995) screened hot pepper germplasms for resistant to *Fusarium* wilt and one of the genotype, cultivar Musalwadi was immune under all conditions. Asahi Yamato, a mid -season Japanese introduction of watermelon and Arka Manik released by Indian Institute of Horticultural Research, Bangalore, India were identified as moderately susceptible whereas, Charleston Gray -122 one of the older cultivars of U.S.A., widely grown in gulf countries and suitable for long distance transport and Pusa Rasal were moderately resistant. However, Smokylee and Calhoun Gray were found highly resistant (Naik, 1990). Madhukar *et al.,* (2002) identified 1HR-3018 as resistant against *Fusarium* wilt of chilli by seed inoculation and rapid-root-dip-transplant method. DS-1 and LCA-206 were moderately resistant and the remaining ones were highly susceptible. Rapid-root -dip-transplanting technique and sick soil technique were employed adequately in identifying the resistant source against wilt of chilli caused by *Fusarium solani* and the genotypes viz., F-112-5-83, SKAU-C-101 and PC-6 were identified as resistant (Devika Rani *et al.,* 2006). Where complete resistance to pathogens exists, the resistance is generally expressed irrespective of inoculum level used in the tests and is usually not largely influenced by environmental changes. This form of resistance is often expressed to soil-invading pathogens which demonstrate host specificity, including members of the genus *Phytophthora* such as *P. sojae, P. vignae* and *P. medicaginis* (Brasier and Hansen, 1992), fungal vascular wilt pathogens, *Pachymetra chaunorhiza* (Croft, 1989) and others. However, resistance to pathogens which show host specificity is not always complete; e.g. certain genotypes of soybean and cowpea show only partial resistance to *P. sojae* and *P. vignae* respectively (Davis *et al.,* 1993). In these examples, the partially resistant genotypes show susceptibility when the hypocotyls is used as an infection court, yet express partial resistance whose expression is inoculum level dependent (Irwin and Langdon. 1982; Davis *et al.*, 1993). For many soil-inhabiting fungi including *Sclerotinia* spp. *Sclerotium rolfsii, Phymatotrichum omnivorum, Rhizoctonia solani, Pythium* spp., naturally occurring resistance in the host species, of interest, is unavailable (Agrios, 1988). While, working with this group of pathogens, the results may be misleading when excessive or inadequate quantity of inoculum is used or conducting test in sterile soil or seedling wilt count instead of adult plant wilt counting. The above limitation should be over come while screening for resistance.

Studies demonstrated that maximum expression of dry root rot required for screening germplasm can be secured by inoculations of seedlings grown in *M. phaseolina* infested soil at the time of moisture stress (Singh and Lodha, 1986; Lodha 1998). This procedure was used to screen 28 elite genotypes rigorously to identify resistant sources. In the absence of complete resistance, least susceptible genotypes RGC 471 and Kutch 8 were categorized as moderately resistant while HG 75 and AG Ill as highly susceptible. These genotypes were used for making three crosses i.e. Kutch 8 x RGC 471; Kutch 8 x HG 75 and HG 75 x AG Ill. Progenies were subsequently advanced to F 5 and F 6 generations (Lodha and Solanki, 1993).

There was a wide range of variation for dry root rot resistance among the F_1s produced by these crosses. The F I from Kutch 8 x RGC 471 showed resistance close to either of the parent. The susceptibility (33.1%) of F_1 was almost equal to the mid- parent value (33.4%) indicating absence of dominance. Scaling test revealed the presence of epitasis in Kutch 8 x ROC 471 and HO 75 x AG 111 crosses and appropriateness of additive dominance model for Kutch x HG 75. In gene effect studies, additive, additive x additive and dominance x dominance were more pronounced in Resistant x Susceptible cross. These findings are in close agreement with those of Rao and Shinde (1985) who reported inheritance of resistance to be polygenic in sorghum for *M. phaseolina* induced charcoal rot and found that non-allelic interaction played an important role. In the absence of complete resistance, only alternative left for the improvement of the population was to select plants possessing more resistance to dry root rot. Thus, few strains combining field resistance as well as high yield were selected (Lodha and Solanki, 1993). Of these, strain CAZG 27-1 ranked first in all India testing in gum content (32.8%). In case of cowpea, of the 33 genotypes evaluated against *M. phaseolina,* 26/4/1, V16, K39, 25/8/2 and CO3 were found moderately resistant (Singh and Lodha, 1986).

However, it is too much to expect in case of soilborne pathogens such as *Fusarium* to have stable resistance because of variability of the pathogen. Nevertheless, resistance in soilborne disease would continue to play an important role for the simple reason that although high resistance may not always be attainable, even moderate resistance may help to make other measures more effective.

Rhizosphere effect- A unique phenomenon in soil borne diseases

The release of nutrient through exudation of seeds and roots stimulates the activity of microbes in the spermatosphere; materials released include large amount of sugars from peas (Flentje and Saksena, 1964) soybean (Brown and Kennedy, 1966), also amino acids and vitamins (Woodcock, 1962). Nutrition of soilborne plant pathogens in the rhizosphere is an important factor in pathogenesis (Toussoun *et al.*, 1960) since they are known to stimulate the germination of chlamydospores (Schroth *et al.*, 1963). There is a strong indication that *Fusarium* population remains constant or even decreases slightly in case of resistant cultivar (Naik *et al.*, 1997b) which lent the possibility to hope that rhizosphere of non-susceptible host crop would lower the inoculum potential. Rhizosphere of garlic, onion, radish and okra were identified as crops reducing *F. oxysporum* population in cucurbits (Naik and Sen, 2002). On the basis of rhizosphere effect in reducing the soil propagules and minimum adverse effect on the crop, garlic, onion and radish were recommended for crop rotation whereas okra and onion for mixed or inter cropping in cucurbit cropping pattern. Cowpea and tomato which are maximum used for mixed cropping in cucurbit fields were identified as symptomless carriers (Naik and Sen, 1999). Similarly brinjal, chilli and cauliflower were also not recommended in the cucurbit cropping pattern (Naik and Sen, 2002). Mixed cropping of pigeon pea with *Crotolaria medicaginea* highly suppressed Fusarium wilt in pigeonpea (Upadhyay and Rai, 1981). A study on the effect of four different cropping systems on Fusarium wilt of pigeonpea influenced *Fusarium udum* population with maximum population noticed in the rhizosphere of pigeonpea continuously intercropped with sorghum and lowest population in pigeonpea intercropped with sorghum followed by sorghum sequenced with safflower in the post rainy season (Naik *et al.*, 1996). However, production systems were not effective in bringing down the wilt incidence mainly because population of *Fusarium udum* was higher than the ETL determined in earlier experiments for the pathogen (Naik and Reddy, 1993a).

Influence of cropping sequences and soil amendments on soil borne diseases

Studies conducted over four years revealed that continuous and sequential clusterbean cropping increased the population of *M. phaseolina* (Lodha *et al.*, 1990a). Significant variations were recorded with sequences involving fallow in the second or third year. In clusterbean-fallow systems, *M. phaseolina* population declined markedly after the fallow. It increased after the second crop of clusterbean but remained less than that under continuous cropping (Lodha 1995a). In sequences involving clusterbean, moth bean and pearl millet, differences among *M. phaseolina* populations were not significant. However, lowest population recorded in the sequences involving moth bean indicated its relative tolerance compared to clusterbean (Lodha and Singh, 1984). During fallow, absence of a crop and increase in soil moisture in the rainy season might have hastened the decomposition of pearl millet residues resulting in a rapid increase of competitive microorganisms which directly or indirectly reduced the population of *M. phaseolina* (Lodha *et al.*, 1990a). This study suggests that crop rotation with less susceptible crop will restrict the increase in *M. phaseolina* population in the soil.

A field experiment was conducted during summer of 1986 for 15 days (6-20 June). The treatments comprised of wet and dry plots with and without mulch. Mulching was done with transparent polyethylene sheet (50µ). In wet plots, one irrigation was given 2 days before mulching. Soil temperatures were recorded at 2 h interval from 10.00 AM to 4.00 PM daily. Propagules of *M. phaseolina* were determined quantitatively, using wet sieve, serial dilution and selective medium of Papavizas and Klag (1975). Studies revealed that polyethylene mulching during hot days considerably increased the soil temperature (Lodha, 1989; Lodha and Solanki, 1992) with maximum soil temperature at 14 hr in mulched plots was 58°C (wet) and 69°C (dry) at 5 cm depth, which in non-mulched plots did not exceed 53°C (wet) and 63°C (dry). In 15 day mulching period, soil temperature reached 65°C or more for atleast 6 days while in other parts of world it never exceeded 65°C in dry mulched soils (Katan *et al.*, 1980; Sheikh and Ghaffar, 1984). These temperature ranges were higher than the thermal death time-temperature (50°C for 9.0 minutes) reported for *M.phaseolina* (Bega and Smith, 1962). Elevated soil temperature and a shift in favour of antagonistic microorganisms reduced the population of *M.phaseolina* from 350 to 7 sclerotia g^{-1} soil at 5 cm soil depth but reduction decreased with increase in soil depth. After 15 days, 149 and 275 viable sclerotia g^{-1} soil were present in wet and dry mulched treatments, compared with 279 and 335 sclerotia g^{-1} soil in the corresponding controls at 130 cm depth (Lodha and Solanki, 1992).

There was a pronounced reduction in the population of *M.phaseolina* also due to solarization in the experiment conducted by Lodha and Burman in 2004. The native population of 54 sclerotia g^{-1} soil declined by 9.7% in non-solarized dry plots, 68.5% in dry solarized and further to 85.2% in wet plus nitrogen plus manure solarized plots. The population also remained at low levels in all the mulched treatments in 1987 and after fallow in 1988. The population in wet plus nitrogen plus manure and wet plus nitrogen non-solarized treatments was significantly lower than those of dry non-mulched plots after fallow. Clusterbean raised in the wet and dry solarized plots remained disease free until harvest, whereas, there were 19 and 23% mortality due to dry root rot in the non-mulched wet and dry plots.

Results of field study have shown that both the amendments combined with one summer irrigation were significantly superior in reducing incidence of dry root rot and wilt on clusterbean and cumin, respectively (Lodha *et al.*, 2001; Mawar and Lodha, 2002). The soil temperature of amended soil after one summer irrigation in June ranged from 38-44°C at 15 cm depth. These temperatures were 0.5 to 5°C higher than those recorded in unamended soil for the same period and 6-16°C higher when amendments were incorporated in July. Merely one summer irrigation in unamended plots reduced dry root rot severity significantly by 34-39% compared with the unamended control without irrigation.

Maximum wilt incidence (26.8%) was recorded in the unamended control plots. A single summer irrigation, or soil amendment with residues either combined with summer irrigation or applied in July at the time of clusterbean planting significantly reduced wilt incidence. Amending the soil with mustard residue or mustard oilcake in summer or with mustard oil-cake in July significantly reduced wilt incidence as compared with summer irrigation alone. Of the residues, mustard oil-cake was significantly more effective than mustard pod straw with a 34% greater reduction in wilt incidence, but differences relating to time of application were not significant.

The microbial make-up in any field at any given time is essentially the product of previous crops and their sequence in the field. Although, disease control through crop rotation (inter of mixed cropping) is a passive process, under actual field conditions some of the benefits may include stimulation of microorganisms that may partially or totally suppress the pathogen. The growing of a succession of different crops in the same field allows time for the soil microbiota to displace or weaken the soilborne pathogens. This might be the reason for reduction of pathogenic *Fusarium* population in treatment with sorghum / pigeon pea intercropping followed by sorghum, safflower sequential cropping. With few notable exceptions soil inhabiting plant pathogens multiplied when furnished access to their preferred host and decline in population (or biomass) when denied access to host. In the Pacific Northwest, a three year rotation (Wheat/peas or lentils/barley) is nearly as effective as soil fumigation for root disease control.

Soil Solarization

Soil solarization or slow soil pasteurization is a soil disinfestations method of disease control. The methodology has been used successfully to reduce population of plant parasitic nematodes, various plant pathogenic fungi, weeds (Horowitz 1980: Katan *et al.*, 1983) and successful control of certain soil borne fungal diseases (Pullman *et al.*, 1981). Soil solarization involves covering soils with clear, thin polyethylene sheets in mid summer, generally for 4 to 6 weeks, before sowing the host crop. Most successful applications of solarization have been achieved as a pre-plant treatment but post-plant treatments have also proved effective (Ashworth and Gaona, 1982; Stapelton and Devay 1982). During solarization, the plastic sheets trap solar energy and prevent heat losses caused by evaporation and convection, thus creating a greenhouse effect. In Mediterranean type of climates temperatures as high as or higher than 60°C are achieved in the surface layers of soil generally representing an increase of 10 to 12°C over uncovered soil. Reports show that many nematode and fungal pathogens are killed by temperatures within the range of 45^{0}C to 55°C (Pullman *et al.*, 1981; Porter and Merriaman, 1983). Consequently, the principal effect of solarization is to raise soil temperatures to levels which are lethal to soilborne pathogens. Katan (1981) has also suggested that merely moistening the soil during summer may result in disease control in certain arid zones. In such regions, this may be termed as summer irrigation.

Solarization technique is especially effective in the regions where climatic conditions favour adequate heating of the soil, since intense solar irradiations and high temperatures are characteristic features of Indian arid regions. Hot arid climate is most appropriate for soil solarization and elevated soil temperatures reduced the viability of a heat-resistant pathogen *M.phaseolina* at top soil layer. The reduction in disease incidence in plant grown in solarized soils, as with any soil treatment, results from the effects exerted on each of the three living components viz., host, pathogen and surrounding microorganisms as well as on the physical and chemical environments which in turn affect the activity and interrelationships of the organisms (Devay and Katan, 1991). Although these microbial processes occur primarily during solarization, they may continue to certain extent and in different ways after the removal of polyethylene sheets and planting. The most pronounced effect of solarization is a physical one i.e. an increase in soil temperatures for several hours daily during the solarization period. However, other accompanying processes such as shifts in the microbial populations, changes in the chemical composition and physical structure of soil, high moisture levels maintained by the polyethylene mulch and changes in gas composition of the soil should also be considered when analyzing mechanisms of disease control. Hence, in the long run, soil solarization may lead to soil suppressiveness reflected by slow reinfestation. If antagonists are adversely affected, the latter resulting in the creation of a biological vacuum. Studies have frequently indicated that solarized soils become hostile and less receptive to pathogen reinfestation. This can be regarded as induced suppressiveness.

Soil solarization effectively reduced the pathogen population of *F. oxysporum* f. sp. *vasinfectum* in the soil and thereby decreased incidence of wilt in cotton plants, improved plant growth and finally cotton yields for as long as three years after treatment. Solarization by covering the soil with clear transparent polyethylene sheet for six weeks during hot summer months showed 80.8% reduction of *F. oxysporum* f. sp. *ciceri*. Solarization of infested soil by mulching with different polyethylene mulches increased soil temperatures in ecosystem lethal to tomato wilt pathogen. Of course, soil temperature decreased with an increase in soil depth; while it drastically reduced pathogen population in soil mulching (25m) for 15 and 60 days and eliminated *Fusarium oxysporum* pathogen from 6 and 24 cm soil depth, respectively resulting in decreased wilt incidence to zero levels. Gamliel and Stapleton (1997) have suggested the incorporation of organic amendments as a non-chemical approach to improving the efficacy and predictability of pathogen control by soil solarization. Pigeonpea wilt causing *Fusarium* propagules were greatly reduced in both dry and irrigated condition immediately after solarization. The population of *Fusarium* remained at low levels in solarised plot through out the growing season (Naik and Reddy, 1993). Mulching with transparent polyethylene for 32 to 49 days effectively checked the population density of *Pseudomonas solanacearum* down to a depth of 15 cm (Chellemi *et al.*, 1994). Enhanced biological activity of beneficial biological control agents in deeper soil profile (15 to 30 cm) also contributed substantially in management of soil-borne plant pathogens (Chen *et al.*, 1991).

The solarization in its totality is a typical integrated management technique mainly because it involves physical, chemical and biological mechanisms and the control is achieved on a variety of target pests such as fungi, bacteria, insects, nematodes and weed density.

Biological control of soil borne diseases

Natural products and chemical compounds discovered as a result of basic research into the molecular mechanism of pathogens and biological control have led to the development of 'biorational' pesticides. However, fewer isolates can suppress plant disease under diverse growing conditions and fewer still have broad-spectrum activity against multiple pathogenic taxa. Nonetheless, intensive screens have yielded numerous candidate organisms for commercial development. Biological control is the reduction of the amount of inoculum or disease producing activity of a pathogen accomplished by or through one or more organism other than inoculum. It is the use of organism, genes or gene product to regulate a pathogen (i) by keeping inoculum below economic threshold level (ETL) (ii) retard or exclude infection (iii) maximizing plant self defense.

Some of the microbial taxa that have been successfully commercialized and are currently marketed as Environment Protection Act-registered (EPA) biopesticides in the United States include bacteria belonging to the genera *Agrobacterium, Bacillus, Pseudomonas,* and *Streptomyces* and fungi belonging to the genera *Ampelomyces, Candida, Coniothyrium, and Trichoderma*. These grow typically towards the target pathogen hyphae.

Of all these, *Trichoderma* spp received attention all over the world because of its ubiquitous distribution. They are the most potent antagonistic fungi and have been extensively used in agriculture of developed countries for the past 25 to 30 years. *Trichodrema* has emerged as a very effective bio control agent as a replacement to many key pesticides so also as an increasing awareness regarding the safety of pesticides for management of particularly soil borne plant pathogens.

The strains of *Trichoderma* spp are opportunistic invaders, fast growing, prolific producers of spores and powerful antibiotic. These are some of ideal characters which make them a very successful bioagent (Kubicek *et al.,* 2002). Many of the uses of these fungi as bio control have been documented recently (Druzhinina *et al.,* 2006). It has several modes of action such as hyperparasitism, antibiosis, reduction in saprophytic ability, induced resistance in the host plant, competition for the nutrients and space, reduction of different volatile and non volatile compounds etc. (Elad and Freemen, 2002; Harman *et al.,*2004). Its mycoparasitism which mainly involves action of antibiosis and cell wall degrading enzymes such as chitinases, β1-3 glucanases and proteases. Since chitin is a major component of fungal cell wall the bio control activity of *Trichoderma* has been attributed to chitinase (Harman, 2000).

Trichoderma harzianum has greater rhizosphere competence and parasitizes the pathogenic fungi. Among seven *Trichoderma* antagonists evaluated under *in vitro* conditions, all of them overgrew on the chilli wilt fungus (*F. solani*) in dual culture. Application of antagonists through farm yard manure at seven days in advance helped in development and proliferation of bioagent compared to simultaneous application and seed treatment of bioagent (Madhukar *et al.*, 2002). Among the different antagonists studied by them, *T. harzianum* PDBCH-10 recorded maximum number and least number of propagules was enumerated in *T. harzianum* PDBCH-8. On the contrary, antagonist reduced the pathogen population considerably and the maximum reduction of *Fusarium* population was recorded at 60 days after application. Among the different methods of application, 7 day prior application recorded maximum reduction of pathogen population. Devika Rani *et al.*, (2006) evaluated several bioagents, among them *Trichoderma viride* (I), *T.viride*-16 and *Pseudomonas fluorescens* (Indigenous) were very effective in inhibiting the growth of *F. solani* both in dual culture and seed coating test. The bio-agents when seed treated on chilli cultivars; highest number of propagules of bio-agent was recovered in *T.viride* (Indigenous) with maximum reduction of pathogen propagule at 60 days after of application. However, an indigenous isolate of *P. flurosences* was identified as an effective inducer of systemic resistance against wilt of chilli. The bioagents, when tested for shelf life, on four carriers under two environmental conditions, highest population of bio-agent was recorded in *T.viride* (Indigenous) in farm yard manure and vermicompost at room temperature whereas liquid broth for *P.fluorescens.*

Antifungal metabolites of *Trichoderma* have been grouped in to pentaketides, threnodes and octaketides. Another hydrolytic enzyme which attracted much attention in the recent times is chitinase. Chitinase enzymes (including N-acetyl glucosaminidase, chitobiase and endochitinase) produced by *T. harzianum* were 26 which were effective in lysing chitin containing fungi. While studying the cytochemical aspects of chitin breakdown by parasitic action of a *Trichoderma* sp. on *F. oxysporum* f. sp *radicis-lycopersici*, rapid hydrolysis of wall bound chitin by extra-cellular chitinase of *Trichoderma* has been investigated. Chitinase negative strains of *T.harzianum* and *T.reesei* were transformed.

Fusarium wilt is a serious problem in cucurbits specially in riverbed cultivation. Muskmelon and watermelon plants of the same genotype escaped infection at a heavily infested site. Screening of rhizosphere microflora of such plants yielded *Aspergillus niger* frequently. Of all the selected antagonistic bacteria and fungi, *A. niger* isolates from rhizosphere of wilt free plants were identified as potential antagonists (Naik, 1990; Sharma and Sen, 1991, Naik and Sen 1994). Strong antagonism by *Aspergillus* sp. against *Fusarium oxysporum* f. sp. *vasinfectum, F. oxysporum* f. sp. *radicis lycopersici, F. solani, F. oxysporum* has been reported (Bora, 1977; Marois *et al.*, 1981; Sharma, 1989; Naik, 1990). Among bacteria, strains of *Pseudomonas* against *F. solani* and *B. subtilis* against *F. oxysporum* f. sp. *melonis, lycopersici, vasinfectum, F. solani* and *F. udum* were claimed to be biocontrol agents. *Aspergillus niger* (A6) and *Bacillus subtilis* (BI) were identified as the best biocontrol agents against *F. oxysporum* isolates of watermelon, both under *in vitro* and *in vivo* since they caused maximum inhibition followed by differential lysis of mycelium in the advanced stage of antagonism. Further, the culture filtrates of both the antagonists inhibited hundred per cent inhibition of germination of micro conidia. These bioagents when coated on the watermelon seeds could sustain the disease pressure. Both *A. niger* (A6) and *B. subtilis* (BI) were effective against a spectrum of nine pathogenic isolates of *Fusarium* sp. (Naik and Sen, 1994). When these two bioagents were applied in six different ways with or without carriers, they improved the germination. *A. niger* AN 27 was developed as potential biocontrol agent against soilborne pathogens (Sen, 2000). The introduced antagonist should survive, proliferate, become active and established itself in a new environment. Hence, proper formulations of antagonist preparations have been evolved. Five formulations were made from AN 27 which included Kalisena SL, Pusa Mrida, Kala Sipahi (capsules) for soil application and Kalisena SD for seed dressing (Sen, 2000).

The work on management of soil borne disease should be based on ecological principles namely inoculum density, economic threshold level, quantification, prediction and agronomic manipulation, since chemical control is arduous, uneconomically and not advisable owing to risk of ground water pollution and death of non target beneficial flora.

Molecular Approaches and techniques in soil borne plant pathogens

Several classical approaches have been developed to detect and identify soil fungal inhabitants for the past several years. Selective media have been devised to exclude the large number of soil organisms and allow growth of target fungi. However the advent of molecular biology has offered a number of revolutionary insights into the detection and enumeration of soil borne fungal pathogens and also has started to provide information on the identification of unknown species from DNA sequences. Recent advances in the detection of plant pathogens using immunological and nucleic acid-based techniques has enabled the major genera and species of disease-causing organisms to be quickly and reliably identified. Methods that are based on the use of monoclonal antibodies or the polymerase chain reaction (PCR) are highly sensitive and specific and have the potential to replace traditional technologies. Detection and identification of pathogens that are soil borne presents a number of difficulties such as sampling and reaction inhibition that are beginning to be overcome. New approaches to antibody production such as that of phage display, the combination of immunoassays with PCR and the development of species-specific PCR primers will allow versatile and wide-scale use of these new techniques in unraveling the several mysteries of soil borne plant pathogens.

Molecular markers have been used to characterize genetic diversity of different isolates of *F. oxysporum* (Edel *et al.*, 1995; Kistler 1997; Alves-Santon *et al.*, 1999) and Kistler (1997) who suggested that certain *F. oxysporum* isolates with in formae speciales may be genetically less similar to each other than those isolates as non-pathogenic strains or those isolate as a group in the other formae species. The complex hierarchal pattern observed within the *F. oxysporum* lineage is consistent with results obtained in earlier studies (Kistler, 1997; Koenig *et al.*, 1997; O'Donnell *et al.*, 1998). The universal rice primers (URP) were initially designed from the repeat sequences of wild rice genome by Kang *et al.*, (2002), who have shown their application in molecular analysis of different organisms across species.

Prasad *et al.*, (2004) studied molecular variability among 45 isolates of *Fusarium* species belonging to eight sections by using universal rice primer's (URPs) designed from the rice micro satellite. Most of the *Fusarium oxysporum* and its formae speciales shared common bands of 1.0, 1.5 and 3.0 kb size. A combined dendrogram of URP markers obtained from 45 *Fusarium* isolates formed five major clusters. Assessment of molecular variability based on URPs indicated clearly that *Fusarium oxysporum* is heterogeneous and most of the formae speciales have close evolutionary relationships. Other species of *Fusarium* viz., *Fusarium oxysporum, F. pallidoroseum* and *F. chlamydosporum* were grouped separately showing low genetic similarity with *F. oxysporum.*

Evaluation of the genetic diversity in *M. phaseolina* isolates from India has been an initial step towards understanding the population structure of the charcoal rot pathogen. The results of the investigation demonstrate the applicability of RAPDs of genomic DNA and ITS, RFLPs in analyzing genetic variation which are particularly useful when combined with *in vitro* and *in vivo* studies (Purkayastha *et al.*, 2006). DNA polymorphism indicated that *M. phaseolina* isolates from the four host species used in their study were closely related but that detectable genetic variation exists among isolates based on different hosts. This was revealed through analysis of the RFLP's of ITS regions 5-8S DNA and part of the 255 r DNA as well as RAPD of genomic DNA. The variation that among the isolates was confined to populations with boundaries defined by the species.

Almeida *et al.*, (2003) recorded genetic variability among Brazilian isolates of *M. phaseolina* from soyabean and showed that one single root can harbor more than one haplotype. RAPD technique can be used for a quick screening of *F. udum* isolates in the laboratory condition (Harris, 1999). Twenty four isolates of *R. solani* collected from soil, root and collar rot or leaf blight of infected plants were analyzed for genetic variation by using 11 primers Four universal rice primers and two inter simple sequence repeat (ISSR) markers and finger print patterns were generated for each isolates. The isolates obtained from same host and same geographical regions showed similarity in DNA finger printing profiles barring few exceptions (Sharma *et al.*, 2005).

One of the most important molecular research areas exciting is the use of disease resistant transgenic plants that express microbial genes involved in the suppression of root infections. Next important research area is the use of transgenic microorganisms as biocontrol agent in which microbe expresses genes from another microorganism whose products are inhibitory to soil borne pathogens. Lastly, the use of nucleic acid hybridization technologies for detection, identification and monitoring of soil borne pathogens and introduced biocontrol agents is another area being explored recently.

Although several fungi such a *Trichoderma* spp. *Cheatomium globosum, Aspergillus niger,* Bacterial strains such a *Pseudomonas fluoresecens* have been commercialized as bioagents, the recent developments in molecular biology specially in understanding the biocontrol mechanism and pathogenesis should lead to recombinant technologies to manage soil borne diseases. Perhaps, the lack of specific genes encoding soil pathogen suppressive trait and present state of inadequate knowledge on ecology of soil borne pathogens are the reasons for not many transgenic plants against these groups of pathogens.

REFERENCES

Agrios, G. N. 1988. Plant Pathology, Academic, New York. pp: 365

Ahmed, N. Tanki, M. I. and Mir, N. M. 1994. Screening of advance breeding lines of chilli and sweet and hot pepper cultivars against Fusarium wilt. *Plant Dis. Res.* 9: 153-154.

Alabouvette, C. 1986. Fusarium wilt suppressive soils from the Chateaurenard region: review of a 10 year study. *Agronomie.* 6: 273-284.

Alabouvette, C., Couteaudier, Y. and Louvet, J. 1982. Comparaison de la receptivite de differents sols et substrats de culture aux fusarioses vascularies. *Agronomie.* 2 : 1-6.

Alabouvette, C., Couteaudier, Y. and Louvet, J. 1984. Dynamique des populations de *Fusarium* spp. et de *Fusarium oxysporum* f.sp. *melonis* dans un sol resistant et dans un sol sensible aux fusarioses vascularies. *Agronomie.* 4 : 729-733.

Alabouvette, C., Rouxel, F. and Louver, J. 1980. Mise en evidence de la specificite de la resistance d' un sol vis-a-vis des fusarioses vascularies. *Ann. Phytopath.* 12: 11-19.

Almedia, A.M.R., Abdelnoor, R.V., Aris, C.A.A., Carvolho, V.P., Ficho, D.S.J., Marin, S.R.R., Benato, L.C., Pinto, M.C. and Carvalho, C.G. 2003. Genotypic diversity among Brazilian isolates of *Macrophomina phaseolina* revealed by RAPD. *Fitopathol. Bras.* 28: 279-285.

Alves Santo, F.M., Benito, E.P., Eslava. A.P. and Diaz Minguez, J.M. 1999. Genetic diversity of *Fusarium oxysporum* isolates from common bean fields in Spain. *Appl. Environ. Microbial.* 65: 3335-3340.

Amir, H. and Alabouvette, C. 1993. Involvement of soil abiotic factors in the mechanisms of soil suppressiveness to Fusarium wilt. *Soil Biol. Biochem.* 25(2): 157-164.

Armstrong, G. M. and Armstrong, J. K. 1968. *Phytopath.* 58: 1242-1246.

Ashworth, L. J. Jr. and Gaona. S. A. 1982. Evaluation of clear polyethylene mulch for controlling Verticillium wilt in an established pistachio nut grove. *Phytopath.* 72: 243.

Baker, R. 1965. The dynamics of inoculum. (*In*:) Baker, K. F. and Synder, W. C. (Eds.). *Ecology of Soilborne Plant Pathogens.* pp: 395-419. Univ. of California Press, Berkeley.

Bega, R. and Smith, R.S. 1962, Time-temperature relationships in thermal inactivation of Sclerotia of Macrophomina phasesolina. *Phytopath.* 52:632-635.

Booth, C., 1971. *The genus Fusarium.* CMI, Kew Survey, London, England, p.237.

Bora, T., 1977. *In vitro* and *In vivo* investigations of the effect of some antagonistic fungi against the damping off diseases of egg plant. *J. Turk. Phytopath.* 6:13-17.

Brasier, C. M. and Hansen, E. M. 1992. Evolutionary biology of *Phytophthora.* Part II: Phylogeny, speciation and population structure. *Ann. Rev. of Phytopath.* 30: 173-200.

Brown, G. I. and Kennedy, B. W. 1966. Effect of oxygen concentration of Pythium seed rot of soybean. *Phytopath.* 56: 407-411.

Bruehl, G. W. 1987. Soil borne plant pathogens. MacMillan Publishing company, New York, Colier Macmillan Publishers, London. pp: 368

Brandes, E. W. 1918. P. R. Agric. Exp. Stn. Rep. 1916.

Buczacki, S. T. 1981. Fungal pathogens in soil and roots. (*In*:) Burchill, R. T. (Ed.). *Methods in Plant Pathology*, Phytopathological Papers, No.26. pp: 17- 21. Commonwealth Mycological Institute, Kew, Surrey, UK.

Buohot, D. 1979. Estimation of inoculum density and inoculum potential: techniques and their value for disease prediction. (*In*:) Schippers. B. and Gams. W. (Eds.). *Soilborne Plant Pathogens* pp-21-34. Academic. London.

Chellemi, D. O., Olson, S. M. and Mitchell, D. J., 1994. Effect of soil solarization and fumigation on survival of soil-borne pathogens of tomato in Northern Florida. Plant Dis. 78: 1167-1172.

Chen, Y., Gamliel, A., Stapleton, J. J. and Avaid, J., 1991. Chemical physical and microbial changes related to plant growth in disinfected soils. (In:) Soil Solarization (Katan, J. and DeVay, J. E. (Eds), pp 103-129. CRC Press, Boca Raton, USA.

Conn, K. L. and Tiwari, J. P., 1990. Survey of *Alternaria* black spot and *Sclerotinia* stem rot in central Alberta in 1989. *Can. Plant Dis. Survey* 70: 66-67.

Chauhan, M. S. 1988. Reaction of varieties/ lines of tomato to wilt. *Indian J. Mycol. Pl. Pathol.* 18:72-73.

Cook, R. J. and Baker, K. F., 1983. The Nature and Practice of Biological Control of Plant Pathogens. *American Phytopathological Society*, St. Paul, 539pp.

Croft, B. J. 1989. A technique for screening sugarcane cultivars for resistance to Pachymetra root rot. *Plant Dis.* 73.651-654.

Davis. R. I., Irwin. J.A.G. and Imrie. B.C. 1993. Development of a glasshouse assay for screening for partial resistance to *P. vignae* in cowpea. *Aust. J. Agric. Res.* 44:755-762.

Devay, J. E. and Katan, J. 1991. Mechanisms of pathogen, control in solarized soils. (*In*:) Katan, Jand Devay, J.E. (Eds.) *Soil Solarization*, CRC, Press, pp-87-101. Boca Raton, FL, USA.

Devika Rani, G.S., Naik, M.K., Patil M.G., Mohan Kumar, H.D., 2006. Screening of chilli varieties against Fusarium wilt, Paper presented in *Annual Meeting and Symposium on Diagnosis and Ecofriendly management of Plant Diseases* held by the Indian Phytopathological Society, at CPCRI, Kerala, 27th & 28th November 2006.

Devika Rani, G.S., Naik, M.K., Raju. K. and Prasad P.S., 2006. Efficacy of a spectrum of bioagents on *Fusarium solani*, causing wilt of chilli, Paper presented in the *National Seminar on New Frontiers in Plant Pathology*, held by the Indian society of Mycology and Plant Pathology, at Kuvempu University, Shimoga, 28-30, September 2006, p113.

Douglas, D. R. 1970. The effect of inoculum concentrations on the apparent resistance of muskmelons to *Fusarium oxysporum* f. sp. *melonis. Can. J. Bot.* 48:687-693.

Dhingra, O. D. and Sinclair, J. B., 1975. Survival of *Macrophomina phaseolina* sclerotia in soil; effect of soil moisture, carbon:nitrogen ratios, carbon sources and nitrogen concentrations. *Phytopathology 65: 236-240.*

Druzhinina, I.S., Schmoll, M., Seiboth, B. and Kubicek, C. P., 2006. Global carbon utilization profiles of wild-type strains, mutants and transformants of *Hypocrea jecorina. Appl Environ Microbiol* 72: 2126-2133.

Edel, V., Sterin Berg, C., Avelange, I., Laguere, G. and Alabouvetee, C., 1995. Comparison of three molecular methods for the characterization of *Fusarium oxysporuim* strains. *Phytopath.* 85: 579-585.

Elad, Y. and Freeman, S., 2002. Biological control of fungal plant pathogens. In: *The Mycota A comprehensive Treatise on fungi as experimental systems for basic and applied research.* K Esser and JW Benett, eds. Vol. XI: Springer Press, Germany pp 93-109.

Erwin, D.C., and Ribeilo, O.K. 1996. Phytophthora-Diseases world-wide. APS Press, Am. Phytopathol. Soc. St. Paul, M.N., USA.

Ferris, H. 1987. Components and techniques of Integrated Pest Management Threshold Determinations for soil- borne pathogens. *Plant Dis.* 71: 452-455.

Flentje, N. T. and H. K. Saksena. 1964. Pre-emergence rotting of peas in South Australia III. Host-pathogen interaction. *Aust. J. Biol. Sci.* 17: 665-675.

Foster. R. C. 1986. The ultra structure of the rhizoplane and the rhizosphere. *Ann. Rev. Phytopath.* 24:211-234.

Gilligan, C.A., 1983. Modeling of soil borne plant pathogens. *Ann. Rev. of Phytopath.* 21: 41-64

Gilligan. C. A. 1985b. Construction of temporal models: ill. Disease progress of soil borne pathogens. *Adv. Pl. Pathol.* 3:66-105.

Gamliel, A. and J.J. Stapleton. 1997. Improvement of soil solarization with volatile compounds generated from organic amendments. *Phytoparasitica,* 25 (Suppl.):3 I S-38S.

Harris, S.A. 1999. RAPD's in systematics- a usefull methodology. (*In:*) Hollin Gswoth, P.M., Bateman, R.M., Gornall, R.J. (Eds.). *Molecular systematics and plant evolution.* pp.211-228, Taylor and Francis Ltd., London, UK.

Harman, G. E., 2000. Myths and dogmas of biocontrol: Changes in perceptions derived from research on *Trichoderma harzianum* T-22. *Pl Dis* 84: 377-391

Harman, G. E., Howell, C. R., Viterbo, A., Chet, I. and Lorito, M., 2004. *Trichoderma species* opportunistic, avirulent plant symbionts. *Nat Rev Microbiol* 2(1): 43-56.

Haymaker, H. H. 1928. Pathogenicity of two strains of the tomato wilt fungus, *Fusarium lycopersici Sacco. J. Agric. Res.* 36:675-719.

Horowitz, M., 1980. Weed research in Israel. *Weed Sci.* 28: 457-460.

Irwin. J. A .G. and Langdon. P.W. 1982. A laboratory procedure for determining relative levels of field resistance in soybeans to *Phytophthora megasperma* f. sp. *glycinea. Aus. J. Agric. Res.* 33:33-39.

Katan, J. (1981). Solar heating (solarization) of soil for control of soil borne pests. *Ann Rev. Phytopath.* 19: 211-236.

Katan, J., Fisher, G. and Grinstein. A. 1983. Short and long term effect of soil solarization and crop sequence on Fusarium wilt and yield of Cotton in Israel. *Phytopath.*73:1215-1219.

Katan, J., Rotem, I., Finke, Y. and Daniel, J. 1980. Solar heating of the soil for the control of pink root and other soil borne diseases in onion. *Phytoparasitica*.80:39-50

Kang, H.W., Park, D. S., Go., S.J. and Eun, M.Y., 2002. Finger printing of diverse genome using PCR with universal rice primers generated from repetitive sequence of koream weedy rice. *Molec.Cells*. 13: 281-287.

Kesavan, G. and Chowdhury, B. 1977. Screening for resistance to *Fusarium* wilt of tomato. *SABRAOJ*. 9:57-65.

Kistler, H.C. 1997. Genetic diversity in the plant pathogenic fungus *Fusarium oxysporum*. *Phytopath*. 87: 3335-3340.

Koeing, R.K., Plotez. R.C. and Kistler, H.C., 1997. *Fusarium oxysporum* f.sp *cubense* consist of small number of diversity and globally distributed clonal linkage. *Phytopath*. 68: 915-923.

Kubicek, C. P., Bissett, J., Druzhinina, I., Kullnig Gradinger, C. and Szakacs, G., 2002. Genetic and metabolic diversity of *Trichoderma:* a case study on Southeast Asian isolates. *Fungal Genet Biol* 38: 310-319.

Lemanceau, P., 1989. Role of competition for carbon and iron in mechanisms of soil suppressiveness to Fusarium wilts. Pages 386-396. (*In*:) E. C. Tjamos and C.H. Beckman, (Eds.) *Vascular Wilt Diseases of Plants-Basic Studies and Control*. Springer Verlag Berlin, NATO ASI Series H28, 590pp.

Lockwood, J. L. 1981. Exploitation competition. Pages 319-350 (*In*:) D.T. Wicklow, G. Carroll, (Eds). The fungal community. Its Organization and Role in the Ecosystem. Dekker, New York.

Lodha, S. (1989). Soil solarization for the control of *Macrophomina phaseolina* (Tassi) Goid. in arid soils. *Indian Phytopath*. 42: 170-171.

Lodha, S. (1995a). Population dynamics of *Macrophomina phaseolina* in aridisols: an overview. *In Proceedings of Global conference on Advances in Research on Plant Diseases and their Management* held at Udaipur from Feb,12-17, 23pp (Abstr.).

Lodha, S. (1998). Effect of sources of inoculum on population dynamics of *Macrophomina phaseolina* and disease intensity in clusterbean. *Indian Phytopath*. 51: 175-179.

Lodha, S. and Burman, U. (2004). Towards managing soil-borne plant diseases in arid region. Central Arid Zone research Institute, Jodhpur, India, Tech Bull, 84pp.

Lodha, S. and Singh, M. (1984). Quantitative determination of *Macrophomina phaseolina* in grass-legume inter cropping. *Annals of Arid Zone*. 23: 259-261.

Lodha, S. and Solanki, K. R. (1992). Influence of solar heating on the control of *Macrophomina phaseolina* and weeds in arid environment. *Indian J.Agril. Sci*. 62: 838-843.

Lodha, S. and Solanki, K. R. (1993). Inheritance of dry root rot resistance in *clusterbean*. *Indian Phytopath*. 45: 430-433.

Lodha, S., Mathur, B.K. and Solanki, K. R. (1990a). Factors influencing population dynamics of *Macrophomma phaseolina* in arid soils. *Plant and Soil.* 125: 75-80.

Lodha, S., Israel, S. and Mawar, R. (2001). Combating diseases for better cumin yield. *Indian Frmg.* 52: 17-20.

Madhukar, H. M., Naik, M. K. and Patil, M. G. 2002. Evaluation of chilli genotypes and bioagents against Fusarium wilt of chilli, *Proceeding of National symposium on Crop Protection and WTO An Indian Scenario*, Feb. 2002. CPCRI, Kassargod.

Marois, J. J., Mitchell, D. J. and Sonoda, R. M. 1981. Biological control of Fusarium crown rot of tomato under field conditions. *Phytopath.* 712:1257-1260.

Mawar, R. and Lodha, S. (2002) Integration of *Brassica* amendments and summer irrigation for the control of *Macrophomina phaseolina* and *Fusarium* in hot arid region. *Phytopathologia Mediterranea.* 41: 45-54.

Menzies, J. D. 1963. The direct assay of plant pathogen populations in soil. *Ann. Rev. Phytopath.* 1:127-142.

Mitchell, J .E. 1979. The dynamics of the inoculum potential of populations of soil- borne plant pathogens in the soil ecosystem. (*In*:) Schippers, B. and Gams. W, (Eds.) *Soilborne Plant Pathogens.* pp-686. Academic, London.

Naik, M. K., 1990. Wilt of Watermelon caused by *Fusarium oxysporum* Schlecht. *Ph.D. Thesis.* Division of Mycology and Plant Pathology, *Indian Agricultural Research Institute*, New Delhi, India.

Naik, M. K. and Reddy, M. V. 1993a. Economic threshold level of inoculum for Fusarium *udum* in pigeonpea. (*In*:) Rajagopal. D, (Eds.). *Soil Organism and Sustainability.* Indian Society of Soil Biol. and Ecol. VAS, GKVK, Bangalore.

Naik, M. K. and Reddy, M. V. 1993b. Relationship between Fusarium wilt incidence and yield loss in pigeonpea. *Int. Pigeonpea Newsltr.* 27:24.

Naik, M. K. and Sen, B. 1994. Effectiveness of biocontrol agents against a spectrum of *Fusarium* isolates causing wilt of watermelon. *Indian J. of Plant Proto.* 21: 19-22.

Naik, M. K. and Sen, B. 1999. Symptomless carriers of Fusarium wilt of watermelon. *J. Soil. Biol and Ecol.*17:21-24.

Naik, M. K., 2003. Challenges and Opportunities for Research in Soil-borne Plant Pathogens with Special Reference to *Fusarium* Species. *J.Mycol. Pl. Pathol.* 33 (1): 1-14.

Naik, M. K. and Sen, B. 2002. Management of Fusarium wilt of watermelon by rhizosphere manipulation of non-curbitaceous crops. *Indian Phytopath.* 55 (1): 95-96.

Naik, M. K., Reddy, M. V. and McDonald, D. 1995. Disease Prediction model study for Fusarium wilt of pigeonpea. (*In*:) Agashe, S. N, (Eds.) *Proceedings of International Conference on Aerobiology.* Malhotra Publishing House, New Delhi.

Naik, M. K., Reddy, M. V., Raju, T. N. and McDonald, D. 1997b Dynamics of *Fusarium udum* population in relation to resistance. (*In*:) Singh, D. V. J (Eds.) *Proceedings of Golden Jublee International Conference of Indian Phytopathological Society on Integrated plant Disease Management for Sustainable Agriculture*, New Delhi, India. Nov. 1997.

Naik, M. K., Reddy, M. V., Raju, T. N. and Rego, T. J. 1996. Pigeonpea production systems in relation to sustainable management of Fusarium wilt. (*In*:) Pathak, P. S. and Goel. B, (Ed.), *Studies in Indian Agroecosystem*: 45- 50. National Institute of Ecology, New Delhi.

Nayeema, J., Ahmed, N. Tanki, M. I. and Dar, G. M. 1995. Screening of hot pepper germp1asm for resistance to usarium wilt *(F. pallidoroseum* [Cooke] Sacc.) *Capsicum and Eggplant, Newsletr*. 14:68- 71.

Nees von Esenbeck, C. G., (1816). "Dan System der pilze and Schwamme". Stahelschen Buchhandlung, Wurzburg.

Nene, Y. L. and Kannaiyan, J. 1982. Screening pigeonpea for resistance to *Fusarium* wilt. *Plant Dis*. 66: 306-307

Nene, Y. L., Haware, M. P. and Reddy, M. V. 1981. Information bulletin, 10. pp-ll. ICRISAT, Patancheru. Andhra Pradesh, India.

Nene, Y. L., Haware, M. P. and Reddy, M. V., Phillips, J. C., Castro, E. L., Kotasthane, F. R., Gupta, D., Singh, G., Ahukla, P. and Sah, R. P. 1989. Identification of broad based and stable resistance to wilt and root rot in chickpea. *Indian Phytopath*. 42: 499-505.

Netzer, D. 1976. Physiological races and soil population levels of *Fusarium* wilt of watermelon. *Phytoparasitica* .4: 131-136.

O'Donnell, K., Cigelink, E., and Nirenberg, H.I. 1998. Molecular sytematics and phylogeography of *Gibberella fujikuroi* species complex. *Mycologica*. 90: 465-493.

Palodhi, P. R. 1975. Studies on the wilt of muskmelon caused by *Fusarium solani* (Mart.) Sacco. pp-45. *M.Sc. Thesis*. Indian Agricultural Research Institute. New Delhi, pp 45.

Palodhi, P. R. and Sen, B. 1983. Perpetuation of cucurbit wilt pathogen in river bed cultivation. *Indian J. Mycol. And Pl. Pathol*. 13: 164-168.

Papaviaz, G.C. and Klag, N.G. (1975). Isolation and quantitative determination of *Macrophomina phaseolina* from soil. *Phytopath*. 65: 182-187.

Pfender, W. F., Rouse. D. I. and Hagedorn. D. J. 1981. A most probable number method of estimating inoculum density of *Aphanomyces eutiches* in naturally infested soil. *Phytopath*. 71:1169-1172.

Porter, I. J. and Merriman, P. R. 1983. Effects of solarization of soil on nematode and fungal pathogens at two sites in Victoria. *Soil Biol. Biochem*. 15:39.

Prasad, R.D., Sharma, T.R., Jana, T.K., Prameela Devi, J., Singh, N.K. and Koundal, K.R., 2004, Molecular analysis of genetic variability in *Fusarium* species using micro satellite markers. *Indian Phytopath.* 57(3):272-279.

Punja, Z. K. 1988. *Sclerotium (Athelia) rolfsii,* a pathogen of many plant speices. pp. 523-534. (*In:*) G. S. Sidhu (Ed.) *Genetics of Plant Pathogenic Fungi.* Vo16. Academic Press. London.

Punja, Z. K. and Damiani, A. 1996. Comparative growth, morphology, and physiology of three *Sclerotium* species. *Mycologia.* 88: 694-706.

Pullman, G. S., De-yay, J. E., Garber, R. H. and Weinhold, A. R. 1981. Soil solarization and thermal death: A logarithmic relationship between time and temperature for four soilborne populations of *Verticillium dahiae, Pythium* spp, *Rhizoctonia solani* and *Thielaviopsis basicola. Phytopath.* 71: 954.

Purkayastha, S., Kaur, B., Dilbaghi, N. and Chaudhury. 2000 Characterization of *Macrophomina phaesolina,* the charcoal rot pathogen of cluster bean using conventional techniques and PCR based molecular markers. *Plant Pathol.* 55: 106-116.

Rao, A. S. and Rao, M. V. 1963. Inoculum potential and *Fusarium* wilt of cotton. *Nature,* 200: 598-599.

Rao, M. V. and Rao, A. S., 1966. *Fusarium* wilt of cotton in relation to inoculum potential. *Trans. Br. Mycol. Sci.*49:403-409.

Rao, D. N. and Shinde, V. K. (1985). Intheritance of charcoal resistance in sorghum. *J. Maharashtra Agric. Uni.* 10: 54-56.

Rouxel, F. Alabouvette, C. and Masson, J. P. 1987. Interet de la modelisation pour l' etude de la receptivite des sols aux maladies d' origine tellurique: exemples d' applications a la hernie des crucifers et aux fusarioses vascularies. C. R. Seances Acad. *Agric. Fr.* 73: 137-149.

Saccardo, P. A. (1910). "Sylloge Fungorum," Vol. XIX. Pira.

Schorth, M. N., Toulsoun, T. A. and Snyder, W. C. 1963. Effect of certain constituents of bean exudates on *germination* of chlamydospores of *Fusarium solani* f. sp. *phaseoli* in soil. *Phytopath.* 53:809-812.

Schneider, R. W. 1982. Suppressive Soils and Plant Disease. American Phytopathological Society. St. Paul, 88pp.

Sen, B. 2000. Biological control: A success story. Jeensanidhi Award Lecture. *Indian Phytopath.* 53:243- 249.

Sharma, J. 1989. Spectrum and application of the antagonist in the biological control of Fusarium wilt of cucurbits. *Ph. D. Thesis.* Division Mycol. Pl. Pathol. IARI, New Delhi. 123+XXVP.

Sharma, M., Guptha, S.K. and sharma, T.K. 2005. Characterization of variability in *Rhizoctonia solani* by using morphological and molecular markers. *Phytopath.* 153: 449-456.

Sharma, J. and Sen, B. 1991. Antagonistic potentials of soil and dung isolates of Aspergillus spp against *Fusarium solani* (Mart.) Sacco causing wilt of cucurbits. *Indian J. Microbial Ecol.* 2:91-97.

Sheikh, A.H. and Ghaffar, A. 1984. Reduction in the viability of sclerotia of *Macrophomina* phaesolina with polythene mulching of soil. *Soil Biol. Biochem.* 16:77-79.

Singh, S. and Lodha, S. (1986). Varietal resistance of cowpea to *Macrophomina phaseolina* (Tassi) Goid, causing dry root rot and its control. *Indian J.Agril. Sci.* 56: 552-555.

Smith, E. F., (1899). U. S. Dep. Agric. Bur. Plant Ind. Bull. 17.

Sneh, B., Dupler, M., Elad, Y. and Baker, R. 1984. Chlamydospore germination of Fusarium *oxysporum* f. sp. *cucumerinum* as affected by fluorescent and lytic bacteria from Fusarium suppressive soil. *Phytopath.* 74: 1115-1124.

Snyder, W. G., and Hansen, H. N. (1940). The species concept in *Fusarium. Amer. J. Bot.* 27, 64-67.

Stapleton, J. J. and Devay, J. E. 1982. Effect of soil solarization on population of selected soilborne microorganisms on growth of deciduous fruit tree seedlings. *Phytopath.* 72:323-326.

Stotzky, G. and Martin, R. T., 1963. Soil mineralogy in relation to the spread of Fusarium wilt of banana in Central America. *Plant Soil.* 18: 317-338.

Stover, R. H. 1962. Fusarial wilt of bananas and other *Musa* species. CMI, *Phytopath.* Pap. 4, 117p.

Stutz, E., Kahr, G. and Defago, G. 1989. Clays involved in suppression of tobacco black root rot by a strain of *Pseudomonas fluorescens. Soil Biol. Biochem.* 21: 361-366.

Toussoun, T. A., Nash, S. M. and Snyder, W. C. 1960. The effect of nitrogen sources and glucose on the pathogenesis of *Fusarium solani* f. sp. *phaseoli. Phytopath.* 50: 137-140.

Toussoun, T. A. 1975. Fusarium suppressive soils. pages 145-151 (*In:*) G.W. Bruehl, (Ed.), *Biology and Control of Soil Borne Plant Pathogens.* American Phytopathological Society, St Paul, 216 pp.

Toussoun, T. A. and Nelson, P. E. 1976. *Fusarium:* A Pictorial Guide for the Identification of *Fusarium* species, pp-43. Pennsylvania State University Press. University Park, Pennsylvania, USA.

Upadhyaya, R. S. and Bharat Rai. 1981. Effect of cultural practices and soil treatments on incidence of wilt disease of pigeonpea. *Plant and Soil.* 62:309-312.

Wellman, F. L. 1939. A technique for studying host resistance and pathogenicity in tomato Fusarium wilt. *Phytopath.* 31:281-283.

Windels, C. E., Burnes, P. M. and Kommedahl, T. 1988. Five year preservation of *Fusarium* spp on silica gel and soil. *Phytopath*. 78:107-109.

Woodcock, W. P. 1962. Influence of aeration of exudation from seeds and on growth of root rot fungi. *Phytopath*. 52: 927.

• • •

Advances in Soil Borne Plant Diseases, 2008
Editors.: Naik and Devika Rani, pp. 33-53
New India Publishing Agency, New Delhi

2

Rhizoctonia solani Kuhn Soil-Inhabiting Nonspecialized Pathogen in Agroecosystem

Arvinder Kaur and H.S. Chaube

INTRODUCTION

Rhizoctonia solani Kuhn (teleomorph: *Thanatephorus cucumeris* (Frank) Donk is cosmopolitan in soils. It is a destructive plant pathogen with an almost unlimited host range. The fungus inhabits in soils and causes damping-off of seedlings, root, crown and stem rots, and occasionally foliar blight (Baker, 1970). *R. solani* is an important symbiont with the saprophytic seedling stage (protocorm) of many orchids (Warcup, 1985). It aggressively colonizes organic matter and is one of the fastest growing fungi. According to available information, the isolates of *R. solani* possess the following-characteristics (Plate 2.1).

(a) Some shade of brown hyphal pigmentation,

(b) Branching near the distal septum of cells in young vegetative hyphae,

(c) Constriction of hyphae and formation of a septa short distance from the point of origin of hyphal branches,

Centre of Advanced Studies in Plant Pathology,
Govind Ballabh Pant University of Agriculture and Technology,
Pantnagar-263145, Uttaranchal, India

(d) Cytoplasm is interconnected through a septal pore (dolipore),

(e) Multinucleate cells in young vegetative hyphae, and

(f) No conidium, except monilioid cells,

(g) The absence of clamp connections, rhizomorph, only sclerotia as soil-borne propagules.

Thanatephorus cucumeris, the sexual form of *R. solani* belonging to phylum Basidiomycota and order Ceratobasidales, is rarely seen except on stems of *Solanum tuberosum* (Adam, 1988). The fruiting thallus is arachnoid to hypochnoid and is composed of short cylindrical holobasidia with stout sterigmata arising in asymmetrical cymes. The basidiospores can germinate repeatedly by forming a sterigma and a ballistospore, particularlary if they are discharged onto hyphae of *R. solani.*

THE SPECIES CONCEPT

In 1815, the genus concept of *Rhizoctonia* was established by the great French mycologist, A. P. De Candolle for the violet root-rot organism. He called *Rhizoctonia* (from the Greek 'death of roots') because it rapidly attacked and killed the roots of phanerogamic plants. As reviewed by Parmeter and Whitney (1970), "the basic characters of the genus, as set forth by De Candolle, were the production of sclerotia of uniform texture with hyphal threads emanating from them and the association of the mycelium with roots of living plants.

Rhizoctonia solani Kuhn also known by the genus name *Moniliopsis* (Moore, 1987), is a common and highly heterogeneous fungal species. Fungi classified as *Rhizoctonia* species represent an amalgam of taxonomically diverse groups that differ in many significant features, including their sexual stages (teleomorph), asexual stages (anamorph) and other characters (Sneh *et al.,* 1991). Because of the importance of many *Rhizoctonia* species as plant pathogens, a variety of useful approaches have been developed for identifying many groups of *Rhizoctonia* spp. based on the information from teleomorph, anastomosis behaviour, and molecular biology. *Rhizoctonia* species may belong to several different orders of basidiomycetes and even ascomycetes (Moore, 1987). The most important *Rhizoctonia* spp. associated with plant diseases belong to three major groups (Ogoshi, 1987): one is the multinucleate *Rhizoctonia,* which has three or more nuclei per cell and teleomorph in the genus *Thanatephorus* Donk. Another is the binucleate *Rhizoctonia,* which has only two nuclei per cell (rarely one or three), and teleomorph in the genus

Ceratobasidium Rogers. The third group includes *R. oryzae* and *R. zeae* , which are multinucleate & have teleomorph in the genus *Waitea* Warcup and Talbot.

R. solani (Thanatephorus cucumeris) is the representative of *Thanatephorus* though other species, *Thanatephorus sterigmaticus* (Talbot, 1965), *Thanatephorus orchidicola* (Warcup & Talbot, 1966) and *Thanatephorus corchorus* (on jute) are also multinucleate (Tu *et al.,* 1977). However, these three species are not commonly isolated and are believed to have limited distribution (Ogoshi, 1987).

ANASTOMOSIS GROUP (HYPHAL FUSION)

R. solani is varied in its pathogenicity, sclerotial morphology, cultural appearance on media, physiological characteristics etc. To understand the diversity in *R. solani,* the most important criteria is anastomosis group and to date 14 anastomosis groups (AGs) are described (Carling *et al.,* 2002). Isolates of *R. solani* are assigned to AG by pairing the isolates with 'tester' strains and observing the hyphal fusion. Isolates may be paired on (a) water agar in petri -dishes (Parmeter *et al.,* 1969; Hyakumachi and Vi, 1987); (b) cellophane/ nitrocellulose/ dialysis membrane placed on agar media (Parmeter *et al.,* 1969; Murray, 1981; Carling *et al.,* 1987); (c) agar - coated slides or cover slips placed on moist filter paper or on agar in petridishes (Tu *et al.,*1969; Murray, 1981) or (d) on a bare slide placed in a moist chamber (Kronald and Stanghellini, 1988).

Anastomosis grouping is most useful method for grouping of *R. solani.* There are three types of hyphal fusion: perfect fusion, imperfect fusion and contact fusion (Plate 2.2). The process of perfect fusion of hyphae of *R. solani* includes hyphal growth, secretion of one or more attracting substances, attraction to the substances, contact of hyphae, cessation of hyphal growth, formation of branch like projections, dissolution of cell walls and connection of protoplasms (Yokoyama *et al.,* 1985). Imperfect fusion differs in that dissolution of cell walls is followed by death of cells (Yokoyama and Ogoshi, 1984). In contact fusion, hyphal contact and apparent contact of wall is seen but there is no membrane to membrane contact. Both mycelia must grow in close contact with each other. If hyphal fusion occurs, the isolates belong to the same AG, usually attraction of other hyphae and death of fused cells are observed. If the attraction, fusion and hyphal death do not occur, these isolates belong to different AGs.

It is important to know the AGs of isolates for studying the pathogenicity of *R. solani.* As a plant pathogen, *R. solani* is regarded as an unspecialized fungus. However, looking at the isolates from the viewpoint of AGs, there is certain host specificity in them, though it is not perfect. For example, a number of isolates of AG-2-1 are from the cruciferae, whereas AG-2-2 have been found infecting the chenopodiaceae (AG-2-2-IV), mainly sugarbeets or the graminae (AG-2-2 III B). Each AG appears to have distinctive morphological and physiological characteristics. Each appears genetically isolated; this is supported by data from serological comparisons (Adams and Butler, 1979), electrophoresis of soluble proteins (Kuninaga and Yokosawa, 1984) and DNA base sequence homology (Vilgalys, 1988). Many anastomosis groups are subdivided into intraspecific groups (ISGs) on the basis of their cultural, virulence, molecular or other characters.

THE PATHOGEN: TYPES OF DISEASES CAUSED BY R. SOLANI

The diseases of plants caused by *R. solani* are many and varied. The adaptability of this fungus to its environment and its ability to survive under a multitude of conditions make it one of the most versatile organisms.

Seed decay, Damping -off of seedlings and Root- rot:

The fungus may invade the seeds while still in the fruit, decaying it there or merely infecting it. The decay process is then resumed after the seed is planted and before germination. The seed may also be invaded by growth of the fungus from infested soil in which it is planted; the greater the amount of inoculum in the soil, more rapidly the decay occurs. The pre-emergence phase of damping-off disease is an extension of seed decay. This delayed attack may result from unfavorable environment or insufficient inoculum for faster action, decay only occurs following germination. The post-emergence phase represents still a further delay in attack, expression of symptoms, or both. Symptoms may develop any time after emergence through the soil surface, until the seedling is past the most susceptible juvenile stage. The fungus attacks underground fleshy roots causing root-rot.

Aerial (Web, leaf, -thread) blights:

When the environment is warm, rainy, humid, *R. solani* may spread through the tops of plants. In some cases, these isolates of the fungus lead an aerial existence independent of the soil (Plate 2.3).

Stem canker

Stem lesions may develop on fully mature plants with well developed tissue. Strands of mycelium and sometimes sclerotia appear on the surface of stem. Roots remain intact until late in the disease, but the tops wilt, turn brown and die.

Bottom rot or Head rot

After development of head in cabbage or lettuce, *R.solani* may invade it from below. Leaves in contact with the soil are invaded, and the fungus spread to the stem and to the leaves above. Initially, the lesions may be sharply defined and brown, they may dry out if the weather becomes drier, turning papery and brown (Plate 2.3).

Crown and Bud rot

Infection occurs in the young leaflets or in leaf bases, and causes petiole decay. The crown of the plant may be killed and the fungus gradually advances into the top of the fleshy root, causing a dry brown decay.

Rot of parts in contact with soil

Any above ground plant parts that come into contact with soil heavily infested with *R. solani* may become infected and finally rot.

Storage rots, black scurf and blemishes

The best-known symptom of *R. solani* is the appearance of black sclerotia on potato tubers. The sclerotia on tubers may germinate in storage and grow through the cloth bags and rot them (Plate 2.3).

SURVIVAL

Rhizoctonia solani is a common soil borne pathogen that has many plant hosts, forms sclerotia in soil, and survives for a long period in the absence of a cultivated host, either as sclerotia or thick walled brown hypha in plant debris. These sources of survival of pathogen can be grouped into three categories as follows:

(i) Infected host (cultivated and collateral as reservoir of inoculum)

(ii) Saprophytic survival outside the host.

(iii) Dormant structure (i.e. sclerotia) in or on the host or soil.

The fungus has a wide host range, which includes crops and common weed species belonging to families such as leguminosae, graminae, solanaceae, brassicaceae, cucurbitaceae, malvaceae, chenopodiaceae, cyperaceae, commelinaceae and caryophyllaceae. Kannaiyan and Prasad (1979) described that *Rosolani* occurs on a number of weed hosts namely *Cyperus rotundus, Digitaria longiflora, Panicum repens* and *Cynodon dactylon.* The isolates of *R. solani* from weed host have been reported to be pathogenic to several crop plants (Bains and Dhaliwal, 1993). *R.solani* causing foliar blight disease of soybean was able to infect several weeds like *Echinochloa crusgalli, Brachiaria platyphylla, Xanthium strumarium, Ipomea hederacea* var *intergriuscola, Sesbania exaltata, Rotthoellia cochinchinesis, Sorghum halepense, Digitaria sunguingalis, Aeschynomene virginica, Sida spinosa, Cyperus rotundus, Melochia corchorifolia, Senna obtusifolia* and *Amaranthus hybridus* (Black *et al.*, 1996).

COLONIZATION AND GROWTH OF R. SOLANI IN SOIL

Saprophytic colonization and substrate exploitation by *R. solani*

According to Garrett (1950, 1956) Competitive Saprophytic Ability (CSA) is the summation of physiological characteristics that makes for success in competitive colonization of dead organic substrates". He further suggested that CSA of a fungus in soil would depend not only on growth rate, production of antibiotics or toxins, enzymatic complement and potential, and tolerance of toxins or antibiotics produced by other micro-organisms, but also on other micro-determinants of the soil ecosystem, including number and variety of antagonists exploiting a substrate. CSA is an intrinsic characteristic and important for substrate exploitation and survival of the parasite in soil. Generally, strong saprophytes are usually able to survive in soil for a longer period of time than weak saprophytes. This concept is true for *R. solani* where degree of saprophytism and longevity of *R. solani* are strongly interrelated. There is evidence that *R. solani* persists in soil primarily as a saprophyte in tissues infected during parasitism or by saprophytically colonizing dead plant tissues in which it can remain dormant or active for long period of time (Boosalis and Scharen, 1959). Interestingly, Sneh *et al.*, (1966) observed a high degree of correlation between the soil infestation level and the saprophytic activity of *R. solani*.

BIOTIC FACTORS AFFECTING PATHOGENICITY:

A. Infective organ

There are some infection sites of *R. solani* on the plant surface such as epidermis, stomata and wounds. The infection structures vary with hosts, isolates and parts of plants. In cotton, *R. solani* penetrated the host epidermis with single hyphae or through infection cushions on host hypocotyls, while it mainly showed cuticle penetration and stomata invasion in cotyledons (Nakayama, 1940). The epidermis of cotton bolls and bracts are penetrated by germinated basidiospores, infection cushion -like structures and hyphae (Luke *et al.*, 1974). These results suggested that there are differences in the behaviour of *R. solani* as a pathogen with situation, age of host plants and characteristics of isolates. It is considered that *R. solani* recognizes some specific signals from the structures or exudates of host surfaces and behaves differently as a reaction to these signals (Akino and Ogoshi, 1995).

Basidiospores of *R. solani* germinate and form appressoria, penetrate host epidermis and form primary lesions. Hyphae developed from the first lesion reinfect host tissue through stomata, and induce secondary lesions (Naito and Sugimoto, 1978). The infection behaviour of hyphae and basidiospores from the same isolate was different. The basidiospore is important in the ecology of *R. solani* because it confers airborne characteristics to the pathogen.

B. Susceptibility and resistance in host plants

There are many reports on the differences in susceptibility of some host cultivars to *R. solani.* Late maturing cultivars are able to escape the disease (Hori *et al.,* 1981). This resistance is attributed to ecological and morphological characteristics of these cultivars. Stockwell and Hanchey (1987) reported the accumulation of lignin and phenolic substances in lesion borders fortified by the pathogen on bean hypocotyls. Cell walls of lesion borders were macerated by cell wall degrading enzymes if they were delignified. This fact suggested that lignifications of the cell wall obstructed the development of lesion areas. Broglie *et al.,* (1991) reported that when the bean chitinase gene was transferred to and expressed in tobacco and canola, their resistance against *R. solani* was increased.

C. Enzymes

Many hydrolytic enzymes play an important role in disease development by *R. solani.* These enzymes digest plant cell walls or intercellular materials. The positive correlation between the production of these enzymes by the fungus and its pathogenicity, and between resistance of plant tissue against fungal enzymes and resistance of plants against a disease has been reported by Baker and Walker (1962).

D. Effect of plant exudates

There are two types of effects of plant exudates on *R. solani:* (a) activation of sclerotia (resting structure), and (b) the induction of infection structures on the host surface. Being a soil borne plant pathogen, *R. solani* usually exists in soil in the saprophytic or resting stages. The activation by host exudates is necessary for the fungus to express its behavior as a plant pathogen (Kohmoto *et al.,* 1995)

Plant exudates activate saprophytic activity of the pathogen in soil and increase the opportunity of infection of the host plant. On potato, maturation and growth of sclerotia and pigmentation of hypha of *R. solani* are affected by volatile or unstable compounds of host exudates (Dijist, 1988). In groundnut, growth of the fungus changed from luxuriant to poor and occurrence of infection cushions from frequent to none as the age of the seedling increased due to a quantitative and qualitative decrease in the exuded compounds (Reddy, 1980). The decrease of exudates with age is due to the decreased permeability of cells with tissue maturation, and this resulted in increasing resistance of the plant with age. Additionally, the plant exudates act as a source of nutrients for mucilage production by the fungus to adhere to the host surface (Armentrout *et al.,* 1987).

E. Differences in anastomosis groups

There are many comparative studies on characteristics of isolates of different anastomosis group (AGs) of *R. solani,* however, differences in pathogenicity within and between AGs is not fully characterized (Ogoshi, 1987). Most of the investigations are mainly of taxonomic significance for detection of variation in each AG; only few reports are available on relationship of AGs to pathogenicity of isolates.

Diversity of the isolates in each AG shows biochemical variation and differences in the structure of the proteins. Abe *et al.,* (1969) demonstrated that the isolates, which reacted with AG-2-2 IV antiserum, were pathogenic on sugar beet, that is; they are *R. solani* AG-2-2 IV isolates. It is also noted that structures on constructions of pectic enzymes varied among AGs and their subgroups in *R. solani* (Cruickshank, 1990).

F. Association with other microorganisms

The effect of nematodes on *R. solani* is indirect, that is they are mediated through the host and, in most cases where their effect is not neutral, nematodes enhance infection (Reynolds and Hansen, 1957). Wiggins and Curl (1979) showed that damage to cotton seedlings caused by *R. solani* could be decreased by the suppressive effect of collembola (Springtails) on the pathogen. Although considered to be unspecialized in their grazing on the fungi, they do show some preferences. *Proisatoma minuta* causes more suppression of *R. solani* than of *S. rolfsii* (Curl *et al.,* 1985). Nitrogen fixing bacteria e.g. *Rhizobium liguminosarum* and *Bradyrhizobium japonicum,* all have decreased infection by *R.solani* (Ehteshamul- Haque and Ghaffer, 1993). Many species of *Streptomyces* are producers of potent antibiotics, which affect soilborne pathogens. For example, *S.hygroscopilus* suppresses *R. solani* on pea (Rothrock and Gittlieb, 1984).

DISEASE DECLINE

Rhizoctonia blight of radish in clay -loam soil increases for three successive plantings and then declines (Henis *et al.*, 1979). The decline occurs only when *R. solani* is present and active. The fungus, *R. solani,* decline does not occur if the soil contains the pathogen but not the host, or the host without the pathogen. Henis and co-workers (1979) explored the changes in microflora associated with such decline and observed an increase in *Trichoderma* spp. As decline intensifies, the number of *Trichoderma* spp. propagules increases. Chet and Baker (1980) continued these studies and established *T harzianum* as the cause of such decline. They further studied *R. solani* infested soil and compared the development of disease suppression in a clay loam at pH levels of 8.1, 6.5 and 5.7. At pH 8.1, disease increased with successive cropping of the three hosts. At the lower pH levels, the disease decreased.

In this case, *T. harzianum* produced conidiophores most abundantly at low pH. Spore germination and growth were favored by acidity. No evidence of antibiotics was involved in the suppression of *R. solani* but direct parasitism was important as hyphae of *T. harzianum* coil around hyphae of *R. solani. Trichoderma harizanum* produced 1,.3 glucanase and chitinase to lyse the wall of *R. solani* (Hadar *et al.*, 1979). The pH optimum for β-1, 3 glucanase from *T. viride* is 4.5 and for its chitinase, 5.3 (Jones and Watson, 1969). This additionally supports the importance of acidity to the effectiveness of *Trichoderma* spp. in antagonistic interactions.

The soil, infested with *R. solani* in radish field had a 7 to 8 fold increase in *Trichoderma* spp. Additionally, *Streptomyces* and bacteria were unchanged (Liu and Baker, 1980). The increase of *T. harzianum* is attributed to the presence of active propagules of *R. solani*. Hypha of *T. harzianum* encircles hypha of *R. solani* and lyses individual cells but do not enter them. Due to lysis, the hyphae loose their structural integrity to the extent that it cannot mobilize reserves leading to its decline.

Decline of sugar beet root rot *(R. solani* A G- 2-21V) in continuous cropping was studied in detail by Hyakunachi and Ui (1982). Continuous cropping in an experimental field resulted in increased root rot until the third year, a leveling off in the fourth year, and decreased root rot in fifth year, after which the cycle of increased and decreased root rot was repeated. This rapid decline in inoculum density in the previous year is believed to be biological in nature.

Jager and Velvis (1983) reported suppressive soils to *R. solani* from potato fields in the Netherlands. They proposed that this suppressiveness in slightly acidic, sandy, Pleistocene soils was conditioned by soil- borne or tuber- borne antagonists.

DISEASE MANAGEMENT

Cultural practices

Leach (1947) carried out extensive studies on the effect of temperature on incidence of damping -off disease of cotton and found that delaying the planting date in spring can improve their chances of escaping disease attack. Temperature and moisture strongly affect *R. solani* diseases through its effect on the biotic and abiotic components of the disease (Katan, 1981). Moisture is important in relation to leaf diseases and damping-off of seedlings, when crops are particularly susceptible to disease at certain stages of their growth e.g. web blight caused by *R. solani* (Palti and Katan, 1997). Singh (1996) had extensively studied control of soil- borne diseases using organic amendments. He found that amendment of soil with wheat straw saw dust and oil cakes reduced population of *R.solani* and black scurf disease of potato. Other cultural practices such as adjustment of planting date, mulching of soil, plant spacing, deep ploughing, use of raised beds, proper timing of irrigation and crop rotation have been found effective in decreasing the incidence and severity of *R. solani* diseases (Hillocks and Waller, 1997).

Soil solarization

Soil solarization is the advanced field technology for control of soil- borne pathogens of nursery beds. It is a simple, less expensive and non-chemical method of soil disinfestations. It involves use of heat as lethal agent by capturing solar energy with traps like white polyethylene. In addition to high temperature, there are other effects like changes in the physico-chemical properties of soil, moisture level, gas composition and shift in microbial population. In Israel, soil solarization has been found effective in controlling *R. solani* on potato and onion. Pullman *et al.,* (1981) demonstrated that solarization with polyethylene sheet effectively controlled the *R.solani* of cotton. Population of *R.solani* was also decreased by solarization in tomato field (Tamietti and Garibaldi 1989).

Use of VAM fungi

Mycorrhiza, representing symbiotic fungal associations with plant roots normally help in uptake of nutrients, have also been known to help in reducing diseases. Studies have clearly shown that the root systems having mycorrhizae are less susceptible to attack by soil-borne pathogens than non-mycorrhizal systems. In *Brassica napus* fields, Iqbal *et al.,* (1977) observed that VA-Mycorrhizal plants suppressed root infection by *R.solani*. Mycorrhizal poinsettia plants also have been shown to tolerate infection by *R. solani* (Stewart and Pfleger, 1977). Similarly, Jalali and Thareja (1981) demonstrated significant reduction in the host infection by *Rhizoctonia* species after mycorrhizal inoculation. Zambolin and Schenck (1983) observed that plants colonized by *Glomus mosseae* appeared to tolerate infection by *R.solani* better than the non-mycorrhizal plants. The infection levels of the pathogen in the roots were similar in both mycorrhizal and non-mycorrhizal plants. Mycorrhizal plants had greater concentrations of phosphorus in the leaves than the non-mycorrhizal plants. Reduced root exudation resulting from improved phosphorus uptake may have an effect on reducing the inoculum density of *R. solani.*

Chemical control

A wide range of fungicides have been evaluated and found effective against *R. solani.* Chet *et al.,* (1982) reported that soil and bulb -borne *R. solani* was reduced by soil treatment with quintozene. Although soil treatment with fungicides is very expensive, yet it has been widely used in glass house, seed-beds and high value commercial crops. This fungicide is commonly used in the control of bean root rot (Abawi, 1989) as a furrow spray (Crossan *et al.,* 1963), and in a mixture of captan, chloroneb and metalaxyl applied as slurry (Abawi, 1989). Choroneb has also proved effective as dry seed dressing in controlling seedling mortality of cowpea (Wiliams, 1975). Benomyl, carbendazim and captafol (Galvez *et al.,* 1989) can give good control of web blight when applied as a foliar spray. For control of *R. solani* in vegetables, the mixture of captan 41 + carbendazim 9 at a rate of 64+ 14 g a.i per 100 l at a dose of 0.5 per plant as drench is effective (Utkhede and Gupta, 1996). In another study, foliar spray of Bavistin (0.05%) along with Raxil (0.2%) or Bavistin (0.2%) was found highly effective in reducing web blight and increasing yield of french bean followed' by Topsin-M and Baycor (Mathew and Gupta, 1996) Flutolanil, a benzanilide systemic compound has been used as a protective and curative fungicide against *R. solani* (Hillocks and Waller, 1997).

Biological control

Several microorganisms including fungi like *Trichoderma* spp., binucleate *Rhizoctonia* spp, *Clostridium* spp., *Aspergillus* spp., bacteria and actinomycetes act as antagonists against *R. solani* (Kwok *et al.,* 1987; Cardoso and Enchandi, 1987; Roy, 1991; Hwang and Benson, 2002). Species of *Trichoderma* are effective against the fungus and inhibit the growth by direct parasitism and antibiosis (Kwok *et al.,* 1987).

Binuleate *Rhizoctonia* (BNR) species have been studied extensively against Rhizoctinia diseases (Ichielevich *et al.,* 1985, Hwang and Benson, 2002).

Possible mechanism of action by BNR includes induction of systemic host resistance and competition for invasion sites or nutrients (Cardoso and Echandi, 1987; Hwang and Benson, 2001). Poromarto and coworkers (1998) demonstrated that pre-colonization of root tissue by BNR was critical to induce host resistance for inhibition of *R. solani* on soybean. Binucleate *Rhizoctonia* isolates were more effective for suppression of stem and root rot than application of *Burkholderia cepcia* (strain 5-5B) (Hwang and Benson, 2002).

Specific strains of *Pseudomonas fluorescens,* which rapidly colonize plants significantly increased yield probably due to reduced disease levels. In canola / rapeseed antibiotic producing strains of *Pseudomonas fluorecens* inhibit the growth of *R. solani* by production of three antibiotics (procyanin, PYJTolnitrin and phenazine carboxamide) and thus are used in the management of root rot disease (Dahiya and Wood, 1987).

Induced resistance

Resistance, in plants being otherwise susceptible to diseases without changing their genetic constitution, can be induced by biotic as well as abiotic agents. Kempe and Sequeira (1983) suggested that resistance to a virulent strain of *Ralstonia solanacearum* was induced in potato by treatment of seed pieces with avirulent or incompatible strain of *R. solanacearum*. Jensen and coworkers (1998) observed good control of *R. solani* in cabbage *(Brassica oleracea)* when seeds imbibed in, benzothiadizole (GGA245704), an activator of plant defense system, was planted directly than dry seeds. Additionally, more number of non-treated seedlings developed symptoms of infection, while activator treatment provided significant protection, even at high inoculum rate used in the experiment.

Integrated Disease Management

R. solani is an unspecialized pathogen and there is little exploitable resistance to it available among crop species. However, there is some pathogenic specialization within the species complexes of *Rhizoctonia*. Cultural factors which favour the vigor of the host and optimal deployment of these resistance mechanism, while also disadvantaging the pathogen e.g. by reducing inoculum potential indicate how the effects of cultural practices can augment the effects of partial resistance. Many forms of the *Rhizoctonia solani* complex show host species specialization, which can permit some control through rotations involving crops, which are not susceptible to the same forms.

Complete control of *R. solani* is rarely achieved by single method of control. In the absence of resistant cultivars, less susceptible cultivars are combined with fungicidal/chemical seed treatment. In the case of raised nursery beds, it is more appropriate to combine soil solarization with organic amendments and rotation with crops, which do not share the pathogen. Integration of banodanil and *Trichoderma harzianum* were found effective for the control of Rhizoctonia pre-emergence damping-off of radish (Lifshitz, *et al.*, 1985). BINAB a mixture of *Trichoderma harzianum* and *T polysporum* has been registered for the control of *Rhizoctonia* damping -off complex (Utkhede and Gupta, 1996).

REFERENCES

Abawi, O.S. 1989. Root rots. *(In:)* Schwartz, H.F. and Pastor-Corrales, M.A. (Eds). *Bean Production Problems in the Tropics. Centro Internacional de Agricultura Tropical*, Cali, Colombia. pp:105-157.

Abe, H., Baba, T. and Takakuwa, T.1969. Serological reactions of root rot pathogen *(Rhizoctonia solani)* of sugar beets. *Ann. Phytopathol. Soc. Japn.* 35: 374.

Adam, G. C. 1988. *Thanatephorus cucumeris (Rhizoctonia solani)*, A species complex of wide host range. *(In:)* O.S. Sidhu. (Eds). *Advances in Plant Pathology: Genetics of Plant Pathogenic Fungi.* Volume 6. Academic Press, London. pp: 535.

Adams, G. C. Jr. and Butler, E. E. 1979. Serological relationships among anastomosis groups of *Rhizoctonia solani. Phytopath.* 69: 629-633.

Akino, S. and Ogoshi, A.1995. Pathogenicity and Host Specificity in *Rhizoctonia solani. (In:)* K. Kohmoto, U. S. Singh and R. P. Singh (Eds). *Pathogenesis and Host Specificity in Plant Diseases.* Volume II. Elsevier Science Ltd., Japan. pp: 37-49

Affilentrout, V.N., Downer, A.J., Grasmick, D.L. and Weinhold, A. R.1987. Factors affecting infection cushion development by *Rhizoctonia solani* on cotton. *Phytopath.* 77: 623

Bains, S. S. and Dhaliwal, H. S. 1993. Development of foci of blighted mung plants due to *Rhizoctonia solani* transmitted from weeds. *Indian Bota. Rept.* 11:11-14.

Baker, K. F. 1970. Types of *Rhizoctonia* diseases and their occurrence. *(In:)* J.R. Paffileter, Jr. (Eds). *Rhizoctonia solani; Biology and Pathology*. Univ. of California Press, Berkeley. pp: 125-148.

Baker, K. R. and Walker, J. C. 1962. Relationship of pectolytic and cellulolytic enzyme production by strains of *Pellicularia filamentosa* to their pathogenicity. *Phytopath.* 52:1119.

Black, B. D., Griffin, J. L., Russin, J. S., Snow, J. P. 1996. Weed host for *Rhizoctonia solani* causal agent for Rhizoctonia foliar blight of soybean *(Glycine max). Weed Tech.* 10: 865-869.

Boosalis, M. G. and Scharen, A. L. 1959. Methods for microscopic detection of *Aphanomyces euteiches* and *Rhizoctonia solani* associated with plant debris. *Phytopath,* 49: 192-198.

Broglie, K., Chet, I., Holliday, M., Cressman, R., Biddle, P., Knowlton, S., Mauvais, C.I. and Broglie, R.1991. Transgenic plants with enhanced resistance to the fungal pathogen *Rhizoctonia solani. Science.* 254: 1194.

Cardoso, J .E. and Echandi, E. 1987. Biological control of *Rhizoctonia* root rot of snap bean with binucleate *Rhizoctonia* like fungus. *Phytopath.* 77: 1548-1551.

Carling, D.E., Leiner, R.H. and Kebler, K. M. 1987. Characterization of a new anastomosis group (AG-9) of *Rhizoctonia solani. Phytopath,* 77:1609-1612.

Carling, D.E., Kuninaga, S. and Brainard, K.A., 2002. Hyphal anastomosis reactions, rDNA- Internal transcribed spacer sequences and virulence levels among subsets of *Rhizoctonia solani* anastomosis group-2(AG-2) and AG-BI. *Phytopath.* 92(1): 43 -50.

Chet, I., Elad, Y., Kalfon, A., Hadar, Y. and Katan, J. 1982. Integrated control of soil- borne and bulb-borne pathogens in Iris. *Phytoparasitica.* 10:229-236.

Chet, J. and Baker, R.1980. Induction of suppressiveness to *Rhizoctonia solani* in soil. *Phytopath.* 70: 994-998.

Crossan, D. F., Biehn, W.L., Morehart, A. L. and Baniecki, J.F.1963. Control of *Rhizoctonia* root rot of snap bean: low volume in-the-furrow versus high volume pre-plant fungicide sprays. *Pl. Dis. Reptr.* 47:109-111.

Cruickshank, R. H. 1990. Pectic zymograms as criteria in taxonomy of *Rhizoctonia. Mycol. Res.* 94:938.

Curl, E. A. R., Gudauskas, R.T., Harper, J. D. and Peterson, C.M. 1985.Effect of soil insects on populations and germination of fungal propagules *(In:)* Parker, C. A., Rovira, A. D., Moore K. J., Wong P. T. W. and Kollmorgen, J. F. (Eds.) *Ecology and Management of Soil-borne Plant Pathogens. Proceedings of the 4th International Congress of Plant Pathology.* Section 5. University of Melbourne. 1983. American Phytopathological Society.

Dahiya, J.S. and Woods, D.L.1987. Control of *Rhizoctonia solani,* causal agent of brown girdling root rot of canola / rapeseed by *Pseudomonas fluorescens. Can. J. Plant Pathol.* 9:275-276.

Dijist, G. 1988. Effect of periderm and water-soluble exudates of potato tubers on black scurf formation before and after halum destruction. *Neth. J Plant Pathol.* 94: 257.

Etheshamul-Haque, S. and Ghaffar, A. 1993. Use of *Rhizobia* in the control of root rot disease of sunflower, okra, soybean and mungbean. *J. Phytopath.* 138: 157-163.

Galvez, G. E., Mora, B. and Pastor-Corrales, M.A.1989. Web blight pp.195-209 in *Tropical,* Cali, Colombia.

Garrett, S.D. 1950. Ecology of the root inhabiting fungi. *Biol. Rev.* 25:220-254.

Garrett, S. D. 1956. Biology of root infecting fungi. Cambridge Univ. Press, London, New York.pp:293.

Hadar, Y., Chet, I. and Henis, Y. 1979. Biological control of *Rhizoctonia solani* damping-off with wheat bran culture of *Trichoderma harzianum. Phytopath,* 69: 64-68.

Henis, Y., Ghaffar, A. and Baker, R. 1979. Integrated control of *Rhizoctonia solani* damping off of radish: effect of successive plantings, PCNB and *Trichoderma harzianum* on pathogen and disease. *Phytopath.* 68: 900-907.

Hillocks, R.J. and Waller, J. M. 1997. *Soil-borne Diseases of Tropical Crops.* CAB International, U.K. pp: 419.

Hori, M., Anraku, M. and Matsumoto, K. 1981. Resistance of rice cultivars from Japan and other countries to sheath blight. *Kinki- Chugoku Agric. Res.* 62: 3

Hwang, J. and Benson, D. M. 2002. Biocontrol of Rhizoctonia stem and root rot of Poinsetta with *Burkholderia cepacia* and Binucleate *Rhizoctonia. Plant Dis.* 86:47-53.

Hwang, J. and Benson, D. M. 2001. Resistance of poinsettia cultivars against *Rhizoctonia* stem rot induced by binucleate *Rhizoctonia. Phytopath.*91: 541.

Hyakumachi, M. and Vi, T .1987. Non-self-anastomosing isolates of *Rhizoctonia solani* obtained from fields of sugar beet monoculture. *Trans. Br. Mycol. Soc.* 89:155-159.

Hyakuma, ti1i, M. and Vi, T. 1982. Decline of Rhizoctonia root rot during sugar beet monoculture. *Mem. Fac. Agric. Hokkaido Univ.* 13: 445-454.

Ichielevich- Auster, M.,Sneh, B., Koltin, Y. and Barash, 1.1985. Supperssion of damping - off caused by *Rhizoctonia* species by a non- pathogenic isolate of *R. solani. Phytopath,* 75:1080-1084.

Iqbal, S.H., Qureshi, K.S. and Ahmad, J.S.1977. Influence of Vescular Arbuscular Mycorrhiza on damping-off caused by *Rhizoctonia solani* in *Brassica napus. Biologia.* 23: 197-208.

Jager, G. and Velvis, H, 1983. Suppression of *Rhizoctonia solani* in potato fields occurrence. *Neth. J Plant Pathol.* 89:21-29.

Jalali, B.L. and Tbareja, M.L.1981. Management of soil-borne plant pathogens by VA Mycorrhizae model system (Abstr.). *Indian Phytopath.* 34: 115 -116.

Jensen, B. D., Latunda-Dada, A. A., Hundson, D. and Lucas, J.A.1998.Protection of *Brassica* seedlings against downy mildew and damping -off by seed treatment with CGA 245704 an activator of systemic acquired resistance. *Pesticides.* 52:63-69

Jones, D. and Watson, D.1969. Parasitism and lysis by soil fungi of *Sclerotinia sclerotiorum* (Lib.) de Bary, a phytopathogenic fungus. *Nature.* 244:287-288.

Kannaiyan, S. and Prasad, N.N. 1979. Sheath blight incidence in weed hosts. *Int. Rice Res. Newsltr.* 4:17.

Katan, J .1981. Solar heating (Solarization) of soil in control of soil-borne pests. *Ann. Rev. Phytopath.* 19:211-236.

Kempe, J. and Sequeira, L. 1983. Biological control of bacterial wilt of potatoes: Attempts to induce resistance by treating tubers with bacteria. *Plant Dis.* 67: 499-503.

Kohmoto,K., Singh, U.S. and Singh, R.P.eds.1995. Pathogenicity and Host Specificity in Plant Diseases. Volume II. Elsevier Science Ltd., U.K. p.43.

Kronland, W.C. and Stanghellini, M.E. 1988. Clean slide technique for the observation of anastomosis and nuclear condition of *Rhizoctonia solani. Phytopath.* 78: 820-822.

Kuninaga, S. and Yokosawa, R. 1984. DNA base sequence homology in *Rhizoctonia solani* Kuehn. IV. Genetic relatedness within AG-4. *Ann. Phytopath. Soc. Japn.* 50:322-330.

Kwok, D. C. H., Gahy, P. C., Hoitink, H. A. J. and Kuter, G. A. 1987. Interaction between bacteria and *Trichoderma hamatum* in suppression of Rhizoctonia damping - off in bark compost media. *Phytopath.* 77: 1206-212.

Leach, L.D. 1947. Growth rates of host and pathogen as factors determining the severity of pre-emergence damping-off. *J Agric. Res.* 75: 161-179.

Lifshitz, R., Lifshitz, S. and Baker, R.1985.Decrease in incidence of Rhizoctonia pre-emergence damping -off by use of integrated chemical and biological control. *Plant Dis.* 69: 431-434.

Liu,S. and Baker,R.1980. Mechanism of biological control in soil suppressive to *Rhizoctonia solani. Phytopath.* 70: 404-412.

Luke, W. J., Pinckard, J.A. and Wang, S.S. 1974. Basidiospore infection of cotton bolls by *Thanatephorus cucumeris. Phytopath.* 64: 107.

Mathew, A. and Gupta, S.K. 1996. Studies on Web blight of French bean caused by *Rhizoctonia solani* and its management. *Indian J. Mycol. Pl. Pathol.* 26(2): 171-177.

Moore, R. T. 1987. The genera of *Rhizoctonia*- like fungi: *Ascorhizoctonia, Ceratorhiza* gen.nov, *Epulorhiza* gen. nov., *Moniliopsis* and *Rhizoctonia. Mycotaxon.* 29: 91-99.

Murray, D.I.L.1981. *Rhizoctonia solani* causing barley stunt disorder. *Trans. Brit. Mycol. Soc.* 76: 383-395.

Naito, S. and Sugimoto, T. 1978. Basidiospore infection and lesion development on sugar beet leaves by *Thanatephorus cucumeris* (Frank) Donk. *Ann. Phytopathol. Soc. Japn.* 44:426.

Nakayama, T .A.1940. A study on the infection of cotton seedlings by *Rhizoctonia solani. Ann. Phytopathol. Soc. Japn.* 10:93.

Ogoshi, A. 1987. Ecology and pathogenicity of anastomosis and intraspecific groups of *Rhizoctonia solani* Kuhn. *Ann .Rev. Phytopathol.* 25:125-143

Palti, J. and Katan,J: 1997. Effect of cultivation practices and cropping systems on soil-borne diseases. (*In:*) Hillocks, R.J. and Waller, J.M. (Eds.) *Soil-borne Diseases of Tropical Crops.* pp. 377-394. CAB International, U.K.

Parmeter, J.R. Jr. and Whitney, H.S. 1970. Taxonomy and nomenclature of the imperfect state. (*In*:) J.R. Parmeter Jr., (Ed.) *Biology and Pathology of Rhizoctonia solani*. pp. 7-19 University of California Press, Berkeley. 255p.

Parmeter, J.R. Jr., Sherwood, R.T. and Platt, W.D.1969. Anastomosis grouping among isolates of *Thanatephorus cucumeris*. *Phytopath*. 59: 1270-1278.

Poromatro, S.H., Nelson, B.D. and Freeman, T.P .1998. Association of binucleate *Rhizoctonia* with soybean and mechanism of biocontrol of *Rhizoctonia solani*. *Phytopath*. 75:1080-1084.

Pullman, G.S., Devay, J.E., Garber, R.H. and Weinhold, A.R.1981. Soil solarization on Verticilliurn wilt of cotton and soil-borne population of *Verticillium dahliae, Pythium* spp., *Rhizoctonia solani* and *Thielaviopsis basicola*. *Phytopath*. 71:954-959.

Reddy, M. N. 1980. Studies on groundnut hypocotyls exudates and behavior of *Rhizoctonia solani* in influencing the disease. *Plant Soil*. 55: 445.

Reynolds, H. W. and Hansen, R.G.1957. Rhizoctonia disease of cotton in presence or absence of the cotton root rot nematode in Arizona. *Phytopath*. 47: 256-269.

Rothrock, C.S. and Gottlieb, D.1984. Role of antibiosis in antagonism of *Streptomyces hygroscopicus* var. *geldanus* to *Rhizoclonia solani* in soil. *Can. J. Microb*. 30: 1440-1447.

Roy, A.K. 1991. Inhibitory effect of *Aspergillus terreus* on *Rhizoctonia solani* causing sheath blight of rice. *J Indian Bot. Soc*.70:95-98.

Singh, R.S.1996. *Introduction to Principles of Plant Pathology*. Third edition. Oxford & IBH Publishing Co. Pvt. Ltd., New Delhi. pp: 377 -384.

Sneh, B.; Burpee, L. and Ogoshi, A.1991. Identification of *Rhizoctonia* species. The American Phytopathological Society Press, St. Paul, MN. pp: 1-5.

Sneh, B., Katan, J., Henis, Y. and Wahl, I. 1966. Methods for evaluating inoculum density of *Rhizoctonia* in naturally infested soil. *Phytopath*. 56: 74-78.

Stewart, E.L. and Pfleger, F.L.1977. Development of poinsettia as influenced by endomycorrhizae, fertilizer and root rot pathogens *Pythium ultimum* and *Rhizoctonia solani*. *Florist's Rev*, 159 (37):79-80.

Stockwell, V. and Hanchey, P. 1987. Lignification of lesion borders in *Rhizoctonia* infected bean hypocotyls. *Phytopath*, 77:589.

Talbot, P.H.B. 1965. Studies of *'Pellicularia'* and associated genera of Hymenomycetes. *Persoonia*, 3:371-406.

Tamietti, G. and Garibaldi, A. 1989. The use of solarization against *Rhizoctonia* under green house conditions in Liguria. *In Fitopatol*. 39: 43-45.

Tu, C.C., Cheng, Y.H., Kimbrough, J. W. 1977. A new species of *Thanatephorus* from jute in Taiwan. *Mycologia*. 69: 400-413.

Tu, C.C., Roberts, D.A. and Kimbrough, J. W. 1969. Hyphal fusion, nuclear conditions and perfect stages of three species of *Rhizoctonia .Mycologia.* 61: 775-783.

Utkhede, R. and Gupta, V.K. 1996. Management of soil-borne Diseases. Kalyani Publishers, New Delhi, India. pp: 100-122.

Vilgalys, R. 1988. Genetic relatedness among anastomosis groups in *Rhizoctonia* as measured by DNA / DNA hybridization. *Phytopath,* 78:698-702.

Warcup, J. H. and Talbot, P. H. B. 1966. Perfect stages of some Rhizoctonias. *Trans. Br. Mycol. Soc.* 49:427-435.

Warcup, J. H. 1985. Pathogenic *Rhizoctonia* and orchids. *(In:)* C.A. Parker, A.D. Rovira, K.J. Moore, P.T.W. Wong and J.K. Kollmorgen (Eds). *Ecology and Management of soil borne Plant Pathogen.* The American Phytopathological Society Press, St. Paul, M.N. pp: 69-70.

Wiggins, E. A. and Curl, E. A. 1979. Interactions of collembola and microflora in cotton Rhizosphere. *Phytopath.* 69: 244-249.

Williams, R.J .1975. Control of cowpea seedlings mortality in southern Nigeria. *Pl. Dis. Reptr.* 59:245-248.

Yokoyama, K. and Ogoshi, A. 1984. Studies on hyphal anastomosis of *Rhizoctonia solani,* IV. Observation of imperfect fusion *Ann. Phytopathol. Soc. Japan.* 50: 398.

Yokoyama, K., Ogoshi, A. and Ui, T. 1985. Studies on hyphal anastomosis of *Rhizoctonia solani* II. The Ultrastructural changes of fungal cells during the perfect fusion. *Trans. Mycol. Soc. Japan.* 26: 199-207.

Zambolin, L. and Schenck, N.C.1983. Reduction of the effects of pathogenic, root infecting fungi on soybean by the mycorrhizal fungus, *Glomus mosseae. Phytopath.* 73: 1402-1405.

• • •

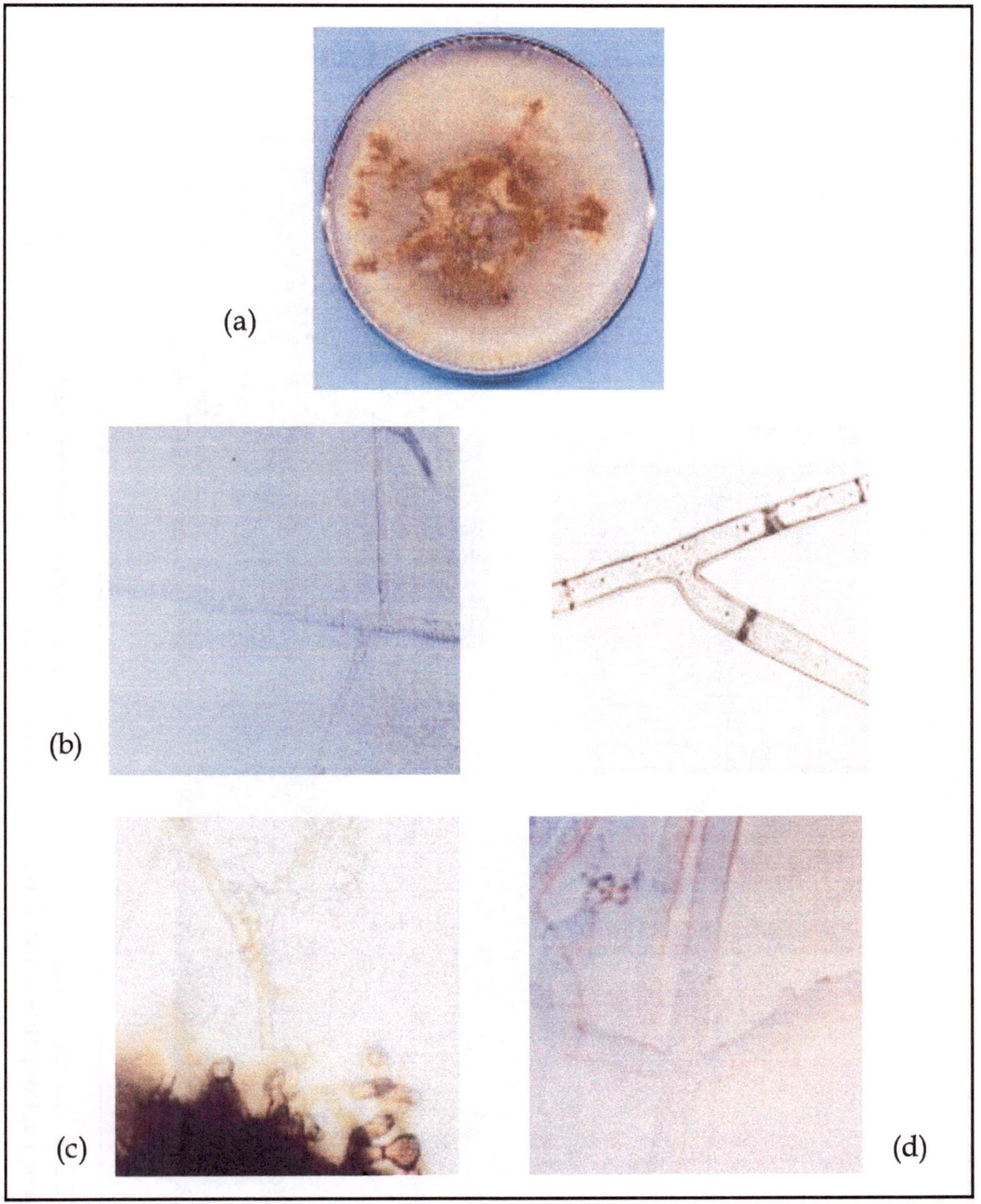

PLATE 2.1: a) 12 days old- Potato dextrose agar culture showing the hyphae and irregular sclerotia of *Rhizoctonia solani*.

b) Hyphae of *Rhizoctonia solani* showing constriction and right angle branching.

c) Monilioid cell of *R.solani*

d) Hyphae showing dolipore septa.

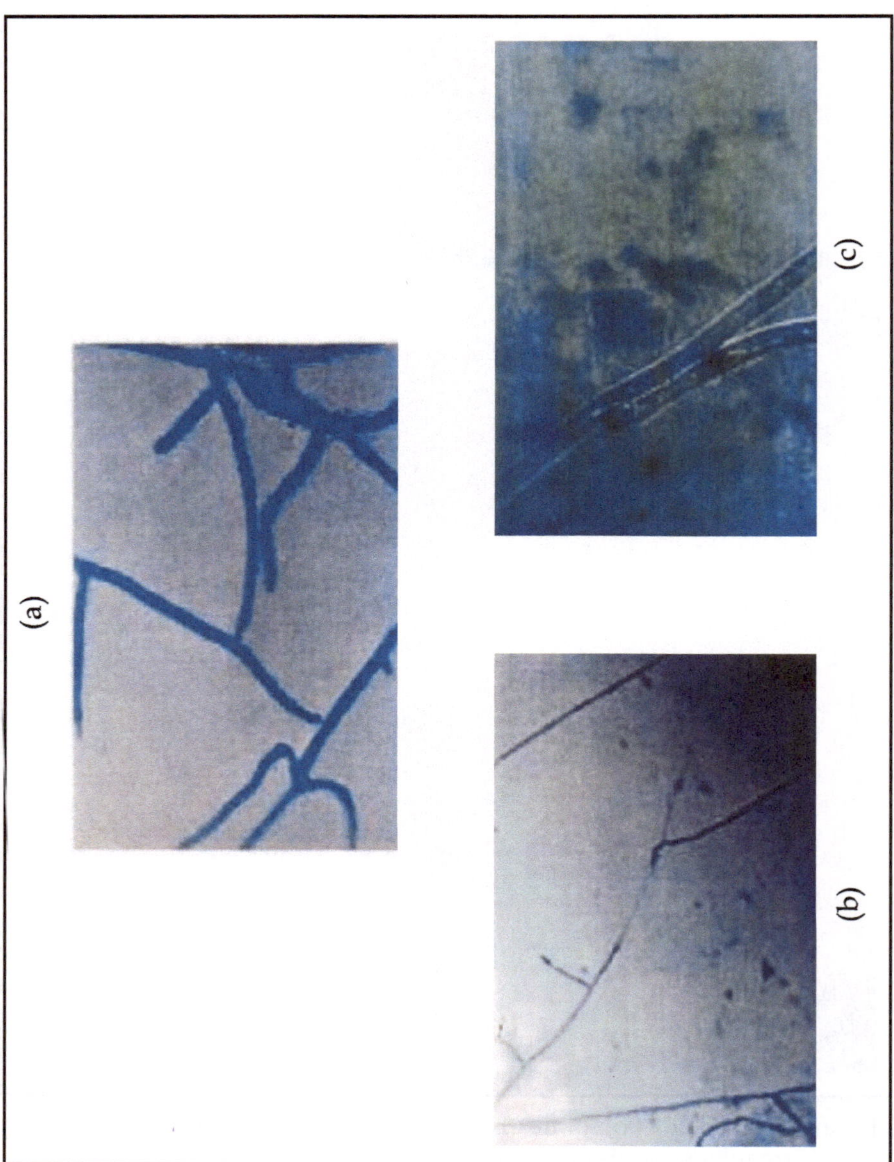

PLATE 2.2: Anastomosis groups of *R.solani*. a) Perfect fusion b) Imperfect fusion c) Contact fusion. (*Courtesy*: Dr. Roopali Sharma, G.B.P.U.A.&T., Pantnagar)

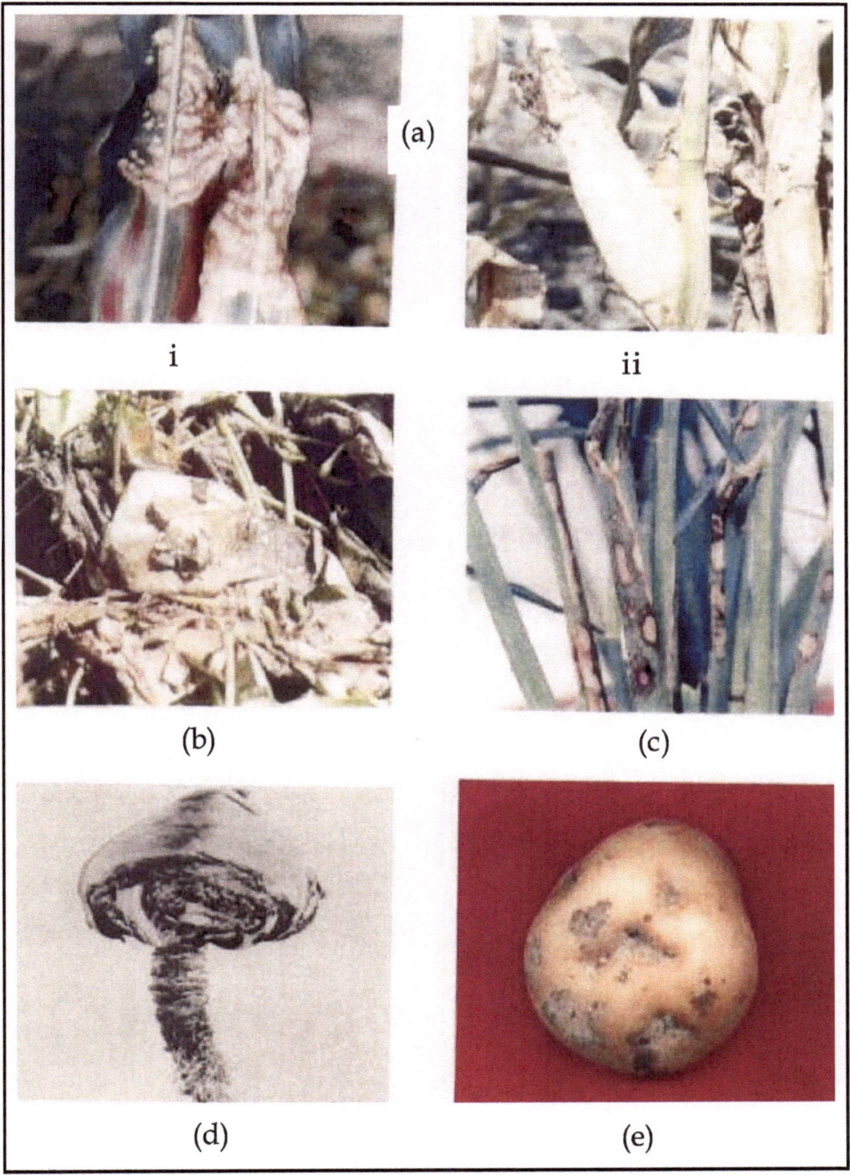

PLATE 2.3: a) Broad leaf and sheath blight of maize showing bands of blighted appearance on i) Leaves ii) Cob

b) Web blight of Urd bean showing web like mycelial growth and sclerotia.

c) Sheath blight of rice.

d) Bottom rot of cabbage; basal leaves have decayed and fallen away.

e) Black sclerotial bodies on potato surface.

Advances in Soil Borne Plant Diseases, 2008
Editors: Naik and Devika Rani, pp. 55-88
New India Publishing Agency, New Delhi

3

Molecular Detection and Characterization of Soil-Borne Fungal Plant Pathogens

T.R. Sharma and T.K. Jana

INTRODUCTION

All plant species grown on this universe are attacked by insect-pests and diseases, resulting huge yield losses in agriculturally important species. Among diseases, extensive yield losses occurred due to soil borne fungal plant pathogens. Soil borne fungal pathogens infect roots or stem bases, their dispersal and survival stages are primarily confined to the soil. Although many soil pathogens may also produce air or water dispersed spores which results in long distance dispersal. Soil borne fungal diseases are specifically those diseases where the soil borne fungi has the capacity to survive in soil for long periods in the absence of their hosts, or survive in the soil outside of seed or host residues, but infection and symptom expression are confined to the subterranean organs of the plant. Soil borne fungi may be divided into two classes: i) *The soil inhabiting fungus* which is characterized by its ability to survive indefinitely as saprophyte (e.g. *Rhizoctonia, Sclerotium*) and ii) *The soil invading or root inhabiting* fungus which is characterized by an expanding parasitic phase and a declining saprophytic phase after the hosts death. The soil invaders exhibit a greater degree of parasitic specialization than the soil

National Research Centre on Plant Biotechnology
Indian Agricultural Research Institute, New Delhi-110012

inhabitor and therefore have a more restricted host range. They may be further sub-divided into three groups: (a) *vascular wilt pathogen* (after initial invasion pathogens are confined to vascular tissue e.g., *Fusarium)*, (b) *ectotrophic root pathogens* (grow extensively on outside of the root and can penetrate the root cortex at different loci, sometimes forms specialized mycelial condensed body or sclerotia, e.g. *Macrophomina*), the (c) *obligate parasites* (specialized pathogens which produced motile zoospores, e.g. *Synchitrium, Phytophthora*). Accurate identification, characterization and early detection of these soil-borne fungal plant pathogens in many crops using morphological criteria are confusing. Traditional methods for the detection of fungal plant pathogen in soil on rhizoplane or in rhizosphere rely on culturing of fungi onto agar plates. The cultural methods are time consuming, taking 3-5 days to determine total viable count and 5-7 days to detect specific fungi. The technology for detection and characterization of soil borne fungal plant pathogens has advanced significantly in recent years and considerable efforts have been focused on the development of DNA based techniques. With the advent of molecular technology, in the early 1980's a major focus has been to identify and characterize various fungal plant pathogens. Molecular detection techniques can rapidly generate accurate results useful for making disease management decisions (Levesque, 1997, 2001). Two major advances occurred over the past 30 years in the molecular detection and characterization of fungal pathogens. i) Detection of plant pathogen by antibodies, particularly by using monoclonal antibodies and enzymes linked immunosorbant assay (ELISA) (Clark and Adams, 1977; Kohler and Milstein, 1975). ii) Molecular characterization by electrophoretic karyotyping, nucleic acid based hybridization (Botstein *et al.*, 1980; Howlett, 1996) and polymerase chain reaction (PCR) (Mullis and Faloona, 1987). Recent reviews on the molecular characterization and detection of genomic variability in filamentous fungi using various techniques provide related information on this topic (Zolan, 1995; Mills *et al.*, 1995; Sharma *et al.*, 1999, 2003). Uses of PCR based method in pathogen detection have already been reviewed. (Bridge *et al.*, 1998; Martin *et al.*, 2000; Bridge, 2002). In this chapter we will briefly illustrate some of the molecular techniques extensively used for the characterization and detection of soil borne fungal plant pathogens.

ELECTROPHORETIC KARYOTYPING

In any living organisms characteristic of genome in terms of number of chromosomes and microscopic morphology of each chromosomes are called karyotype. In fungi, these chromosomes are too small and very difficult to study. Recent advances in gel electrophoresis technology enabled to separate the chromosome or its parts in intact form known as electrophoretic karyotyping (EK). EK variability can be effectively used to detect genomic rearrangements of a particular soil borne fungal strain within and between species as taxonomic tool. In 1984, there was a major breakthrough in the development of pulse field gel electrophoresis (PFGE) where intact chromosomes migrate through a pulsed field in agarose gel matrix with a velocity dependent on their size and three dimensional structure (Schwartz and Cantor, 1984). Using rare-cutting restriction endonucleases which cleave the fungal genome infrequently, it is possible to generate small number of large DNA fragments to form macro-restriction profiles characteristic for one or more isolates (Kempken and Walz, 1996). Satisfactory separation of large DNA fragments can only be achieved in an alternating electrical field. Comparison of macro-restriction profiles for different isolates will indicate whether they are indistinguishable and thus potentially useful. PFGE has proved to be invaluable in investigation of outbreaks of soil-borne infection caused by phytopathgenic fungi (Howlett, 1996). The technique requires highly skilled personnel but is very reproducible and has the advantage of generating genetic data which can be statistically analyzed.

NUCLEIC ACID HYBRIDIZATION BASED TECHNIQUES

In molecular biology, the first molecular marker that has been used for the characterization of genetic variability in different organism is restriction fragment length polymorphism (RFLP). This analysis is based on action of restriction endonucleses; enzymes that cleave double stranded DNA at specific target sequences. Analysis of DNA (the DNA may be nuclear, ribosomal or mitochondrial) by RFLPs is one of the several molecular techniques that are used to determine genetic variation in soil-borne plant pathogenic fungi (Michelmore and Hulbert, 1987). In fungi, the mtDNA is small and present in high copy number and shows abundant polymorphisms resulting from mutations (Bruns *at al.*, 1991; Talyor, 1986), which makes it suitable for restriction enzyme analysis. Because of the rapid rate of evolution of mtDNA, which is estimated in animals to be 10 to 100 times faster than that of nuclear DNA (Brown *et al.*, 1982), analysis of mtDNA may provide a more sensitive measurement of genetic divergence on population than other methods.

Nucleic acid hybridization involves cutting of target DNA with restriction enzymes, electrophoretic separation of individual cut DNA fragments and Southern blotting on a nitrocellulose membrane and hybridization with DNA probes. The DNA probe complementary to the target DNA fragments is hybridized to the membrane. The hybridized product can be directly detected if the probe is labeled with a radio-active or fluorescent dye. Alternatively, it can be detected indirectly by using an enzyme linked immunosorbant assay

(ELISA). Preparation of non-radioactive DNA probes for plant pathogen detection has already been explained by Sharma *et al.* (2002).

Another approach for differentiating among strains of a fungal species is by using random clone hybridization probes. In this case the genomic DNA of the fungal strains digested, and the resulting restriction fragments are cloned in a suitable vector. Clones from this library are then tested as hybridization probes against digested DNA from other strains. For example, Elias *et al.* (1993) produced a genomic library of one isolate of *F. oxysporum* f.sp. *lycopersici* to estimate the genetic diversity within forma-speciales of *Fusarium oxysporum*. They tested 50 clones from a large collection of isolates of the same and the other forma-speciales. This screening allowed them to identify 10 repetitive DNA fragments among the 50 tested clones, suggesting that 20% of the genome of *F. o.* sp. *lycopersici* is composed of repetitive DNA. Manicom and Baayen (1993) described a similar approach where one probe, generating multiple banding pattern with *F. oxysporum* strains was obtained for further investigations at the intra-specific level by using RFLPs. These methods facilitate the molecular characterization and observing variations below species level in many soil borne fungal species.

An alternative hybridization technique is a sandwich assay involving the immobilization of a DNA probe to a membrane, which then serves as a receptor/capture site for hybridization with target DNA from the test pathogens. A second labeled receptor probe is then added which increases the sensitivity and specificity of the technique.

POLYMERASE CHAIN REACTION BASED TECHNIQUES

Polymerase chain reaction methods developed by Kerry Mullis (Saiki *et al.*, 1985) are now widely used for the detection and characterization of genetic variability in fungi. In this technique, the DNA is extracted from fungal tissue and used as template for amplification. Short sequences of DNA known as primers are annealed to the complementary DNA target. The primers are then extended across the target sequence using a heat-stable DNA polymerase (generally derived from *Thermus aquaticus* bacterium) and deoxynucleoside triphosphates (dNTP's), resulting in the doubling of the starting target material. Multiple repeats of the denaturation, annealing and extension steps result in an exponential increase in the amount of target DNA. The PCR product is separated on agarose gel and detected by staining with ethidium bromide.

The use of an intermediate step (modification) is necessary when PCR techniques are used for the detection of fungal plant pathogens in soil or soil amended composts. Though PCR amplification significantly enhances the sensitivity of these techniques, a problem which frequently encountered in the application of PCR to detect the pathogen from soil samples, disease roots or stem parts or complex environment samples due to the presence of inhibitors in the soil or in plant compounds which may interfere with the DNA isolation used in PCR reaction (De Boer *et al.*, 1995; Cullen and Hirsch, 1998; Tien *et al.*, 1999; Pennanen *et al.*, 2001). A clean separation or filtrations of soil samples are necessary before the isolation of DNA. The use of column (matrix based) or polyvinylpolypyrrolidene spin column based extraction system seems to be fruitful in this case (Berthelet *et al.*, 1998; Garci-Pedrajas *et al.*, 1999). The DNA can be extracted from fungal pathogen by using a phenol: chloroform: isoamyl alcohol extraction procedure. A simplified procedure for isolating fungal DNA from infested soil, which can be used in PCR detection, was developed by Garcia-Pedrajas *et al.* (1999). The DNA is then amplified in a PCR assay with primers specific to the target fungi and the amplified product is subsequently detected by gel electrophoresis. A number of modification and improvement of PCR techniques such as internal transcribed spacer (ITS) RFLP, random amplified polymorphic DNA (RAPD), amplified fragment length polymorphism (AFLP), simple sequence repeats (SSR), variable number of temdem repeats (VNTR) and real-time PCR can be used for identification, characterization, detection and epidemiological studies of fungal plant pathogens.

The ITS regions of rDNA have proven particularly useful for separation of fungal taxa at the species to genus level, because of the rate of accumulation of mutations in these regions often approximates to the rate of speciation (Bruns *et al.*, 1991; Gardes and Burn, 1996; Pasquer and Brygo, 2000). The commonly used earliest PCR-based fingerprinting technique is RAPD (Williams *et al.*, 1990) or arbitrarily primed-PCR (AP-PCR) (Welsh and McClelland, 1990). This technique has been extensively used in molecular characterization of fungal pathogens (Sharma *et al.*, 1999; Bridge and Arora, 1998; Sharma *et al.*, 2003).

Another more commonly used DNA fingerprinting primers are based on repetitive DNA sequences. The most extensively used primers with both bacteria and fungi, are relatively complex sequences of around 15 - 25 bases that occurs frequently in the genome. These sequences are often considered as universal primers, since primers derived from bacteria can also be used to amplify DNA from lowest eukaryotes i.e. fungi. The primers designed from protein II gene sequences of M13 bacteriophage are commonly used (Meyer *et al.*, 1992). VNTR derived oligonucleotides as PCR primers were applied for population studies of several filamentous fungi (Bridge *et al.*, 1997).

The second types of repetitive sequences used for fingerprinting are the microsatellites or SSR. The SSR's are tandem arrays of short DNA sequences composed of 1 -6 base pair (bp) motifs. These motifs are usually repeated at least five times in the genome. They are found specifically in eukaryotic genomes, including fungi. Due to their ubiquity, co-dominant inheritance, high polymorphism and their use in PCR, SSR's are very powerful genetic makers, especially for the study of closely related fungi (Hantula *et al.*, 1996). The third types of repetitive sequences are larger than microsatellites called minisatellite. The individual minisatallite unit typically varies from 7-100 bp, and the repeat ranging from 0.5 kb to several kilobases (Anderson and Nilsson-Tillgreen, 1997).

The AFLP is another technique used for fingerprinting fungal genomes (Zhang and Steffenson, 2000; Majer *et al.*, 1996). In this method total genomic DNA is extracted and digested with restriction enzymes. Adaptors containing short oligonucleotides are then attached to the ends of cut fragments. Primers complementary to the adaptors part of the oligonucleotide sequence are then used to amplify a proportion of the digestion fragments. This technique is considered to produce many more fragments than repetitive sequence fingerprinting.

Beside above mentioned methods, a few important approaches are now available for detailed high throughput analysis of the distribution of root pathogens in soils. One of the difficulties with PCR is that the final detection of the PCR product using gel electrophoresis is time consuming and cumbersome, and detection and identification of the amplified products usually requires electrophoretic and blot hybridization analysis for interpretation. Recent research has focused on colorimetric detection of PCR product as an alternative technique to gel electrophoresis (Rowhani *et al.*, 1998). The development of colorimetric-PCR assays for the detection of soil borne fungal plant pathogen, which involves the use of 16 S/23S rRNA intergenic spacer and ITS region as genomic target. Now an immuno-enzymatic method, PCR-ELISA, is available for sensitive detection of DNA or RNA (Alard *et al.*, 1993; Roche Molecular Biochemicals, 2001). In PCR-ELISA the PCR products are non-radioactively labeled with digoxygenin-11-UTP (DIG) during amplification. This is followed by immobilization of a biotin-labeled oligonucleotide (capture probe) of complementary sequence to the target DNA strand in the wells of a streptavidin-coated microtiter plate. The DIG-labeled amplified DNA is then hybridized to the capture probe in the microtiter well, and the DIG is then detected immunologically in an ELISA like assay (Roche Molecular Biochemical, 2001). Using the ribosomal ITS 1 as the target molecule for PCR-ELISA, species of *Fusarium, Phytophthora* are reliably identified (Grimm and Geisen, 1998; Bailey *et al.*, 2002).

Differences in primary sequence composition of an organism can be detected under partially denaturing conditions using denaturing or temperature-gradient gel electrophoresis [(DGGE and TGGE). Muyzer and Smalla, 1998; Smit *et al.*, 1999; Vainio and Hantula, 2000] or by single-strand conformation polymorphism [(SSCP) Schwieger and Tebbe, 2000]. Alternatively, DNA length polymorphisms can be assayed directly, by amplified ribosomal intergenic spacer analysis [(ARISA)] or by digesting amplified products with selected restriction enzymes and then characterizing the resulting fragment length polymorphism (RFLPs). Difference in fungal community structure in soil can be detected by analyzing complete mixtures of restriction fragments. However it is usually advantageous to visualize only the terminal restriction fragments (TRFs) by fluorescently labeled one of the two PCR primers (Liu *et al.*, 1997). Because only a singly band is visualized from each distinct amplification product. The number of TRFs in a T-RFLP profile indicates the species richness in the amplified sequences. T-RFLP analysis can be used to study diversity in various soil fungi at different taxonomic levels (Buchan *et al.*, 2002). When applied to ribosomal gene sequences, RFLP analyses termed as amplified ribosomal DNA (rDNA) restriction analysis [ARDRA (Tiedje *et al.*, 1999)] or fluorescently tagged ARDRA [FT-ARDRA (Livak *et al.*, 1995; McSpadden-Gardener and Weller, 2001)]. The choice of approach depends on the degree of sequence diversity in the amplified product. FT-ARDRA is particularly useful to distinguish among larger taxonomic fungal groups such as genera within a domain because the amplified fragments are likely to differ at multiple restriction sites. DGGE are appropriate for diversity studies on a more limited scale because the sequence differences that distinguish taxa may not correspond to most restriction sites.

The obligonucleotide sequences used as amplification primers and the amount and quality of nucleotide sequence determine the differences at fungal community level. Previously, most DNA-based community profiling methods rely on analyses of ribosomal DNA because they contain substantial phylogenetic information (Woese *et al.*, 1990). New primers are designed from the conserved sequences that allow the detection of multiple taxa in a single PCR reaction (Liu and Stahl, 2002). Profiles based on amplified ribosomal sequence have become the bases for the description of soil borne fungal communities. However, the precision with which taxonomic assignments can be made is limited by the number of nucleotides sequenced and the potential for the generation of sequence hetrogeneity during PCR amplification (Qiu *et al.*, 2001; Speksnijder *et al.*, 2001) because of the complexity of soil borne fungal pathogens. Many researchers find convenient to use partial sequencing (250-

500 bp) of a large number (>100) of rDNA clones to characterize the populations present in their clone banks (Liesack *et al.*, 1997). Besides the sequencing RFLP and / or T-RFLP analyses of individual clones can be used with limited success. For the analyses of soil borne pathogens PCR with domain-specific primers, FT-ARDRA and partial sequencing of cloned 16S sequences can be used for phylogenetic analysis at the genus level (Dunbar *et al.*, 2001).

For molecular typing of soil-borne plant pathogens the single strand conformation polymorphism (SSCP) technique is also a fast molecular typing method that can differentiate between fungal isolates according to sequence differences in single genes (Myer *et al.*, 1985; Orita *et al.*, 1989). SSCP can detect any base pair variation in a DNA sequence, without prior sequence information. The technique depends on the fact that single nucleotide changes can alter the three-dimensional conformation of single stranded DNA thus affecting its electrophoretic mobility resulting sequence polymorphism. The polymorphic bands can be excised from the gel, sequenced, and more specific PCR primers can be designed to amplify the particular polymorphic fragments of the alleles. This technique is called sequence characterized amplified region (SCAR) analysis. Population differences detected by PCR-based approaches must be confirmed by an independent method because of the limited quantitative power of profiling methods and the inherent uncertainties associated with amplification products derived from mixed templates. For quantitative estimation of plant pathogenic fungi or microbial diversity in soil can be performed by other approaches such as dilution extinction PCR (McSpadden Gardener *et al.*, 2001) or real-time PCR (Heid *et al.*, 1996; Belgrader *et al.*, 2001; Lees *et al.*, 2002). Real-time PCR is the important development in quantitative diagnostic methods and promises to be useful at very low levels of infection or infestation (Bach *et al.*, 2003). The most frequently employed application of this technique makes use of TaqMan chemistry (Parkin Elmer Applied Biosystems. Foster city, CA) in conjunction with a sequence detection system (SDS) (RBI 7700; Perkin-Elmer). In this method the fluorogenic TaqMan probe labeled on opposite ends with a reporter dye and a quencher dye, anneals between the PCR products, the 5'-3' exonuclease activity of *Taq* DNA polymerase cleaves annealed probe molecules and release reporter dye which results in an intense fluorescent signal that is measured by SDS during each cycle of the development of extremely specific, sensitivity and accurate assays to quantify pathogen infection in seeds (Zhang *et al.*, 1999; Bates and Taylor., 2001) roots (Boehm *et al.*, 1994) and foliage (Winton *et al.*, 2002), or even plants with its latent stage of infection (Bach *et al.*, 2003) . In the future, a third approach to characterizing fungal diversity in rhizosphere will involve hybridization of amplified sequence to DNA arrays (microarrays) of known sequences. All of these approaches can also be adapted to examine metabolically active populations within soil-borne microbial communities by using extracted RNA to generate DNA templates by reverse-transcriptase-mediated PCR (RT-PCR) (Gibson *et al.*, 1996; Nogales *et al.*, 2001).

The work on identification and characterization of some important soil-borne fungal plant pathogens i.e. *Phytopthora* (Oomycetes), *Fusarium* (Ascomycetes), *Rhizoctonia* (Basidiomycetes), and *Macrophomina, Sclerotinia* and *Sclerotium* (Deuteromycetes) using molecular techniques have been discussed in the succeeding paragraphs.

Phytophthora

Phytophthora is a major genus of soil borne plant pathogens within the diploid, Oomycete fungi. Currently, the genus is assigned to the order pythiales and phyllum Oomycota within the group of heterokont, biflagellate (Hawksworth *et al.*, 1995). There are about 60 described *Phytophthora* species (Erwin and Ribiero, 1996). Most of these species are primary invaders of healthy plant tissue with limited saprotrophic ability. Many are responsible for serious diseases of economically important crops. Conventionally, *Phytophthora* species were discriminated on the basis of structure of sporangium (non papillate, semi papillate and papillate) the form of antheridium (amphigymous or paragynoms) and whether the taxon is homothallic or heterothallic (Waterhouse, 1963). Since late eighties, population genetic studies based on molecular methods have improved our ability to both identify *Phytophthora* species units and to determine their natural relationships.

Direct restriction analysis without hybridization of purified DNA has been used to reveal variations in *Phytophthora* species. Restriction analysis of total genomic DNA or its components such as nuclear, ribosomal or mitochondrial DNA in *Phytophthora* species was studied by different workers (Carter *et al.*, 1990; Forster and Coffey, 1993; Forster *et al.*, 1995). Differences in digestion patterns (RFLP) of *P. infestans* were found useful diagnostic character upto the level of strains. Carter *et al.* (1990) used differences in mtDNA restriction profiles combined with other markers to characterize *P. fragrance* and investigated its taxonomic position. Foster and Coffey (1993) also described a technique to pre-screen numerous isolates of *Phytophthora* on the basis of mtDNA polymorphism. After digestion with a restriction enzyme, they identified polymorphisms in mtDNA from total DNA, since mtDNA bands were distinguished on gels against the nuclear background. This allowed them to compare the mtDNA patterns from large number of isolates. In *Phytophthora*, phylogenetic analysis through RFLP's of mitochondrial and nuclear DNA has successfully resolved the issues within groups of species (Forster *et al.*, 1995). Besides RFLP, different molecular methods used for isolate characterization of *P. infestans* were RFLP-PCR, RAPD, AFLP and SSR (microsatellited) (Maurand *et al.*, 1995; Griffith and Shaw, 1998; Van der Lee *et al.*, 1997; Knapora *et al.*, 2001; Knapora and Gissi, 2002).

Amplification of rDNA by PCR and sequence analysis of the large and small submits rDNA has confirmed *Phytophthora* as chromist (Van de Peer *et al.*, 1996). Later, sequence analysis of ITS regions has confirmed its value in distinguishing morphologically similar *Phytophthora* species and phylogenies of species (Foster *et al.*, 1995; Crawford *et al.*, 1996; Cooke and Duncan, 1997). Cooke *et al.* (2000) characterized and assessed phylogenetic relationships among 50 *Phytophthora* species on the basis of ITS sequence variation of genomic rDNA. Recently, besides genomic rDNA, sequence variation of intergeneric mitochondrial DNA spacers (mtDNA-IGS) of *P. infestans* and related species was also reported (Wattier *et al.*, 2003). Bohm *et al.*, (1999) developed a real time quantitative PCR for monitoring *P. infestans* and *P. citricola* in their respective host plant roots. The examples of molecular characterization and identification of *Phytophthora* species is given in Table 3.1.

Fusarium

The genus *Fusarium* with a large number of species is a ubiquitious soil-inhibiting fungus has world-wide distribution. They comprise phytopathogenic and saprophytic species. The pathogenic *Fusarium* species can be further subdivided into groups each with a distinctive host range. Traditionally, each of these groups is treated as an intra-specific taxon called forma-species. The generic concept of *Fusarium* was first mooted by Link in 1809 (Link, 1809) by taking into consideration the presence of distinctive canoe- or banana- shaped canidia. Differences in the shape of the macroconidia are central to the identification of many *Fusarium* species, although there are other characters that also serve to differentiate the genus into species. Unfortunately, like other fungi *Fusarium* also has limited morphological characters which are also influenced by natural selection and environmental conditions.

With the introduction of molecular technologies, *Fusarium* taxonomy is once again an area of great controversy (Leslie *et al.*, 2001). The taxonomy of *Fusarium* has undergone a number of changes over the last 100 years, and consequently the concept of a species within the genus has varied greatly between broader and narrower concepts. Various methods of molecular biology have been introduced in the field of *Fusarium* research that is outlined in Table 3.1. In the begining molecular characterization of the genus *Fusarium* is based on RFLPs of nuclear or mitochondrial or ribosomal DNA. Variation of mitochondrial DNA (mtDNA) has been used extensively to assess inter- and intra-species relationships in *F. oxysporum* (Jacobson and Gordon, 1990; Kim *et al.*, 1991; Kistler *et al.*, 1987). Molecular variability of mtDNA is commonly revealed by RFLP analysis after hybridization of total DNA with mtDNA probes. For example, RFLP in mtDNA were useful for studying relationships between populations of *F. oxysporum* and to detect genetic exchange among populations. This fungal species includes both non-pathogenic and phytopathogenic strains, the latter being divided in to formae-speciales and races on the basis of their host-specificity (Kistler *et al.*, 1987; Jacobson and Gordon, 1990; Kim *et al.*, 1991; Perez-Artes *et al.*, 1995). The genetic relatedness of strains among different forma-specials causing wilt in cucurbits (Kim *et al.*, 1992) or crucifers (Kistler *et al.*, 1987) was investigated by RFLP analysis of mt DNA. In *F.oxysporum,* variation in mtDNa have also been used to detect genetically isolated populations within a particular forma-speciales (Jacobson and Gordon, 1990). Perez-Artes *et al.* (1995) differentiated different races (0-6) of 45 *F.o.*f. sp. *ciceri* isolates of California, India and Spain using mt DNA RFLP analyis. RFLP analyis of mtDNA has been shown to yield valid taxonomical characters in formae-speciales and races of *F. oxysporum* (Table 3.1). However, now a days PCR-based techniques i.e. ITS, IGS-RFLP, RAPD, AFLP and real-time PCR have been used extensively to differentiate between different species of *Fusarium,* formae-speciales under *F. oxysporum, F. solani,* and different races under formae-speciales of *F.o.* isolates (Alves-Santos *et al.*, 2001, 2002; Assigbetse *et al.*, 1994; Bentley *et al.*, 1995, 1998; Bridge *et al.*, 1993 Chiocchetti *et al.*, 1999; Filon *et al.*, 2003; Gherbawy, 1999; Jimenez-Gasco *et al.*, 2001; Jimenez-Gasco and Jimenez-Diaz, 2003; Kelly *et al.*, 1994, 1998; Kistler, 1997; Lodwig *et al.*, 1999; Mukherjee *et al.*, 1997; Namiki *et al.*, 1998; Oullette and Seifert, 1993; Paavanen-Huhtala *et al.*, 1999; Steenkamp *et al.*, 2000; Suga *et al.*, 2002; Wang *et al.*, 2001; Woo *et al.*, 1996; Yli-Mattila *et al.*, 1996, 1997). We have also performed RAPD analysis for molecular characterization of different isolates of *Fusarium oxysporum* and other *Fusarium* species (Jana *et al.*, unpublished). Grouping of different formae speciales and races of *F. oxysporum* is depicted in Plate 3.1.

Rhizoctonia

Rhizoctonia solani (teleomorph: *Thanatephorus cucumeris*) is an important soil borne pathogenic fungus causing diseases of many agricultural and horticultural crops world-wide (Sneh *et al.*, 1996; Ogoshi, 1996). Anastamosis grouping of *Rhizoctonia* isolates has remained the most frequently used method for differentiating anastomosis groups (AGs) despite being time consuming and requiring some degree of skill (Anderson, 1982). Later, with the use of molecular marker techniques it was proved that the AGs seem to be genetically different and regarded as a complex fungus comprising of several species.

Anastomtosis group is a convenient method for classification but sometime anastomosis reaction can be difficult to identify thus warrant other methods of identification of *R. solani* isolates (Carling *et al.*, 1999, 2002). A variety of molecular methods have been used for characterization and grouping of *R. solani* isolates such as DNA / DNA hybridization (Kuninaga and Yokosawa, 1985; Vilgalys, 1988; Kuninaga, 1996) electrophoretic karotyping through PFGE (Keijer *et al.*, 1996), analysis of ribosomal rDNA ((Vilgalys and Gonzales, 1990; Liu and Sinclair, 1993; Schneider *et al.*, 1997) and PCR- fingerprinting using RAPD-PCR (Toda *et al.*, 1999a, b), ERIC and REP-PCR (Toda *et al.*, 1999a), UP-PCR (Lubeck and Poulsen, 2001) and AFLP (Julian *et al.*, 1999). Kuninaga (2002) reviewed the current situation of the taxonomy of the genus *Rhizoctonia* and the *R. solani* species. All the molecular techniques which are used for characterization of *Rhizoctonia* starting from species to AGs level are listed in Table 3.1. Recently, Lees *et al* (2002) developed quantitative real-time PCR assays for the detection and identification of *R. solani* AG-3 from soil samples.

Macrophomina

The genus *Macrophamina* is represented by a single species *phaseolina* commonly found in more than 500 plants over a vast geographical area (Mihail and Taylor, 1995). Various studies have been directed at analyzing variation in the species using chlorate utilization, pycnidial formation and fractal dimension analysis of hyphal branching patterns (Mihail *et al.*, 1994). Analysis at the genome level has been performed by characterizing the prevalence and size distribution of minichromosomes (Jones and wang., 1997). This study demonstrated that minichromosome patterns are highly variable and do not allow for grouping of any population of *M. phaseolina.* The molecular techniques like internal transcribed spacer (ITS) region comparison, showed limited variation among the isolates within the species (Chase *et al.*, 1994; Su *et al.*, 2001). Variability was also observed when comparing patterns obtained by RAPD analysis of *M. phaseolina* isolates (Su *et al.*, 2001). Vandemark *et al.*, (2000) used AFLP analysis to assess the genetic variability of a number of isolates of *M. phaseolina.* Recently, we have characterized different isolates of *Macrophomina phaseolina* obtained from a vast number of plants with the aim to study genetic variation from diverse hosts and geographical locations with molecular markers (Jana *et al.*, unpublished). DNA fingerprints with RAPD could successfully be used to distinguish the genetic variability of *M. phaseolina* populations from various parts of the world and one of such fingerprint patterns is given in Plate 3.2.

Sclerotinia species

Sclerotinia

Sclerotinia sp. is an important soil-borne nectrotrophic plant pathogen that causes rot and wilt resulting about 95% yield loss in crops like oilseed rape, bean, tomato, sunflower (Conn and Tiwari, 1990; Boland and Hall, 1994). *Sclerotinia,* unlike many fungal genus have been taxonomically well defined by morphological, cytological and physiological features (Kohn 1979; Le Tourneau, 1979; Cruishank, 1983). However, rational and systematic definition of intra-specific grouping based on factors such as pathogenicity, virulence or geographic distribution has been difficult because of problems in assay procedures.

A variety of molecular criteria for examining genetic diversity within and between mycelial compatibility groups (MCG's) of *S. sclerotiorum* have been used (Kohn *et al.*, 1990). Kohn *et al.* (1991) initially used three molecular makrers such as RFLP of genomic DNA, smaller mt rRNA gene, larger mtRNA gene with homologus, nuclear DNA or heterologus DNA probes. All confirmed that each MCG's in the sample was genotypically unique. Kohli *et al.* (1992) examined genetic heterogeneity of distantly located field populations of *S. scleotiorum*. They observed a complex fingerprint profile for 33 of the 36 MCG's which allowed them to determine that some clones were distributed over large geographic distances, and that each field population of *S. sclerotiorum* contained more than one clone. Polymorphism in DNA amplified by PCR yield valuable information especially when comparing large group of strains. PCR amplification with ITS showed no variation in size among the 58 strains of *S. sclerotiorum* (Kohn *et al.*, 1991). Later, Errampalli and Kohn (1996) have also shown chromosome polymorphism through electrophoretic karyotyping (EKs) of 83 isolates of *S. sclerotiorum.* However, their study confirms that EK's of 83 isolates were variable within agricultural and natural populations, as well as among the species such as *S. sclerotiorum, S. minor and S. trifoliarum.* Similarly, cloning of *S. sclerotiorum* showed that the number of chromosome varies from 8 to 10 and hybridization with homologous and heterologous probes confirmed existence of four linkage groups (Cubeta *et al.*, 1997). Apart from these markers, assessments of polymorphism using microsatellite and SSCP have also been reported in *Sclerotinia* species (Sirjusingh and Kohn, 2001; Carbone *et al.*, 1999).

Sclerotium rolfsii

Sclerotium (teleomorph *Athelia*) *rolfsii* is a soil-borne plant pathogen that is distributed from temperate to tropical regions and cause diseases on a wide range of plant species (Punja, 1988). Genetic relationships of different MCG of *S. rolfsii* in a specific geographic region have been reported from the studies conducted in Texas (Nalim *et al.*, 1995), Japan (Okabe and Matsumoto, 2000) and South Africa (Cillier *et al.*, 2000). It was also found that several MCG were represented among isolates recovered from a single field. [Punja and Grogan (1983)]. The isolates of *S. rolfsii* from a MCG displayed variation in morphological characteristics, such as in number and size of sclerotia in culture, indicating that phenotypic differences are common among members of a group. Similarly, difference in growth, morphology and physiology of three *Sclerotium* species was reported by this group (Punja and Damiani, 1996). This was again substantiated by the RAPD data, which showed that banding patterns differences occurred among isolates from the same MCG. Harlton *et al.*, (1995) demonstrated that isolates of *S. rolfsii* within the same MCG also displayed genetic variation in ITS region of ribosomal DNA using RFLP. Okabe *et al.*, (1998) using same method classified Japanese isolates into five ITS-RFLP groups which had distinct geographical distribution patterns. It was assumed that isolates with dissimilar ITS-RFLP grouping from the same MCG could have undergone genetic changes through selection, mutation or genetic drift, while at the same time conserving the alleles governing somatic compatibility. They suggested that genetic exchange among isolates within a MCG could give rise to new genetic combinations. Later, they have observed the ITS heterogeneity (at nucleotide level) within different isolate of a single strain of *S. rolfsii* (Okabe *et al.*, 2001). Okabe's group (Okabe and Matsumoto, 2000; Okabe's *et al.*, 2001) and Nalim *et al.*, (1995) reported that members of MCG were predominantly clonal in origin. Contrary to these observation, recently, Punja and Sun (2001) using a larger sample size demonstrated that clonality could be detected in some MCG of *S. rolfsii* using RAPD, but not all. Cillier *et al.*, 2000 reported similar findings using AFLP analysis of isolates of *S. rolfsii* from South Africa. Species level differences in *S. rolfsii* can be ascertained by molecular techniques (Table 3.1). Harlton *et al.*, (1995) distinguished two species using ITS-RFLP. Punja and Sun (2001) observed that RAPD patterns in *S. delphini* were very distinct from those of *S. rolfsii.*

CONCLUSION

Molecular characterization and detection of fungal pathogens by using chromosome or its small unit i.e. nucleic acid based techniques has several advantages over other methods. For instance, serological methods have been

TABLE 3.1. Molecular techniques used for the detection and characterization of important soil borne fungal plant pathogens

Pathogen	Molecular technique	Target Sequence	Reference
Phytophthora species	EK, RFLP	Whole genome mt DNA rDNA	Forster *et al.*, 1989; Goodwin *et al.*, 1992; Tooly and Carras, 1992; Cooke and Duncan, 1997
	PCR, RFLP	ITS and IGS region of rDNA	Forster *et al.*, 1995
	DAF	Whole genome	Drenth *et al.*, 1993
	RAPD	Whole genome	
	AFLP	Whole genome	
	ITS1 and ITS2 sequencing	ITS	Forster *et al.*, 1995; Cooke and Duncan,1997
	r DNA sequencing	rDNA	Crawford *et al.*, 1996. Cooke *et al.*, 2000
P. cinnamomi	RAPD, Dot blot PCR	Whole genome, species-specific DNA probe	Dobrowolski and O Brien, 1993; Judelson and Mes senger-Routh, 1996
	RFLP, RAPD, microsatellite	Whole genome	Goodwin *et al.*, 1992; Drenth *et al.*, 1993;
			Fraser *et al.*, 1999; Dobrowolski *et al.*, 2003
P.infestans	RFLP-PCR, RAPD, AFLP, SSR, PCR with specific primer PINF and ITS-5. Real-time-PCR; Sequencing	Whole genome, rDNA ITS, mt DNA-IGS	Carter *et al.*, 1990; Van der Lee *et al.*, 1997; Griffith and Shaw, 1998; Brurberg *et al.*, 1999; Bohm *et al.*, 1999; Cooke *et al.*, 2000; Knapova and Gisi, 2002; Wattier *et al.*, 2003
P. megakarya	RAPD	Whole genome	Nyasse *et al.*, 1999
P. megasperma	RFLP	Whole genome	Whisson *et al.*, 1992
P. megasperma	mt DNA RFLP	mt DNA	Forster et al., 1989
P. parasitica var. nicotianae		Whole genome	Zhang *et al.*, 2003
P.sojae	RAPD	Whole genome	Meng *et al.*, 1999

Contd...

Table 3.1. Contd...

Fusarium species	Beta tubulin, ITS, IGS, rDNA RFLP, UP-PCR, RAPD, RFLP, rDNA sequencing	Whole genome, rDNA ITS, IGS region	Yli-Mattila *et al.*, 1997 a, b, 2002
F. avenaceum	RAPD, UP-PCR, rDNA, RFLP	Whole genome, rDNA	Yli-Mattila *et al.*, 1996,
F. culmorum	Microsatellite	Whole genome	Giraud *et al.*, 2002
F. graminearum	RAPD	Whole genome	Ouellet and Seifert, 1993
F. moniliforme	RFLP, PCR	Whole genome	Murillo *et al., 1998*
F. oxysporum	EK, PCR-RFLP, RAPD, AFLP	Whole genome, IGS of rDNA	Appel and Gordon, 1995; To-Anum *et al.*, 1995; Kistler, 1997; Paavanen-Huhtala *et al.*, 1999; Gherbawy, 1999; Bao et *al.*, 2002
F.o. f. sp. *bassilici*	RAPD	Whole genome	Chiocchetti *et al.*, 1999
F.o. f. sp. *ciceri*	RAPD, RFLP, microsatellite	Whole genome, repeat rich region	Kelly *et al.*, 1994, 1998; Mukherjee *et al.*, 1997; Perez-Artes *et al.*, 1995; Barve *et al.*, 2001; Jimenez-Gasco *et al.*, 2001. Jimenez-Gasco and Jimenez-Diaz, 2003
F.o. f. sp. *cubense*	EK, RAPD, DAF, PCR-RFLP	Whole genome	Mia, 1990; Bentley *et al.*, 1995, 1998; Thomas *et al.*, 1994; Lodwig *et al.*, 1999
F. o. f. sp. *cucumerium*	RAPD, RFLP	Whole genome	Wang *et al.*, 2001
F.o. f. sp. *cucuribitae*	RAPD	Whole genome	Namiki *et al.*, 1994
F.o. f. sp. *glycinea*	RFLP	Whole genome	Whisson *et al.*, 1992
F.o. f. sp. *luffae*	RAPD	Whole genome	Wang *et al.*, 2001
F.o. f. sp. *lycopersici*	RFLP	Whole genome	Elias *et al.*, 1993
F.o. f. sp. *gladioli*	RFLP	Whole genome	Mess *et al.*, 1994
F.o. f. sp. *melonis*	RFLP	Whole genome	Jacobson and Gordon, 1990; Namiki *et al.*, 1998

Contd...

Table 3.1. Contd...

F.o. f. sp. *nivenum*	EK, RFLP	Whole genome	Kim *et al.*, 1991, 1992
F.o. f. sp. *pisi*	RAPD	Whole genome	Grajal-Martin *et al.*, 1993
F.o. f. sp. *phaseoli*	RFLP, RAPD	Whole genome	Woo *et al.*, 1996; Alves-Santos *et al.*, 2001
F.o. f. *sp. vasinfectum*	RAPD, rDNA and mt DNA RFLP, nitrate reductase and phosphate permease gene, Small unit of genome	Whole genome, rDNA, mt DNA	Bridge *et al.*, 1993; Assigbetse *et al.*, 1994; Smith *et al.*, 2001.
F. solani	EK	Whole genome	Nazarreth and Bruschi, 1994; Suga *et al.*, 2002
F. solani f. sp. *glycines*	RAPD	Whole genome	Li *et al.*, 2000
F. solani f.sp. *phaseoli*	Real-time PCR	Whole genome	Filon *et al.*, 2003
F. subglutinans	PCR	Whole genome	Steenkamp *et al.*, 2000
Rhizoctonia solani	EK	Chromosome	Keijer *et al.*, 1996
	DNA/DNA hybridization	Specific DNA region	Carling and Kuninaga, 1990
	rDNA RFLP	rDNA	Vilgalys and Gonzalz, 1990
	r DNA ITS spacer	ITS of rDNA	Liu and Sinclair, 1993
	RAPD	Whole genome	Duncan *et al.*, 1993; Yang and Kharbanda, 1996; Toda *et al.*, 1999a,b; Neeraja *et al.*, 2002
	ERIC and REP-PCR	Whole genome	Toda *et al.*, 1999a
	AFLP	Whole genome	Julian *et al.*, 1999
	UP-PCR	Whole genome	Lubeck and Poulsen, 2001
	PCR, Sequencing of ITS	rDNA ITS	Kuninaga *et al.*, 1997, 2000
	Realtime-PCR		Lees *et al.*, 2002
M. phaseolina	EK	Chromosome	Jones and wang., 1997

Contd...

Table 3.1. Contd...

	r DNA ITS-RFLP	Whole genome, rDNA,	Chase *et al.*, 1994; Su *et al.*, 2001
	RAPD	Whole genome	Su *et al.*, 2001
	AFLP	Whole genome	Vandemark *et al.*, 2000
Sclerotinia sclerotirum	EK	Chromosome	Erampalli and Kohn, 1996
	RFLP	Whole genome	Kohn *et al.*, 1988
	RFLP	Whole genome	Kohn *et al.*, 1991
	RFLP	Whole genome	Kohli *et al.*, 1992
	PCR of ITS, AP-PCR	ITS	Kohn *et al.*, 1991
	microsatellite, SSCP	Flanking region between two repeat rich region of the genome.	Carbone *et al.*, 1999, Sirju singh and Kohn, 2001
	PCR based detection	Specific region of genome	Freeman *et al.*, 2002
Sclerotium rolfsi	RAPD	Whole genome	Harlton *et al.*, 1995
	RAPD, ITS-RFLP	ITS region of rDNA	Okabe and Matsumoto, 2000; Okabe *et al.*, 1998, 2001
	AFLP	Whole genome	Cilliers *et al.*, 2000
	RAPD	Whole genome	Punja and Sun, 2001

found to be more difficult, since the fungal antigens are complex and variable at different growth stages. The presence or absence of spore bearing structures and slow growing nature of certain fungal pathogens will not affect their detection by nucleic acid based techniques, since all fungal cells contain DNA which can be used for analysis. However, DNA probes can only be used in those fungi, which are generally not amenable to morphological identification methods.

REFERENCES

Alard, P., Lantaz, O., Sebagh, M., Calvo, C.F., Well D., Chavanel, G., Senik, A., and Charpentier, B. 1993. A versatile ELISA-PCR assay for mRNA quantitation from a few cells. *Bio. Tech.* 15: 730-736

Alves-Santos, F., Ramas, B., Garcia-Sanchez, M.A., Eslava, A.P., and Diaz-Minguez, J.M. 2001. A DNA based procedure for *in planta* detection of *Fusarium oxysporum* f. sp. *phaseoli. Phytopath.* 92: 237-244.

Alves-Santos, F.M., Cordeiro-Rodrignes, L., Sayages, J.M., Martin-Domingnez, R., Gracia-Benavides, P., Crespo, M.C., Diaz-Minguez, M., and Eslava, A.P. 2002. Pathogenicity and race characterization of *Fusarium oxysporum* f.sp. *phaseoli* isolates from Spain and Greece. *Plant Pathol.* 51: 605-611.

Anderson, N.A. 1982. The genetics and pathology of *Rhizoctonia solani*. *Ann. Rev. Phytopath.* 65: 994-998.

Anderson, T.H., and Nilsson-Tillgren, T. 1997. A fungal minisatellite. *Nature.* 386:771.

Appel, D.J. and Gordon, T.R. 1995. Intra-specific variation within populations of *Fusarium oxysporum* based on RFLP analysis of the intergenic spacer region of the rDNA. *Exp. Mycol.* 19: 120-128.

Assigbetse, K.B., Fernandez, D., Dubois, M.P., and Geiger, J.P. 1994. Differentiation of *Fusarium oxysporum* f. sp. *vasinfectum* races on cotton by random amplified polymorphic DNA (RAPD) analysis. *Phytopath.* 84: 622-626.

Bach, H. J., Jesson, I., Schloter, M. and Munch, J.C. 2003. A TaqMan-PCR protocol for quantification and differentiation of phytopathogenic *Clavibacter michiganensis* subspecies. *J. Microbiol. Meth.* 52:85-91.

Bailey, A.M., Mitchell, D.J., Manjunath, K.L., Nolasco, G., and Niblett, C.L. 2002. Identification to the species level of the plant pathogens *Phytophthora* and *Pythium* by using unique sequences of the ITSI region of ribosomal DNA as capture probes for PCR ELISA. *FEMS Microbiol.* 207: 153-158.

Bao, J.R., Fravel, D.R., O'Neil, N.R., Lazarovitz, G., and Van Berkum, P. 2002. Genetic analysis of pathogenic and non pathogenic *Fusarium oxysporum* from tomato plants. *Can. J. Bot.* 80: 271-279.

Barve, M.P., Haware, M.P., Sainani, M.N., Ranjekar, P.K., and Gupta, V.S. 2001. Potential microsatellites to distinguish four races of *Fusarium oxysporum* f. sp. *ciceri* prevalent in India. *Theo. Appl. Genet.* 102: 138-147.

Bates, J.A. and Taylor, E.J.A. 2001. Scorpion ARMS primers for SNP real-time PCR detection and quantification of *Pyrenophora teres*. *Mol. Plant Pathol.* 2: 275-280.

Belgrader, P., Young, S., Yuan, B., Primeau, M., Christel, L.A., Pourahmadi, F., and Northrup, M.A. 2001. A battery powered notebook thermal cycler for rapid multiplex real time PCR analysis. *Anal. Chem.* 73: 286-289.

Bentley, S., Pegg, K.G., and Dale, J.L. 1995. Genetic variation among a world-wide collections of isolates of *Fusarium oxysporum* f. sp. *cubense* analysed by RAPD-PCR fingerprinting. *Mycol. Res.* 99: 1378-1384.

Bentley, S., Pegg, K.G., Moore, N.Y., and Davis, R.D. 1998. Genetic variation among vegetative compatibility groups of *Fusarium oxysporum* f. sp. *cubense* analysed by DNA fingerprinting. *Phytopath.* 88: 1283-1293.

Boehm, E.W.A., Ploetz, R.C., and Kistler, H.C. 1994. Statistical analysis of electrophoretic karyotype variation among vegetative compatibility groups of *Fusarium oxysporum* f. sp. *cubense. Mol. Plant Microbe. Interact.* 7: 196-207.

Bohm, J., Hahn, A., Schubert, R., Bahnweg, G., Adler, N., Nechwatal, J., Oeklmann, R.M., and Obwald, W. 1999. Real-time quantitative PCR: DNA determination in isolated spores of the mycorrhizal fungus *Glomus mosseae* and monitoring of *Phytophthora infestans* and *Phytophhtora citricola* in their respective host plants. *J. Phytopathol.* 147: 409-416.

Boland, G.J. and Hall, R. 1994. Index of plant hosts of *Sclerotinia sclerotiorum. Can. J. Plant Pathol.* 16: 93-108.

Botstein, D., White, R.L., Skolnick, M., and Davis, R.W. 1980. Construction of genetic linkage map in man using restriction fragment length polymorphisms. *Am. J. Hum. Genet.* 32: 314-331.

Bridge, P. 2002. Biochemical and molecular techniques. (*In*:) J.M Waller, J.M Lenne and S.J Waller (Eds.) *Plant Pathologist's Pocketbook.* Wallingford: CABI Publishing, pp 229-244.

Bridge, P.D., and Arora, D.K. 1998. Interpretation of PCR methods for species definition. (*In*:) P.D. Bridge, D.K. Arora, R.P. Elander, and C.A. Reddy, (Eds.) *Application of PCR in Mycology*. CAB International, Wallingford, UK, pp 64-83.

Bridge, P.D., Couteaudier, Y., and Clarkson, J. 1998. Molecular variability of fungal pathogens. Oxford: CAB International. Wallingford, UK.

Bridge, P.D., Ismail, M.A., and Rutherford, M.A. 1993. An assessment of aesculin hydrolysis, vegetative compatibility and DNA polymorphisms as criteria for characterising pathogenic races within *Fusarium oxysporum* f. sp. *vasinfectum. Plant Pathol.* 42: 264-269.

Bridge, P.D., Pearce, D.A., Rivera, A., and Rutherford, M.A. 1997. VNTR derived oligonucleotides as PCR primers for population studies in filamentous fungi. *Lett. Appl. Microbiol.* 24: 426-430.

Brown, W.M.E.M., Pranger, A., Wang, A., and Wilson, C. 1982. Mitochondrial DNA sequences of primates: tempo and mode of evolution. *J. Mol. Evol.* 18: 225-239.

Bruns, T.D., White, T.J., and Taylor, J.W. 1991. Fungal molecular systematics. *Ann. Rev. Ecol. Syst.* 22: 525-564.

Brurberg, M.B., Hannukkala, A., and Harmansen, A. 1999. Genetic variability of *Phytophthora infestans* in Norway and Finland as revealed by mating type and fingerprint probe. *Mycol. Res.* 103: 1609-1615.

Buchan, A., Newell, S.Y., Moreta, J.I.L., and Moran, M.A. 2002. Analysis of internal transcribed spacer (ITS) regions of rRNA genes in fungal communities in a south-eastern US salt. Microbial. *Ecology*. 43: 329-340.

Carbone, I., Anderson, J.B., and Kohn, L.M. 1999. Patterns of descent in clonal lineages and their multilocus fingerprints are resolved with combined gene genealogies. *Evolution* 53:11-21.

Carling, D.E., and Kuninaga, S. 1990. DNA base sequence homology in *Rhizoctonia solani* Kuhn: Inter-and intragroup relatedness of anastomosis group-9. *Phytopathol.* 80: 1362-1364.

Carling, D.E., Kuninaga, S., and Brainard, K.A. 2002. Hyphal anastomosis reactions, and other rDNA-ITS sequences and virulence levels among subsets of *Rhizoctonia solani* anastomosis group-2 (AG-2) and AG-BI. *Phytopath.* 92: 43-50.

Carling, D.E., Pope, E.J., Brainard, K.A., and Carter, D.A. 1999. Characterization of mycorrhizal isolates of *Rhizoctonia solani* from an orchid, including AG-12, a new anastomosis group. *Phytopath.* 89: 942-946.

Carter, D.A., Archer, S.A., Buck, K.W., Shaw, D.S., and Shattock, R.C. 1990. Restriction fragment length polymorphism of mitochondrial DNA of *Phytophthora infestans. Mycol. Res.* 94: 1123-1128.

Chase, T.E., Jiang, Y., and Mihail, J. 1994. Molecular variability in *Macrophomina phaseolina. Phytopath.* 84: 1149. (Abstr.)

Chiocchetti, A., Ghignone, S., Minuto, A., Gullino, M.L., and Migheli, Q. (1999). Identification of *Fusarium oxysporum* f. sp. *basillici* isolated from soil, basil seed and plants by RAPD analysis. *Plant Dis.* 83: 576-581.

Cillier, A.J., Herselman, L., and Pretorius, Z.A. 2000. Genetic variability within and among mycelial compatibility groups of *Sclerotium rolfsii* in South Africa. *Phytopathol.* 90:1026-1031.

Clark, M.F., and Adams, A.N. 1977. Characteristics of the microplate method of enzymel linked immunosorbent assay (ELISA) for the detection of plant viruses. *J. Gen. Virol.* 34: 475-483.

Conn, K.L., and Tewari, J.P. (1990). Survey of *Alternaria* blackspot and *Sclerotinia* stem rot in central Alberta in 1989. *Can. Pl. Dis. Survey.* 70: 66-67.

Cooke, D.E.L., and Duncan, J.M. 1997. Phylogenetic analysis of *Phytophthora* species based on the ITSI and ITS2 sequences of ribosomal DNA. *Mycol. Res.* 101: 667-677,

Cooke, D.E.L., Drenth, A., Duncan, J.M., Wagels, G., and Brasier, C.M. 2000. A molecular phylogeny of *Phytophthora* and related oomycetes. *Fungal Genetics Bio.* 30: 17-32.

Crawford, A.R., Bassam, B.J., Drenth, A., MacLean, D. J., and Irwin, J.A.G. 1996. Evolutionary relationships among *Phytophthora* species deduced from rDNA sequence analysis. *Mycol.* 100: 437-443.

Cruishank, R.H. 1983. Distribution between *Sclerotinia* species by their pectic zymograms. *Trans. Br. Mycol. Soc.* 80: 117-119.

Cubeta, M.A., Cody, B.R., Kohli, Y., and Kohn, L.M. 1997. Clonality in *Sclerotina sclerotiorum* on infected cabbage in eastern North Carolina. *Phytopath.* 87: 1000-1004.

Cullen, D.W., and Hirsch, P.R. 1998. Simple and rapid method for direct extraction of microbial DNA from soil for PCR. Soil Biol. Biochem. 30: 983-993.

De Boer, S.H., Ward, L.J., Le, X., and Chittaranjan, S. 1995. Attenuation of PCR inhabitation in presence of plant-compounds by addition of BLOTTO. *Nucleic Acid Res.* 23: 2567-2568.

Dobrowolski, M.P., and O'Brien, P.A. 1993. Use of RAPD-PCR to isolate species-specific DNA probe for *Phytopthora cinnamoni. FEMS Microbiol. Lett.* 113: 43-48.

Dobrowolski, M.P., Tommerup, I.C., Shearer, B.L. and OBrien, P.A. 2003. Three clonal linages of *Phytophthora cinnamomi* in Australia revealed by microsatellites. *Phytopath.* 93: 695-704.

Drenth, A., Goodwin, S.B., Fry, W.E., and Davidse, L.C. 1993. Genotypic diversity of *Phytophthora infestans* in the Netherlands revealed by DNA polymorphism. *Phytopath.* 83: 1087-1092.

Dunbar, J., Tecknor, L.O. and Kuske, C.R. 2001. Phytogenetic specificity and reproducibility and new method for analysis of terminal restriction fragment profiles of 16S rRNA genes from bacterial communities. *Appl. Environ. Microbiol.* 67: 190-97

Duncan, S., Barton, J.E., and O'Brien, P.A. 1993. Analysis of variation in isolates of *Rhizoctonia solani* by random amplified polymorphic DNA assay. *Mycol. Res.* 97: 1075-1082.

Elias, K.S., Zamir, D., Lichtman, P.T., and Katan, T. 1993. Population structure of *Fusarium oxysporum* f. sp. *lycopersici*: Restriction fragment length polymorphisms provide genetic evidence that vegetative compatibility group is an indicator of evolutionary origin. *Mol. Pl. Microbe Interact.* 6: 565-578.

Errampalli, D., and Kohn, L.M. 1996. Electrophoretic karyotypes of *Sclerotinia sclerotiorum. Appl. Environ. Microbiol.* 62: 4247-4251.

Erwin, D.C., and Ribeiro, O.K. 1996. *Phytophthora Diseases* World-wide. APS Press, Am. Phytopathol. Soc. St. Paul, M.N., USA.

Filon, M., St-Arnaud, M., and Jabaji-Hare, S.H. 2003. Quantification of *Fusarium solani* f. sp. *phaseoli* in mycorrhizal bean plants and surrounding mycorrhizosphere soil using real-time polymerase chain reaction and direct isolation on selective media. *Phyopath.* 93: 229-235.

Forster, H., and Coffey, M.D. 1993. Molecular taxonomy of *Phytophthora megasperma* based on mitochondrial and nuclear DNA polymorphisms. *Mycol. Res.* 97: 1101-1112.

Forster, H., Kinscherf, T.G., Long, S.A., and Maxwell, D.P. 1989. Restriction fragment length polymorphisms of the mitchondrial DNA of *Phytophthora megasperma* isolated from soybean, alfalfa and fruit trees. *Can. J. Bot.* 67: 529-537.

Forster, H., Learn, G., and Coffey, M.D. 1995. Towards a better understanding of the evolutionary history of species of the genus *Phytophthora* using isozymes, DNA

RFLPs and ribosomal spacer sequences. *In:Proc. of the Eur. Assoc. for Photo Res. Conf. Phytophthora infestans* Dowley, L.J., Bannon, E., Cooke, L.R., Keane, T., and O'Sullivan, Eds, 42-54.

Fraser, D.E., Shoemaker, P.B., and Ristaino, J.B. 1999. Characterization of isolates of *Phytophthora infestans* from tomato and potato in North Carolina from 1993 to 1995. *Plant Dis.* 83:633-638.

Freeman, J., Ward, E., Calderon, C., McKartney, A. 2002. A polymerase chain reaction (PCR) assay for the detection of Inoculum of *Sclerotinia sclerotiorum. Eur. J. Plant Pathol.* 108: 877-886.

Garcia-Pedrajas, M.D., Bainbridge, B.W., Heale, J.B., Perez-Artes, E., and Jimenez-Diaz, R.M. 1999. A simple PCR-based method for the detection of the chickpea-wilt pathogen *Fusarium oxysporum* f. sp. *ciceris* in artifical and natural soils. *Eur. J. Plant Pathol.* 105: 251-259.

Gardes, M., and Burns, T.D. 1996. ITS-RFLP matching for identification of fungi. *Methods Mol. Biol.* 50: 177-186.

Gherbawy, Y.A.M.H. (1999). RAPD profile analysis of isolates belonging to different forma speciales of *Fusarium oxysporum*. *Cytologia.* 64: 269-276.

Gibson, U.E.E.M., Heid, C.A., and Williams, P.M. 1996. A novel method for real time quantitative RT-PCR. *Genome Res.* 6: 995-1001.

Giraud, T., Fournier, E., Vautrin, D., Solignac, M., Vercken, E., Bakan, B., and Brygoo, Y. (2002). Isolation of eight polymorphic microsatellite loci using an enrichment protocol, in the phytopathogenic fungus *Fusarium culmorum. Mol. Ecol. Notes.* 2: 121-123.

Goodwin, S.B., Spielman, L.J., Matuszak, J.M., Bergeron, A.N., and Fry, W.E. 1992. Clonal diversity and genetic differential of *Phytophthora infestans* populations in Northern and Central Mexico. *Phytopath.* 82: 955-961.

Grajal-Martin, M.J., Simon, C.J., and Muehlbauer, F.J. 1993. Use of random amplified polymorphic DNA to characterise race 2 of *Fusarium oxysporum* f. sp. *pisi. Phytopath.* 8: 612-7614.

Griffith, G.W., and Shaw, D.S. 1998. Polymorphism in *Phytophthora infestans*: four mitochondrial haplotypes are detected after PCR amplification of DNA from pure cultures or from host lesions. *Appl. Environ. Microbiol.* 64: 4007-4014.

Grimm, C. and Geisen, R. 1998. A PCR-ELISA for the detection of potential fumonisin producing *Fusarium* species. *Lett. Appl. Microbiol.* 3: 33-35.

Hantula, J., Dusabenyagasani, M., and Hamelin, R.C. 1996. Random amplified microsatellites (RAMS) a novel method for characterizing genetic variation within fungi. *Eur. J. Plant Pathol.* 26: 159-166.

Harlton, C.E., Levesque, C.A., and Punja, Z.K. 1995. Genetic diversity in *Sclerotium (Athelia) rolfsii* and related species. *Phytopath.* 85: 1269-1281.

Hawksworth, D.L. Kirk, P.M., Sutton, B.C., and Pegler, D.N. 1995. Ainsworth and Bisby's Dictionary of the Fungi, 8th Ed. CAB International, Wallingford, Oxon, UK.

Heid, C.A., Stevens, J., Livak, K.J., and Williams, P.M. 1996. Real-time quantitative PCR. *Genome Res.* 6: 986-994.

Howlett, B.J. 1996. Application of pulsed field gel electrophoresis to the study of genomes of fungal pathogens of plants. *Aust. J. Plant Pathol.* 25: 219-225.

Jacobson, D.J., and Gordon, T.R. 1990. Variability of mitochondrial DNA as an indicator of relationships between populations of *Fusarium oxyporum* f. sp. *melonis. Mycol. Res.* 94: 734-744.

Jimenez-Gasco, M.M., Perez-Artes, E., Jimenez-Diaz, R.M. 2001. Identification of pathogenic races 0, 1B/C, 5, and 6 of *Fusarium oxysporum* F.sp. *ciceris* with Random Amplified Polymorphic DNA (RAPD). *Eur. J. Plant Pathol.* 107 (2): 237-248.

Jiminez-Gasco, M.M., and Jimenez-Diaz, R.M. 2003. Development of a specific polymerase chain reaction-based assay for the identification of *Fusarium oxysporum* f. sp. *ciceris* and its pathogenic races 0, 1A, 5 and 6. *Phytopathol.* 93: 200-209.

Jones, R.W., and Wang, H. 1997. Immunolocalization of a â-1, 4-endoglucanase from *Macrophomia phaseolina* expressed in plants. *Can. J. Microbiol.* 43: 491-495.

Judelson, H.S., and Messenger-Routh, B. 1996. Quantitation of *Phytophthora cinnamomi* in avcado roots using a species-specific DNA probe. *Phytopath.* 86: 763-768.

Julian, M.C., Acero, J., Salazar, O., Keijer, J., and Rubio, V. 1999. Mating type correlated molecular markers and demonstration of heterokaryosis in the phytopathogenic fungus *Thanatephorus cucumeris* (*Rhizoctonia solani*) AG 1-IC by AFLP DNA fingerprint analysis. *J. Biotechnol.* 67: 49-56.

Keijer, J., Houterman, P.M., Dullemans, A.M., and Korsman, M.G. 1996. Heterogeneity in electrophoretic karyotype within and between anastomosis groups of *Rhizoctonia solani. Mycol. Res.* 110: 789-797.

Kelly, A., Alcala Jiminez, A.R., Bainbridge, B.W., Heale, J.B., Perez-Artes, E., and Jiminez Diaz, R.M. 1994. Use of genetic fingerprinting and random amplified polymorphic DNA to characterize pathotypes of *Fusarium oxysporum* f. sp. *ciceris* infecting chickpea. *Phytopath.* 84: 1293-1298.

Kelly, A.G., Bainbridge, B.W., Heale, J.B., Perez-Artes, E. and Jimenez-Diaz, R.M. 1998. *In-planta* polymerase chain reaction detection of the wilt-inducing pathotype of *Fusarium oxysporum* f.sp. *ciceris* in chickpea (*Cicer arietinum* L.). Physiol. *Mol. Plant Pathol.* 52: 397-409.

Kempken, F. and Walz, M. 1996. The use of rare cutting endonuclease in electrophoretic karyotyping of fungal genomes. *Fungal Genet. Biol.* 20: 89-92.

Kim, D.H., Martyn, R.D., and Magill, C.W. 1991. Mitochondrial DNA restriction fragment length polymorphisms in *Fusarium oxysporum* f. sp. *niveum. Phytoparasitica.* 19: 211-223.

Kim, D.H.R.D., Martin, C., and Magill, W. 1992. Restriction fragment length polymorphism groups and physical map of mitochondrial DNA from *Fusarium oxysporum* f.sp. *nivem. Phytopath.* 82: 346-353.

Kistler, H.C., Bosland, P.W., Benny, S., Leong, S., and Williams, P.H. (1987). Relatedness of strains of *Fusarium oxysporum* from crucifers measured by examination of mitochondrial and ribosomal DNA. *Phytopath.* 77: 1289-1293.

Kistler, H.C. 1997. Genetic diversity in the plant pathogenic fungus *Fusarium oxysporum. Phytopath.* 87: 474-479.

Knapova, G., and Gisi, U. 2002. Phenotypic and genotypic structure of *Phytophthora infestans* populations of potato and in France and Switzerland. *Plant Pathol.* 51: 641-653.

Knapova, G., Tenzer, I., Gessler, C., and Gisi, U. 2001. Characterization of *Phytophthora infestans* from potato and tomato with molecular markers. Biodiversity in plant pathology. *Proceedings of 5th congress of the European Foundation for Plant Pathology,* Taormina, Pisa, Italy: SIPV, 6-9.

Kohler, G., and Milstein, C. 1975. Continuous culture of fused cells secreting antibodies of predefined specificity. *Nature.* 56: 495-497.

Kohli, Y., Morrall, R.A.A., Anderson, J.B., and Kohn, L.M. 1992. Local and *trans*-Canadian clonal distribution of *Sclerotinia sclerotiorum* on canola. *Phytopath.* 82: 875-880.

Kohn, L.M., Carbone, I., and Anderson, J. B. 1990. Mycelial interactions in *Sclerotinia sclerotiorum. Exp. Mycol.* 14: 225-267.

Kohn, L.M., Petsche, D.M., Baile, S.R., Norak, L.A., and Anderson, J.B. 1988. Restriction fragment length polymorphisms in nuclear and mitochondrial DNA of *Sclerotinia* species. *Phytopath.* 78:1047-1051.

Kohn, L.M., Statsovski, E., Carbone, I., Royer, J., and Anderson, J.B. 1991. Mycelial incompatibility and molecular markers identify genetic variability in populations of *Sclerotinia sclerotiorum. Phytopath.* b: 480-484.

Kohn, L.M. 1979. Delimitation of economically important plant pathogenic *Sclerotinia* species. *Phytopath.* 69: 881-886.

Kuninaga, S., and Yokosawa, R. 1985. DNA base sequence homology in *Rhizoctonia solani* Kuhn. IV. Genetic relatedness among seven anastomosis groups. *Ann. Phytopathol. Soc. Jpn.* 51: 127-132.

Kuninaga, S., Carling, D.E., Takeuchi, T. O., and Yokosawa, R. 2000. Comparison of rDNA-ITS sequences between potato and tomato strains in *Rhizoctonia solani* AG-3. *J. Gen. Plant Pathol.* 66: 2-11.

Kuninaga, S., Natsuaki, T., Takeuchi, T., and Yokosawa, R. 1997. Sequence variation of the rDNA ITS regions within and between anastomosis groups in *Rhizoctonia solani. Curr. Genet.* 32: 237-243.

Kuninaga, S. 1996. DNA base sequence complementary analyses. (*In*:) B. Sneh, S. Jabaji-Hare, S. Neate and G. Dijst (Eds.). *Rhizoctonia species: Taxonomy, Molecular Biology, Ecology, Pathology and Disease Control.* Kluwer Academic Publishers, Dordrecht, The Netherlands, pp 73-80.

Kuninaga, S. 2002. Current situation of the taxonomy of the genus *Rhizoctonia* and the *R. solani* species complex. *Jpn. J. Phytopathol.* 68:3-20.

Lees, A.K., Cullen, D.W., Sullivan, L., and Nicolson, M.J. 2002. Development of conventional and quantitative real-time PCR assays for the detection and identification of *Rihizoctonia solani* AG-3 in potato and soil. *Plant Pathol.* 51: 293-302.

Leslie, J.F., Zeller, K.A., and Summerell, B. 2001 Icebergs and species in populations of *Furarium. Physiol. and Mol. Pathol.* 59: 107-117.

Le Tourneau, D. 1979. Morphology, cytology and physiology of *Sclerotinia* species in culture. *Phytopath.* 69: 887-890.

Levesque, C.A. 1997. Molecular detection tools in integrated disease management: overcoming current limitations. *Phytoparastica.* 25: 3-7.

Levesque, C.A. 2001. Molecular methods for detection of plant pathogens -What is the future ? *Can. J. Plant Pathol.* 24: 333-336.

Li, S., Tam, Y.K., and Hartman, G.L. 2000. Molecular differentiation of *Fusarium solani* f. sp. *glycines* from other *F. solani* based on mitochondrial small subunit rDNA sequences. *Phytopathol.* 90: 491-497.

Liesack, W., Janssen, P.H., Rainey, F.A., Ward-Rainey, N.L., and Stackenbrandt, E. 1997. Microbial diversity in soil: the need for a combined approach using molecular and cultivation techniques. (*In*:) J.D van Elsas, J.Trevors, M.Wexler (Eds.) Modern Soil Microbiology, New York: Marcel Dekker 375-440.

Link, H.E. 1809. Observations in ordines plantrum naturals. Disseratio I. Mag. Grs. Naturf. Freunde. Berlin 3: 3-12.

Liu, W., Marsh, T.L., Cheng, H., and Forney, J.L. 1997. Characterization of microbial diversity by determining terminal restriction fragment length polymorphism of genes encoding 16s rRNA. *Appl. Environ. Microbiol.*63: 4516-22

Liu, W.T., and Stahl, D. (2002). Molecular approaches for the measurement of density, diversity, and phylogeny. *Appl. Environ. Microbiol.* 82: 14-34

Liu, Z.L., and Sinclair, J.B. 1993: Differentiation of intraspecific groups within anastomosis group 1 of *Rhizoctonia solani* using ribosomal DNA internal transcribed spacer and isozyme comparison. *Can. J. PlantPathol.* 15: 272-280.

Livak, K.J., Flood, S.J.A., Marmaro, J., Giusti, W., and Deetz, K. 1995. Oligonucleotides with fluorescent dyes at opposite ends provide a quenched probe system useful for detecting PCR product and nucleic acid hybridization. *PCR Methods App.* 5: 357-362.

Lodwig, E.M., Bridge, P.D., Rutherford, M.A., Kung'u, J., and Jeffries, P. 1999. Molecular differences distinguish clonal lineages within east African populations of *Fusarium oxysporum* f. sp. *cubense*. *J. Appl. Microbiol.* 86: 71-77.

Lubeck, M., and Poulsen, H. 2001. UP-PCR cross-blot hybridization as a tool for identification of anastomosis groups in the *Rhizoctonia solani* complex. *FEMS Microbiol. Lett.* 201: 83-89.

Majer, D., Mithen, R., Lewis, B.G., Vos, P., and Oliver, R.P. 1996. The use of AFLP fingerprinting for the detection of genetic variation in fungi. *Mycol. Res.* 100: 1107-111.

Manicom, B., and Baayen, R.P. 1993. Restriction fragment length polymorphisms in *Fusarium oxysporum* f.sp. *dianthi* and other fusaria from *Dianthus* species. *Plant Pathol.* 42: 851-857.

Martin, R.R., James, D., and Levesque, C.A. 2000. Impacts of molecular diagnostic technologies on plant disease management. *Ann. Rev. Phytopath.* 38: 207-239

Maurand, R., Archer, S.A., Buck, K.W., Shaw, D.S., and Shattock, R.C. 1995. The use of RAPD markers in genetic studies of *Phytophthora infestans*. (*In*:) L.J. Dowley, E. Bannan, L.R. Cooke, T. Kaene, E. O'Sullivan, (Eds.) *Phytopthora infestans*. Dublin: Boole Press Ltd, pp 150.

McSpadden Gardener, B.B., and Weller, D. M. 2001. Changes in population of rhizosphere bacteria associated worth take all of wheat. *Appl. Environ. Microbiol.* 67: 4414-25

McSpadden Gardener, B.B., Mavridi, D.V., Thomashow, L.S., and Weller, D.M. 2001. A rapid PCR based assay characterizing rhizosphere populations of 2, 4-DAPG-producing bacteria. *Phytopath.* 91: 44-54.

Meng, X.Q., Shoemaker, R.C., and Yang, X.B. 1999. Analysis of pathogenicity and genetic variation among *Phytophthora sojae* isolates using RAPDs. *Mycol. Res.* 103:173-178.

Mess, J.J., Van Doorn, J., Roebroeck, E.J.A., Van Egmond, E., Van Aarbrijk, J., and Boonekamp, P.M. 1994. Restriction fragment length polymorphisms, races and vegetative compatibility groups within a world-wide collection of *Fusarium oxysporum* f. sp. *gladioli*. *Plant Pathol.* 43:362-370.

Meyer, W., Morawetz, R., Borner, T., and Kubicek, C.P. 1992. The use of DNA fingerprint analysis in the classification of some species of *Trichoderma* aggregate. *Curr. Genet.* 21: 27-30.

Mia, V.P.W. 1990. Using karyotype variability to investigate the origin and relatedness of isolates of *Fusarium oxysporum* f. sp. *cubense*. (*In*:) R.C. Ploefz, (Ed.) Fusarium Wilt of Banana. The Am. Phytopathol. Soc. St. Paul, MN, pp 55-62.

Michelmore, R.W., and Hulbert, S.H. 1987. Molecular markers for genetic analysis of phytopathogenic fungi. *Ann. Rev. Phytopath.* 25: 383-404.

Mihail, J.D., and Taylor, S.J. 1995. Interpreting variability among isolates of *Macrophomina phaseolina* in pathogenicity, pycnidium production, and chlorate utilization. *Can. J. Bot.* 73: 1596-1603.

Mihail, J.D., Obert, M., Taylor, S.J., and Bruhn, J.N. 1994. The fractal dimesion of young colonies of *Macrophomina phaseolina* produced from microsclerotina. *Mycologia.* 86: 350-356.

Mills, D., McCluskey, K., Russel, B.W., and Agnan, J. 1995. Electrophoretic karyotyping: Methods and Application. (*In*:) R.P. Singh and U.S. Singh (Eds.) *Molecular Methods in Plant Pathology*, Michigan: Lewis Pub. Inc. pp 81-86.

Mukherjee, P.K., Bhagwat, A., Mitra, J., Mitra, R., and Murthy, N.B.K. 1997. Restriction analysis of ribosomal DNA: a tool for differentiating between races of *Fusarium oxysporum* f.sp. *ciceris.* *(In:) Indian Phytopathological society-Golden Jubilee, Int. Conf. Integrated Plant Disease Management in Sustainable Agriculture,* New Delhi, India, pp 43 (abstr.).

Mullis, K.B., and Faloona, F.A. 1987. Specific synthesis of DNA *in vitro* via a polymerase catalyzed chain reaction. *Meth. Enzymol.* 155: 335-350.

Murillo, I., Cavallarin, L., and San Segundo, B. 1998. The development of a rapid PCR assays for detection of *Fusarium moniliforme. European J. Plant Pathol.* 104: 301-311.

Muyzer, G., and Smalla, K. 1998. Application of denaturing gradient gel electrophoresis (DGGE) and temperature gradient gel electrophoresis (TGGE) in microbial ecology. Antonie van Leeuwenhoek. *J. Microbiol. Serol.* 73: 127-41

Myer, R.M., Fischer, S.G., Lerman, L.S., and Maniatis, T. 1985. Nearly all single base substituation in DNA fragments joined to a GC-clamp can be detected by denaturing gradient gel electrophoresis. *Nucleic Acid Res.* 13: 3131-3145.

Nalim, F.A., Starr, J.L., Woodard, K.E., Senger, S., and Keller, N.P. 1995. Mycelial compatibility groups in Texas peanut field populations of *Sclerotium rolfsii. Phytopath.* 85: 1507-1512.

Namiki, F., Shiomi, T., Kayamura, T. and Tsuge, T. 1994. Characterisation of the formae-speciales of *Fusarium oxysporum* causing wilts of cucurbits by DNA fingerprinting with nuclear repetitive DNA sequences. *Appl. Environ. Microbiol.* 60: 2684-2691.

Namiki, F., Shiomi, T., Nishi, K., Kayamura, K., and Tsuge, T. 1998. Pathogenic and genetic variation in the Japanese strains of *Fusarium oxysporum* f. sp. *melonis. Phytopath.* 88: 804-810.

Nazareth, S.W., and Bruschi, C.V. 1994. Electrophoretic karyotype of *Fusarium solani. Microbiology,* 140: 1373-1375.

Neeraja, C.N., Vijayabhanu, N., Shenoy, V.V., Reddy, C.S., and Sarma, N.P. 2002. RAPD analysis of Indian isolates of rice sheath blight fungus *Rhizoctonia solani. J. Plant Biochem. Biotech.* 11:43-48.

Nogales, B., Moore, E., Llobet-Brossa, E., Rosselo-Mora, R., Amann, R., Timmis, N. 2001. Combined use of 16S ribosomal DNA and 16S rRNA to study the bacterial community of polychorinated biphenyl-polluted soil. *Appl. Environ. Microbiol.* 67: 1874-84.

Nyasse, S., Grievet, L., Risterucci, A.M., Blaha, G., Berry, D., Lanand, C., and Despreaux, D. 1999. Diversity of *Phytophthora megakarya* in central and west Africa revealed by isozyme and RAPD markers. *Mycol. Res.* 103: 1225-1234.

Ogoshi, A. 1996. Introduction-The genus *Rhizoctonia* (*In*:) B. Sneh, S. Jabaji-Hare, S. Neate and G. Dijst (Eds.). *Rhizoctonia* species: Taxonomy, *Molecular Biology, Ecology, Pathology and Disease Control*. Kluwer Academic Publishers, Dordrecht, The Netherlands, pp 1-9.

Okabe, I., and Matsumoto, N. 2000. Population structure of *Sclerotium rolfsii* in peanut fields. *Mycoscience.* 41: 145-148.

Okabe, I., Arakawa, M., and Matsumoto, N. (2001). ITS polymorphism within a single strain of *Sclerotium rolfsii. Mycoscience.* 42: 107-113.

Okabe, I., Morikawa, C., Matsumoto, N., and Yokoyama, K. 1998. Variation in *Sclerotium rolfsii* isolates in Jpn. *Mycoscience.* 39: 399-407.

Orita, M., Iwahana, H., Kanazana, H., Hayashi, K., and Sekiya, T. 1989. Detection of polymorphisms of human DNA by gel electrophoresis as single-strand conformation polymorphisms. *Proc. Natl. Acad. Sci.* USA 86: 2766-2770.

Ouellet, T., and Seifert, K.A. 1993. Genetic characterization of *Fusarium graminearum* strains using RAPD and PCR amplification. *Phytopath.* 83: 1003-1007.

Paavanen-Huhtala, S., Hyvonen, J., Bulat, S.A., and Yli-Mattila, T. 1999. RAPD-PCR, isozyme, rDNA RFLP and rDNA sequence analysis in identification of Finnish *Fusarium oxysporum* isolates. *Mycol. Res.* 103: 625-634.

Pasquer, A., and Brygo, Y. 2000. Diversity of soil fungi studied by PCR-RFLP ITS. *Mycol. Res.* 104: 1027-1032.

Penananen, T., Paavolainen, L., and Hantula, J. 2001. Rapid PCR-based method for the direct analysis of fungal communities in complex environmental samples. *Soil Biol. Biochem.* 33: 697-699.

Perez-Artes, E., Roncero, M.I.G., and Jiminez Diaz, R.M. 1995. Restriction fragment length polymorphism analysis of the mitochondrial DNA of *Fusarium oxysporum* f. sp. *ciceris. J. Phytopathol.* 43: 105-109.

Punja, Z.K., and Damiani A. 1996. Comparative growth, morphology, and physiology of three *Sclerotium* species. *Mycologia.* 88: 694-706.

Punja, Z.K., and Grogan, R.G. 1983. Basidiocarp induction, nuclear condition, variability, and heterokaryon incompatibility in *Athelia rolfsii. Phytopath.* 73: 1273-1278.

Punja, Z.K., and Sun, L.J. 2001. Genetic diversity among mycelial compatibility groups of *Sclerotium rolfsii* (teleomorph *Athelia rolfsii*) and *S. delphini*. *Mycol. Res*. 105: 537-546.

Punja, Z.K. 1988. *Sclerotium rolfsii,* a pathogen of many plant species. *Adv.Plant Pathol*. 6: 523-534.

Qiu, X., Wu, L., Huang, H., McConel, P.E., Palumbo, A.V. 2001. Evaluation of PCR generated chimeras mutations, and heterduplexes with 16S rRNA gene based cloning. *Appl. Environ. Microbiol*. 67: 880-87

Roche Molecular Biochemicals 2001. PCR ELISA (DIG Detection). Semiquantitative Detection of DIG-labeled Products by Hybridzation-based Microtiter Plate Assay. Instruction Manual Version 3 May 2001. Available at http://biochem.roche.com/pack-insert/1636111A.pdf.

Rowhani, A., Biardi, L., Routh, G., Daubert, S.D., and Golani, D.A. 1998. Development of a sensitive colorimetric-PCR assay for detection of viruses in woody plants. *Plant Dis*. 82: 880-884.

Saiki, R.K., Gelfand, D.H., Stoffel, S., Scharf, S., Higuchi, R.H., Horn, G.T., Mullis, K.B., and Elrich, H.A. (1985). Primer directed enzymatic amplification of DNA with a thermo stable DNA polymerase. *Science*. 239: 487-492.

Schneider, J.H.M., Salazer, O., Rubio, V., and Keijer, J. 1997. Identification of *Rhizoctonia solani* associated with field-grown tulips using ITS rDNA polymorphism and pectic zymograms. Eur. *J. Plant Pathol*. 103: 607-622.

Schwartz, D.C., and Contor, C.R. 1984. Separation of yeast chromosome-sized DNA's by pulse field gel electrophoresis. *Cell*. 37: 67-75.

Schwieger, F., and Tebbe, C.C. 2000. Effect of field inoculation with *Sinorhizobium meliloti* communities in rhizospheres of a target plant (*Medicago staiva*) and a non target plant (*Chenopodium album*) - linking of 16s rRNA genebased single strand conformation polymorphism community profiles to the diversity of cultivated bacteria. *Appl. Environ. Microbiol*. 66: 3556-65

Sharma, T.R., Jana, T.K., and Arora, D. K. 2003. Molecular characterization and genomic variability of filamentous fungi. (*In*:) Arora, D. K, Kachatourians, G. (Eds.), *Applied mycology and biotechnology*, Vol. 4. Elsever Press, In press.

Sharma, T.R., Prachi, and Singh, B.M. 1999. Application of polymerase chain reaction in phytopathogenic microbes. *Indian J. Microbiol*. 39:79-91.

Sharma, T.R., Tewari, J.P., and Singh, B.M. 2002. Use of polymerase chain reaction for the synthesis of non-radioactive DNA probes in phytopathogenic fungi. (*In*:) G.S. Shekhavat, B.P. Singh and R.A. Singh (Eds). *Molecular Application Plant Disease Management*. Malhotra Publishing House, New Delhi, India, pp 54-62.

Sirjusingh, C., and Kohn, L.M. 2001. Characterization of microsatellites in the fungal plant pathogen, *Sclerotinia sclerotiorum*. *Mol. Ecol. Notes*. 1: 267-269.

Smit, E., Leeflang, P., Glandorf, B., Van Elsas, J.D., and Wernars, K. 1999. Analyses of fungal diversity in the wheat rhizosphere by sequencing of cloned PCR-amplified genes enco ding 18S rRNA and temperature grandient gel electrophoresis. *Appl. Environ. Microbiol.* 65: 2614-2621.

Smith, S.N., DeVay, J.E., Hsieh, W.H., and Lee, H.J. 2001. Soil-borne populations of *Fusarium oxysporum* f. sp. vasinfectum, a cotton wilt fungus in California fields. *Mycologia.* 94: 737-743.

Sneh, B., Jabaji-Hare, S., Neate, S., and Dijst, G. 1996. *Rhizoctonia species: Taxonomy, molecular biology, ecology, pathology and disease control.* Kluwer Academic Publishers, Dordrecht, The Netherlands.

Speksnijder, A.G.C.L., Kowalchuck, G.A., DeJong, S., Kline, E., Stephen, J.R., and Laanbroek, H.J. 2001. Microvariation artifacts introduced by PCR and cloning of closely related 16S rRNA gene sequences. *Appl. Environ. Microbiol.* 67: 469-72

Steenkamp, E., Britz, H., Coutinho, T., Wingfeld, B., Marasas, W., and Wingfeld, M. 2000. Molecular characterization of *Fusarium subglutinans* associated with mango malformation. *Mol. Plant Pathol.* 1:187-193.

Su, G., Suh, S.O., Schneider, R.W., and Russin, J.S. 2001. Host specialization in the charcoal rot fungus, *Macrophomrina phaseolina. Phytopath.* 91: 120-126.

Suga, H., Ikeda, S., Taga, M., Kageyama, K., and Hyakumachi, M. 2002. Electrophoretic karyotyping and gene mapping of seven formae speciales in *Fusarium solani. Curr. Genet.* 41: 254-260.

Taylor, J.W. 1986. Fungal evolutionary biology and mitochondrial DNA. *Exp. Mycol.* 10: 259-269.

Thomas, V., Rutherford, M.A., and Bridge, P.D. 1994. Molecular differentiation of two races of *Fusarium oxysporum* f. sp. *cubense. Lett. Appl. Micriobiol.* 18: 193-196.

Tiedje, J., Asuming-Brempong, S., Nusslein, K., Marsh, T., Flynn, S. 1999. Opening the black box of soil microbial diversity. *Appl. Soil Ecol.*13: 109-22

Tien, C.C., Chao, C.C., and Chao, W.L. 1999. Methods for DNA extraction from various soils: a comparison. *J. Appl. Microbiol.* 86: 937-943.

To-Anum, C., Nelson, H., and Ouchi, S. 1995. Electrophoretic karyotyping of *Fusarium oxysporum. Ann. Phytopathol. Soc. Jpn.* 61:350.

Toda, T., Hyakumachi, M., and Arora, D.K. 1999a. Genetic relatedness among and within different *Rhizoctonia solani* anastomosis groups as assessed by RAPD, ERIC and REP-PCR. *Microbiol. Res.* 154:247-258.

Toda, T., Hyakumachi, M., Suga, H., Kageyama, K., Tanaka, A., and Tani, T. 1999b. Differentiation of *Rhizoctonia* AG-D isolates from turfgrass into subgroups I and II based on rDNA and RAPD analysis. *Eur. J. Plant Pathol.* 105:835-846.

Tooly, P.W., and Carras, M.M. 1992. Separation of chromsomes of *Phytophthora* species using CHEF gel electrophoresis. *Exp. Mycol.* 16: 188.

Vainio, E.J., and Hantula, J. 2000. Direct analysis of wood-inhabiting fungi using denaturing gradient gel electrophoresis of amplified ribosomal DNA. *Mycol. Res.* 104: 927-936.

Vandemark, G., Martinez, O., Pecina, V., and Alvardo, M., de J. 2000. Assessment of genetic relationships among isolates of *Macrophomina phaseolina* using simplified AFLP technique and two different methods of analysis. *Mycologia* .92:656-664.

Van de Peer, Y., Van der Auwera, G., and De Wachter, R. 1996. The evolution of stramenopiles and alveolates as derived by „substitution rate calibration" of small ribosomal subunit RNA. *J. Mol. Evol.* 42: 201-210.

Van der Lee, T., De Witte, I., Drenth, A., Alfonso, C., and Govers, F. 1997. AFLP linkage map of Oomycete *Phytopthora infestans. Fungal. Genet. Biol.* 321: 278-291.

Vilgalys, R., and Gonzalez, D. 1990. Ribosomal DNA restriction fragment length polymorphisms in *Rhizoctonia solani. Phytopath.* 80: 151-158.

Vilgalys, R. 1988. Genetic relatedness among anastomosis groups in *Rhizoctonia solani* as measured by DNA/DNA hybridization. *Phytopath.* 78: 698-702.

Wang, P., Lo, H.S., and Yeh, Y. 2001. Identification of *F.o. cucumerium* and *F.o. luffae* by RAPD generated DNA probes. *Lett. Appl. Microbiol.* 33: 397-401.

Waterhouse, G.M. 1963. Key to the species of *Phytophthora* de Bary. Mycol. Paper No. 92, 22 pp. CMI Kew, UK.

Wattier, R.A.M., Gathercole, L.L., Assinder, S.J., Gliddon, C.J. Deahl, K.L., Shaw, D.S and Mills, D.I. 2003. Sequence variation of intergeneric mitochondrial DNA spacers (mtDNA-IGS) of *Phytophthora infestans* (Oomycetes) and related species. *Mol. Ecol. Notes* 3:136-138.

Welsh, J., and McClelland, M. 1990. Fingerprinting genomes using PCR with arbitrary primers. *Nucl. Acids Res.* 18: 7213-7218.

Whisson, S.C., Maclean, D.J., Manners, J.M., and Irwin, J.A.G. 1992. Genetic relationships among Australian and North American isolates of *Phytophthora magasperma* assessed by multicopy DNA probes. *Phytopath.* 82: 863-868.

Williams, J.G.K., Kubelik, A.R., Livak, K.J., Rafalski, J.A., and Tingey, S.V. 1990. DNA polymorphisms amplified by arbitrary primers are useful as genetic markers. *Nucl. Acids Res.* 18: 6531-6535.

Winton, L., Stone, J.K. Watrud, L.S., and Hansen, E.M. 2002. Simultaneous one-tube qualification of host and pathogen DNA with realtime polymerase chain reaction *Phytopath.* 92: 112-116.

Woese, D.R., Kandler, O., and Wheelis, M.L. 1990. Towards a natural system of organisms: proposal for the domains Archae, Bacteria, and Eurarya. *Proc. Natl. Acad. Sci.* USA 87:285-93

Woo, S.L., Zonia, A., Del Sorbo, G., Lorito, M., Nauni, B., Scala, F., and Noviello, C. 1996. Characterization of *Fusarium oxysporum* f. sp. *phaseoli* by pahtogenic races, VCGs, RFLP's and RAPD. *Phytopath.* 86: 966-973.

Yang, J., and Kharbanda, P.D. 1996. Characterization, virulence and genetic variation of *Rhizoctonia solani* AG-9 in Alberta. *Plant Dis.* 80: 513-518.

Yli-Mattila, T., and Hyvonen, J. 1996. RAPD-PCR analysis of *Fusarium* strains cladistic evaluation of the results. *Sydowia.* 48: 184-195.

Yli-Mattila, T., Mironenko, N.V., Alekhina, I.A., Hannukkale, A., and Bulat, S.A. 1997a. Universally primed polymerase chain reaction analysis of *Fusarium avenaceum* isolated from wheat and barley in Finland. *Agr. Food Sci. Fin.* 6:25-36.

Yli-Mattila, T., Paavanen, S., Hannukkala, A., Parikka, P., Tahvonen, R., and Karjalainen, R. 1996. Isozyme and RAPD-PCR analysis of *Fusarium avenaceum* strains from Finland. *Plant Pathol.* 45:126-134.

Yli-Mattila, Y., Paavanen-Huhtala, S., Bulat, S.A., Alekhina, I.A., and Nirenberg, H.I. 2002. Molecular, morphological and phylogenetic analysis of *Fusarium avenaceum, F. arthrosporioides, F. tricinctum* species complex - a polyphasic approach. *Mycol. Res.* 106: 655-669.

Yli-Mattila, Y., Paavanen-Huhtala, S., Hannukkala, A., Alekhina, I.A., Hyvonen, J., and Bulat, S.A. 1997b. RAPD-PCR, UP-PCR, rDNA RFLP, rDNA sequence and isozyme analysis of *Fusarium* isolates in Finland. *Cereal Res. Comm.* 25:629-630.

Zhang, A.W., Hartman, G.L., Currio-penny, B., Pederson, W.L., and Beacker, K.B. (1999). Molecular detection of *Diaporthe phaseolorum* and *phomopsis longicolla* from soyabean seeds. *Phytopath.* 89: 796-804.

Zhang, S., and Steffenson, B.J. 2000. A simple and sensitive silver staining method for detecting AFLP markers in fungi. Fungal. *Genet. Newsletr.* 47: 101-102.

Zhang, X.G., Sun, W.X., Guo, L., Yu, J.F., and Chang, C.J. 2003. Genetic and apathogenic variation among tobacco black shank strains of *Phytoph thora parasitica* var. *nicotianae* from the main tobacco growing in China. *J. Phytopath.* 151: 259-266.

Zolan, M.E. 1995. Chromosome-length polymorphism in fungi. *Microbiol. Rev.* 59: 686-698.

• • •

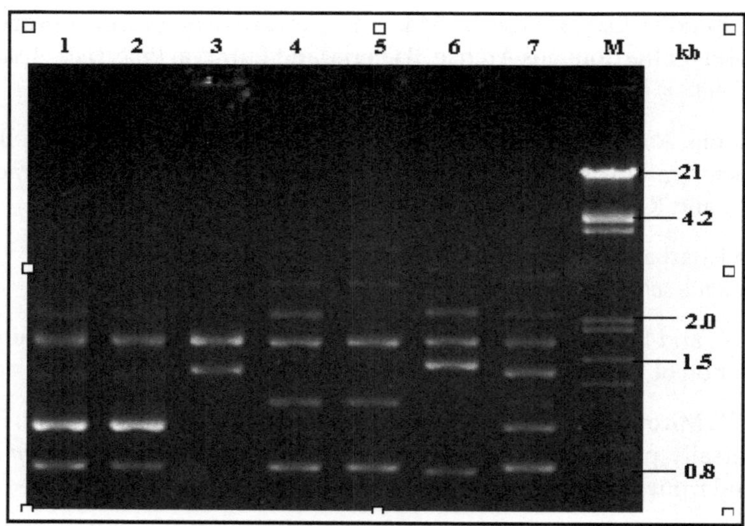

PLATE 3.1. RAPD profiles of different isolates of *Fusarium oxysporum* obtained with primers OPA-13. M = Molecular weight markers (Lambda DNA cut with *Eco* RI and *Hind* III) (Bangalore Genei, Bangalore, India). Lanes 1 to 7, are isolates of *Fusarium oxysporum*. PCR products were resolved on 1.4% agarose gel in TAE (0.5 X) buffer.

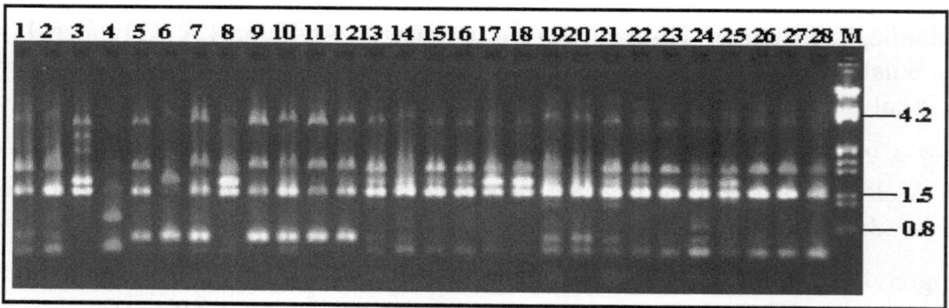

PLATE 3.2. RAPD profiles of different isolates of *Macrophomina phaseolina* obtained with prime OPC-18. M = Molecular weight markers (Lambda DNA cut with *Eco* RI and *Hind* III) (Bangalore Genei, Bangalore, India). Lanes 1 to 28, are isolates of *M. phaseolina*. PCR products were resolved on 1.4% agarose gel in TAE (0.5 X) buffer.

Advances in Soil Borne Plant Diseases, 2008
Eds.: Naik and Devika Rani, pp. 89-127
New India Publishing Agency, New Delhi

Advances in Plant Diseases Caused by *Sclerotium rolfsii* and their Management

R.D. Prasad and M.K. Naik*

INTRODUCTION

The form genus *Sclerotium* was described by Saccardo (Saccardo, 1913) to include fungi that produce sclerotia and had sterile mycelia. Several fungi were included in this genus, and all were characterized by production of small, dark brown or black spherical sclerotia that were internally differentiated into a rind, cortex and medulla. Several plant pathogenic fungi with this sclerotial type and with no known sexual or conidial state were placed in the genus *Sclerotium*. However, as the teleomorphic states were discovered, affinities to the Basidiomycotina (Punja, 1988) or Ascomycotina (Krause and Webster, 1972) were established for some members of the form genus.

One of the best-known members of the genus *Sclerotium* is *S. rolfsii Sacco* (teleomorph: *Athelia rolfsii* (Curzi) Tu & Kimbrough) is a soil borne plant pathogenic fungus which is prevalent in warm temperatures and sub tropical regions of the world (Aycock, 1966; Punja, 1985). The pathogen has host range of over five hundred plant species belonging to dicotyledonous plants

Directorate of Oilseeds Research, Rajendranagar, Hyderabad-560 024, India
* Department of Plant Pathology, University of Agricultural Sciences, College of Agriculture, Raichur- 584 101, Karnataka, India

and a few monocotyledonous species (Aycock, 1966). Two related species, *S.delphinii* and *S.coffeicola*, which are reported to occur on ornamental bulbs (delphinium, iris, and lily) and coffee, respectively, produce larger sclerotia that are lighter brown to orange in color (Aycock, 1966; Stevens, 1931). It is not clear whether these differences in host occurrence and morphology of sclerotia are sufficient to warrant a separate species designation for these isolates, within *S. rolfsii*, field isolates also may exhibit mycelial incompatibi1ity by forming barriage zones when paired in culture (Punja & Grogan, 1983a), which had led to the establishment of mycelial compatibility groups (MCGs) within the species (Punja, 1988; Punja & Grogan, 1983b).

Another species, *S. bataticola*, which causes diseases in various hosts, including ashy stem blight of bean and soybean, charcoal rot of sorghum and corn, stem rot on watermelon, and root and wood rot on citrus, produces numerous small, black, irregular sclerotia" usually in the pith of the stem or stalks of infected plant and occasionally, also conidia in pycnidia of *Macrophomina* type. A third *Sclerotium* species, *S. cepivorum*, which causes white rot disease of onion and garlic, in addition to sclerotia, also produces occasional sterile conidia on sporodochia (Agrios, 1988).

MORPHOLOGICAL AND GENETIC VARIATIONS IN *SCLEROTIUM* SPECIES

Isolates of *S. coffeicola*, *S. rolfsii* (*Corticium rolfsii*) and *S. delphinii* from diverse geographical areas show differences in morphology (colony characteristics and sclerotial formation), growth response to different temperatures and media (radial growth, dry weight and sclerotial production), and the ability to produce oxalic acid and pectinase enzymes. The best medium for colony growth was V8 agar for *S. coffeicola* and *S. rolfsii* and PDA for *S. delphinii*. All species formed clamp connections on leading hyphae, and with an average number of 2 nuclei in each species. The basidial stage was only induced in *S. rolfsii* on PDA containing charcoal and was not observed in the other two species. The sclerotia of all 3 species were comprised of 3 layers: an outer rind, a middle cortex and an inner medulla. The cortical tissue was 15-20 cell layers thick in *S. coffeicola* compared with 4-10 layers in the other 2 species. Proteins were also distributed in both the cortex and medulla of *S. rolfsii* and *S. delphinii*, but were more concentrated in the medulla of *S. coffeicola*, and occurred in the walls and the cytoplasm. Phenolic compounds were observed in the rind cells of all sclerotia, with *S. coffeicola* containing the least amount, and were localized to the walls. All species produced oxalic acid, pectinase and polyphenoloxidase in culture, with the highest production in *S. rolfsii*, followed by *S. delphinii* and *S. coffeicola*. These species can be differentiated by colony characteristics, such as morphology and size of sclerotia, by differences in histochemical staining and composition of sclerotia, and by growth responses to temperature (Punja and Damiani, 1996).

Field isolates of *S. rolfsii* from California that formed basidia on PDA containing 2% activated charcoal (C-PDA) single-basidiospores (S1) strains were obtained that showed pronounced variability in growth rate, sclerotial production and frequency of clamp formation, suggesting that the fie1d iso1ates were either heterokaryotic or diploid. Of these 100 S1 strains, 85 formed the basidial state in culture. An additional 62 field isolates from various hosts and geographic areas also showed variability; of these 36 were previously induced in fruit on C-PDA. The hyphal tip cells of field isolates and S1 strains were multinucleate, and discharged basidiospores contained 2 nuclei. Antagonism zones formed in pairings among sibling or non-sibling S1 strains, but isolations made from the zone of interaction of 120 of these pairings yielded 6 morphologically distinct colonies, 2 of which were shown to be heterokaryotic by fruiting and progeny analysis.

Formation of the antagonism zone in pairings between S1 strains. (Heterokaryon incompatibility) is not a complex barrier to formation of heterokaryons (Punja and Grogan, 1983b). When a large variety of isolates tested to study the impact of several environmental factors on teleomorph induction in *S. rolfsii,* and a comparison of morpholpgical features of each teleomorph induced, an isolates of the teleomorph stage induced were *Athelia rolfsii.* A corn-leaf-culture method is recommended for rapid, dependable teleomorph production (Tu *et al.,* 1992). Significant difference in the surface morphology of sclerotia, *viz.,* 'small' and 'large', of S. *rolfsii* was noticed, while the internal tissue organization, in general, was the same in both the types consisting of a thick skin, rind, cortex and medulla. The difference in size of sclerotia appears to be due to difference in the volume of the medullary region. The isozyme pattern of GPI-1 of both sclerotial types was same but significant difference was observed in the relative mobility, of PEP-1 (Prithiviraj *et al.,* 1996). The presence of different races of S. *rolfsii* in Cuba was confirmed by Hernandez-M'edina and Aleman (1997). A series of cultural, morphological and pathogenic characteristics were evaluated in 8 S. *rolfsii* isolates obtained from *Phaseolus vulgaris* and sunflower. Variability was found in their mycelial growth rate, number and diameter of the sclerotia, mycelial density, the presence of rhizomorphs, duration of sclerotial formation and pathogenicity on sunflower and *P.vulgaris.* In Japan, sixty-seven isolates of the southern blight fungus were divided into 5 groups on the basis of ITS-RFLP analysis of nuclear rDNA by Okabe *et al.,* (1998). Morphological characters of sclerotia varied between groups. Three groups were reidentified as S. *rolfsii,* and 2 resembled S. *delphinii* in RFLP patterns and/or in having large sclerotia and relatively low optimal growth temperature (28°C). Hyphal anastomosis was observed between different ITS-RFLP groups, as well as between different isolates belonging to the same groups. These results indicate that populations of this fungus consist of several different subgroups, although morphological differences are not always evident. S. *rolfsji* isolates can also be classified based on sclerotial formation types, mycelial compatibility groupings and sclerotial fine structure as observed by Scanning Electron Microscopy (SEM). Chen *et al.*, (1998) divided isolates of stem rot pathogen (S. *rolfsii*) of periwinkle into A, R and middle types on the basis of colony morphology. Mature sclerotia of A type, R type and middle type are composed of an outer melanized rind 2 to 4 cell layers thick, an underlying cortical layer and an innermost medullary region comprising loosely interwoven hyphae. Mazzanti *et al.*, (1997) confirmed variability in pathogenicity of S. *oryzae (Magnaporthe salvinii)* rice.

SCLEROTIAL BIOGENISIS

The biogenesis of sclerotia in S. *rolfsii* was associated with lipid peroxidation. There was a direct relationship between the number of sclerotia formed and lipid peroxidation levels in the mycelial colonies. Lipid peroxides of possible membranous and cytoplasmic origin were traced in sclerotial exudate. Sclerotia biogenisis may be associated with oxidative stress caused by growth conditions (Georgiou, 1997). Three stages viz., sclerotial initials, development and mature stages in sclerotial development were documented by Zarani and Christias (1997) in sclerotial biogenesis of S. *rolfsii* using scanning electron microscope. Initials developed from primordia, small tufts of mycelial hyphae held together very loosely. Each tuft grew in size and increased in hyphal density. Mature sclerotia were compact, spherical and had completely shed the outer layer of aerial hyphae. The skin surface was rough and scaly. With increasing age of culture and maturation of sclerotia, the levels of lipid peroxidation products and melanin increased. Lipid peroxidation and melanin formation may proceed in parallel during sclerotial biogenesis and maturation. Both lipid peroxidation and melanin formation are strongly affected by Fe_2^+ and by antioxidant vitamins (ascorbic acid), microelements (selenium) and mercapto compounds (glutathi.one). Myceliogenic germination and lytic activity via melanin production could be affected by (anti)oxidants, which could potentially be used for controlling sclerotia producing fungi without using traditional toxic fungicides.

Six days after inoculation sclerotial initials appeared in *S. cepivarum*, formed by hyphae branching and looping. From 6 to 8 days, the number and size of initials increased, and mucilagenous material appeared. By day 9, hyphal bundles were formed in the mycelium. Between 9 and 11 days, spherical forms developed and the sclerotia grew. By day 12, an acellular matrix appeared, and to day 18, this matrix progressively obscured the surface hyphae and became black. A lack of ovoid rind cells developed at the surface. In general, the rind of sclerotia aged in dry conditions had a broken, irregular appearance versus fresh sclerotia or sclerotia aged under moist, axenic conditions. Sclerotia aged dry developed 1 to 4 layers of rind cells, while sclerotia kept moist developed only 1 or 2 layers.

MYCELIAL COMPATIBILITY GROUPS (MCG)

Ponja and Grogan (1983b) showed that *S. rolfsii* isolates can be placed in mycelial compatibility groups (MCG) based on myceilial interactions. They identified 25 groups from 72 isolates. When isolates from the same MCG were paired, hyphae intermingled with little or no cell death, whereas isolates from different groups formed antagonistic zones that were accompanied by plasmolytic killing of hyphal cells. Twenty five mycelial compatibility groups (MCG) were identified from 366 isolates of *Sclerotium rolfsii* collected from Texas groundnut fields based on the formation of an antagonism zone between incompatible mycelia of paired isolates. The same MCG were often detected in several geographically distant countries in Texas. DNA amplification patterns resulting from the use of the 18-base oligonucleotide primer NK2 and from restriction digests of the internal transcribed spacer (ITS) region of the rDNAin a subset of 80 isolates from four fields to determine the genetic similarity of isolates within and between I2 MCG showed three NK2-amplified DNA patterns in genomic DNA and four MboI restriction digest patterns of the ITS region. All isolates within a MCG gave identical patterns for each marker, and some MCG shared the same ITS and NK2 patterns. MCG sharing the same DNA patterns were often from the same field. *S. rolfsii* isolate Sf-2 from black bean and isolates Sg-1, Sg-3, Sg-5, Sg-6, Sg-7, Sg-8 and Sg-9 from sunflower originating from different areas in Cuba and different soil types were placed into 5 groups with different antagonistic capacities or showing compatibility reactions (Hernandez and Aleman, 1997). Amplified fragment length polymorphism (AFLP) analysis was used by Cilliers *et al.*, (2000) to assess genetic variability between isolates and MCGs of *S. rolfsii*, the causal organism of stem rot or southern blight of groundnut. For preliminary genetic analysis, 10 isol ates were selected from each of two MCGs and compared with each other using the restriction enzymes EcoRI and Msel and 4 primer pairs. The number of polymorphisms ranged from 10 to 36 per primer combination, with an average of 22.5. AFLP analysis clearly showed genotypic differences (22%) among MCGs B andC, with a maximum variation of 6.41% between any two isolates per group using four primer pairs. Certain isolates could not be distinguished from each other. A more in-depth study of 10 isolates from MCG B, using 8 additional primer pairs, showed small genetic differences (maximum of 4.2% an minimum of 0.2%) between isolates. These results suggested that DNA could be pooled for comparison of MCGs. Pooled DNA from isolates within groups using 20 primer pairs confirmed differences between 9 MCGs. This technique effectively differentiated MCGs of *S. rolfsii* from each other and also detected differences between isolates within a single MCG.

Pairings among 119 isolates of *S. rolfsii*, 11 of *S. delphinii* and 2 of *S. coffeicola* on potato-dextrose agar to establish mycelial compatibility groups (MCGs) revealed 49,3 and 2 MCGs, respectively, in a worldwide collection. Within an MCG, isolates were often from the same host and geographical area; however, widely diverse isolates were also grouped within the same MCG. The host of origin of the isolate was not correlated with the MCG except in *S. delphinii*. Many MCGs were comprised of only 1 isolate. Variation in nuclear rDNA (internal transcribed spacer (ITS) regions) was examined following restriction enzyme digests. Restriction fragment length polymorphisms (RFLPs) were obtained and could distinguish amongst the 3 *Sclerotium* spp. and 3 *Athelia* spp. (an outgroup). Combined banding patterns for the 4 enzymes were used to characterize intraspecific variation in 3 *Sclerotium* spp. There were 12 subspecific groupings in *S. rolfsii*, 1 in *S. delphmii* and 2 in *S. coffeicola;* some of these groupings correlated with their MCG. However, isolates within an MCG could show different ITS RFLP patterns and certain patterns were also dispersed among different MCGs. Furthermore, segregation of 2 MboI restriction patterns in the ITS region among 29 single-basidiospore strains, derived from 4 parental field isolates of *S. rolfsii* with 1 pattern, suggested the presence of 2 distinct rDNA types in the field isolates. The rDNA types may reflect a heterokaryotic nuclear condition in field isolates of *S. rolfsii*. The restriction maps and phylogenetic analyses supported a close affinity of the 3 *Sclerotium* spp., which may be more appropriately designated as subspecific varieties of *S. rolfsii* (Harlton *et al.*, 1995).

S. cepivorum isolates (146) from Canada and other locations (23) were evaluated for mycelial compatibility in agarculture and by transmitted light microscopy. In compatible interactions, hyphae of paired colonies anastomosed without cell deterioration or death at the point of fusion. In some incompatible interactions, hyphae of paired colonies were initially similar to those of compatible interactions but were later followed by deterioration or death of the fused cells. Incompatible interactions were associated with a red-brown colour that, in mass, produced a dark line at the junction of the 2 colonies. In other incompatible interactions, hyphae bypassed or repelled each other without fusing. Compatible and incompatible reactions were used to define 9 mycelial compatibility groups (MCGs). No incompatibility responses occurred among isolates from individual fields but 2 MCGs occurred among isolates from different. Among isolates from other regions, 7 additional MCGs occurred that consisted of 1-8 isolates. Two isolates were variable in reaction and were not assigned to any MCG. Three isolates from Switzerland were compatible with MCG-l isolates from Ontario; and 1 isolate from Switzerland, 2 from England, L from New Zealand and 2 from Australia were all intercompatible. It is suggested that mycelial compatibility can be used to reflect genetic heterogeneity among isolates of *S. cepivorum*. Three field populations of *Sclerotium cepivorum* from Ontario and Quebec in Canada, one field sample from the Northisland of New Zea1and, as well as isolates from other onion- and garlic-growing areas were classified in mycelial compatibility groups (MCGs) and further characterized by DNA fingerprints and DNA sequence polymorphisms at six genomic regions. MCGs in the population samples were genetically distinct clones. Members of a clone belonged to a unique MCG with an associated multilocus genotype and DNA fingerprint. The structure in each field population was predominantly clonal based on the repeated sampling of genotypes over a wide geographic area. Only three clones were identified within each of the population samples. The hypothesis that these clones were introduced to the three agricultural production areas is supported by their recovery from additional locations. Some evidence for past recombination was detected when genotypes from the three population samples and from other locations were pooled.

MODE OF INFECTION

S. rolfsii attacks tissues directly. However, it produces considerable mass of mycelium and kills and disintegrates tissues by secreting oxalic acid, and also pectinolytic, cellulolytic, and other enzymes before it actually penetrates the host. Once the fungus becomes established in the plants, its subsequent advance and production of mycelium and sclerotia are quite rapid, especially during conditions of high moisture and high temperature. The pathogen grows. survives, and attacks plants best near the soilline, perhaps because of more favorable temperatures there, more plentiful supply of organic substances the fungus uses for food, and perhaps, 1ess competition or antagonism by other soil organisms (Agrios, 1988).

A young hypha of *S. cepivorum* penetrated the plant cuticle. Cytochemical techniques gave evidence of phenoloxidase and esterase activities in microconidia and thus confirmed their infectious potential. Massive production of microconidia was induced by immersion of the fungus in water.

These elements suggest that in high moisture conditions microconidia of *S. cepivorum* might play crucial role in survival and dissemination of the fungus. Metcalf and Wilson (1999) observed that *S. cepivorum* hyphae penetrated the epidermis and hypodermis of onion roots and grew into the cortex. Immediately following penetration only the cells through which *S. cepivorum* grew were lysed, but subsequently cells were killed and cell walls disintegrated ahead of the infection hyphae. *S. cepivorum* produced 2 polygalacturonases (PG) and-2 pectinesterases (PE) in culture. The spatial distribution of these enzymes was associated with cell death and cell wall degradation. The epidermis, hypodermis, endodermis and vascular tissues were more resistant to hydrolysis than the cortex, but only the endodermis and cells within it retained nuclei following hydrolysis of the surrounding cortical tissues. The cavity within the root cortex became filled with swollen, vacuolate *S. cepivorum* hyphae.

SYMPTOMATOLOGY

When seedlings are attacked, the fungus invades all the parts of the seedling, and the seedlings die quickly. When the fungus attacks grown up plants, it grows into the cortex and girdles the plants, which eventually die. Usually the infection begins on the succulent stem as a dark brown 1esion just below the soil-line. The first symptoms appears as yellowing or wilting of lower leaves or drying back of leaves from the tips downward and then symptoms progress to the upper leaves. The fungus grows upwards in the plant covering stem lesions with white cottony mycelium. The fungus spreads more rapidly into roots and destroys them. On all infected tissues, and even on the nearby soil, the fungus produces numerous small sclerotia.

SURVIVAL AND PERPETUATION

The fungus overwinters mainly as sclerotia. It is spread by moving water, infested soil, contaminated tools, infected transplant seedlings, infected vegetables and fruits, and in some hosts, as sclerotia mixed with the seed.

Selective isolation of Sclerotium

A selective medium containing KH_2PO_4, 1g; $MgSO_4.7H_2O$, 0.5 g; KNO_3, 2g; thiamine HC1, 1 mg; minor element solution, 10ml; gallic acid, 160mg; potassium oxalate, 10g; agar 20g; distilled water, 1 l was used for the selective isolation of *S. rolfsii* from field sclerotia contaminated with other organisms and from organic matter. Selectivity is based on tolerance and utilization of oxalate, low pH (4.2) and presence of gallic acid (Backman and Rodriguez Kabana, 1976).

Sampling techniques for recovery of sclerotia of Sclerotium

A screening-flotation method, which enables the comparison of inoculum levels of S. *oryzae (Leptosphaeria salvinii)* infecting rice in the soil was devised by Krause and Webster (1972). Mature sclerotia are hydrophobic, buoyant and quite uniform in size (c. 270 mm Dia.); most are produced in lesions on stem and 1eaf tissue of infected rice plants at water level and deposited in or on the soil during harvest and tillage; and no increase has been demonstrated in sclerotial numbers on rice residue incorporated with soil. Soil samples collected from such fields were air dried, then soaked in water, blended and washed on to 100-mesh screens from which sclerotia and soil were washed and the floating sclerotia separated.

A flotation technique using an aqueous solution of blackstrap molasses was developed by Rodriguez Kabana *et al.,* (1974) for rapid extraction of sclerotia of *Sclerotium rolfsii* from soil. Briefly, a 50-ml (70g) sub sample of air-dried, sieved (4-mm mesh) soil was placed in a 600 ml beaker and enough extracting solution added to provide a final volume of 350 ml the mixture was stirred with an electric stirrer (16,00 rpm) for 30s and ,allowed to settle for 30s. The1iquid was decanted through a 250 mm (60 meshes) screen of 7.5cm diameter. And the sclerotia were counted and plated on selective oxalate gallic acid medium to determine viability. The extracting solution consisted of 200 ml blackstrap molasses (sp. gr. 1.00, 25°C), 800 ml tap water, and 12.5 mg/ml separan NP 10 (a flocculating agent). This extracting solution had a sp. gr. value of 1.073 (25 °C). More than 80% of sclerotia of S. *rolfsii* were recoverable by the method, which detected as few as 1 sclerotium/50 ml soil. No sclerotia were detected in fields under maize which had been under groundnut the previous year, while 3.33-sclerotia/250 ml soil was found in groundnut fields, which had been under maize the previous year. The distribution of sclerotia in fields under groundnut mono culture indicated heterogeneity with a pronounced row effect where plots were contigeuous.

Another simple method of direct counting of viable sclerotia of S. *rolfsii* in soil was developed by Toribio (1977). Sub samples of 30 g air-dried, sieved soil were placed in 135 nm diameter dishes and watered to saturation. The dishes were sealed in plastic bags and incubated at 30°C for 48h. The bags were opened and the germinating sclerotia were counted. Recovery of viable sclerotia was 95-100%. The method can be used to determine sclerotial populations in natural soils. Sampling and extraction procedures as described above to estimate numbers, spatial pattern, and temporal distribution of sclerotia of S. *rolfsii* in soils were tried by Punja *et al.,*(1985). The mean inoculum density of S. *rolfsii* in 6 commercial fields ranged from 0.3 to 53.7 sclerotia/ 300 cm^3 of soil. The range of sclerotial density in the least and most heavily infested field was 0-5 and 2-225, respectively. The means and ranges varied with time of sampling and were influenced by previous cropping history and cultural practices. The frequency distribution of sclerotia among soil samples from the 6 fields at all sampling dates was best described by the negative binomial probability distribution. Values of the dispersion parameter, k, were 0.15-4.4 and variance-to-mean ratios were significantly greater than unity for all samples, indicating a clustering of inoculum. The range of sclerotia l density, frequency distribution of sclerotia and extent of variability among samples obtained from one field were influenced by sample probe size. Samples taken with a 10.2-cm-dia. probe inserted 7.5cm deep had a wide and variable range of inoculum density and the distribution of inoculum was described by the negative binomial. Bulked samples obtained with a 18.8cm diam. soil auger inserted 12cm deep had a smaller and less variable range of inoculum density, and the distribution was best described by the Poisson. Samples taken along simulated diagonal, parallel or diamond-shaped paths in each or 3 non uniformly infested fields provided similar mean inoculum density values, all of which were within 5% of the estimated population mean determined by the quadrate method. Total numbers of samples required were considerably reduced with these sampling patterns, however. Maximum recovery of laboratory or soil- produced sclerotia from artificially infested soil samples was high (91-95%) with a wet- sieving assay method and intermediate (77-86%) with a flotation-sieving method. A methanol assay procedure provided a low recovery (28:331110) of viable soil-produced sclerotia and intermediate recovery (57-59%) of viable laboratory-produced sclerotia.

MANAGEMENT OF DISEASES CAUSED BY SCLEROTIUM

Chemical methods

Chemical control of two important *Sclerotium* diseases *viz.*, S. *rolfsil* and S. *cepivorum* are discussed at length. Chemical fungicides effective in controlling various *Sclerotium* diseases are listed out in Table 4.1.

The application of either PCNB (quintozene) lOG, quintozene- chlorpyrifos 10 2G, ethoprop (ethoprophos) -quintozene 3 10G, or quintozene 10G + ethoprophos 10 G in narrow bands (10 or 15 cm) centered over the row at half the recommended rate per ha in many treatments resulted in significant increases in yield and *Sclerotium rolfsii* control. Narrow band width applications of the foliar-applied flutolanil 50WP and diniconazole 25WP tended to give increased yie1ds and better control of diseases than wider band width applications (Csinos, 1989). Fungicide treatments (benodanil 1.6-13.1 g a.i./ m, flutolanil 3.3 and 13.1 g a.i./m, quintozene 14.7 g a.i./m, Nor-Am SN 596 3.3 and 6.5 g a.i./m) were applied at planting to field-grown iris to control

S. *rolfsii*. After 9 months, 2 and 98-100% of the untreated and treated iris respectively, were healthy (Chastagner *et al.*, 1990). Studies on effect of tebuconazole, an ergo sterol biosynthesis inhibitor on S. *rolfsii* mycelium and sclerotia revealed that mean ED50 values for inhibition of mycelial growth of S. *rolfsii* was 0.08 g/ml, respectively, for tebuconazole, compared with 3.9 mg/ml for the pentachloronitrobenzene (PCNB) standard. The ED50 values for the inhibition of sclerotia formation by S. *rolfsii* were 0.13 and 4.99 mg/ml for tebuconazole and PCNB, respectively. ED50 values for inhibition of sclerotial initials were 0.14 and 2.75 mg/ml for tebuconazole and PCNB, respectively. In the field, tebuconazole provided good control of southern stem rot *(S. rolfsii)* of groundnut when applied 7 times as a foliar spray at rates of 188-250 g/ha (Brenneman *et at.*, 1991). The efficacy of foliar- applied flutolanil, diniconazole, tebuconazole and propiconazole for control of southern stem rot S. *rolfsii)* and effects on yield of groundnuts was compared with that of granular PCNB. Numbers of stem rot loci were significantly decreased and yields were increased in plots treated with banded and broadcast applications of flutolanil, diniconazole and tebuconazole. While these fungicides increased yields to a greater extent than quintozene (Hagan *et at.*, 1991). PCNB 10G (quintozene) and quintozene + ethoprop applied on 10 cm (narrow) or 25 cm (wide) band widths, were evaluated for control of southern stem rot *(S. rolfsii)* and impact on yield of groundnut. Hagan *et al.*,(1991) in another study showed that narrow band-width applications of quintozene at 5.6 a.i./ha, significantly reduced disease and increased groundnut yield over years Narrow and wide band width applications of quintozene + ethoprop at 5.6 + 1.7 kg a.i./ha and 11.2 + 3.3 kg a.i./ha reduced southern stem rot incidence and increased yields. Disease control and yield response following treatment with quintozene + ethoprop were similar to that in plots treated with quintozene on both narrow and wide band widths. Diniconazole, applied twice at 0.28 kg a.i./ha, gave better disease control, and higher yields than quintozene or quintozene + ethoprop in some years. Yield losses in groundnut *(cv.* Florunner) caused by southern stem rot *(S. rolfsii)* were evaluated in growers' fields at different levels of infestation and applications of pesticides, including PCNB (quintozene), chlorpyrifos, ethoprop, diniconazole, propiconazole, tebuconazole and fonofos. In general (based on an average of results over the 7 years) treatment with any of the fungicides increased yield by an average of 10.6% and decreased average disease incidence by 52.6% compared with untreated plots. Plots treated with a combination of insecticide and quintozene produced the highest yields and had the lowest disease incidence (5.5% more and 45.1% less than control plots, respectively) (Bowen *et al.*, 1992). In field tests to determine the effects of cyproconazole(0, 123,24.6, 49.1,73.9 and 98.7 g/ha) and Chlorothalonil

(0, 210, 420 and 630 g/ha), applied alone and in tank mixes on southern stem rot and groundnut yields, quadratic curvilinear reductions in the incidence of stem rot with increasing rates of cyproconazole were observed in both years. Culbreath *et al.,* (1995) found that chlorothalonil had no effect on incidence of stem rot. Yield increase was linear or curvilinear with increasing rates of cyproconazole, with yields converging with the higher rates of cyproconazole within all rates of ch1orothalonil. Drenches of lynx 3.6F (tebuconazole), ProStar 50W (flutolanil), Fluazinam 500F (fluazinam), Terraguard 50 W (triflumizole), and Curalan DF (vinclozolin) were compared with Terraclor 75W (PCNB) for the control of southern blight on container-grown aucuba *(Aucuba japonica)* by Hagan and Olive (1999). ProStar 50W and Fluazinam 500F across a range of rates completely protected aucuba from attack by the causal fungus, S. *rolfsii.* Also, both fungicides gave better disease control than the current industry standard Terraclor 75W. Lynx 3.6F at the 0.54 g a.i./litre rate proved as effective in controlling southern blight as ProStar SOW and Fluazinam 500F. Terraclor 75W, which was applied at over twice the labelled drench rate, gave inconsistent disease control. The insecticide, chloropyrifos suppress southern stem rot, caused by S. *rolfsii,* on groundnut as effectively as the fungicide terraclor (quintozene). Hagan *et al.,* (1989) suggested that granular chloropyrifos can be an attractive alternative to quintozene on dryland groundnuts because of its lower sclerotia/g it was necessary to treat garlic cloves with iprodione at 1000 g/100 kg and the soil at 1-2 g/m2. Treatment of cloves only was sufficient in areas with < 0.2 sclerotia/g. Combined treatments of quintozene at 6 g / ml or formaldehyde at 9.5-76 with iprodione on the cloves appeared to be promising, particularly for the most infested areas (Resende, 1987). Among post-emergence fungicides tested for their efficiency in controlling S. *cepivorum* in an onion monoculture dicarboximides procymidone and myclozolin were more effective. From the azote group, triadimefon, triadimenol and ethyltrianol gave the best control. The optimum time for application was when symptoms first became visible; hence, winter onions, most affected in the test region, needed 2 treatments, in mid-October and mid-April (Krauthausen and Chietinger, 1989). The effects of different methods of applying procymidone, either alone or combined, were evaluated for disease control in onions which showed that seed treatments combined with sprays to the soil surface gave the most effective control of white rot, reducing disease from 78 to 16% and increasing yields from 5.7 to 14.6 t/ha in brown onions. Stem base sprays applied two formulations of procymidone, -Sumisclex 500 (50% a.i.) and ns Flocol (27.5% a.i.), were equally effective (Porter *et al.,* 1991). Corbaz and Dorsaz (1992) observed that triadimefon and cyproconazole were the most promising sterol demethylation inhibitor (DMI) fungicides for control of S. *cepivorum* on onions in seedbed and field trials. Tebuconazole used as seedling dip (2S ml/l), followed by 2 sprays (750 ml/Fed) after transplanting at 6 week intervals, and significantly reduced white rot of onion (EI Shehaby *et al.*, 1992). Bugaret *et al.*,(1996) achieved good control of the white rot disease with vinclozolin, procymidone, diethofencarb + carbendazim + procymidone.

Natural products

Most of the studies on use of natural plant extracts/products for control of the fungi under genus *Sclerotium* are at laboratory1evel and none of those studies lead to development of commercial product for field use. Various plant extracts tried for control of fungi under genus *Sclerotium* and found inhibitory includes Datura *(D. stramonium)*, River-red gum *(Eucalyptus camaldulensis)*, Limon scented spotted gum *(E. citriodora)* and Osqar *(Calotropis procera)*, *Brachiaria humidicola, Crotalaria paulinG, Eucalyptus citriodora* and *Brassica oleracea* var. *capitata*, Neem oil, *Adhatoda vasica, Aegle marmelo, Anisomeles vata, Azadirachta indica, Cymbopogon flexuosus*, rhizomes *of Curcuma amada* and resin of *Ferula lostida, Adhatoda vasica, Moringa oleifera, Azadirachta indica, Lawsonia inermis, Sapindus trifoliatus, Rauvolfia serpentina, Pongamia glabra (P. pinnata)* and *Lantana camara* (Singh and Dwivedi, 1990; Daya *Ram et al.*, 1994; Das *et al*, 1997; Pinto *et al.,1998;* Yossry *et al.*, 1998).

Cultural practices for management of *Sclerotium* diseases

Effect of chemical fertilizer amendments

Sclerotial germination can be significantly inhibited by 10 mM soil amendments of urea, Ca cyanamide, ammonium su1fate, ammonium nitrate, ammonium chloroide, sodium nitrite or potassium nitrite. Sclerotia were killed by urea, Ca cyanamide, and nitrites. Amendments of ammonium compounds at 10 mM also promoted colonization of sclerotia by soil microorganisms (Fang *et al.*, 1988). Ahmed and Baker (1989) observed that disease incidence due to S. *rolfsii* in wheat was lower in plots heavily fertilized with inorganic N (urea) at 100 and 120 kg/ha, and higher in those with organic amendments (mustard oil cake and well rotted cow dung). Potash application significantly reduced *Sclerotium oryzae (Magnaporthe salvinii)* incidence with a corresponding increase in grain and straw yields (Jayaraj *et al.*, 1991). In the field, urea (80 kg/ha) and ZnSO4 (10 kg/ha) initially increased the foot rot (S. *rolfsii)* of soybeans population followed by a slight decrease at a later stage. A reduction in the disease was recorded with all the fertilizers used andZnSO4 was the most effective. Urea (80 kg/ha) and ZnSO4 (10 kg/ha) stimulated plant growth which was reflected by increased plant height and yield (Deb and Dutta, 1992).

Effect of crop rotation

The value of cotton *(Gossypium hirsutum* cv. Deltapine 90) in rotation with peanut *(Arachis hypogaea* cv. F1orunner) for the management of root-knot nematode *(Meloidogyne arenaria)* and southern blight (S. *rolfsii)* was studied for 6 years in a field in southeast Alabama by Rodriguez Kabana *et al.,* (1991). Peanut yields following either 1 or 2 years of cotton (C-P and C-C-P, respectively) were higher than those of peanut monoculture without nematicide [P(-)] At-plant, application of aldicarb to continuous peanut [P(+)] averaged 22.1% higher yields than those for P(-) over the 6 years of the study. The use of aldicarb in cotton and peanut in the C-C-P rotations increased yields of both crops over the same rotations without the nematicide. When the nematicide was applied to both crops in the C-P rotation, peanut yields were increased in only '2 of the possible 3 years when peanut was planted. Application of aldicarb to cotton only in the C-P rotation did not improve peanut yie1ds over those obtained with the rotation without nematicide. Juvenile populations of M. *arenaria* determined at peanut-harvest time were lowest in plots with cotton. Plots with C-P or C-C-P had lower populations of the nematode than those with either P(-) or P(+). The incidence of S. *rolfsii* in peanut was lower in plots with the rotations than in those with peanut monoculture. Aldicarb application had no effect on the occurrence of southern blight. In another study, Rodriguez Kabana *et al.,* (1991) studied the effect of crop rotation of 'Florunner' peanut *(Arachis hypogaea)* with the soyabean cv. Kirby for 6 years to evaluate long-term effects on management of *Meloidogyne arenaria* and S. *ro1fsii* in peanut production. Plantings included peanut monoculture, soyabean every other year, and soyabean 2 of every 3 years; for each, aldicarb was applied at-plant every year or was never applied. After 1 or 2 years of soyabean, densities of J2 of *M. arenaria* at peanut harvest were less than 10% of those in peanut control plots. However, populations usua1ly resumedhigh 1eve1s (>150 J2/100 cm3 soil) after a subsequent peanut crop. Aldicarb increased average yield from peanut mono culture by 20%. In the first 4 years, peanut yields of I-year and 2-year rotation were comparable to but no better than those from peanut mono culture receiving aldicarb. During the 6-year study, however, monocultured peanut yield decreased by 35% whereas yields from rotation plots remained high. Regeneration of yield in monocultured peanut was associated with a sustained high population of *M arenaria* and a progressive increase in incidence of southernblight. Thus, long-term rotation of peanut with Kirby soyabean postponed peanut yield decline but did not reduce population densities of *M. arenaria* below economically damaging levels. Brenneman *et al.,* (1995) had grown florunner groundnut after 1, 2 or 3 years of Tifton 9 bahia grass *(Paspalum notatum)* and in alternating years with bahia grass. Continuous groundnut was grown in non-treated plots and in plots treated with flutolanil (4.48 kg/ha). In continuous groundnut, stem rot *(S. rolfsii)* incidence was 4, 18, 19 and 44% during 1990-93, respectively, without flutolanil, and 0, 4, 10 and 17% with flutolanil. In 1993, stem rot incidence was 39, 29, 17 and 23% in the-3rd, 2nd, 1st and alternating year of groundnut, respectively. Taylor and Rodriguez Kabana (1999) found that southern blight (white mold) fungus *(S. rolfsii)* of peanuts *(Arachis hypogaea)* can be suppressed by rotation with cotton.

Physical methods

Soil solarization

Grinstein *et al.,* (1979) tried mulching moistened soil with transparent polythene sheets to raise soil temperature due to solar heat for the control of weeds and soil borne diseases in peanut field soils naturally infected with S. *rolfsii.* Soils mulched for 6 wks in July-August and then sown with groundnuts the following spring showed significant decreases in the percentage of diseased plants, rotten pods and weeds and an increase of 52.8% and 123.5% in total yield and crop quality, respectively. Tiwari (1997b) studied the effect of soil solarization on the sclerotia of S. *rolfsii,* causal agent of root rot of chickpea, at 3 different soil depths (5, 10 and 15 cm) in 2 different soil types (sandy clay loam and clay loam) in pot experiments during the hot summer in Madhya Pradesh, India. Pots covered with transparent polythene sheet attained markedly higher temperatures compared with the uncovered control. Average soil temperature exceeded 50°C and 45°, at 5 and 10 cm depths, respectively, in solarized soils. Clay loam soil had higher temperatures than sandy clay loam at 5 and 10 cm depths. Solarization significantly decreased weed flora and the chickpea crop sown in the solarized pots was completely free from sclerotial root rot, when sclerotia were placed at 5 cm depth. The chickpea crop was also disease free where sclerotia were placed at 5 and 10 cm depths in pots with moist soil.

The effect of soil solarization using a transparent polyethylene film (35 mm thick) covering soil for 34 and 55 days was compared with application of vinclozolin (75 mg/m of planting row) on the control of S. *rolfsii* on *Phaseolus vulgaris* cultivar IAC-carioca. S. *rolfsii* inoculum multiplied in rice seeds (360 g/plot) was used for soil inoculation. In solarized soil, maximum temperatures of 41^{0}C and 35^0 at 10 and 20 cm depths, respectively, were attained, while in non solarized soil, temperatures of 35°C and 31°C were attained, respectively. During two successive P. *vulgaris* cultivations, solarization promoted yield increase and a decrease in the number of recovered and viable sclerotia. The emergence, weight and height of plants were not affected by solarization. There was no difference between the fungicide application and solarization for the characteristics evaluated (Ghini *et al.,* 1997).

Matrod *et al.,* (1991) studied the effect of moist and dry plots covered with 100 or 180-200 mm thick transparent polyethylene sheeting, 400 mm transparent polyvinyl chloride sheeting or 60 mm black polyvinyl chloride sheeting during July and August on S. *cepivorum* of garlic (125 sclerotia/100 g dry soil). In moist soils, the viable number of sclerotia decreased by 75.2-83.2% in plots covered by clear plastic mulches compared with a decrease-of 49.6;:59:2% with the b1ack p1astic mulch (12% in the uncovered control). In dry soils, the reduction in sclerotial number ranged from 1.6 to 10.4% in the covered plots (1.6% in the control). It has been observed that soil solarization during the summer before a reliable and practical method of control of white rot of garlic. Reduction to undetectable levels of S. *cepivorum* in the upper 20 cm layer of soil, even in heavily infested soils, after solarization for 8-11 weeks. White rot progress curves in subsequent crops of garlic indicated disease onset 4 months after planting. Rates of disease progress and final incidence of dead plants were greatly reduced in solarized plots, with yield increments of 40.6-155.5% greater than the unsolarized control plots. However, gar1ic crops in the second year after solarization had disease levels and yield reductions that were unacceptable to the growers; it is suggested that this is attributable to the high incidence of white rot of garlic that can be induced by low inoculum densities in the soil. Disease progress curves in the unsolarized plots suggested that secondary infections occur.

Soil solarization significantly reduced S. *cepivorum* sclerotial viability in two separate trials in New Zealand to 40.2 arid-53.3%, respectively when soil was covered with clear 50 mm thick polythene for 4 weeks. Sclerotial viability further decreased in two New Zealand sites; Cantetbury (to 8.7%) and Blenheim (shallow silt loam soil) (to 0%) when the soil was solarized for 8 week period. Solarization increased the soil temperature by 6-7°C. Microorganisms isolated from the recovered sclerotia included species of *Trichoderma, Verticillium, Mucor, Aspergillus* and four unidentified bacterial species (McLean *et al.,* 2001).

Mulching of *Sclerotium oryzae* infested soil (moist or dry) with polyethylene sheets during hot summer days of May and June increased the soil temperature at) cm from 36 to 48°C (wet) and from 44 to 52°C (dry), and at 20 cm from 32 to 38°C (wet) and from 35 to 39°C (dry). In artilicially-infested soil the sclerotia were not eradicated, but 95-100% loss of viability was observed at 5 cm by mulching for 1 week and at 20 cm by mulching for 8 weeks. Mulching effects were not influenced by the moisture content of the soil or by amendments with lucerne or wheat straw. Mulching of naturally- infested soil at a second site did not eradicate S. *oryzae,* but reduced sclerotial viability by 93% (Haroon Usmani and Ghaffar, 1982). Effects of soil solarization on natural and/or artificially established populations of the fungi were examined by Mihail and Alcorn (1984) using 51 mm clear polyethylene tarplor-2, 4 or 6 wk during spring, summer and autumn trials. Complete control of buried, precounted sclerotia of S. *rolfsii* was achieved to depths of 15 cm in the summer out only to 1 cm in the autumn. Sharma *et al.,* (1984) observed that the viability of sclerotia of *S. oryzae (Magnaporthe salvinii)* in soil samples from depths of 0-5 cm was controlled by a maximum of 49.6, 73.7, 83 and 93.4% after 3, 6,-9 and 12 days respectively, of mulching with 0.5 mm sheeting. This technique for the control of rice stem rot is less costlier than fumigation, is nonhazardous and leaves no residues.

Host Plant Resistance

Sclerotium cepivorum

Utkhede and Rahe (1978) evaluated 64 North American and European onion cultivars, including dry bulb, sweet Spanish, pickling and bunching types, for resistance to S. *cepivorum.* Infection in the field was least in Ailsa Craig, Dako and Wolska. These cultivars were also among those least infected in thelaboratory. The same cu1tivats have shown resistance to white rot in other trials also (Utkhede and Rahe, 1981; Rahe and Utkhe, 1982). Onion cultivars Hybrid-3 and IIHR Yellow were moderately resistant to S. *cepivorum* in India (Somkuwar, *et al.,* 1997).

Sclerotium oryzae

The initial screening of 55 types of 17 *Oryza* species for resistance to S. *oryzae (Magnaporthe salvinii),* and the subsequent reevaluation of the 24 most resistant entries, revealed that significant differences exist in resistance. Eleven types were found to be significantly more resistant than Calusa, California's most resistant cultivar. Resistant wild sources were crossed with M101 (susceptible) and advanced generations were analysed. The F2 revealed transgressive segregation towards resistance in six of the eight populations studied. Overdominance at one or more loci, or dominance at two or more loci, for resistance is suggested. All broad-sense heritability estimates were above 50%, and for the cross M101 X *Oryza rufipogon* 'Al00912' the estimate was 91%. Resistance appears to be of the horizontal type (Figoni, 1981).

By a new method of inoculation S. *oryzae* was cultured on oatmeal agar and a portion of the medium with 10 sclerotia was introduced between the sheath and the stem. Natural infection was produced by transplanting the genotypes in a field where infected straw had been added to the soil for 4 yr. A significant interaction was found between 2 isolates and the varieties inoculated. Of 20 Italian varieties only CRM 357/55, CRM 293/687, Raffaello and Padano were moderately resistanti Reactions of the genotypes were the same under natural and artificial infection (Moletti *et al.*, 1983).

S. *oryzae-resistant* germplasm line (PI566666), originating from the cross *L201/0ryza rufipogon* Al 00912//80H3793/3/82Y5 I, was released in 1993. In yield trials 87Y550 produced 10.4 t/ha, compared with 10.6 for L202, its sister line. It headed about the same time as L202 and was somewhat shorter (70 versus 76 cm). It has glabrous leaves and long grains (7.8 mm length and 2.3 mm width for brown rice grains) with light brown pericarp. The endosperm amylose content of 87Y550 was 230 g/kg and the spreading value was intermediate (17 g/litre KOH solution). On a scale where I = immune and 10 = dead, the S. *oryzae* resistance score of 87Y550 was 4.1, compared with 6.3 for L202.

Sclerotium rolfsii

The potato cultivars Centennial Russet, White Rose and Kennebec and advanced breeding lines AD7267-1, AD7277-1, BCg 289-1, WC672-2 and WC 285-18 (a sibling of Centennial Russet) were screened for resistance to S. *rolfsii*. Line BC9289-1 showed the highest incidence of both stem and tuber infection. All cultivars and lines except Centennial Russet and BC9289-1 showed very little tuber infection (Voss *et al.*, 1984). When 75 cultivars and elite breeding lines of rice were tested for resistance to seedling blight *(S. rolfsii)* by Bhardwaj and Singh (1984) in India, no blight occurred in IR 3941 or ES45 Plp 2B.

Stem rot is a widespread, devastating disease of the cultivated groundnut *(Arachis hypogaea)* caused by the soil borne pathogen *Sclerotium rolfsii* and resistant cultivars are needed to reduce cost of groundnut production. The Valencia market types of groundnut were significantly more susceptible to S. *rolfsii* infection than Spanish, runner or Virginia types. Toalson, Tifrun, Pronto and Sunbelt Runner had the fewest disease loci, and Early Bunch, Tifrun, Sunbelt Runner, Floriglant, NC 7 and GK 3 had the highest yields (Branch and Csinos, 1987). Among 43 groundnut cultivars screened against S. *rolfsii*, RSG- 130 and RSG-132 were resistant and RSG-191 was moderately resistant (Chitale *et al.*, 1990). ICGV87157 is also more resistant than the popular Indian cultivar JL24 to the *S. rolfsii* [Nigam *et al.*, 1991]. A total of 859 groundnut germplasm accessions and breeding lines were screened in field trials for resistance to stem and pod rot caused by *S. rolfsii* at the ICRISAT Asia Center, Patancheru, India. Of the 20 selected genotypes, 7 interspecific hybrid derivatives (326, 988, 1019, 1024, 1065, 1267 and 1364) consistently exhibited stable resistance to both stem and pod rot. Nine breeding lines (ICGV 86034, 86124, 86252, 86388, *86590*, 86606, 86635, *87160* and 87359) showed low susceptibility to stem and/or pod rot (Mehan, *et al.,* 1995). Krishnakanth *et al.,* (1999) evaluated ten Spanish bunch cultivars along with resistant and susceptible controls over two seasons using artificial inoculation for resistance to *S. rolfsii.* Two known resistant genotypes, ICGV 87165 and ICGV 8650, were rated as resistant while most genotypes were susceptible. Dh 8 was comparable or superior to the resistant controls. Fifteen peanut genotypes of diverse eco-geographical regions were evaluated for their genetic resistance to *S. rolfsii* under field conditions in Pakistan in 1995. The most resistant line was PG-480 while the most promising variety, in terms of yield and yield components potential, was Cina. Variety Cina is recommended for general cultivation because of its relative field resistance to the pathogen, promising yielding potential and shorter maturity duration in the Swat valley (Muhammad *et al.,* 1998). 'Toalson' and 'Southerp Runner' are 2 genetically diverse cultivars with partial stem rot resistance; whereas, 'Florunner' and 'Sunbelt Runner' are two stem rot susceptible cultivars. Mass-selected populations derived from different combinations of crosses among these 2 resistant and 2 susceptible cultivars were evaluated in Georgia Coastal Plain Experiment Station, Tifton, Georgia and USA. The resistant cultivar Toalson produced 40% higher pod yields and had 50% fewer stem rot disease loci than the other partially resistant cultivar Southern Runner. Mass-selected populations derived from the Sunbelt Runner x Toalson cross combination also had significantly higher yield and better stem rot resistance than mass-selected population derived from the Florunner x Southern Runner cross combination (Branch and Brenneman, 1999).

Of 48 brinjal (aubergine) ctiltivars evaluated for resistance to S. *rolfsii* in India, 4 cultivars (BL-I, BWR 54, PBR 91-2 and Pusa Bhairav) were immune, 2 (local brinjal and selection 10-2-56) were resistant, With a disease incidence of <20%, 3 (ManjariGota, perennial brinjal and SM-6-: 7 (PF)) were moderately susceptible (disease incidence 21 -50%) and 39 cultivars were highly susceptible (Sugha *et al.,* 1990).

Developed in a breeding programme with a tomato bulk F2 progeny derived from a cross between Step 54 and Southland, these breeding lines have in common resistance to southern blight caused by S. *rolfsii*. The original source of resistance is not known, but F11 Weslaco breeding lines W268MS56 and W27JM 856 were common throughout all 6 pedigrees and were derived from the Step 54 and Southland cross. Lines 5635M and 5707M additionally have *Lycopersicon* pimpinellifolium in their pedigrees. In screening trials with isolates Scr 20, 24, 30 and 32 of *S. rolfsii* artificially inoculated onto the plants, the lines showed a survival response equivalent to that of the resistant L. *pimpinellifolium* line PII26432 over 2 years (Leeper *et al.,* 1992).

When fifteen cotton cultivars, including commercial varieties belonging to the *Gossypium* species *G. herbaceum, G. arboreum, G. hirsutum* and *G. barbadense,* were screened *for* resistance to *Sclerotium rolfsii,* the G. *herbaceum* genotypes Jayadhar and SM88 and in the hybrids DDH2 and DHHll showed greater resistance (Prabhu *et al.,* 1997).

Biological control of *Sclerotium* diseases

As with many other diseases, those caused by genus *Sclerotium* continue to receive considerable attention with regard to the development of biological control strategies. Many reports are now available snowing potential of fungi, mycorhizae and bacteria as bioagents on various diseases caused by *Sclerotium* (Table 4.2).

Mechanism of biological control of *Sclerotium*

Fungal biological control agents may directly parasitize sclerotia or mycelium of *Sclerotium.* S. *rolfsii* produces 1ectins which serves as recognition factors for fungal biological control agents. The ability of different isolates of the *Trichoderma* spp. to attack S, *rolfsii* was correlated with the agglutination of conidia of *Trichoderma* by S. *rolfsii* (Barak *et al.,* 1985). Penetration of the sclerotia of *Sclerotium rolfsii* by *Trichoderma* will follow a direct and an indirect approach. Hyphae of *Trichoderma* penetrate the rind and cortex, lyse the medullar tissue, and produce chlamydospores inside and conidia outside the sc1erotia. Degraded sclerotia became dark, soft and empty, and disintegrated under slight pressure. Penetration capacity is an important but not the only property required for *Trichoderma* isolates to be efficient bio-control agents (Henis *et al.,* 1983). Isolates of T. *harzianum,* T. *hamatum* and T. *koningii* could penetrate the sclerotial wall, establish and sporulate inside the sclerotium of S. *rolfsii* (Desai and Schlosser, 1999).

Indirectly the fungal bioagents may produce hydrolytic enzymes and degrade sclerotia and mycelia of *Sclerotium.* The ability of *Trichoderma harzianum* to attack *Sclerotium rolfsii* is apparently connected with the production of chitinase and *B-(1,3)-* glucanase inside the attacked sclerotia during parasitism. SEM and TEM micrographs show that the mycoparasite degraded the walls of sclerotial cells, and the attacked cells lost their cytoplasmic contents. It is assumed that T. *harzianum* utilizes sclerotial cell contents, thus enabling it to sporulate intensively on the sclerotial surface and inside the digested cells (Elad *et al.,* 1984). Motto *et al.,* (1986) observed that hyphae of T. *harzianum* developed in the medulla of S. *rolfsii* sclerotia, growing on the inside of the cell wall and in the cell lumen. The cytoplasm of penetrated ce1ls rapidly degenerated. The hyperparasite passed from cell to cell by lytic perforation of the walls. Germ tubes of medullar cells were also parasitized, with wall lysis and digestion of cell cytoplasm.

Benhamou and Chet (1996) studied the interaction between T. *harzianum* and sclerotia of the soil borne plant pathogen S. *rolfsii* by SEM and TEM to assess the potential role of enzymatic hydrolysis in the antagonistic process. SEM investigations revealed that hyphae of T. *harzianum* grew aoundantly on the sclerotial surface forming a dense branched mycelium that appeared to establish contact with the outer host cells through thin mucilage. Cross-sections of parasitized sclerotia by light microscopy showed that hyphae of the antagonist multiplied on the sclerotial surface and displayed the ability to penetrate the rind. Growth of antagonist in the rind layer was mainly intracellular, and host-wall penetration was achieved by means of constricted hyphae. Ultra structural observations showed that *Trichoderma* growth and development co- incided with extensive host cell alterations. Cell breakdown due to host cell-wall disruption was observed most frequently in inner rind cells adjacent to medullary cells. Ingress of T *harzianum* hyphae in the medulla was characterized mainly by a change in the mode of growth from intra to intercellular. *Trichoderma* hyphae did not penetrate the medullary cells, although the latter showed pronounced alterations. The use of wheat germ agglutinin/ovomucoid-gold complex for localization of chitin monomers resulted in regular labeling of both host and antagonist cell walls, even when sclerotia were massively colonized. There was no indication of cell-wall disorganization in either the host or the antagonist, as shown by the regular distribution of labeling, even in zones of penetration by constricted *Trichoderma* hyphae. In the medulla, gold labeling was regularly distributed over the thick host cell walls, even when the medullary cells showed obvious signs of disorganization. When ultra thin sections of parasitized sclerotia were incubated with the gold-complexed beta-1,3-glucanase for localization of beta-1,3- glucans, a regular distribution of gold particles was observed over the walls of both outer and inner rind cells. Incubation with gold-complexed lipoprotein lipase yielded a pattern of labeling similar to that obtained with gold-complexed beta-1, 3-glucanase. The culture filtrates of *T harzianum* contain various antibiotic substances that inhibit the growth of S. *rolfsii* (Upadhyay and Mukhopadhyay, 1983).

Ten wild-type strains and 2 benomyl-tolerant mutants of *Talaromyces flavus* were examined for their ability to secrete the cell wall-degrading enzymes chitinase, beta- 1,3-glucanase and cellulase, to parasitize sclerotia of *S. rolfsii,* to reduce bean stem rot caused by *S. rolfsii.* The benomyl-tolerant mutant BenRTFl-R6 overproduced extracellular enzymes and exhibited enhanced antagonistic activity against *S. roffsii* compared with the wild-type strains and other mutants. Correlation analyses between the extracellular enzymatic activities of different isolates of *T.flavus* and their ability to antagonize *S. rolfsii* suggested that mycoparasitism by *T. flavus* and biological control of S. *rolfsii* were related to the chitinase activity of *T. flavus* (Madi *et al.*, 1997).

Biological control of *S. cepivorum*

T. harzianum and *Bacillus subtilis* were able to decrease white rot infection in the field (Abd EI Moity and Shatla, 1981; Utkhede and Rahe, 1983). Rotations of Trifolium/sorghum, Trifolium/maize, wheat/sorghum, faba bean/sorghum, faba bean/maize and lentil/cotton decreased sclerotial numbers of *Sclerotium cepivorum* in soil in the presence of *T. hariianum* but wheat/maize had no effect. In the absence of *T. harzianum* rotations of lentil/cotton and wheat/soyabean decreased *S. cepivorum* in soil but other rotations had no effect. Onion cultivation increased numbers of *S. cepivorum* sclerotia and *T. harziamum* populations in soil more than the other crops (Abdel Rahman *et al.*, 1990). Bacterial antagonists produce antibiotic substances, which inhibit growth of the sclerotial fungi. *Pseudomonas aeruginosa* produces pyocyanin which inhibits sclerotial germination.

Bilogical control of *Sclerotium rolfsii* in various crops

Integration of quintozene and *T. harzianum* significantly reduced disease incidence (76% control) and increased the root, green foliage and sucrose yield/ha of sugarbeet (Upadhyay and -Mukhopadhyay, 1986). Papavizas and Lewis (1989) when used several wild-type strains and mutants of G. *virens, T. hamatum, T. harzianum* and *T. viride* against *S. rolfsii* in the greenhouse, ten strains of G. *virens* and 4 strains of *T. harzianum* suppressed damping-off of snapbean *(Phaseolus vulgaris)* by 30-50% and blight by 36-74%. G. *virens* strains were more effective in suppressing disease than strains of *T. harzianum*. A wheat bran preparation of *T. harzianum* enriched with nutrient solution was evaluated against S. *rolfsii* infected sugarbeet by Ciccarese *et al.*, (1992). Application of this preparation at 10 g/m2 soil at the same time or 1 year after soil inoculation with S. *rolfsii* at 1.5 g/m2, significantly reduced sugarbeet root rot and increased yield. The same preparation retained its protective effect against S. *rolfsii* the next year, but a second treatment with *T. harzianum* increased the protective value.

Several strains of *G. virens* and *T. harzianum* used alone were equal to or more effective than double and triple mixtures of such strains in disease suppression. Two simple formulations of an antagonistic strain of *T. koningii* were employed against southern blight disease caused by S. *rolfsii* in seedling, potted outdoor and field-grown tomatoes by Latunde Dada (1993). Corncob germling inoculum and mycelium powder of *T. koningii* significantly controlled damping off, blight and wilting in tomato. The populations of the antagonist increased from an initial 1 X 10^4 to about 1 X 10^6 c.f.u./g soil in the protected plants. Moreover, sclerotial counts decreased significantly in these soils and those sclerotia found had been parasitized by T. *koningii. Trichoderma*-protected plants were more vigorous than those in the other treatment categories. Lewis *et al.*, (1993) investigated the effects of Pyrax dust (pyrax), alginate pellets and activated vermiculite/bran(VBB)formulations containing *G. virens* on pre- and postemergence damping-off of *Phaseolus vulgaris* caused by S. *rolfsii.* All 3 formulations added to soil to provide 103-104 c.f.u. of G. *virens/g* soil significantly prevented damping-off, but no treatment yielded a plant stand similar to that in noninfested soil (73-85%) after 35 d of growth. Pyrax, alginate pellets and activated VBB gave final stands of 59-73, 57-71 and 62- 77%, respectively. The addition of the Formuiatlons toloatriy sandin the field resulted in a 1000-fold increase in propagule numbers of *G. virens* within 11 d of application.

A bacterium, identified as *Serratia marcescens* was found to be the best biocontrol agent of *S. rolfsii* of bean (up to 75% disease reduction) by Ordentlich *et al.*, (1987). The drench and drip application of S. *marcescens* suspension were more effective in controlling S. *rolfsii* than spraying, mixing in soil or seed coating.

Rangeshwaran and Prasad (2000b) tested 300 rhizosphere bacteria for antagonism against *Sclerotium rolfsii,* the causal organism of root/collar rot in sunflower. Eleven rhizobacteria antagonistic to *S. rolfsii* were identified which included *Pseudomonas fluorescens* (PDBCAB 1, 2, 15 and 26), *Pseudomonas* spp. (PDBCAB 7,8, 15), *P. putida-(PDBCAB,* 19), *Streptomyces* sp.(PDBCAB23), *Bacillus pantothenticus (Virgibacillus pantothenticus)* (PDBCAB 28) and *Alcaligenes odorans* (PDBCAB 12). Highest disease suppression was exhibited by *P. fluorescens (PDBCAB-2)* and P. *putida* (PDBCAB 19) (60-63% disease-free plants). The results indicate that selection of antagonistic bacteria must be based on greenhouse performance instead of relying on laboratory results. Inoculation with plant growth-promoting strains of *P. putica, P. fluorescens* and *P. alcaligenes* reduced the incidence of disease caused by *Sclerotium rolfsii* in bean (*Phaseolus vulgaris*) (Gamliel and Katan, 1993).

Laha *et al.*, (1996) found that 16 isolates of fluorescent pseudomonads out of 48 isolates from cotton seedling rhizospheres were antagonistic to *Sclerotium rolfsii,* the causal agent of cotton wilt. Isolate FP-47 was most effective in significantly reducing the disease.

Integrated management of *Sclerotium* diseases

Sclerotium cepivorum

Rushdi *et al.*, (1974) reported that planting onion on 15 October, planting in ridges and the application of pentacnloronitrobenzene (quintozene) (23.8 kg / ha) reduced infection by white rot *(Sclerotium cepivorum)* from 80 to 2%. Infection increased with increasing number of irrigations. Treatment of garlic cloves with tebuconazole (at 1 nil of Folicur 25% 1-1) achieved a significant reduction in the rate of disease progress and the final incidence of plant death by *Sclerotium cepivorum* garlic yiel ds were improved. Although soil solarization provided the best control of garlic white rot, bringing soil populations of S. *cepivorum* to negligible levels, similar levels of disease control and garlic yields were achieved when tebuconazole was sprayed to stem bases of plants grown from cloves also treated with tebuconazole. This double treatment almost doubled the yield compared with untreated plants and significantly increased bulb quality under high disease pressure conditions. Soil solarization was also highly effective in a second consecutive crop of garlic, with significant improvements in yield and garlic quality. The effect of vermicompost, soil solarization, *T. harzianum* and *B. subtilis* incorporated before or after soil solarization was evaluated in soil naturally infested by the garlic pathogen S. *cepivorum* by Pereira *et al.*, (1996). After incorporation of vermicompost and antagnists, the moist soil was covered with a transparent polyethylene film of 0.1 mm for 60 d. *T harzianum* and *B. subtilis* were inoculated after removal of the plastic cover, and at 30, 60 and 90 d, evaluations were carried out based on the percentage of viable sclerotia in culture media. The results showed that soil solarization significantly reduced sclerotial viability of S. *cepivorum*. Incorporation of *T. harzianum* in soil after soil solarization increased percentage control from 79 to 98%. Incorporation of *T. harzianum* in soil after solarization effectively controlled S. *cepivorum* in soil. When the fungal antagonists were inoculated after solarization, a significant growth of populations of both antagonists occurred, attaining 10^8 c.f.u./g of soil.

Onion and garlic rotation with groundnut crops, application of *Trichoderma harzianum* and soil drenching with 0.02% carbendazim reduced stem rot by 44-60 % and pod rot by 27-60% in groundnuts infected by *Sclerotium rolfsii*. These treatments reduced soil S. *rolfsii* populations and increased populations of *T. harzianum* and also increased pod yields by 17-47% and seed Wt by 29-33% (Asghari and Mayee, 1991). Soil solarization and wheat + groundnut were effective in reducing incidence of S. *rolfsii* (Bicici *et al.*, 1994). Combined application of *T. harzianum, Rhizobium* and carbendazim reduced groundnut root rot caused by *S. rolfsii* (Muthamilan and Jeyarajan, 1996).

The effect of different inoculum densities, soyabean cultivars (Bragg, PK 262, PK 308, PK 327 and PK 416) and integrated management practices on the incidence of seed and seedling rot caused by *S. rolfsii* were investigated by Uma Singh *et al.,* (1998). Seeds inoculated with S. *rolfsii* had 17.5% germination with only 5% normal seedlings as compared with 77.5% germination and 70% normal seedlings in the control. In seed treatment tests with the antagonists *Trichoderma harzianum* and *Gliocladium virens,* and the fungicides thiram, bavistin (carbendazim) and vitavax (carboxin + thiram), both pre- and post-emergence seedling rot was effectively managed by seed treatment with carboxin + thiram plus *T. harzianum* or G. *virens.*

The fertilizers ammonium sulphate $(NH_4)_2SO_4$, ammonium nitrate (NH_4NO_3), diammonium phosphate $(NH_4)_2HPO_4$, or urea applied to soil at a field rate of 135 kg/ha, 15 cm deep of nitrogen (N) (0.09 mg of N/g) or *Gliocladium virens* (GI-3) biomass at a rate of 7.5 kg/ha, 15 cm deep (0.05 mg/ g) did not reduce the viability of sclerotia of *S. rolfsii* (Sr-l) of snap bean *(Phaseolus vulgaris)* when each was applied alone. However, treatment with fertilizer together with the low rate of GI-3 biomass significantly reduced the sclerotial viability. The treatments that were effective in reducing the viability by more than 75% were the application of $(NH_4)_2S0_4$ or $(NH_4)_2$ HPO_4 and the low rate of GI-3 biomass. In fumigation studies the conidia or-isolates Gl-3, Thm-4 of *Trichoderma hamatum,* and Tv-l of *Trichoderma viride* were used together with metham sodium (vapam), at l7.la.i/g of soil reduced the viability of sclerotia more than each individual component. The results of this study suggest the feasibility of effective disease reduction with an approach utilizing biological control in combination with fertilization or fumigation (Hoynes *et al.,* 1999).

Deep ploughing combined with no-layby cultivation effectively reduced southern blight *(S. rolfsii)* on processing carrots. Layby cultivations negated the beneficial effects of deep ploughing. Fungicides used for control of S. *rolfsii* were effective for lay by carrots *in* deep-ploughed soil and for both 1ayby and no-layby carrots in disced soil (Gurkin and Jenkins, 1985).

The effect of simultaneous applications of *T. viride* with PCNB (quintozene), captan or Aldrex T (thiram + aldrin) on the basal stem rot *(S. rolfsil)* of tomato was studied by Wokocha (1990) in artificially inoculated field plots. The results showed that a combination of *T. viride* with PCNB and *T. viride* alone gave good disease control during both dry and wet seasons. In another study soil solarization in combination with the introduction of G. *virens* was found to be a potential disease management strategy for control of southern blight caused by *S. rolfsii* on tomatoes in North California. Highest

TABLE 4.1: Chemical fungicides effective against *Sclerotium* diseases

Disease	Crop	Chemical/Dosage	Reference
Stem rot *(S. rolfsii)*	Groundnut 30 lb a.i./ac	Potassium azide	Rodriguez Kabana, *et al.*, 1972
Stem rot *(S. rolfsii)* White rot *S. cepivorum)*	Groundnut Onion	PCNB Benomyl@1-21b/ga1 and Thiophanate-methyl 1974 @I-4 lb/gal as root dip treatments	Sturgeon, 1973 Maloy and Machtmes, 1974
Seedling blight *(S. rolfsii)*	Apple	Brassicol Nursery bed application (120 g/bed)	–
Southern blight *(S. rolfsii)*	Tomato	Vapam	McCarter *et al.*, 1976
Sclerotial wilt *(S. rolfsii)*	Betelvine *(Piper betle* L.)	Merculine at 0.3% and 0.5 1/ft^2 row Rizolex-S-3349 (tolclofos-methyl)	Bartaria *et al.*, 1982 Deshpande and Tiwari, 1991
White rot (S. *cepivorum* Sclerotium rot (S. *ro/fsii)*	Onion, garlic Potato	Vinclozolin Soil treatments with PCNB (quintozene) at 25 kg/ ha and seed piece. treatments with 2% Dithane M-45 (mancozeb)	Jennrich, 1983 Sclerotium rot Bisht, 1982 ,
Sclerotium root rot (S. *ro/fsii)*	Sugarbeet *(Beta vulgaris)*	Brassicol (quintozene), with 10 kg/ha	*Waraitchetal.*, 1986
Foot rot *(sclerotiumro/fsii)*	Soybean Sugar beet	Furazolidone @100 mg/ml Brassicol (quintozene), Demosan (chloroneb)	Deb and Dutta, 1989 Root rot Shamla *et al.*, 1990
Stem rot (S. *oryzae)*	Rice	Carbendazim,	Sunder *et at.*, 1991

TABLE 4.2. Biological control agents of *Sclerotium*

Bio-agent	Disease	Crop	Reference
Trichoderma harzianum and *Bacillus subtilis*	Collar rot *(S. rolfsii)*	Lentil	Agrawal *et al.*, 1977

Contd...

Table 4.2. Contd...

Coniothyrium minitans	White rot *(S. cepivorum)*	Onion	Ahmed and Tribe, 1977
T. viride and *T. harzianum*	Root rot *(S. rolfsii)*	Sugarbeet	Mathur and Sarbhoy, 1978
Sporidesmium Sclerotivorum	White rot *(S. cepivorum)*	Onion	Ayers and Adams, 1979
Bacillus and *Pseudomonas* spp.	White rot *(S. cepivorum)*	Onion	Wong and Hughes,1986
P. fluorescens	Stem rot *(S. rolfsii)*	Peanut	Ganesan and Gnanamanickam, 1987
T. harzianum (Soil application)	Root rot *(S. rolfsii)*	Sugarbeet	Ciccarese *et al.*, 1990
T. harzianum,. penicillium pinophilum	Root rot *(S. rolfsii)*	Sesamum	
T. harzianum, T. koningii and T aureoviride	Root rot *(S. rolfsii)*	Tomato	Monaco *et al.*, 1991
T. harzianum	Foot rot *(S. rolfsii)*	Chilli	Sariah Meon, 1991
T. viride T harzianum and *T koningii*	Foot rot	Soybean	Deb andDutta; 1991
B. subtilis cepivorum)	White rot *(Sclerotium*	Onion	Reddy *et al.*, 1992
P. fluorescens	Stem rot *(S. oryzae)*	Rice 1992	Elangovan and Gnanamanickam,
P. fluorescens, B. pumilus, B. cereus, B. subtilis	Damping-off	Snapbeans	Lewis *et al.*, 1993
T. viride, T. harzianum and *Gliocladium roseum*	Collar rot *(S. rolfsii) Cicer arietinum)*	Gram	Sugha *et al.*, 1993
G. *virens* and *T longibrachiatum*	Root and stem rot diseases *(S. rolfsii)*	Groundnut	Sreenivasaprasad, and Manibfiushanrno, 1993

Contd...

Table 4.2. Contd...

Trichoderma spp.	Pre-emergence damping off *(S. ralfsii)*	Bean	Silveira *et al.*, 1994
G. *virens*	Southern blight *(S. rolfsii)*	Carrot and tomato	Ristaino *et al.*, 1994
T. aureoviride	Stem rot *(S. rolfsii)*	Soybean *(Glycine mar)*	Dharmaputra and Retnowati, 1994 *Chaetomium globosum,*
White rot	Onion	Kay and Stewart,	*(S. cepivorum)* 1994
Trichoderma sp.			
T. harzianum,	Collar rot *(S. rolfsii)*	Lentil	Iqbal *et al.*, 1995
T. koningii and			
T. viride			
Trichoderma spp.	Root rot (S. *rolfsii)*	String bean	Roberti *et al.*, 1996
		(Phaseolus vulgaris)	
Rhizobium	Stem rot (S. *ro!fSii)*	Peanut	Salui and
(Jcg-9 strain of the cowpea group)			Bhattacharyryya, 1998
Glomus fascicu/latum (mycorrhizal fungi)	Stem rot *(S. rolfsii)*	Groundntit	Kulkarni *et al.*, 1997
Trichoderma and *Gliocladium* species	Collar rot *(S. rolfsii)*	Sunflower	Prasad *et al.*, 1999
T. pseudokoningii	Rhizome rot *(S. rolfsii)*	Ginger	Mukhergee *et al.* 1999
T. harzianum, T. viride and *P. fluorescens*	Stem rot *(S. rolfsii)*	Tomato	Thribhuvanamala *et al.*, 1999
G. *catenulatum*	White rot *(S. cepivorum)*	Onion	Tsigbey *et al.* 1999
T. harzianum	Foot rot *(S. rolfsii)*	Wheat	Kalappanavar and Patil, 2000.
T. harzianum	Collar rot *(S. rolfsii)*	Sunflower	Mesta and Arnaresh, 2000

Contd...

Table 4.2. Contd...

T. harzianum	Stem rot *(S. rolfsii)*	Groundnut	Biswas *et al.*, 2000
T. harzianum *T virens*	White rot *(S. cepivorum)*	Onion	McLean and Stewart, 2000

soil temperature measured during the solarization period in 1988 and 1990 were 41 and 49°C, respectively, whereas maximum differences in soil temperature between plots that were solarized and plots that were not solarized were 9°C in 1988 and 14°C in 1990. G. *virens* significantly reduced numbers of sclerotia of S. *rolfsil* by 100, 96 and 56% to depths of 30 cm in 1988, 1989 and 1990, respectively, whereas solarization alone reduced numbers of sclerotia by 62% in 1990. When disease pressure was high in 1988, disease incidence in plots planted immediately after treatment was reduced by 49% in solarized soils amended with G. *virens* before solarization compared with untreated controls (Ristaino *et al.*, 1991). Mulching with a transparent plastic sheet along with incorporating green manure was also effective in contro-ling *Sclerotium rolfsii* on tomato (Tu *et al.*, 1991). Ramirez *et al.*, (1998) investigated the efficacy of fungicides and humus for the control of white rot of tomato, caused by *S. rolfsii* Combined application of vinclozolin, chlorothalonil, liquid humus, and worm humus, which resulted in 91.56% healthy plants and a 86% reduction in the number of sclerotia (Melero Vara, 2000).

The timing of solarization (with clear plastic mulch) in relation to the planting of pepper *(Capsicum annuum)* and the timing of soil amendment with a bran prill formulation of *G. virens* were evaluated for the control of southern blight and the survival of sclerotia of the pathogen S. *rolfsii* in bell pepper in the field. Solarization during crop growth increased the incidence of southern blight and *G. virens* was not effective under the mulch. In addition, pepper yields were low when the soil was solarized during crop growth. In contrast, the solarization of fallow soil in raised beds for 6 weeks prior to crop growth significantly reduced disease incidence in the pepper crop. In addition, in 2 years, G. *virens* alone reduced southern blight in non-solarized soils and reduced the survival of sclerotia of S. *rolfsii* to depths of 30 cm at all locations in soil in both years (Ristaino *et al.*, 1996).

REFERENCES

Abd El Moity, T. H. and Shatla, M. N. 1981. Biological control of white rot disease of onion *(Sclerotium cepivorum)* by *Trichoderma harzianum. Phytopath.* Z.100: 29-35.

Abdel Rahman, T. M., Mohamed, M. S., El Shabrawy, A. M. and Rushdi, M. H. 1990. Effect of different crop sequences on biological control of onion white rot disease by *Trichoderma harzianum* Rifai. *Assiut J. Agril. Sci.* 21: 149-164.

Agrawal, S. C., Khare, M. N. and Agrawal, P. S. 1977. Biological control of *Sclerotium rolfsii* causing collar rot of lentil. *Indian Phytopath.* 30:176-179.

Agrios, G.N. 1988. Plant Pathology, Academic Press, New York, 803p.

Ahmed, A. H. M. and Tribe, H.T. 1977. Biological control of white rot of onion, (*Sclerotium copivorum*) by *Coniothyrium minitans. Plant Pathol,* 26: 75-78.

Ahmed, F. and Bakr, M. A. 1989. Effect of organic and inorganic nitrogenous amendments on foot and root rot of wheat. *Bangladesh J. Pl Pathol.* 5:85-86.

Asghari, M. R. and Mayee, C. D. 1991. Comparative efficiency of management practices on stem and pod rots *of* groundnut. *Indian Phytopath.* 44: 328-332.

Aycock, R., 1966. Stem rot and other diseases caused by *Sclerotium rolfsii.* North Carolina Agriculture Experiment Station, *Technical Bulletin,* 194.

Aers, W. A. and Adams, P. B. 1979. Mycoparasitism of sclerotia of *Sclerotinia* and *Sclerotium* species by *Sporidesmium sclerotivorum. Can J.Microbiol.* 25: 17-23.

Backman, P. A. and Rodriguez Kabana, R 1976. Development of a medium for the selective isolation of *Scierotium rolfsii. Phytopath.* 66: 234-236.

Bartaria, A. M., Tomar, K. S. and Joshi, L. K. 1982. Fungicidal control of sclerotial wilt of pan (*Piper betle* L.). *Pesticides.*16: 7-10.

Barak, R, Elad, Y., Mirelman, D. and Chet, I. 1985. Lectins: a possible basis for specifi recognition in the interaction of *Trichoderma* and *Sclerotium rolfsii. Phytopath.* 75: 458-462.

Benhamou, N. and Chef, I. 1996. Parasitism of sclerotia of *Sclerotium rolfsii* by *Trichoderma harzianum:* pultrastructural and cytochemical aspects of the interaction. *Phytopath.* 86: 405-416.

Bhardwaj, C. L. and Singh, B. M. 1984. Evaluation of some promising rice cultivars and breeding lines against *Sclerotium rolfsii. lndian.J. Mycol.Pl.Pathol.* 14: 166.

Biswas, K. K., Chitreswar Sen and Sen,C. 2000. Management of stem rot of groundnut caused by *Sclerotium rolfsii* througn -*Trichoderma harzianum. 1ndian Phytopath,* 53: 290-295.

Bisht, N. S. 1982. Control of Sclerotium rot of potato. *Indian Phytopath.* 5: 148- 149.

Bicici, M., Cinar, O. and Erkilic, A. 1994. Cultural, chemical, physical, and biological methods,to control stem rot, *Sclerotium rolfsii Sacc.* in peanut. *Turkish J. Agril. For.* 18: 423-435.

Bowen, K. L., Hagan, A. K. and Weeks, R. 1992. Seven years of *Sclerotium rolfsii* in peanut fields: yield losses and means of minimization. *Plant Dis.* 76: 982-985.

Branch, W. D. and Csinos, A. S. 1987. Evaluation of peanut cultivars for resistance to field infection by *Sclerotium rolfsii. Plant Dis.* 71: 268-270.

Branch, W. D and Brenneman, T.B. 1996. Pod yield and stem rot evaluation of peanut cultivars treated with tebuconazole. *Agron J.* 88: 933-936.

Branch, W. D. and Brenneman,T. B. 1999. Stem rot disease evaluation of mass-selected peanut populations. *Crop Protection.* 18: 127-130.

Brenneman, T. B., Murphy, A. P. and Csinos, A. S. 1991. Activity of tebuconazole on *Sclerotium rolfsii* and *Rhizoctonia solani,* two soil borne pathogens of peanut. *Plant Dis.* 75: 744-747.

Brenneman, T. B., Summer, D. R., Baird, R. E., Burton, G. W. and Minton, N. A. 1995. Suppression of fo1iar and soil borne peanut diseases in bahiagrass rotations. *Phytopath.* 85: 948-952.

Bugaret, Y., Sauris, P. and Marin, H. 1996. White rot and rust of garlic. Mastering chemical control, it is possible. *Phytoma.* 487: 45-47.

Chastagner, G. A., Staley, J. M. and Riley, K. 1990. Control of *Sclerotium rolfsii* on bulbous tris and lilies with in-furrow fungicide application. *Acta Horti.* 266: 457-467.

Chen CheinWei., Lin CheinYi., Bsu ZhenChuan., Chen, C. W., Lin, C. Y. and Hsu, Z. C. 1998 Occurrence of periwinkle diseases in Taiwan. *Buletin of-Taichung strict agricultural. Improvement Station.* 54: 47-57 (Periwinkle stem rot disease).

Chitale, K., Tyagi, R. N. S. and Singh, R. D. 1990. Reaction *of* chickpea cultivars to collar rot disease induced by *Sclerotium rolfsii. Indian J.Mycol. Pl. Pathol.* 20: 262-263.

Ciccarese, F., Frisullo, S., Amenduni, M. and Caponero, A. 1990. *Trichoderma harzianum* Rifai in biological control of sugarbeet root-rot caused by *Sclerotium rolfsii* Sacc. *Informatore Fitopatologico.*40: 61-63.

Ciccarese, F.,.Frisullo, S., Amenduni, M. and Cirulli,M. 1992.Use in the open field *of Trichoderma harzianum* Rifai in the biological control of sugarbeet root rot caused by *Sclerotium ro1fsii* Sacc. *Informatore Fitopatologico,* 42: 63-64.

Cilliers, A. J., Berselman, L. and Pretorius, Z. A. 2000. Genetic variability within and among mycelial compatibi1ity groups of *Sclerotium rolfsii* in South Africa. *Phytopath.* 90: 1026-1031.

Corbaz, R. and Dorsaz, L. 1992. Chemical control of onion white rot *(Sclerotium cepivorum). Revue Suisse de Viticulture, d'Arboriculture et d'Horticulture.* 24: 147-149.

Csinos, A. S. 1989. Targeting fungicides for control of southern stem rot on peanut. *Plant Dis.* 73:723-726.

Culbreath, A. K., Brenneman, T. B., Bondari, K. Reynolds, .K. L., and McLean, B. S. 1995. Late leaf spot, southern stem rot, and peanut Yield responses to rates of cyproconazole and chlorothalonil applied alone and in combination. *Plant Dis.* 79: 1121-1125.

Das, S. R., Pani, B. K. and Kar, S. 1997. Evaluation of some plant extracts against *Sclerotium rolfsii* causing stem rot in tuberose. *Envi.Ecol.*15: 975-976.

Daya Ram, Tewari, V. P. and Ram, D. 1994. Control of chickpea collar rot *(Sclerotium rolfsii)* with soil application of plant products. *J.Appld Biol.*4: 38-40.

Deb, P. R. and Dutta, B. K. 1989. Nitrofurans as possible plant chemotherapeutants for the control of foot rot of soybean caused by *Sclerotium rolfsii* Sacco. *J.Phytopath.* 124: 309-315.

Deb, P. R. and Dutta, B. K. 1991. Studies on biological control of foot rot disease of soybean caused by *Sclerotium rolfsii* 'Sacc. *Zeitschrift fur Pflanzerikrankheiten und flanzenschutz.*98: 539-546.

Deb, P. R. and Dutta, B. K. 1992. Effect of inorganic fertilizers on the development of foot rot disease of soybean caused by *Sclerotium rolfsii. Indian Phytopath.* 5: 89-97.

Desai, S. and Schlosser, E. 1999. Parasitism of *Sclerotium rolfsii* by *Trichoderma. Indian Phytopath.*52: 47-50.

Deshpande, A. L. and Tiwari, R. K. S .1991. A new fungicide Rizolex against collar rot of betel vine caused by *Sclerotium rolfsii. IndianPhytopath.* 44: 348-352.

Dharmaputra, O. S. and Retnowati, I. 1994. The possibility of controlling *Sclerotium rolfsii* on soybean *(Glycine max)* using *Trichoderma* and tebuconazole. *Biotropia.* 7: 18-29.

Elad, Y., Barak, R. and Chet, I. 1984. Parasitism of sclerotia of *Sclerotium rolfsii* by *Trichoderma harzianum. Soil Biology and Biochemistry.* 16: 381-386.

Elangovan, C. and Gnanamanickam, S. S. 1992. Incidence of *Pseudomonas fluorescens* in rhizosphere of rice and their antagonism toward *Sclerotium oryzae. Indian Phytopath.*45: 358-361.

El Shehaby, A. I., Radwan, I. A., Moustafa, M. S.B., and Abd ElHalim, A. A. 1992. Control of white rot disease of onion by fungicides. *Egyptian J.Agricul. Reseach.* 70: 1065-1076.

Fang, H., C., Liu, T. M. E. and Tu,C. C. 1988. Effect of chemical fertilizers and nitrogenous compounds on the sclerotia of *Sclerotium rolfsii* in soil. *Plant Protection Bulletin, Taiwan.* 30: 101-110.

Figoni, R. A. 1981. The inheritance of resistance to stem rot *(Sclerotium oryzae)* in rice. *Dissertation Abstracts International,* B. 42:903B.

Gamliel, A and Katan, J. 1993. Suppression of major and minor pathogens by fluorescens pseudomonads in solarized and non solarized soils. *Phytopath.* 83: 68-75.

Ganesan, P and Gnanamanickam, S. S. 1987. Biological control of *Sclerotium rolfsii* Sacc.in peanut by inoculation with *Pseudomonasfluorescens. Soil Biol.Biochem.* 19: 35-38.

Georgiou, D. C. 1997. Lipid peroxidation in *Sclerotium rolfsii:* A new look into the mechanism of sclerotia biogenesis in fungi. *Mycol. Res.* 101: 460-464.

Ghini, R., Bettiol, W. and Caldari Junior, P. 1997. Control of *Sclerotium rolfsii* in beans by soil solarization. *Summa Phytopathologica.* 23:143-145.

Grinstein, A., Katan, J., Razik, A. A., Zeydan, O. and Elad, Y. 1979. Control of *Sclerotium rolfsii* and weeds in peanuts by solar heating of the soil. *PlantDis. Reptr.* 63:1056-1059.

Gurkin, R. S. and Jenkins, S. F. 1985. Influence of cultural practices, fungicides, and inoculum placement on Southern blight and Rhizoctonia crown rot of carrot. *Plant Dis.* 69: 477-481.

Hagan, A. K., Weeks, J. R and Reed, R B. 1989. Southern stem rot suppression on peanuts with lorsban insecticide. *Down to Earth.* 45: 22-24.

Hagan, A. K., Weeks, J. R and Bowen, K. 1991. Effects of application timing and method on control of southern stem rot of peanut with foliar-applied fungicides. *Peanut Science.* 18: 47-50.

Hagan, A. K. and Olive, J. W. 1999. Assessment of new fungicides for the control of southern blight on Aucuba. *J.Envir. Horti.* 17: 73-75.

Harlton, C.E., Levesque, C. A. and Punja, Z. K. 1995. Genetic diversity in *Sclerotium (Athelia) rolfsii* and related species. *Phytopath.* 85: 1269-l-1.

Haroon Usmani, S. M. and Ghaffar, A. 1982. Polyethylene mulching of soil to reduce viability of sclerotia of *Sclerotium oryzae. Soil Biol. Biochem.* 14: 203-206.

Henis, Y., Adams, P. B., Lewis, J. A. and Papavizas, G. C. 1983. Penetration of sclerotia of *Scler-Otium rdlfsii* by *Trichoderma* spp. *Phytopath,* 73:1043-1046.

Hernandez, C. A. and Aleman, R. 1997. Mycelial interaction and antagonism between isolates of *Sclerotium rolfsii* Sacco from different sources. *Centro Agricola.* 24: 87-88.

Hoynes, C. D., Lewis, J. A., Lumsden, R. D. and Bean, G. A. 1999. Biological control agents in combination with fertilization or fumigation to reduce sclerotial viability of *Sclerotium rolfsii* and disease of snap beans in the greenhouse. *J.Phytopath.* 147: 175-182.

Iqbal, S. M., Bakhsh, A., Hussain, S. and Malik, B. A. 1995. Microbial antagonism against *Sclerotium rolfsii,* the cause of collar rot of lentil. *Lens Newsltr.* 22: 48- 49.

Jayaraj, J., Ramanujam, K. and Ramabadran, R. 1991. Effect of potash nutrition on the stem rot incidence and yields of rice. *Journal of Potassium Res.* 7: 62-66.

Jennrich, H. 1983. Results of tests with vinclozolin in the control of *Sclerotium cepivorum* in *Allium* species. *Mededelingen van de aculteit Landbouwwetenschappen Rijksuniversiteit Gent.,* 48: 551-557.

Kalappaqavar, I. K. and Fatil, N. K. 2000. Biological control of *Sclerotium rolfsii* Sacco a causal agent of foot rot of wheat. *Karnataka J. Agricul Sci.* 13: 478-479.

Kay, S. J. and Stewart, A. 1994. Evaluation of fungal antagonists for control of onion rot in soil box trials. *Plant Pathol.* 43: 371-377.

Krause, R. A. and Webster, R. K. 1972. Sclerotial production, viability determination and quantitative recovery *of Sclerotium oryzae* from soil. *Mycologia.* 64:1333- 1337.

Krauthausen, H. J. and Schietinger, R. 1989. Studies on the control of white rot *(Sclerotium cepivorum* Berk.*)* in onIons. *Zertscntift -fur-Pflanzerikrarikheiten und Pflanzenschutz,* 96: 604-614.

Krishnakanth, A., Gowda, M. V. C. and Motagi, B. N. 1999. Response of Spanish groundnuts to stem and pod rots caused by *Sclerotium rolfsii -Sacc. lntr. Arachis Newsltr.*19: 27-28.

Kulkarni, S. A., Srikant Kulkarni., Sreenivas, M. N., Kulkarni, S. 1997 Interaction , between vesicular-arbuscular (VA) mycorrnizae and *Sclerotium rolfsii* -Sacco in groundnut. *Karnataka J. Agricul Sci.* 10: 919-921.

Laha, G. S., Verma, J. P. and Singh, R. P. 1996. Effectiveness of fluorescent pseudomonads in the management of sclerotial wilt of cotton. *Indian Phytopath,* 49: 3-8.

Latunde Dada, A. O. 1993. Biological control of southern blight disease of tomato caused by *Sclerotium rolfsii* witn simplified mycelial formu1ations of *Trichoderma koningii. Plant Path.* 42: 522-529.

Leeper, P. W. Phatak, S.C., Bell, D.K., George, B. F., Cox, E.L., Oerther, G. E. and Scully, B. T. 1992. Southern blight-resistant tomato breeding lines: 5635M, 5707M, 5719M, 5737M,5876, and 5913M. *HortScience.* 27: 475-478.

Lewis, J. A., Papavizas, G. C. and Hollenbeck, M. D. 1993. Biological control of damping-off of snap beans causea by *Sclerotium rolfsji* in the greennouse and field with formulations of *Gliocladium virens. Biol. Control.* 2: 109-115.

Madi, L., Katan, T., Katan, J. and Henis, Y. 1997. Biological control of *Sclerotium rolfsii* and *Verticillium dahliae* by *Talaromyces flavus* is mediated by different mechanisms. *Phytopath.*87: 1054-1060.

Maloy, O. C. and Machtmes, R. 1974. Control of onion white rot by furrow and root-dip application of fungicides. *Plant Dis. Reptr.* 58: 6-9.

Mathur, S. B. and Sarbhoy, A. K. 1978. Biological control of Sclerotium root rot of sugarbeet. *Indian Phytopath.* 31: 365-367.

Matrod, L., Faddoul, J., Elmeamar, A.and AI Chaabi, S. 1991. The use of solar energy for controlling white rot disease of garlic. *FAO Plant Production and Protection Paper,* 109: 118-128.

Mazzanti de CastAnon, M. A., Mazza de Daiad, S. M. and Gutierrez de Arriola, S. A. 1997. Pathogenic variability of *Sclerotium oryzae* in Argentina. *Fitopatologia.* 32: 58-63.

McLean, K. L. and Stewart, A. 2000. Application strategies for control of onion white rot by fungal antagonists. *New Zealand Journal of Crop and Horticultural Science.* 28: 115-122.

McLean, K. L., Swaminathan, J. and Stewart, A. 2001. Increasing soil temperature to reduce sclerotial viability of *Sclerotium cepivorum* in New Zealand soils. *Soil Biol. Biochem.* 33: 137-143.

Mehan, V. K., Mayee, C. D., McDonald, D., Ramakrishna, N. and Jayanthi, S. 1995. Resistance in groundnut to *Sclerotium rolfsii-caused* stem and pod rot. *Intr.J. Pest Manag.* 41: 79-83.

Melero Vara J. M., Prados Ligero, A. M. and Basallote Ureba, M. J. 2000. Comparison of physical, chemical and biological methods of controlling garlic white rot. *European J. Plant Pathol.* 106: 581-588.

Mesta, R. K. and Amaresh, Y. S. 2000. Biological control of *Sclerotium* wilt of sunflower. *Plant Dis. Res.* 5: 202-203.

Metcalf, D. A. and Wilson, C. R. 1999. Histology of *Sclerotium cepivorum* infection of onion roots and the spatial relationships of pectinases in the infection process. *Plant Pathol.* 48: 445-452.

Mihail, J. D. and Alcorn, S. M. 1984. Effects of soil solarization on *Macrophomina phaseolina* and *Sclerotium rolfsii. Plant Dis.* 68: 156-159.

Moletti, M., Mazzini, F., Villa, B. and Baldi, G. 1983. Evaluation of the resistance of rice genotypes to *Sclerotium oryzae* Cart. by means of artificial and natural infection. *Informatore Fitopatologico.* 33: 59-62.

Monaco, C., Perello, A., Alippi, H. E. and Pasquare, A. 0.1991. *Trichoderma* spp.: a bio-control agent of *Fusarium* spp. and *Sclerotium rolfsii* by seed treatment. *Adv.Horti. Sci.* 5:92-95.

Muhammad, A., Ayub, K. and Malik, N. J. 1998. Evaluation of peanut Arachis hypogaea genotypes for ne1d resistance to *Sclerotium rolfsli* and gen'etic yield potential. *Sarhad J. Agricl.* 14: 597-601.

Mukherjee, P. K., Sherkhane, P. D. and Murthy, N. B. K. 1999. Induction of stable benomyl-tolerant- phenotypic mutants of *Trichoderma pseudokoningii* MTCC 3011, and their evaluation for antagonistic and biocontrol potential. *Indian J.Expt. Biol,* 37:710-712.

Muthamilan, M. and Jeyarajan, R. 1996. Integrated management of Sclerotium root rot of groupdnut involving *Trichoderma harzianum, Rhizobium* and carbendazim. *Indian J Mycol.Pl Pathol.* 26: 204-209.

Mutto, S. D., Ambra, V. and Ferrata, M. 1986. Ultrastructural aspects of the parasitism of *Trichoderma harzianum* on sclerotia of *Sclerotium rolfsii. Phytopath Medit.* 25: 10-18.

Nigam, S. N., Reddy, L. J., Subrahmanyam, P., Reddy, A. G. S., McDonald, D. and Rodriguez Kabana R, Robertson, D.G., Wells, L., Weaver, C. P. and King, P. S. 1991. Cotton as a rotation crop for the management of *Meloidogyne arenaria and Sclerotium rolfsii* in peanut. *J. Nematol*, 23: 652-657.

Okabe, I., Morikawa, C., Matsumoto, N. and Yokoyama, K. 1998. Variation in *Sclerotium rolfsii* isolates in Japan. *Mycoscience*.39: 399-407.

Ordentlich, A., Elad, Y. and Chet, I. 1987. Rhizosphere colonization by *Serratia marcescens* for the control of *Sclerotium rolfsli. Soil Biol.Biochem*. 19: 747-751.

Papavizas, G. C. and Lewis, J. A. 1989. Effect of *Gliocladium* and *Trichoderma* on damping-off and blight of snapbean caused by *Sclerotium rolfsii* in the greenhouse. *Plant Pathol*. 38: 278-286.

Pereira, J. C. R., Chaves, G. M., Zambolim,L., Matsuoka, K., Acuna, R. S., Vale, F. X. R. do. And DoVale, F. X. R. 1996. Control of *Sclerotium cepivorum* by the use of vermicompost, solarization, *Trichoderma harzianum* and *Bacillus subtilis Summa Phytopathologica*, 22: 228-234.

Pinto, C. M. F., Maffia, L. A., Casali, V. W. D. and Cardoso, A. A. 1998. *In vitro* effect of plant leaf extracts on mycelia1 growth and sclerotia1 germination of *Sclerotium cepivorum. J. Phytopath*. 146: 421-425.

Porter, I. J., Maughan, J. P. and Towers, G. B. 1991. Eva1uation of seed, stem and soi1 applications of procymidone to control white rot *(Sclerotium cepivorum* Berk.) of onions. *Aust.J. Expert. Agric*. 31: 401-406.

Prabhu, H. V., Hiremath, P. C. and Patil, M. S. 1997. Evaluation of cotton genotypes against collar rot caused by *Sclerotium rolfsii* Sacco .*Karnataka J.Agricul. Sci*. 10: 929-930.

Prasad, R. D., Rangeshwaran,R. and Kumar, P. S. 1999. Biological control of root and collar rot of sunflower. *J. Mycol.Pl. Pathol*. 29: 184-188.

Prithiviraj, B., Singh, U. P., Kobayashi, K., Sugawara, K., Nishimura, R., Sakai Wada, A. 1996. Variability in sclerotial morphology in *Sclerotium rolfsn:* SEM and isozyme studies. *Zeitschrift fur Pflanzenkrankheiten und Pflanzenschutz*. 103: 495-501.

Punja, Z. K. 1985. The biology, ecology and control of *Sclerotium rolfsii. Ann. Rev.Phytopath*. 23: 97-127.

Punja, Z. K. 1988. *Sclerotium (Athelia) rolfsii*, a pathogen of many plant speices. pp. 523-534. (*In*:) G. S. Sidhu (Ed.) *Genetics of Plant Pathogenic Fungi*. Vo16. Academic Press. London.

Punja, Z. K. and Grogan, R. G. 1983a. Basidiocarp induction, nuclear condition, variability and heterokaryon incompatibility in *Athelia (Sclerotium) rolfsii. Phytopath*. 73: 1274-1278.

Punja, Z. K. and Grogan, R. G. 1983b. Hypha1 interactions and antagonism among fie1d isolates and single-basidiospore strains of *Athelia (Sclerotium) rolfsii. Phytopath*. 73: 1279-1284.

Punja, Z. K., Smith, V. L., Campbell, C. L. and Jenkins, S. F. 1985. Sampling and extraction procedures to estimate numbers, spatial pattern, and temporal distribution of sclerotia of *Sclerotium rolfsii* in soil. *Plant Dis*, 69: 469-474.

Punja, Z. K. and Damiani, A. 1996. Comparative growth, morphology, and physiology of three *Sclerotium* species. *Mycologia*. 88: 694-706.

Rahe, J. E. and, Utkhede, R. S. 1982. Resistance to white rot in commercial cliltivars of *Allium cepa. Canadian J.Pl. Path*, 4: 310.

Ramirez, R., Santos, R., Bracho, F., Sandoval, L. and Castro de Rincon, C. 1998. Control of *Sclerotium rolfsii* Sacco with fungicides and humus. Revista de la Facultad de Agronomia, Universidad del Zulia, 15: 534-544.

Rangeshwaran, R and Prasad, R. D. 2000b. Biological control of *Sclerotium* rot of sunflower. *Indian Phytopath*, 53: 444-449.

Reddy, M. S., Rahe, J. E. and Levesque, C. A. 1992. Influence of onion seed bacterization on germination and mycosphere microflora of *Sclerotium cepivorum* sclerotia. *Can J. Microbiol*. 38: 1135-1143.

Resende, M. L. V, de., Zambolim, L. and Cruz~ilho J da. 1987. Effectiveness of fungicides in controlling white rot of garlic *(Allium sativum)*, in relation to the level of sclerotia of *Sclerotium cepivorum* in the soil. *Fitopatologia Brasi1eira*. 12: 71-78.

Ristaino, J. B., Perry, K. B. and Lumsden, R. D. 1991. Effect of solarization and *Gliocladium virens* on sclerotia *of Sclerotium rolfsli*, soil microbiota, and the incidence of southern blight of tomato. *Phytopatho*. 81: 1117-1124.

Ristaino, J. B., Lewis, J. A. and Lumsden, R. D. 1994. Influence of isolates of *Gliocladium virens* and delivery systems on biological control of southern blight on carrot and tomato in the field. *Plant Dis*. 78: 153-156.

Ristaino, J. B., Perry, K. B. and Lumsden, R. D. 1996. Soil solarization and *Gliocladium virens* reduce the incidence of southern blight *(Sclerotium rolfsii)* in, bell pepper in the field. *Biocontrol Science and Technology*. 6: 583-593.

Roberti, R. and Flori, P. and Pisi, A. 1996. Biological control of soilborne *Sclerotium rolfsii* infection by treatment of bean seeds with species of *Trichoderma. Petria*. 6: 105-116.

Rodriguez Kabana, R., Backman, P. A. and Wiggins, E. A. 1974. Determination of sclerotial populations of *Sclerotium rolfsii* in soil by a rapid flotation-sieving technique. *Phytopath*. 64: 610-615.

Rodriguez Kabana, R., Robertson, D. G., Wells, L. and King, P. S. 1991. Soybean-peanut rotations for the management of *Meloidogyne arenaria* and *Sclerotium rolfsii. Nematropica*. 21: 147-154.

Rodriguez Kabana, R, Backman, P. A., Ivey, B. and Farrar, L. L. 1972. Effect of post-emergence application of potassium azide on nematode populations and development of *Sclerotium rolfsii* in a peanut field. *Pl.Dis. Reptr*. 56:362-367.

Saccardo, P. A. 1913. *Sclerotium rolfsii.* Sylloge fungorum XXII. Pavia, Italy. 1500pp.

Salui, M. and Bhattacharryya, P. 1998. Competition of Rhizobium with the root rot pathogen *(Sclerotium rolfsii)* on groundnut rhizoplane. *J.Mycopathgical Research,* 6: 73-80.

Sariah Meon. 1991. Hyphal interactions of *Trichoderma* sp. and *Sclerotium rolfsii,* the causal pathogen of foot-rot of chilli.*Pertanika.* 4: 167-169.

Sharma, S. K., Kaur, J. and Mehrotra, R. S. 1984. Effect of polyethylene mulching on the viability of sclerotia of *Sclerotium oryzae* in soil. *Indian J.Pl Path.* 2: 33-38.

Sharma, B. S., Pathak, V. N., Kalpna Bhatnagar, Bhatnagar, K. 1990. Fungicidal management of root rot of sugar beet induced by *Sclerotium rolfsii Sacc. Indian J.Mycol.Pl.Pathol.* 20:207-210.

Silveira, N. S. S., Michereffi, S. J., Menezes, M. Campos Takaki, G. M. 1994. Potential of *Trichoderma* spp. isolates on the control of *Sclerotium rolfsii* on beans. *Summa Phytopathologica.* 20: 22-25.

Singh, R. K. and Dwivedi, R. S. 1990. Fungicidal properties of neem and blue gum against *Sclerotium rolfsii* -Sacc., a foot rot pathogen of barley. *Acta Botanica Indica,* 18: 260-262.

Sugha, S. K., Sharma, B. K. and Tyagi, P. D. 1990, Factors affecting development of collar rot of gram *(Cicer arietinum)* caused by *Sclerotium rolfsii. Indian J.Agricul.Sci.* 63: 382-385.

Somkuwar, R. G., Gowda, R. V. and Singh, T. -H. 1997. Varietal response to white rot in onion. *Mad. Agril. J.* 84: 343-344.

Sreenivasaprasad, S. and Manibhusbanrao, K. 1993. Efficacy of *Gliocladium virens* and *Trichoderma longibrachiatum* as biological control agents of groundnut root and stem rot diseases. *Int J.Pest Manag.* 39: 167-171.

Stevens, F. L. 1931. A comparative study of *Sclerotium rolfsii* and *Sclerotium delphinii. Mycologia.* 23: 204-222.

Sturgeon, R. V. Jr. 1973. Evaluation of methods of applying soil fungicides for control of soil fungi on Spanish peanuts. *Journal of the American Peanut Research and Education Association, Inc.,* 5: 175-177.

Sunder, S.,Dodan, D. S., Ahuja,S. C. and Singh, R.1991.Spray schedule for control of rice stem rot. *Oryza.* 28: 421-423.

Taylor, C. R and Rodriguez Kabana, R. 1991. Optimal rotation of peanuts and cotton to manage soil-borne organisms. *Agril Systems.* 61:57-68.

Thiribhuvanamala, G., Rajeswari, E. and Duraiswamy, S. 1999. Biological control of stem rot of tomato caused by *Sclerotium rolfsii* Sacco. *Mad. Agril J.* 86: 30-33.

Toibio, J. A. 1977. A simple method of direct counting of viable sclerotia of *Sclerotium rolfsii* in soil. *Annales dePhytopathologie,* 9: 177-182.

Tsigbey, F. K., Nutsugah, S. K. and Ritchie, B. J. 1999. *Gliocladium catenulatum* in association with *Sclerotium cepivorum* on onions leaves in Ghana. *Plant Dis.* 83: 198.

Tu, C. C., Cheny, Y. H. and Hwang, S. C. 1991. Effects of solar energy and green manures on the control of southern blight of tomato. *Plant Protection Bulletin Taipei.* 33: 95-102.

Tu, C. C., Hsieh, T., Tsai, W. H. and Kimbrough, J. W. 1992. Induction of basidia and morphological comparison among isolates of *Athelia (Sclerotium) rolfsii. Mycologia.* 84: 695-704.

Uma Singh, Tliapliyal, P. N. and Singh, U. 1998. Effect of inoculum density, host cultivars and seed treatment on the seed and seedling rot of soybean caused by *Sclerotium rolfsii. Indian Phytopath.* 51: 244-246.

Upadhyay, J. P. and Mukhopadhyay, A. N. 1983. Effect of non-volatile and volatile antibiotics of *Trichoderma harzianum* on the growth of *Scierotium rolfsii. Indian J.Mycol. Pl. Pathol.* 13: 232-233.

Upadhyay, J. P. and Mukhopadhyay, A. N.1986. Biological control of *Sclerotium rolfsii* by *Trichoderma harzianum* in sugar beet. *Trop. Pest Manag.* 32: 215-220.

Utkhede, R. S. and Rahe, J. E. 1978. Screening commercial onion cultivars for resistance to white rot. *Phytopath.* 68: 1080-1083.

Utkhede, R. S. and Rahe, J. E. 1981. Development of resistance to white rot in bulb onions. *Phytopath.* 71: 910.

Utkhede, R. S. and Rahe, J. E. 1983. Effect of *Bacillus subtilis* on growth and protection of onion against white rot. *Phytopath. Z.* 106: 199-203.

Voss, R. E., Weinhold, A. R., Bowman, T., Dennett, C. and Guerard, J. 1984. Potato cultivars differ in susceptibility to *Sclerotium rolfsii. American Potato J.* 61: 541.

Waraitch, K. S., Kanwar, R. S., Bipen Kumar and Kumar, B. 1986. Fungicidal control of sclerotium root rot of sugarbeet *(Beta vulgaris)* caused by *Sclerotium rolfsii. Indian Phytopath.* 39: 100-102.

Wokocha, R. C. 1990. Integrated control of *Sclerotium rolfsii* infection of tomato in the Nigerian Savanna: effect of *Trichoderma viride* and some fungicides. *Crop Protection.* 9: 231-234.

Wong, W. C. and Hughes, I. K. 1986. *Sclerotium cepivorum* Berk. In onion *(Allium cepa* L.) crops: isolation and characterization of bacteria antagonistic to the fungus in Queensland. *J.Appld Bacteriol.* 60: 57-60.

Yossry, A. A., El All, S. M. A., and EI Imery, S. M. 1998. Fungitoxic properties of some plant extracts against growth of soil borne disease fungi. *Annals of Agricultural Science* Cairo, Special Issue 3: 891-909.

Zarani, F. and Christias, C. 1997. Sclerotial biogenesis in the basidiomycete *Sclerotium rolfsii.* A scanning electron microscope study. *Mycologia.* 89: 598-602.

• • •

Advances in Soil Borne Plant Diseases, 2008
Editors: Naik and Devika Rani, pp. 129-142
New India Publishing Agency, New Delhi

5

Plant Growth Promoting Rhizobacteria Mediated Strategy for Disease Management

Jayashree Gopalakrishnan

INTRODUCTION

Biological control involves the suppression of pests/organisms with other organisms. It is an eco-friendly method for management of plant pathogens which can be done exploiting the fungi (*Ampelomyces, Trichoderma, Coniothyrium, Candida* etc.) or bacteria (*Agrobacterium, Bacillus, Pseudomonas* and *Streptomyces*) from rhizosphere, as well as phylloplane and endophytes (Jayaraman and Verma, 2002). The rhizosphere microfloras are predominantly composed of gram negative pseudomonads. The fluorescent pseudomonads constitute a large proportion of the rhizosphere microflora presumably due to their nutritional variability being able to use a large number of organic substances (Stolp and Gadkari, 1981). However earlier workers used fluorescent pseudomonads extensively they now prefer *Bacillus* sp. due to their spore forming ability and heat tolerance which make its formulation easier.

The term PGPR is used to refer to beneficial rhizosphere microorganisms which readily colonize roots of several crops and promote plant growth and yield. Kloepper *et al.,* (1980) first demonstrated the Plant growth promoting

Division of Plant Pathology, Indian Agricultural Research Institute
New Delhi-110012

activity of PGPR is due to their ability to produce siderophores. Mutant PGPR which were defective in fluorescent siderophore production could not promote plant growth (Kloepper and Scroth, 1981). Suppression of plant disease by PGPR involves the production of siderophores, antibiotics and or aggressive root colonization.

PGPR used as biocontrol agents in India has been listed by Gnanamanickam and Podile (2003). Review articles on chemical and biological control of bacterial disease worldwide was done by Burr (2001) and India by Jayaraman and Verma (2002). Review of rhizobacteria for bacterial disease management was given by Dube (2001) and for diseases of plantation crops by Sarma *et al.*, (2003). The importance of PGPR can be seen by the fact that a PGPR Newsletter published in USA and 6 International workshops have been held (2003, in Calicut India). The 7th Workshop was held in Netherlands in 2004.

RHIZOSPHERE BACTERIA IN THE MANAGEMENT OF PLANT DISEASES

Bacterial plant diseases

Fluorescent pseudomonads have been exploited in the biocontrol of bacterial diseases of plants. Kloepper *et al.*, (1980) reported that four strains of Plant Growth promoting rhizobacteria (PGPR) belonging to *Pseudomonas* sp. isolated from potato periderm or celery roots were antagonistic both *in vitro* and filed to *Erwinia carotovora* subsp. *carotovora* (*Ecc*) . Liao (1989) also demonstrated that eight isolates of *P.fluorescens* and 16 of *P. putida* were antagonistic to *Ecc*. Out of this one strain PP-22 of *P.putida* was antagonistic to 24 strains of soft rot bacteria (*Erwinia, Pseudomonas, Xanthomonas* and *Cytophaga*; 1 strain of *Ralstonia solanacearum*, 4 pathovars of *Xanthomonas campestris* and 5 pathovars of *Pseudomonas syringae)*. Cruz *et al.*, (1992) showed the biocontrol activity of *Pseudomonas* against *Clavibacter michiganensis* subsp. *sepedonicus*). *Bacillus subtilis* has been reported to be an excellent antagonist for the control of *Xam* (Safiazov *et al.*, 1995). The protection afforded by the antagonists (by seed bacterisation) was effective upto cotyledonary stage but could not afford protection beyond 2-3 leaf stage (Pain, 1991). It appears that if epiphytic or rhizosphere bacteria have to be used as input for integrated management of bacterial blight of cotton (BBC) one has to include both seed bacterisation with spray of antagonists at early leaf stages, i.e. repeated application of biocontrol agent. Chauhan and Chopra (1993) showed that the isolates 6D and 12D of antagonists used by them were less effective than streptomycin sulphate and copper oxychloride. Fluorescent Pseudomonads were isolated from cotton rhizosphere (Laha, 1994). Out of 48 isolates only 9 were antagonistic to *Xam*. *P. fluorescens* has been found to inhibit growth of *Xam in vitro* and also reduced disease intensity significantly when seed treatment was done followed by foliar spray (Arya and Parashar, 1997). Fluorescent pseudomonads act through a variety of secondary metabolites. At IARI a *P. fluorescens* strain PfK-26 was identified as most effective against *Xcm*. Through TLC four compounds (2 fluorescent, 2 non fluorescent) were isolated and all the four compounds have been shown to inhibit the susceptible reaction on pre-inoculation (0-8) hours to challenge inoculation of *Xcm* (Mondal *et al.*, 1998). The four compounds possessed Rf values of 0.72, 0.64, 0.51 and 0.46 were referred to as compounds I, II, III and IV. Compounds I and IV were fluorescent. The compounds II and IV suppressed the growth of *Xcm in vitro* experiments (Mondal and Verma, 1998; Mondal *et al.*, 1998). The antagonism of the or suppression of susceptible reaction of *Xcm* by cotton rhizosphere strain Pfk-26 of *P. fluorescens* is due to the production of siderophore and cyanic acid. Other secondary metabolites may also be involved. The isolate possessed the maximum inhibitory activity against Race 32 of *Xcm in vitro*. Seed bacterization with the strain improved seed germinability (by 10%), increased percentage of normal seedlings and also caused maximum increase in plant height (52.5 cm) and dry weight (147.5 g). When this strain was applied on *Xcm* inoculated seeds the percentage of germination increased by 20% and 17%., reduced the number of plants with cotyledonary infection by 54% and 40% and gave highest protection of *Xcm* induced cotyledonary infection by 75% and 69% under sterilized and unsterilised soil respectively (Mondal, 1998).

21 rhizobacterial isolates were collected by Dutta (2001). Three isolates (MRb-1, MRb 13 and MRB 19) were found to inhibit the mungbean bacterial leaf spot pathogen (*Xanthomonas axonopodis* pv. *vignaeradiateae, Xav*) under *in vitro* conditions. MRb-1 isolate was characterized and tentatively identified as belonging to *P. fluorescens*. Bacterisation with MRb-1 caused the maximum increase in percentage of germination (6.82%), dry weight (by 32%), leaf area (20.5%) and chlorophyll content (by 8.6%) over the non-bacterised water treated control. MRb-1 could also reduce the cotyledonary infection by 71.42%-88.95% and trifoliate infection by 28.62-52.635. The bacterium was found to produce a moderate level of HCN (5-20 nmol/ml of medium).

Pseudomonas fluorescens inhibited *Xanthomonas oryzae* pv. *oryzae* (Sivamani *et al.*, 1987; Anuratha and Gnanamanickam, 1987). *Pseudomonas fluorescens* is also a plant growth promoting bacteria and increased plant height to 27% over control (Sakthivel *et al.*, 1986). *P.fluorescens* is also reported to be antagonistic to citrus canker pathogen (*Xanthomonas campestris* pv. *citri*) by Unnamalai and Gnanamanickam (1984) and *Ralstonia solanacearum* in banana, eggplant and tomato by Anuratha and Gnanamanickam (1990).

Little is known about the genetics of the mechanism of action of the PGPR. Saha *et al.,* (2000) studied the plasmid borne determinants of colony morphology, pigmentation, antibiotic resistance and antibiosis in *Pseudomonas* sp. antagonistic to *Xanthomonas axonopodis* pv. *malvacearum. In vitro* antibiosis of wild (containing 8.5 kb plasmid) and cured strains of *Pseudomonas* sp. were tested against *Xam* by dual culture method and it was found that the wild strain gave a clear halo zone depicting antibiosis while the cured strain could not exhibit a similar property. *In vivo* antibiosis was checked against the pathogen and the cured strain could not protect cotton from *Xam* infection (as seen by water soaked reaction) while the wild type strain (in presencee of *Xam*) gave protection which was exhibited by HR (hypersensitive reaction). Transformation of cured cells with plasmid restored the antibiosis and *in vivo* HR (in presence of *Xam*) in susceptible cotton.

Rhizosphere Bacteria in the Management of Fungal Diseases

Since 1976 a number of fungal diseases have been reported to be suppressed by fluorescent pseudomonades. (Table 5.1)

Most studies appear to be *in vitro* or in greenhouse conditions. Vidhyasekharan and Muthamilan (1995) developed formulation for seed treatment and soil application for soil management of chickpea wilt. According to them seed treatment with antagonist formulation effectively controlled chickpea wilt disease in the field trials and increased the yield. When seed treatment was followed by root zone application the efficacy of the *P.fluorescens* formulation increased. Sivakumar (1999) found in field trials that when *P.fluorescens* was applied by combination approach viz. seed (20g/ kg seed), soil (2.5 kg peat formulation/ ha 25 days after sowing) and foliar spray (6g peat formulation/litre applied 42 and 48 days after sowing) reduced the percent disease reduction of Banded leaf and sheath blight disease in maize by 75 over single application method as seed treatment, soil application and spray application in case of Pusa Composite variety of maize. Endophytic *Pseudomonas fluorescens* Endo PR 8 isolate was reported to reduce the damping off by *Sclerotium rolfsii* and *R.solani* to the extent of 76% and 84% respectively (Bhowmik *et al.,* 2002). Some of the mixed cropping systems of plantation crops where PGPR have scope and potential is given in Table 5.2. (Source Sarma *et al.,* 2003).

Mechanism of action of rhizobacteria:

Secondary metabolites are produced by substrates provided by previous metabolism. They are not essential for growth. Some of the secondary metabolites are involved in pathogen suppression. For example siderophores, hydrogen cyanide, diacetylphloroglucinals, pyoluteorin, pyrollnitrin etc. Rhizobacteria especially non-pathogenic *Pseudomonas* strains induce systemic resistance in plants. Currently novel volatile organic compounds have been identified not to only promote the growth of *Arabidopsis thaliana* but also to protect it against *Erwinia carotovora* subsp. *carotovora* (Ryu *et al.*, 2003). These compounds have been identified to be 2, 3 butanediol and acetoin. Some of the secondary metabolites identified to be involved in disease suppression are given in Table 5.3.

VESICULAR ARBUSCULAR MYCORRHIZAE AND PGPR INTERACTION

Beneficial bacteria such as PGPR (Burr and Caeser, 1984, Mayer and Lindermann, 1986) somehow work in tandem with mycorrhizal fungi to give not only additive or synergistic growth enhancement of the host plant but also increase the mycorrhizal colonization possibly due to increased spore generation (Mayo *et al.*, 1986). Gaekwad (1994) observed that application of rhizobacteria with VAM showed synergistic interaction as reflected in better plant growth, P assimilation and increased grain yield in bajra. The isolates used were antagonistic to *Sclerotium rolfsii* and the indirect growth promotion may have been due to the PGPR antagonism or displacement of native deleterious microflora. The synergistic action of both the VAM and PGPR is increasingly being exploited for disease management by applying the two together as consortia.

GENETIC ENGINEERING OF BACTERIA FOR BIOLOGICAL CONTROL

Biological control of plant diseases uses micro-organisms including bacteria as biocontrol agents. The commercialization of the use of Plant Growth Promoting Rhizobacteria (PGPR) or bacteria in general as biocontrol agents is impeded by their inconsistent field performance. The molecular genetic aspects have to be, therefore, used to develop a strategy to engineer bacteria which are not only having plant growth promotion genes but also biocontrol genes under variable set of environmental conditions. Such genes have already been identified and / or cloned: eg. genes which regulate production of insecticidal proteins (Wabiko, *et al.*, 1985; Hofte *et al.*, 1986) and antibiotics (Gutterson *et al.*, 1986). The antibiotic 2, 4 -diacetylphloroglucinol (Phl) was involved in the biocontol of various soil-borne plant pathogens by fluorescent pseudomonads. Phl-specific probes and primers based on DNA sequence of *PhlD* biosynthetic gene were used to distinguish Phl-producing strains among bacterial isolates from roots of wheat grown in soil that exhibit natural biocontrol of take all decline (Raaijmakers and Wellers, 1998). Thomashow *et al.*, (1998) have cloned genes for the synthesis of phenazine-1-carboxylic acid (PCA) from *Pseudomonas fluorescens* strain 2-79 and inserted into Phl producing strain *P.fluorescens* strain Q8rl-96. The transgenic derivative strains producing both Phl and PCA, colonized roots aggressively, retained activity against take all and in addition controlled *Rhizoctonia* and *Pythium* root rot better than parent strain. Plant growth hormones (Comai and Kosuge, 1982) and phosphate solubilisation genes (Goldstein and Liu, 1987) have also been cloned/identified. These genes could use to develop improved bacterial genes or to engineer genes directly into the plants. Molecular probes could also be developed to facilitate the selection for better strains.

Binding of the bacterial cells to the roots may prevent the interaction between plants and plant pathogenic microbes. Binding genes have been identified in *Agrobacterium radiobacter, Azospirillum* spp. and *Rhizobium* spp. (Lambert and Joos, 1989). By manipulating or transferring such genes to strains of interest, their association with plants may be strengthened. There is evidence that binding genes from *Agrobacterium* spp. and *Rhizobium* spp. were functionally interchangeable (Dylan *et al.*, 1986). Lambert and Joos (1989) suggested that an elegant way of improving the association between plants and selected bacteria would be engineering positive or negative selection genes in plants. For example, positive selection could be obtained by introducing opine synthesis genes into plants and introducing opine catabolism gene into the bacterium of interest. The opine in the root exudates of the engineered plant might favor the proliferation of the engineered bacteria. Genes that encode for bacteriocides could exert negative selection pressure on the general rhizosphere communities and favor proliferation of bacteria engineered to be bactericide resistant. A combination of negative and positive selection could then be used to further improve the colonization level. Some of the genes associated with PGPR traits are shown in Table 5.4.

CONCLUSION

There is a growing demand for biopesticides among both conventional and organic growers suggesting that a potential market is going to be there in the future. The success of biopesticides will depend on development of sound research base, product marketing and extension. The industry, government and academia will need better linkage to improve the funding for research and development of biopesticides.

In crops where the pesticides are frequently applied for control of disease as combined spray the use of PGPR is limited unless it is compatible with the pesticides. Mathew (2003) observed that in rice the *Pseudomonas fluorescens*

TABLE 5.1. Use of fluorescent pseudomonades in the management of fungal diseases

Disease	Pathogen
Take all of wheat	*Gaeumanomyces graminis* pv. *tritici*
Flax wilt	*Fusarium oxysporum* fsp. *lini*
Sheath rot of rice	*Sarocladium oryzae*
Sheath blight of rice	*Rhizoctonia solani*
Cucumber wilt	*Fusarium oxysporum*
Damping off of maize	*Pythium* spp.
Carnation wilt	*Fusarium oxysporum* fsp. *lini*
Seedling blight of cotton	*Rhizoctonia solani*
Wilt of cotton	*Fusarium oxysporum*
Root rot of cotton and damping off	*R.solani*
Cucumber damping off	*Pythium aphanidermatum*
Radish wilt	*F.o.* f.sp. *raphani*
Tomato wilt	*F.o.* f.sp. *lycopersici*
Bean root rot	*Sclerotium rolfsii*
Pea wilt	*F.o.* f.sp. *pisi*
Root rot	*Aphanomyces euteichus*
Pea damping off	*Pythium ultimum*
Sugar beet root rot	*Aphanomyces cochlioides*
Chickpea wilt	*Fusarium oxysporum* f.sp. *ciceris*
Pigeon pea wilt	*Fusarium oxysporum* f.sp. *udum*
Banded leaf and sheath blight of maize	*Rhizoctonia*

TABLE 5.2. Mixed cropping systems where PGPR show potential

Cropping System	Pathogen Profile
Coconut and Pepper, Ginger and turmeric	*Phytopthora palmivora, P.capsici, Ganoderma lucidum, Thielaviopsis paradoxa, Pythium aphanidermatum*
Coconut and Cacoa	*P.citropthora, P. parasitica*

Contd...

Table 5.2. Contd...

Coconut and Vanilla	*P.meadii, Fusarium oxysporum* f.sp. *vanillae*
Arecanut and Pepper	*P.meadii, P.capsici*
Arecanut and Cocoa	*P.meadii, P.capsici*
Coffee+Pepper	*F.oxysporum* f.sp. *coffeae, P.capsici, R.similes, M.incognita, Fomes noxius, Rosselina*
Coffee+pepper +cardomom	*Rhizoctonia solani, Pythium vexans*
Tea+Pepper	*Poria lateritia, Phomopsis theae*

TABLE 5.3. Secondary metabolites identified to be involved in disease suppression.

Bacterium	**Disease**	**Metabolite**	**Reference**
Pseudomonas fluorescens	Pythium damping off	Siderophores	Kloepper *et al.*, 1980
Pseudomonas aeruginosa	*Sclerotium rolfsii, Fusarium solani, Xanthomonas axonopodis* pv., *citir, Erwinia carotovora*	Siderophores	Leong, 1986
P.fluorescens	*Gaeumannomyces graminis*	Phenazine 1 -carboxylic acid	Hamden *et al.*, 1991, Thomashow and Weller, 1990
P.fluorescens strain CHAO	*Thielaviopsis basicola*	2, 4 diacetyl phloroglucinol	Keel *et al.*, 1990
P.fluorescens strain Pf-5	*Pythium ultimum, Rhizoctonia solani, Erwinia carotovora* subsp. *atroseptica*	2, 4 diacetyl phloroglucinol	–
P.fluorescens strain PFM-2	*Septoria tritici*	2, 4 diacetyl phloroglucinol	–
P.fluorescens strain	*Xanthomonas oryzae* pv.*oryzae*	Siderophores, HCN	–
P.fluorescens strain	*Ralstonia solanacearum*	Siderophore, antibiotic	–

Contd...

Table 5.3. Contd...

P.fluorescens	*Pythium ultimum*	Antibiotic	–
P.fluorescens strain CHAO	*Thielaviposis basicola*	HCN	Ahl *et al.*, 1986
P.fluorescens strain P-11	*Thielaviposis basicola*	HCN	Pitrowska-Seget, 1995
P.fluorescens strain CRb-26	*Xanthomonas axonopodis pv. malvacearum*	Siderophore, HCN, Phenolic compounds	Mondal, 1998
Bacillus subtilis strain MRb	*Xanthomonas axonopodis pv. vignaeradiateae*	Phenolic compounds	Datta, 2001

TABLE 5.4. Some of the genes associated with PGPR traits.

Gene	**Competency**	**Trait**
Rhi	Rhizosphere competency	Biocontrol
Att	Rhizosphere competency	Biocontrol/ Plant growth
*Chv*A, *Chv*B	Flagellar synthesis, motility and chemotaxis genes	Rhizosphere competence
CYP79B2, *CYP79B3*, *nit* 1, *nit* 4	IAA	Plant Growth Promotion
Ipt, *Zog*1, *Zox* 1	Cytokinin	Plant Growth Promotion
ga 3-1, *ga* 3-2	Gibberellins	Plant Growth promotion
*Phl*A to *Phl*F	2 acetyl phloroglucinol synthesis	Biocontrol
*plt*A to *plt*G, *plt*L, *plt*M and *plt*R	Pyoluteorin	Biocontrol
*hcn*ABC	HCN	Biocontrol
bph	Degradation of polychlorinated biphenyl	Bioremediation
*Cin*R, *Cin*I	N-acetyl homoserine lactone regulation pathway	Gene regulation

was compatible with 9 pesticides (Quinalphois, Triazophos, Imidacloprid, carbaryl 50 WP, chlorpyriphos, Estofenprox, Mancozeb, Hexaconazol, carbendazim). Similarly Laha and Venkatraraman (2001) have reported that *P. fluorescens* is compatible with carbendazim while studying sheath blight management in rice. Hence in future formulations the compatibility with pesticides and botanicals should be studied in order to use PGPR in IDM.

Formulation of the PGPR is a problem and various workers have used talc, peat, chitosan etc. The multiplication of bacteria on Kings B medium is expensive and Renymol (2002) suggested coconut water as a suitable medium for *Trichoderma* while Mathew (2003) suggested use of coconut water for growth of *P.fluorescens*.

Genetically, there are options to modify the PGPR to develop strains which are having good biocontrol properties such as early production of antifungal or antibacterial metabolites. However it is critical to keep in mind public concerns on the use of genetically modified microorganisms. Organizations like Mahyco envisage the future to belong to NISC/Metabugs. Metabugs is a concept which envisages a core competent bug with a number of capabilities including those related to rhizosphere competence, survival related function in which genes for biostimulation, biofertilisation and biocontrol are stacked. Alternatively the NISC (Near Isogenic Strain Complex) may be more viable proposition. This involves the use of near isogenic lines differing only for the set of genes and their regulatory system involved in the specific additional function of bio-stimulation or biocontrol or biofertilisation or bioremediation etc (Shenoy and Kalagudi, 2003).

REFERENCES

Ahl, P., Voisard, C. and Defago, G. 1986. Iron bound siderophores, cyanide and antibiotic involved in suppression of *Thielaviopsis basicola* by a *Pseudomonas fluorescens* strain. *J. Phytopathol.* 116: 121-134.

Anuratha, C. S. and Gnanamanickam, S. S. 1987. *Pseudomonas fluorescens* suppresses development of bacterial blight symptoms. *Int. Rice Res. Newsltr.* 12: 17.

Anuratha, C. S. and Gnanamanickam, S. S. 1990. Biocontrol of bacterial wilt caused by *Pseudomonas fluorescens* in India with antagonistic bacteria. *Plant. Soil.* 124: 109-116.

Arya, S. and Parashar, R. D. 1997. Biological control of bacterial blight of cotton by phylloplane antagonists. *Int. Conf. on Integrated Disease Management for Sustainable Agriculture,* 10-15 November, 1997, New Delhi, India. pp:195 (Abstr.).

Bhowmik, B., Singh, R. P., Jayaraman, J. and Verma, J. P. 2002. Population dynamics of cotton endophytic *Pseudomonas,* their antagonism and protective action against the major pathogens of cotton. *Indian Phytopath.* 55: 124-131.

Burr, T. J. 2001. Future development of chemical and biological controls for bacterial diseases of plants. (*In:*) De Boer, S.H. (Ed.) *Plant Pathogenic Bacteria.* Kluwer Academic Publishers, Netherlands.

Burr, T. J. and Caesaer A. J. 1984. Beneficial Plant Bacteria. CRC *Crit. Rev. Plant Sci.* 2: 1-20.

Comai, L. and Kosuge, T. 1982. Cloning and characterisation of a virulence determinant of *Pseudomonas savastonai. J. Bacteriol.* 149: 40-46.

Cruz, A. R., De, L. A. , Poplawsky, A. R. and Weise, M. V. 1992. Biological suppression of ring rot by fluorescent pseudomonads. *Appl. Environ. Microbiol.* 58: 1986-1991.

Chauhan, M. S. and Chopra, B. L. 1993 Biological control of bacterial blight of cotton in the Northern Zone. *J. Cotton Res. Develop.* 7: 307-312.

Datta, S. 2001. Enzymatic response of mungbean cultivars to *Xanthomonas axonopodis* pv. *vignaeradiateae* and biotic/ abiotic elicitors. *Ph. D. Thesis,* Indian Agricultural Research Institute, New Delhi.

Dube, H. C. 2001. Rhizobacteria in biological control and plant growth promotion. *J. Mycol.Pl. Pathol.* 31: 9-21.

Dylan, T., Telpi, L., Stanfield, S., Kashyap, L., Douglas, C., Yanofsky, M., Nester, E., Helinski, D.R. and Ditta, G. 1986. *Rhizobium meliloti* genes required for nodule development are related to chromosomal virulence genes in *Agrobacterium tumefaciens. Proc. Natl. Acad. Sciences,* U.S.A. 83: 4403-4407.

Gaekwad, S. H. 1994. Role of rhizobacteria on mycorrhizae formation in cereals. *Ph. D. Thesis,* Indian Agricultural Research Institute, New Delhi.

Gnanmanickam, S. S. and Podile, A. R. 2003. An overview of PGPR work in India. *Molec. Plant Microbe Interac.* 20: 43-46.

Goldstein, A. H. and Liu, S. T. 1987. Molecular cloning and characterisation of a mineral phosphate solubilising gene from *Erwinia herbicola. Bio/Technology.* 5: 72-74.

Gutterson, N. I., Layton, T. J., Zingle, J. S. and Warren, G. J. 1986. Molecular cloning of genetic determinants for inhibition of fungal growth by a fluorescent Pseudomonad. *J. Bacteriol.* 165: 696-703.

Hamden, H., Weller, D. M. and Thomashow, L. S. 1991. Relative importance of fluorescent siderophores and other factors in biological control of *Gauemannomyces graminis* var. *tritici* by *Pseudomonas fluorescens* 2-79 and m4-80R. *Appl. Environ. Microbiol.* 57: 3270-3277.

Hofte, H., De Greve, H., Seurinck, J., Janssens, S., Mahillon, J., Ampe, C., Vandekerkhove, J., Vanderbruggen, H., Montagu, M.V., Zabeau, M. and Vaeck, M. 1986. Structural and functional analysis of a cloned endotoxin of *Bacillus thuriengenesis berliner* 1715 *Eur.J.Biochem.* 161: 273-280.

Jayaraman, J. and Verma, J. P. 2002. Current Scenario of bacterial diseases in India and their management. *Ann. Rev. Plant Pathol.* 1: 27-50.

Keel, C., Wirthner, P., Oberhansli, T., Voisard, C., Burger, U., Haas, D. and Defago, G. 1990. Pseudomonads as antagonists of plant pathogen in rhizosphere: role of antibiotic 2, 4 diacetylphloroglucinol in the suppression of black root rot of tobacco. *Symbiosis.* 9: 327-341.

Kloepper, J. W. and Schroth, M. N. 1981. Relationship of *in vitro* antibiosis of plant growth promoting rhizobacteria to plant growth and the displacement of root microflora. *Phytopath.* 71:1020-24.

Kloepper, J. W., Leong, J., Teintze, T. and Schroth, M. 1980. *Pseudomonas* siderophore: A mechanism explaining disease suppressive soils. *Curr. Microbiol.* 4:317-320.

Laha, G. S. 1994. Role of fluorescent pseudomonads in reducing the deleterious microorganisms associated with seeds and seedlings of cotton. *Ph. D. Thesis,* Indian Agricultural Research Institute, New Delhi.

Laha, G. S. and Venkataraman, S. 2001. Sheath blight management in rice with biocontrol agents. *Indian Phytopath.* 54: 461-464.

Lambert, B. and Joos, H. 1989. Fundamental aspects of rhizobacterial plant growth promotion research. *Trends in Biotech.,* 7: 215-219.

Leong, J. 1986. Siderophores: their biochemistry and possible role in biocontrol of plant pathogens. *Annu. Rev. Phytopathol.* 24: 187-209.

Levy, E., Gough, F. J., Berlin, K. D., Guiana, P. W. and Smith, J. T. 1992. Inhibition of *Septoria tritici* and other phytopathogenic fungi and bacteria by *Pseudomonas fluorescens* and its antibiotics. *Plant Pathol.* 41: 335-41.

Liao, C. H. 1989. Antagonism of *Pseudomonas putida* strains PP22 to phytopathogenic bacteria and its potential use as biocontrol agent. *Plant Dis.* 73: 223-226.

Mathew, A. V. 2003. *Pseudomonas fluorescens*-antagonism, compatibility with pesticides and alternate media for mass multiplication *Molec.Plant Microbe Interac.* 20:159-164

Mayer, J. R. and Linderman, R. G. 1986. Response of subterranean clover to dual incouation with VA mycorrhizal fungi and PGPR bacterium *Pseudomonas putida. Soil Biol. Biochem.* 18: 185-190.

Mayo, K., Davis, R.E. and Motta, J. 1986. Stimulation of germination of spores of *Glomus versiforme* by spore associated bacteria . *Mycologia.* 78: 426-431.

Mondal, K. K. 1998. Secondary metabolite of bacteria associated with cotton rhizosphere and their role in bacterial disease of cotton. *Ph.D. Thesis,* Indian Agricultural Research Institute, New Delhi.

Mondal, K. K. and Verma, J. P. 1998. Role of secondary metabolites of *Pseudomonas fluorescens* in the biocontrol of plant pathogens. *(In:)* Singh S.P. and Hussaini, S. S. (Eds.) *Biological Suppression of Plant Diseases , Phytoparasitic Nematodes and Weeds* pp:70-80. Project Directorate of Biological Control, Bangalore, India.

Mondal, K. K., Dureja, P., Singh, R. P. and Verma, J. P. 1998. Secondary metabolites of *Pseudomonas fluorescens* in the suppression of Bacterial Blight of Cotton induced by *Xanthomonas axonopodis* pv. *malvacearum*. *7th International Congress on Plant Pathology*, Edinburgh, August 9-16, 1998, 5.2.79 (Abstr.).

Raaijmakers, J. M. and Weller, D. M. 1998. Natural plant protection by 2, 4 Diacetylphloroglucinol produced by *Pseudomonas* spp. In Take all decline soils. *Molec.Plant.Microbe Interac.* 11: 144-152.

Renymol. 2002. Standardisation of techniques for commercial multiplication of *Trichoderma viride*. A biocontrol fungus. Project report submitted to Bharathiar University for award of degree of M.Sc. in Microbiology.

Ryu, C. M., Farag, M. A., Hu, C. H., Reddy, M. S., Pare, P. W. and Kloepper, J. W. 2003. Volatiles produced by PGPR elicit plant growth promotion and induced resistance in arabidopsis. *Molec..Plant Microbe Interac.*20:93-100.

Pain, A. 1991. Exploitation of Rhizobacteria in the Management of Bacterial Blight of Cotton. *M. Sc. Thesis*, Division of Plant Pathology, Indian Agricultural Research Institute, New Delhi.

Pitrowska-Seget, Z. 1995. The effect of cyanogenic strains of *Pseudomonas fluorescens* on plant growth. *Acta Microbiologica Polonica.* 44: 161-170.

Safiazov, J. S., Manners, R. N. and Sattarova, R. K. 1995. The use of the best antagonists for the control of cotton diseases. *Field Crops Res.* 43: 51-54.

Saha, S., Singh, R.P., Verma, J.P. and Jayaraman, J. 2000. Plasmid-borne determinants of colony morphology, pigmentation, antibiotic resistance and antibiosis in *Pseudomonas* sp. antagonistic to bacterial blight of cotton. *Curr.Sci.* 79:1384-85.

Sakthivel, N., Sivamani, E., Umnamalai, N. and Gnanamanikam, S. S. 1986. Plant growth promoting rhizobacteria in enhancing plant growth and suppressing plant pathogen. *Curr. Sci.* 55: 22-25.

Sarma, Y. R., Krishnakumar, V. and Anand Raj, M. 2003. Scope and Role of PGPR in cropping systems in plantation crops and spices in relation to their agronomy and pathology. *Molec. Plant Microbe Interac.*20:3-20

Shenoy, V. V. and Kalagudi, G. M. 2003. Metabug and Near isogenic strain consortia concepts for plant growth promoting rhizobacteria. *Molec. Plant Microbe Interac.*20: 108-114.

Sivakumar.G. 1999. Management of Banded Leaf and Sheath Blight of Maize caused by *Rhizoctonia solani* fsp. *sasakii* using *Pseudomonas fluorescens. PhD. Thesis.* Indian Agricultural Research Institute, New Delhi.

Sivamani, E., Anuratha. C. S. and Gnanamanickam, S. S. 1987. Toxicity of *Pseudomonas fluorescens* toward bacterial pathogen of banana (*Pseudomonas solanacearum*) and rice (*Xanthomonas campestris pv. oryzae*). *Curr. Sci.* 56: 547-548.

Stolp, H. and Gadkari, D. 1981. Non pathogenic members of the genus *Pseudomonas*. (*In*:) Starr, M.P., Stolp, H., Truper, H.G., Balows, A. and Schlegel, H.G. (Eds.). *The Prokaryotes: A Handbook on Habitats, isolation and identification of bacteria*. Vol.1, pp: 719-741. Springer Verlag, Berlin.

Thomashow, L. S. and Weller, D. M. 1990. Role of antibiotics and siderophores in the biocontrol of take all disease of wheat. *Plant Soil*. 1129: 93-99.

Thomashow, L. S., Mavrodi, D. V., Huang, Z., Raaijmakers, J. M. and Weller, D. M. 1998. Molecular approaches to improve the efficacy of bacterial biocontrol agents. *7th International Congress on Plant Pathology*, Edinburgh, August 9-16, 1998, 3.5.2S (Abstr.).

Unnamalai, N. and Gnanamanickam, S. S. 1984. *Pseudomonas fluorescens* is antagonistic to *Xanthomonas citri* (Hasse) Dye, the incitant of citrus canker. *Curr Sci*. 53: 703-704.

Vidhyasekharan, P. and Muthamilan, M. 1995. Development of formulation of *Pseudomonas fluorescens* for control of chickpea wilt. *Plant Dis*. 79:782-786.

Wabiko, H., Held, G.A. and Bulla, L.A. Jr. 1985. Only a part of the protoxin gene of *Bacillus thuringiensis* subsp. *berliner* 1715 is necessary for insecticidal activity. *Appl. Environ.Microbiol*. 49:706-708.

• • •

Advances in Soil Borne Plant Diseases, 2008
Editors : Naik and Devika Rani, pp. 143-164
New India Publishing Agency, New Delhi

Genomics of *Pseudomonas fluorescens* and Biological Control of Soil Borne Plant Diseases

Dhananjay Naik, M.S. Rao, Abdul Rahiman Bantwal# and M.K. Naik*

INTRODUCTION

Throughout the history of agriculture, humans have struggled to reduce the adverse effects of plant disease on their crops. Early agriculturalists realized the benefits of cultural practices such as crop rotation and the use of organic soil amendments in promoting plant productivity. It is now well established that many of these effects are achieved by promoting the natural microbiological processes that keep plant disease in check. Cultural practices were mainstays of traditional agricultural systems and still provide the primary approaches for management of many soilborne diseases today. For example, disease suppressive soils, into which pathogen(s) can be introduced without causing the expected levels of disease severity, can result from alterations in cropping patterns or other cultural practices. Recently, *Pseudomonas* spp. that produces a specific antibiotic has been associated convincingly with soil

Division of Entomology and Nematology, Indian Institute of Horticultural Research (ICAR), Hessaraghatta Lake P.O., Bangalore 560 089 India

\# Department of Biotechnology, Jnanasahyadri, Kuvempu University, Shankaraghatta Shimoga- 577 451 India.

* Department of Plant Pathology, University of Agricultural Sciences, College of Agriculture Raichur, 580 101

suppressiveness. This compelling evidence for the role of *Pseudomonas* spp. in natural processes of biological control has built even greater enthusiasm for the development of these bacteria as biological control agents

The potential of specific strains of *Pseudomonas* spp. for suppression of plant disease in agriculture has been demonstrated in hundreds, of experiments worldwide. Typically, a collection of strains isolated at random from plant surfaces are inoculated individually onto seeds, roots, or aerial plant tissues and inoculated plants are compared to non-inoculated plants in disease assays. Over and over again, *Pseudomonas* spp. has been identified for their suppressive effects on plant pathogens in these studies. For example, in greenhouse studies in which a random collection of culturable bacteria were screened directly for suppression of Pythium damping-off, fluorescent pseudomonads comprised a large proportion (33% to 100%) of the effective biocontrol strains (Loper, 1988). There is no question that strains representing diverse genera of gram-negative and gram-positive bacteria can suppress plant disease, but fluorescent pseudomonads are typically among the most effective antagonists selected for suppression of both soilborne and aerial diseases of plants (Loper *et al.*, 1997). Approximately one third of the articles coded for the subjects of biological control and plant disease in the Agricola database involve *Pseudomonas* spp. Therefore, the genus has been the focus of ecological and mechanistic research evaluating biological control of plant disease.

Pseudomonas fluorescens belongs to Plant Growth Promoting Rhizobacteria (PGPR), the important group of bacteria, which play a major role in the plant growth promotion, induced systemic resistance, biological control of pathogens etc. PGPR are known to enhance plant growth promotion and reduce severity of many fungal diseases (Hoffland *et al.*, 1996 and Wei *et al.*, 1996). These PGPR employ different mechanisms to arrest the fungal growth. Production of lytic enzymes like chitinase and β-1, 3-glucanase by the PGPR strains is considered a major antagonistic property of strains. These lytic enzymes act on chitins and glucans, which are the major constituents of the cell wall of the majority of the fungi except the oomycetes group. Hydrolytic action of these enzymes results in degradation of the fungal cell wall. Antifungal effects of chitinase produced by many bio-control agents like *Trichoderma* spp. and *Pseudomonas* spp. on plant pathogens have been reported (Haran *et al.*, 1996; Krishnamurthy, 1996; Krishnamurthy *et al.*, 1999).

Most research dedicated to improve crop yields and plant growth with microbial inoculants has focused on symbiotic bacteria. However, free-living soil micro-organisms, such as *Azospirillum* and various Pseudomonads (Plate 6.1), also might be used to promote plant growth. PGPR are rhizobacteria that have the ability to promote the growth of plants following inoculation onto seeds or subterranean plant parts. PGPR mediate improved plant growth either directly by stimulation of the plant or indirectly through disease control of pathogens or triggering and inducting of host defense mechanism.

ANTIBIOTICS PRODUCED BY *P. FLUORESCENS*

Pseudomonas fluorescens produces the antibiotics 2,4-diacetylphloroglucinol (DAPG), phenazine-1-carboxylic acid (PCA), phenazine (Phe), antimicrobial metabolites such as pyoluteorin (Plt), hydrogen cyanide (HCN), siderophores pyoverdine (Pvd), salicylic acid (Sal) and pyochelin (Pch) which inhibit a broad spectrum of plant pathogenic fungi, bacteria, nematode (Raaijmakers *et al.*, 1997; Schmidli *et al.*, 1997; Boruah and Kumar, 2002) and control a variety of root and seedling diseases. There are many evidences substantiating the importance of DAPG in biological control. The production of DAPG is governed by Phl gene. The population size of DAPG producers in the rhizosphere is correlated with the disease suppressiveness of the soil and *in situ* antibiotic production. The diverse DAPG producing *Pseudomonas* spp. has been isolated from the rhizosphere of various crop plants and their role in promoting plant growth and inhibiting root diseases are the subjects of ongoing investigations worldwide.

PHL GENE AND 2, 4-DIACETYLPHLOROGLUCINOL (DAPG) PRODUCTION IN *P. FLUORESCENS*

DAPG is the major determinant in the biological control of a range of plant pathogens by many fluorescent *Pseudomonas* spp. Phl A, B, C, D genes are responsible for the production of DAPG in *P. fluorescens* (Abbas *et al.*, 2002). A 4.8 - Kb chromosomal DNA region from *P. fluorescens* carrying Phl biosynthetic gene is used as a probe to determine Phl biosynthetic locus conserved within Phl producing *Pseudomonas* strains of worldwide origin. The Phl producers have considerable phenotypic and genotypic diversity (Plate 6.2). Two phenotypically distinct groups are detected. The first group produces Phl, pyoluteorin, and hydrogen cyanide and consisted of 13 strains from almost all locations sampled in the United States, Europe, and Africa. The second produced only Phl and HCN and consisted of 32 strains from the U.S. and European soils.

To characterize the abundance and diversity of these functionally important bacterial populations the reliable technique is gene based polymerase chain reaction. The Phl-D gene is predicted to encode a polyketide synthase that synthesis the monoacetyle phloroglucinol (MAPG), the immediate precursor of 2, 4-DAPG antibiotic.

Molecular tools for the selection of *P. fluorescens* strains based on Phl gene and chitinase productions are of paramount importance. This is reliable, faster and easy to screen the best strains for disease management (Plate 6.3 and 6.4). This review highlights the gene-based detection of *P. fluorescens* for antibiotic production and chitinase production and mechanism involved in the biological control of soil borne diseases by a versatile antagonistic agent *P. fluorescens*.

Interest in biological control of plant pathogens has been stimulated in recent years by trends in agriculture towards greater sustainability and public concern about the use of hazardous pesticides. There is now unequivocal evidence that antibiotics play a key role in the suppression of various soil borne plant pathogens by antagonistic micro organisms. The significance of antibiotics in biocontrol, and more generally in microbial interactions, often has been questioned because of the indirect nature of the supporting evidence and the perceived constraints to antibiotic production in rhizosphere environments. Reporter gene systems and bio-analytical techniques have clearly demonstrated that antibiotics are produced in the spermosphere and rhizosphere of a variety of host plants. Several abiotic factors such as oxygen, temperature, specific carbon and nitrogen sources, and micro elements have been identified to influence antibiotic production by bacterial biocontrol agents. Among the biotic factors that may play a determinative role in antibiotic production are the plant host, the pathogen, the indigenous microflora, and the cell density of the producing strain (Raijmakers *et al.*, 2002).

Screening of *Phl* gene positive *Pseudomonas fluorescens*, encoding for 2-4- diacetylphloroglucinol (dapg), a broad spectrum antibiotic

Pseudomonas fluorescens produces a broad-spectrum antibiotic 2, 4-Diacetylphloroglucinol (DAPG). This is a major determinant in the biological control of a range of plant pathogens by many fluorescent *Pseudomonas* spp. Production of DAPG is governed by *Phl* gene. The *Phl*-A, B, C and D genes are responsible for the biosynthesis of the antifungal and antibacterial metabolite 2, 4-DAPG (*Phl*) by the bio-control strain *P. fluorescens* (Abbas *et al.*, 2002).

Phl producing fluorescent *Pseudomonas* spp. are the key components of the natural biological control that operates in take-all-suppressive soils. The production of antifungal compounds plays an important role in the suppressive properties of *P. fluorescens* strain CHA0 and it is suggested that over production of these compounds may have deleterious effects on plants. Haas *et al.*, (1991) and Maurhofer *et al.*, (1992) concluded that, depending on the host-pathogen system, enhanced antibiotic production by *P. fluorescens* may result in improved disease suppression or in contrast a toxic effect on the plant.

Antibiotic DAPG and hydrogen cyanide (HCN) individually contribute to the suppression of black root rot. The GacA gene is directly followed by a uvrC gene which suggested that GacA, perhaps is activated by conditions of restricted growth, thus functions as a global regulator of secondary metabolism in *P. fluorescens* (Laville *et al.*, 1992) for the production of DAPG.

P. fluorescens strain CHA0 produces the antimicrobial metabolites 2, 4-DAPG (*Phl*), pyoluteorin (Plt) and hydrogen cyanide, siderophores pyoverdine (Pvd), salicylic acid (Sal) and pyochelin (Pch) which are important for disease suppression in plants (Schmidli-Sacherer *et al.*, 1997).

Role of antibiotics phenazine-1-carboxylic acid (PCA) and *Phl* in the bio-control activity of this fluorescent *Pseudomonas* spp. support the earlier evidence that mechanisms in addition to PCA are responsible for suppression of take-all by strain 2-79. Other workers carried out similar studies and the results revealed that the ability of certain fluorescent *Pseudomonad* strains to suppress plant fungal diseases has been linked to production of secondary metabolites such as 2, 4-DAPG (*Phl*). Genetic modification of bio-control strains for increased production of antifungal metabolites has allowed the development of strains with improved bio-control capabilities (Lohrke *et al.*, 1997).

DAPG producing *P. fluorescens* F113 is proposed as a potential bio-control inoculant for the protection of potato crops against the potato cyst nematode, *Globodera rostochiensis* (Cronin *et al.*, 1997; Raaijmakers and Weller, 1998).

Sharifi-Tehrani *et al.* (1998) reported that fluorescent *Pseudomonads* producing the antimicrobial compound 2, 4-DAPG (*Phl*) are being studied for use as bio-control agents of soilborne fungal diseases. Some of them can produce pyoluteorin (Plt) in addition to *Phl*, whereas others synthesise only *Phl*. There was no correlation between protection *in vivo* and inhibition of the corresponding fungal pathogen on plates. They suggested that promising bio-control *Pseudomonads* may be identified based on the ability to produce *Phl* and/or specific amplified ribosomal DNA restriction analysis (ARDRA) based fingerprints. Analysis of restriction patterns of genomic DNA obtained after hybridization with the *Phl* gene probe and cluster analysis of restriction patterns of amplified DNA coding for 16S rRNA (ARDRA) and randomly amplified polymorphic DNA (RAPD) markers indicated that the strains that produce both *Phl* and pyoluteorin were genetically highly similar (Keel *et al.*, 1996).

Transposon mutagenesis of strains *P. fluorescens* F113 and W81 has demonstrated that the bio-control abilities of these strains were mediated by 2, 4-DAPG (*Phl*) or lytic enzyme production, respectively (Dunne *et al.*, 1998).

Duffy *et al.*, (1998) studied the effect of fusaric acid on the production of DAPG by *P. fluorescens*. He reported that, fusaric acid (produced *by F. oxysporum*) repressed the production of the antibiotic 2, 4-DAPG (*Phl*) by some but not all *P. fluorescens* bio-control strains.

Duffy and Defago (1999) studied the liquid culture assays to identify several minerals and carbon sources, which had a differential influence on the production of the antibiotics 2, 4-DAPG, pyoluteorin (Plt), the siderophores, salicylic acid and pyochelin by the model strain CHA0.

Raaijmakers *et al.*, (1999) gave the protocol for high pressure liquid chromatography/mass spectrometry that allowed specific identification and quantification of the antibiotic 2,4-DAPG (*Phl*) produced by naturally occurring fluorescent *Pseudomonas* spp. on roots of wheat (cv. *Penewawa*) grown in a soil suppressive to take-all (*Gaeumannomyces graminis* var. *tritici*) of wheat. Bonsall *et al.*, (1997) developed a protocol to readily isolate and quantify *Phl* from broth and agar cultures and from the rhizosphere environment of plants.

Boruah and Kumar (2002) tested the production of antibiotics phenazine (Phe), 2, 4-DAPG (*Phl*), and siderophore pyoverdin (Pyo) in standard King's B and succinic acid media, respectively. They studied the *in vitro* antibiosis of the metabolites, their antibacterial and antifungal activity against bacterial test organisms *Corynebacterium* spp., *Mycobacterium phlei*, and *M. smegmatis*, and test fungi *F. moniliforme* and several other *Fusarium* spp.

Genetically modified *Pseudomonas putida* WCS358r, are known to have improved activity against soil borne pathogens (Bakker *et al.*, 2002). Two genetically modified derivatives carried the *Phz* or the *Phl* biosynthetic gene loci and constitutively produced either the antifungal compound phenazine-1-carboxylic acid (PCA) or the antifungal and antibacterial compound 2, 4-DAPG. Effects of the genetically modified bacteria on the rhizospheric fungi and bacteria were analysed by using amplified ribosomal DNA restriction analysis (ARDRA). Introduction of the genetically modified bacterial strains caused a transient change in the composition of the rhizosphere microflora. However, introduction of the genetically modified organisms (GMOs) did not affect the several soil microbial activities that were investigated in the study.

Molecular detection of *phl* gene using gene specific primers

The antibiotic 2, 4-DAPG (*Phl*) is an important factor in the biological control of many soil borne diseases by fluorescent *Pseudomonas* spp. A 6.5 kb genomic DNA fragment from *P.fluorescens* Q2-87 conferred production of *Phl* and of a red pigment distinct from *Phl,* but which typically is present when *Phl* is produced, upon all 13 *Phl* nonproducing recipient *Pseudomonas* strains into which it was introduced. The size of the genomic region involved in *Phl* production and the consistency and specificity, with which these sequences transferred *Phl* biosynthetic capability, support the conclusion that the 6.5-kb fragment contains the *Phl* biosynthetic locus (Bangera and Thomashow, 1996).

McSpadden *et al.,* (2001) developed the new primers for the easy detection of *Phl-D* +ve strains based on cloning and sequence analysis of *Phl-D* open reading frame from five genotypically distinct strains of *P. fluorescens.*

Isolation, cloning and sequence analysis of *phl* gene

Tn5 mutagenesis and complementation analysis were used to clone a 6-kb genomic fragment required for biosynthesis of 2, 4-DAPG (*Phl*) from *fluorescent Pseudomonas* spp. str. F113. A recombinant plasmid, pCU203, containing this region partially complemented a *Phl* production-negative mutant (F113G22) derived from str. F113. Genetically modified strain of *P. fluorescens,* M114. Str. M114 (pCU203) showed enhanced antagonism towards *Pythium ultimum in vitro* and significantly increased the emergence of sugarbeet seeds in the same soil compared with emergence induced by the parent strain M114 (Fenton *et al.,* 1992).

Maurhofer *et al.,* (1995) reported that introduction of the recombinant cosmid pME3090 into *P. fluorescens* strain CHA0 increased 3 to 5 fold the production of the antibiotic metabolites pyoluteorin (Plt) and 2, 4-DAPGl (*Phl*) *in vitro.*

A derivative of *P. fluorescens* strain CHA0 with a mutation in the global regulator gene (GacA), which is unable to produce *Phl,* Plt and HCN, failed to protect cucumber against damping-off caused by *Pythium ultimum.* A functional gacA gene was necessary for the protection of dicotyledons against root diseases, but not for that of Gramineae. Results also indicated that Pch and/or Sal were involved in the ability of the GacA- Pvd- mutant of CHA0 to suppress root diseases in *Poaceae* (Schmidli-Sacherer *et al.,* 1997).

The rifampicin resistant bio-control strain of *P. fluorescens* CHA0-Rif and its derivative CHA0-Rif/pME3424, has improved the bio-control activity and thus enhanced production of the antibiotics 2,4-DAPG (*Phl*) and pyoluteorin (Plt). The introduced strains did not significantly affect the abundance of dominant genotypic groups of culturable bacteria discriminated by restriction analysis of amplified 16S rDNA of 2500 individual isolates. In micro organisms with a synthetic bacterial community, both bio-control strains reduced the population of a strain of *Pseudomonas* but did not affect the abundance of other bacterial strains including 2 highly antibiotic-sensitive isolates (Natsch *et al.*, 1998).

CHARACTERIZATION OF CHITINASE ISOFORMS PRESENT IN LOCAL COLLECTIONS OF *PSEUDOMONAS FLUORESCENS*

Viswanathan and Samiyappan (2001) reported that the cell free culture filtrate of *P. fluorescens* grown on chitin-amended medium significantly inhibited the mycelial growth of *Colletotrichum falcatum*. This cell free culture filtrate contained 3-7 polypeptides and Western blotting analysis revealed that 3-5 chitinase isoforms are present in the filtrate.

Rangaraj *et al.*, (2002) reported the 35 kDa chitinase induced by *P. fluorescens* in rice plants purified and tested the antifungal activity against mycelium and sclerotia of *Rhizoctonia solani*. The production of antibiotics, antifungal compounds, bacterial lytic enzymes and the induction of plant chitinase were correlated with antifungal activity against sclerotia of *R. solani*.

Ehteshamul *et al.*, (1997) reported the significant reduction in the infection of *Meloidogyne javanica* on chickpea in pot experiments in the soil amendment with crustacean chitin. Chen-Chong Shun *et al.*, (2000) reported the effect of purified chitinase extracted from *Vigna unguiculata* sub spp. *sesquipedalis* on *F. oxysporum* f.sp. *niveum*. He reported the greatest inhibition of two forms of *F. oxysporum* at 10µg/ml.

The antagonistic (antifungal) activity of crude chitinase was found to be located in a low molecular weight fraction of the enzyme, which does not possess chitinase activity. Both crude and purified chitinase were able to lyse the cell walls of intact mycelium. Low molecular weight non enzymatic substances largely determine antagonistic activity of *Bacillus* spp. against micromycetes, whereas the role of chitinase is to utilize chitin, which is ubiquitously present in soil (Melent *et al.*, 2001).

Trichoderma spp. has been described as antagonistic agents against various phytopathogenic fungi. This ability makes *Trichoderma* a healthier, safer and environmentally friendlier biological choice, compared with polluting chemical pesticides, usually employed in agriculture. Mycoparasitism is one of the mechanisms involved in antagonistic processes, based on activity of extracellular lytic enzymes produced by *Trichoderma* and has been proven as a good biological agent. Molecular characterization of lytic isoenzymes was achieved by electrophoretic separation in renaturalizing conditions.

Nematophagous fungi *Pochonia chlamydosporia* and *P.suchlasporium* growing on colloidal chitin produced increasing chitinolytic activity over time. Both fungi displayed higher chitinolytic activity in cultures supplemented with yeast extract. Major endochitinases with optimum pH of 5.5 for both fungi were separated by affinity chromatography on a macroporous cross linked chitin matrix (Tikhonov *et al.*, 2001).

Frankowski *et al.*, (2001) conducted the study to determine enzymatic properties of the endochitinase and the N-acetyl-beta-1, 4-D-hexoseaminidase from *Serratia plymuthica*. They correlated the effect of temperature and pH on the efficacy of chitinase in inhibiting the pathogen.

Dana *et al.*, (2001) reported the regulation of expression of chit 33 using a 1.4-kb fragment of the 5'-upstream sequences of chit 33 in *Aspergillus niger* goxA (glucose oxidase encoding) gene. Ramanathan *et al.*, (2002) studied the inhibitory effect and growth promoting effect of talc based *P. fluorescens* on the mycelial growth of ragi (*Eleusine coracana*) blast fungus, *Magnaporthe grisea*. Shternshis *et al.* (2002) reported the 4-fold decrease in the severity of midge blight by the spray of 1% chitinase. He also reported the significant suppression of the independent spur blight by chitinase and phytoverm.

The study of seed-specific chitinases has contributed significantly to current knowledge of this ubiquitous protein family, including antifungal properties, structure, specificity, and catalytic mechanism. Indeed, the first plant chitinase for which a three-dimensional structure was solved had been isolated from barley seeds. Moreover, the finding that a chitinase could rescue a somatic embryo mutant was the first evidence of a non-defensive function. Several lines of evidence have substantiated the biotechnological potential of chitinases to counter plant fungal disease. Specific chitinases accumulate in seeds of many species as part of their normal developmental programme. Some chitinases can also be induced in developing and germinating seeds in response to microbial attack. All known seed isoforms belong to classes I, II, IV and VII, which are encoded by *Chi*-genes, as well as to the more divergent class III encoded by *Chi*-b genes (Gomez *et al.*, 2002).

Takikawa *et al.*, (2002) studied *Enterobacter cloacae* based on chitinase gene transformation and lytic infection by virulent bacteriophages; a phylloplane strain of *E. cloacae* was isolated from tomato cv. *Moneymaker* leaves and transformed with a chitinase gene. The chitinase gene was used as a reporter gene to construct a new system for easy and rapid monitoring of transgenic strains of *E. cloacae* released in the environment, in combination with specific recognition by virulent bacteriophages.

Nawani *et al.*, (2002) purified and characterized the chitinase from *Microbispora* spp.by gel filtration chromatography. They reported that the chitinase had a molecular weight of 35 kDa and showed maximum activity towards p-nitrophenyl-beta-D-N,N'-diacetylchitobiose, indicating a chitobiosidase activity. The chitinase from *Microbispora* spp. may have potential applications in the recycling of chitinous wastes, particularly due to its thermophilic and acidophilic character.

Liu *et al.*, (2002) studied the chitinolytic activities of *Bacillus thuringiensis* through SDS-PAGE and Western blotting and showed that the chitinase produced by some *B. thuringiensis* strains had a molecular weight of about 61 kDa. The bioassay results indicated that the chitinase producing *B. thuringiensis* strains could enhance the insecticidal activity of *B. thuringiensis* strain DL5789 against *Spodoptera exigua* larvae, with an enhancing ratio of 2.35-fold. Chitinase was widely produced in *B. thuringiensis* strains and some of the strains could enhance the toxicity of active *B. thuringiensis* strain. Chitinase produced by *B. thuringiensis* might play a role in the activity of the biopesticide.

Phenyl alanine lyase (PAL), chitinase and beta-1, 3-glucanase activities increased significantly from 3 days after *P. fluorescens* treatment and reached the maximum 6 days after treatment. Challenge inoculation with *F. oxysporum* further increased the enzyme activities. These results suggest that the enhanced activities of defence enzymes and elevated content of phenolics may contribute to bioprotection of banana plants against *F. oxysporum* (Thangavelu *et al.*, 2003).

GENOMICS OF *P. FLUORESCENS*-PF5, A PARADIGM STRAIN

The selection of *P. fluorescens* Pf-5 as the subject for genomic sequencing was based upon the importance of Pf-5 as a biological control organism, its rhizosphere competence, the large spectrum of antibiotics and other secondary metabolites that it produces, and its status as model environmental strain for studies of gene regulation. Strain Pf-5 was described early in the history of biological control research (Howell and Stipanovic 1979), and it was the first biological control agent for which the chemical basis of disease suppression was known.

Soil borne diseases suppression by *P. fluorescens*- strain PF5.

Pf-5 was first described for its capacity to suppress soilborne diseases of cotton caused by *Rhizoctonia solani* (Howell and Stipanovic, 1979) and *Pythium ultimum* (Howell and Stipanovic, 1980). These pathogens are the major cause of seed rot and seedling death throughout much of the cotton Belt in the

USA. Seedling diseases caused an estimated loss of 828,000 bales of cotton nationwide in 1995, with similar losses in previous years. *R. solani* and *P. ultimum* are widespread pathogens with broad host ranges that constrain food and fiber production worldwide (Martin and Loper, 1999). Since it was first described, Pf-5 has been shown to suppress these pathogens on plant hosts including cucumber, pea, and maize (Kraus and Loper, 1992; Henkels and Loper, unpublished). Furthermore, Pf-5 suppresses a number of other soilborne or residue-borne fungal pathogens. When inoculated onto wheat straw residue, Pf-5 suppresses ascocarp formation by the tan spot pathogen of wheat, *Pyrenophora tritici-repentis,* thereby decreasing inoculum available for infection of the subsequent wheat crop (Pfender *et al.*, 1993). On turfgrass, Pf-5 suppresses dollar spot caused by *Sclerotinia homoeocarpa* and leaf spot caused by *Dreschslera poae,* which are destructive and widespread diseases affecting golf courses, home lawns, and amenity turf areas (Rodriguez and Pfender, 1997). Pf-5 also suppresses Fusarium crown and root rot of tomato, caused by *Fusarium oxysporum* f. sp. *radicis-lycopersici* (Sharifi-Tehrani *et al.*, 1998) and seed piece decay of potato caused by the bacterial pathogen *Erwinia carotovora* (Xu and Gross, 1986).

Because of the wide variety of diseases suppressed by Pf-5, it is well-recognized by biological control researchers around the world. Pf-5 commonly is included as a reference strain in studies of biocontrol agents (as examples, Keel *et al.*, 1996; Kloepper, 1991; McSpadden-Gardener *et al.*, 2000; Sharifi-Tehrani *et al.*, 1998; and Xu and Gross, 1996).

The power of genomics is being applied to biological control by sequencing the complete genome of the biological control organism *Pseudomonas fluorescens* Pf-5. Strain Pf-5 was selected as a model biocontrol strain for a genomic sequencing project because of its success as a biocontrol agent, its use as a standard biocontrol strain by many researchers, and the body of knowledge that exists on the mechanisms by which Pf-5 suppresses soil borne plant disease. Pf-5 was first described for its capacity to suppress soil borne diseases of cotton caused by *Rhizoctonia solani* (Howell and Stipanovic 1979) and *Pythium ultimum* (Howell and Stipanovic 1980). Since it was first described, Pf-5 has been shown to suppress these pathogens on plant hosts including cucumber, pea, and maize. Pf-5 also suppresses a number of other soil borne or residue-borne pathogens such as *Pyrenophora tritici-repentis, Sclerotinia homoeocarpa* and *Dreschslera poae, Fusarium oxysporum* f. sp. *radicis-lycopersici* and *Erwinia carotovora*. Pf-5 produces a suite of antibiotics including pyrrolnitrin (Howell and Stipanovic 1979), pyoluteorin (Howell and Stipanovic 1980; Nowak-Thompson *et al.* 1999), and 2, 4-diacetylphloroglucinol (Nowak-Thompson *et al.* 1994). Each of these compounds has a different spectrum of activity against plant pathogens, and their roles in biological control have been established in various biological control organisms. Pf-5 is remarkably similar to *P. fluorescens* strain CHA0, a rhizosphere bacterium isolated in Switzerland, in its biological control activities and its antibiotic production. Strains Pf-5 and CHAO have been developed as model organisms for the study of biological control, rhizosphere biology, and antibiotic regulation in gram-negative bacteria, and the collective knowledge accumulated on these strains is unsurpassed among environmental isolates of *Pseudomonas* spp. Genomic sequencing of Pf-5 began in March of 2002 at The Institute of Genomics Research, under the direction of Dr. Ian Paulsen, and was completed in September of 2003. Currently, the genome is being annotated to assign functions to as many genes in the genome of Pf-5 as possible. Completion of genome annotation is expected in spring of 2004. The genome of Pf-5 is estimated at 7.3 Mega bases, which is 0.6 to 1.3 Mega bases larger than genomes of other *Pseudomonas* spp. that have been sequenced to date. As expected from the spectrum of antibiotics produced by Pf-5, many biosynthetic gene clusters for secondary metabolite production are present in the genome. The prevalence of large gene clusters for secondary metabolite production must contribute to the large size of the Pf-5 genome. Genes known to contribute to biological control of Pf-5 are distributed throughout the genome; they are not clustered in groups analogous to the pathogenicity islands found in pathogenic *Pseudomonas* spp. There is no evidence for the presence of genes involved in pathogenicity (i.e., phytotoxins, Type III secretion systems, effectors, etc.) in the genome of Pf-5; such genes are found in the genomes of phytopathogenic *P. syringae*. Therefore, there are significant differences in the genomes of phytopathogenic and biocontrol strains of *Pseudomonas* spp. Twenty one genes encoding outer-membrane receptors for siderophores or other iron-uptake systems are found in the genome of Pf-5, suggesting the importance of iron-acquisition in the ecology of this bacterium. Post-genomics technology offers great promise for enhancing knowledge of biological control. To this end, we have developed a low-density micro array in which factors influencing the expression of genes required for biological control are being evaluated. Knowledge of factors influencing the expression of biological control genes is needed to understand the currently-unexplained variation in biological control that can occur from field to field. Variable efficacy poses a serious obstacle to the development of biocontrol products by industry and the use of these products by growers. More reliable biocontrol products and processes can be developed if sources of variation are identified and strategies for decreasing variability are implemented.

Broad spectrum antibiotic production by Pf-5

Pf-5 produces a broad spectrum of antibiotics. Pf-5 produces a suite of antibiotics including pyrrolnitrin, pyoluteorin, and 2, 4-diacetylphloroglucinol; it also produces hydrogen cyanide and the siderophores pyochelin and pyoverdine. Each of these compounds has a different spectrum of activity against plant pathogens, and their roles in biological control have been established in various biological control organisms. The spectrum of antibiotics produced by Pf-5 is remarkably similar to that produced by another well-known biological control strain, *P. fluorescens* CHA0, which was isolated in Switzerland (Stutz *et al.*,1986).

The genetics of pyoluteorin production has been investigated in strain Pf-5. Pyoluteorin is composed of a resorcinol ring linked to a bichlorinated pyrrole moiety. It inhibits the growth of Oomycetes, including plant pathogens in the genus *Pythium* (Howell and Stipanovic, 1980). Pyoluteorin is derived from proline, the primary precursor to the dichloropyrrole ring, and three molecules of acetyl-coA, which are incorporated into the resorcinol ring via a polyketide mechanism of biosynthesis (Nowak-Thompson, *et al.*, 1997). 24-kb of the pyoluteorin gene cluster in strain Pf-5 has been sequenced and the deduced protein products of identified ORFs correspond logically to the proposed biochemical transformations required for pyoluteorin synthesis (Nowak-Thompson *et al.*, 1997, 1999). Halogenation of the pyrrole ring in pyoluteorin appears to involve the pyoluteorin biosynthetic genes pltA, pltD, and pltM. These genes encode members of a newly-described class of enzymes that chlorinate secondary metabolites, a function previously ascribed to haloperoxidases.

Unlike the synthases that participate in the well-described pathways for polyketide synthesis in actinomycetes, a combination of mono- and multi-functional enzymes appears to be responsible for the synthesis of the resorcinol moiety of pyoluteorin (Nowak-Thompson *et al.*, 1997, 1999). Thus, the biosynthetic gene cluster for pyoluteorin offers some striking contrasts to the general classes of polyketide synthases and chlorinating enzymes established for gram-positive micro organisms.

Regulation of antibiotic production in Pf-5

Regulation of antifungal metabolite production by biological control strains of *Pseudomonas* spp. is controlled by complex regulatory networks that respond to environmental and density-dependent signals and are coupled to the physiological status of the bacterium. Loci that regulate the production of antifungal metabolites in Pf-5 and the closely-related strain CHA0 include a two-component regulatory system encoded by gacS and gacA (Corbell and Loper, 1995; Gaffney *et al.*, 1994; Laville *et al.*, 1992), the sigma factors S (Sarniguet *et al.*, 1995) and 70 (Schnider *et al.*, 1994), the cofactor PQQ (Schnider *et al.*, 1995), the anaerobic regulator anr (Laville *et al.*, 1998) the protease Lon (Whistler *et al.* ,2000) and ptsP, a paralog of sugar phosphotransferase enzyme I (Whistler *et al.*, 2000). Many of these loci control multiple phenotypes including stress response in *P. fluorescens* (Sarniguet *et al.*, 1995; Whistler *et al.*, 1998, 2000), indicating that regulation of antibiotic production is intricately enmeshed in the physiology of the bacterial cell.

Pf-5 is representative of other important biological control agent

Biological control strains of *P. fluorescens* have been isolated from soils and plant surfaces all over the world, and they comprise a diverse group of environmental isolates. Nevertheless, biological control strains do fall into groups defined by biochemical and molecular criteria. Studies evaluating collections of biocontrol strains by BOX- and ERIC-PCR (McSpadden-Gardener *et al.*, 2000) or restriction analysis of amplified spacer ribosomal DNA (Keel *et al.*, 1996; Sharifi-Tehrani *et al.*, 1996) consistently place Pf-5 in a group with CHA0 and other pyoluteorin and phloroglucinol-producing biocontrol strains.

MAINTENANCE OF EFFICIENT STRAIN OF *P. FLUORESCENS* IN FIELD CONDITION

Two diverse sequence areas were found through DNA sequencing of 16S-23S rDNA spacer regions in five closely related *Pseudomonas fluorescens* and *P. putida* biocontrol bacteria. The wild type *Pseudomonas fluorescens* biocontrol bacteria can be separated from all others by using a specific primer (5TCGCGATTAAACTCAC3') and the PCR method. We can now use this method to follow the biocontrol bacterium in the soil. Its population can be detected using the most probable number PCR method.

REFERENCES

Abbas, A., Morrissey, J. P., Marquez, P. C., Sheehan, M. M., Delany, I. R. and O' Gara, F. 2002. Characterization of interactions between the transcriptional repressor PhlF and its binding site at the phlA promoter in *Pseudomonas fluorescens* F113. *J. Bacteriol.* 184: 11, 3008-3016.

Bakker, P.A.H.M., Glandorf, D. C. M., Viebahn, M., Ouwens, T.W.M., Smit, E., Leeflang, P. Wernars, K., Thomashow, L. S., Thomas, Oates, J. E., Loon, L.C, van-Loon-L. C. and Veen, J.H. van (Ed.); Laanbroek, H. J. (Ed.); Vos-W.M-de 2002. Effects of *Pseudomonas putida* modified to produce phenazine-1-carboxylic acid and 2, 4-

diacetylphloroglucinol on the microflora of field grown wheat. *Proceedings of the 9th International Symposium on Microbial Ecology*, Amsterdam, Netherlands, August 2001. *Antonie van Leeuwenhoek*. 81: 1-4, 617-624.

Bangera, M. G. and Thomashow, L. S. 1996 . Characterization of a genomic locus required for synthesis of the antibiotic 2, 4-diacetylphloroglucinol by the biological control agent *Pseudomonas fluorescens* Q2-87. *Molecular Plant Microbe interactions*. 9: (2) 83-90.

Bonsall, R. F., Weller, D. M., and Thomashow, L. S. 1997. Quantification of 2, 4-diacetylphloroglucinol produced by fluorescent *Pseudomonas* sp. *in vitro* and in the rhizosphere of wheat. *Applied and Environmental Microbiology*. 63:951-955. Vol. 93, No. 1, 2003 63.

Boruah, H. P. D. and Kumar, B. S. D., 2002. Biological activity of secondary metabolites produced by a strain of *Pseudomonas fluorescens*. *Folia Microbiologica*. 47 (4) 359-363.

Chen Chong Shun, Zhu Xue Feng, Yu Zhi Fang, Chen, C. S., Zhu-X. F. and Yu-Z. F.2000. Effect of a purified chitinase from *Vigna sesquipedalis* on fungi. Acta-*Phytophylacica Sinica*. 27 (4) 375-376.

Corbell, N. and J. E. Loper. 1995. A global regulator of secondary metabolite production in *Pseudomonas fluorescens* Pf-5. *J. Bacteriol*. 177:6230-6236.

Cronin, D., Moenne-Loccoz, Y., Fenton, A., Dunne, C., Dowling, D. N. and O'-Gara, F.1997. Role of 2,4-diacetylphloroglucinol in the interactions of the biocontrol pseudomonad strain F113 with the potato cyst nematode *Globodera rostochiensis*. *Applied and Environmental Microbiology*. 63 (4):1357-1361.

Dana, M., Benitez, T., Kubicek, C.P., Pintor, Toro, J. A. and Elad, Y. (ed.); Freeman, S. (ed.); Monte, E. 2001 . Chitinase 33 gene expression in *Trichoderma harzianum* during mycoparasitism. IOBC-WPRS Working Group "Biological Control of Fungal and Bacterial Plant Pathogens". *Proceedings of the sixth meeting, Biocontrol Agents: Mode of Action and Interaction with Other Means of Control*, Sevilla, Spain, November 30 - December 3, 2000. Bulletin-OILB-SROP. 2001, 24: 3, 363.

Duffy, B. K., Defago, G. and Duffy, B., Rosenberger, U. (Ed.); Defago-G 1998. A *Fusarium* pathogenicity factor blocks antibiotic biosynthesis by antagonistic pseudomonads. *Molecular approaches in biological control*. Delemont, Switzerland, 15-18 September, 1997. Bulletin-OILB-SROP. 1998. 21(9): 145-148.

Duffy, B. K. and Defago, G.1999. Environmental factors modulating antibiotic and siderophore biosynthesis by *Pseudomonas fluorescens* biocontrol strains. *Applied and Environmental Microbiology*. 65 (6): 2429-2438.

Dunne,C., Cronin, D., Moenne-Loccoz, Y., O'-Gara, F.and Duffy,B. Rosenberger, U.; Defago, G. 1998. Biological control of phytopathogens by phloroglucinol and hydrolytic enzyme producing bacterial inoculants. *Molecular approaches in biological control*. Delemont, Switzerland, 15-18 September, 1997. *Bulletin-OILB-SROP*. 1998, 21(9) 19-25.

Ehteshamul-Haque, S., Sultana, V., Ara, J., Qasim, R. and Ghaffar, A. 1997. Use of crustacean chitin and plant growth promoting bacteria for the control of *Meloidogyne javanica* root knot nematode in chickpea. *Pakistan. J. Nematolo.* 15 (1-2) 89-93.

Fenton, A. M., Stephens, P.M., Crowley, J., O'-Callaghan, M. and O'-Gara, F. .1992. Exploitation of gene(s) involved in 2,4-diacetylphloroglucinol biosynthesis to confer a new biocontrol capability to a *Pseudomonas* strain. *Applied and Environmental-Microbiology*. 58 (12): 3873-3878.

Frankowski, J., Berg, G., Bahl, H. and Elad,Y. (ed.); Freeman, S. (Ed.); Monte, E. 2001. Purification and properties of two chitinolytic enzymes of the biocontrol agent *Serratia plymuthica* C48. IOBC-WPRS Working Group "Biological Control of Fungal and Bacterial Plant Pathogens". *Proceedings of the sixth meeting, Biocontrol Agents: Mode of Action and Interaction with Other Means of Control*, Sevilla, Spain, November 30 - December 3, 2000. Bulletin-OILB-SROP. 24 (3): 319.

Gaffney, T. D., S. T. Lam, K. Gates, A. Frazelle, J. Di Maio, S. Hill, S. Goodwin, N. Torkewitz, A. M. Allshouse, H. J. Kempf, and J. O. Becker. 1994. Global regulation of expression of anti-fungal factors by a *Pseudomonas fluorescens* biological control strain. *Mol. Plant Microbe Interact.* 7:455-463.

Gomez, L., Allona, I., Casado, R. and Aragoncillo, C .2002. Seed chitinases. *Seed Science Research.* 2002. 12 (4): 217-230

Haran, S., Schickler, H., Pe'er, S., Longemann, S., Oppenheim, A. and Chet, I. 1996. Increased constitutive chitinase activity in the transformed *Trichoderma harzianum. Biol.Control.* 3: 101-103.

Haas, K., Keel, C., Laville, J., Maurhofer, M., Oberhansli, T., Schnider, U., Voisard, C., Wuthrich, B., Defago,G. and Hennecke, H. (Ed.); Verma, D.P.S. 1991. Secondary metabolites of *Pseudomonas fluorescens* strain CHA0 involved in the suppression of root diseases. *Advances in molecular genetics of plant-microbe interactions.* Vol. 1. *Proceedings of the 5th International Symposium on the molecular genetics of plant-microbe interactions*, Interlaken, Switzerland, September 9-14, 1990-1991, 450-456.

Hoffland, E., Hakulinen, J. and Van Pelt, J. A.1996.Comparision of systemic resistance induced by avirulant and nonpathogenic *Pseudomonas* species. *Phytopath.* 86:757-762.

Howell, C. R. and R. D. Stipanovic . 1979. Control of *Rhizoctonia solani* in cotton seedlings with *Pseudomonas fluorescens* and with an antibiotic produced by the bacterium. *Phytopath.* 69:480-482.

Howell, C. R. and R. D. Stipanovic. 1980. Suppression of *Pythium ultimum* induced damping-off of cotton seedlings by *Pseudomonas fluorescens* and its antibiotic pyoluteorin. *Phytopath.* 70:712-715.

Krishnamurthy, J. 1996. Studies on *Trichoderma* chitinases effective against rice sheathblight pathogen *Rhizoctonia solani* Kuhn, *M.Sc. (Agri.) Thesis*, Tamil Nadu Agricultural University,Coimbatore, 79 pp.

Krishnamurthy, J., Samiyappan, R., Vidhyasekaran, P., Nakkeeran, S., Rajeswari, E. and Balasubramanian, P. 1999.Mechanism involved in mycoparasitism by *Trichoderma* spp. on *Rhizoctonia solani,* the rice sheath blight pathogen. *J.BioSci.* 24: 207-213.

Keel, C., D. M. Weller, A. Natsch, G. Defago, R. J. Cook, and L. S. Thomashow. 1996. Conservation of the 2, 4-diacetylphloroglucinol biosynthesis locus among fluorescent *Pseudomonas* strains from diverse geographic locations. *Appl. Environ. Microbiol.* 62:552-563.

Kloepper, J. W. 1991. Development of *in vivo* assays for prescreening antagonists of *Rhizoctonia solani* on cotton. *Phytopath.* 81:1006-1013.

Kraus, J. and J. E. Loper. 1992. Lack of evidence for a role of antifungal metabolite production by *Pseudomonas fluorescens* Pf-5 in biological control of Pythium damping-off of cucumber. *Phytopath.* 82:264-271.

Lohrke, S., Moenne-Loccoz, Y., McCarthy, J., Powell, J., Higgins, P., Joyce, H. and O'-Gara, F.1997. Experiences in the molecular ecology of the rhizosphere with microbial releases. JIRCAS-*International-Symposium-Series.* 1997, No. 5: 91-98.

Laville, J., C. Voisard, C. Keel, M. Maurhofer, G. Defago,and D. Haas. 1992. Global control in *Pseudomonas fluorescens* mediating antibiotic synthesis and suppression of black root rot of tobacco. *Proc. Natl. Acad. Sci. USA.* 89:1562-1566.

Laville, J., C. Blumer, C. Von Schroetter, V. Gaia, G. Defago, C. Keel, and D. Haas. 1998. Characterization of the hcnABC gene cluster encoding hydrogen cyanide synthase and anaerobic regulation by ANR in the strictly aerobic biocontrol agent *Pseudomonas fluorescens. CHA0. J. Bacteriol.* 180:3187-3196.

Loper, J. E. 1988. Role of fluorescent siderophore production in biological control of *Pythium ultimum* by a *Pseudomonas fluorescens* strain. *Phytopath.* 78: 166-172.

Loper, J. E., Nowak-Thompson, B., Whistler, C. A., Hagen, M. J., Corbell, N. A., Henkels, M. D, and Stockwell, V. O. 1997. Pages 73-79 *(In:)* Ogoshi, A., Kobayashi., K., Homma, Y., Kodama, F., Kondo, N., and Akino, S. *Biological control mediated by antifungal metabolite production and resource competition: An overview. Proceedings of the Fourth International Workshop on Plant Growth Promoting Rhizobacteria.*

Liu-M., Cai-Q. X., Liu-H. Z., Zhang-B. H., Yan-J. P. and Yuan-Z. M .2002. Chitinolytic activities in *Bacillus thuringiensis* and their synergistic effects on larvicidal activity. *Journal of Applied Microbiology.* 93 (30): 374-379.

Martin, F. N. and J. E. Loper. 1999. Soilborne plant diseases caused by *Pythium* spp.: Ecology, epidemiology, and prospects for biological control. *Critical Reviews in Plant Sciences.* 18:111-181.

Maurhofer, M., Keel, C., Schnider, U., Voisard, C., Haas, D. and Defago, G. 1992. Influence of enhanced antibiotic production *in Pseudomonas fluorescens* strain CHA0 on its disease suppressive capacity. *Phytopath.* 82(2): 190-195.

Maurhofer, M., Keel, C., Haas, D. and Defago, G. 1995. Influence of plant species on disease suppression by *Pseudomonas fluorescens* CHA0 with enhanced antibiotic production. *Plant Pathol.* 44(1): 40-50.

Melent'-ev, A. I., Aktuganov, G. E. and Galimzyanova, N. F. 2001. The role of chitinase in the antifungal activity of *Bacillus* sp. 739. *Microbiology Moscow.* 70 (5): 548-552

McSpadden-Gardener, B. B., K. L. Schroeder, S. E. Kalloger, J. M. Raaijmakers, L. S. Thomashow, and D. M. Weller. 2000. Genotypic and phenotypic diversity of phlD-containing *Pseudomonas* strains isolated from the rhizosphere of wheat. *Appl. Environ Microbiol.* 66:1939-1946.

Nawani, N. N., Kapadnis, B. P., Das, A. D., Rao, A. S. and Mahajan, S. K . 2002 . Purification and characterization of a thermophilic and acidophilic chitinase from *Microbispora* spp. V2. *J.Appl.Microbiol.* 93: 6, 965-975.

Natsch, A., Keel, C. and Hebecker, N .1998. Impact of *Pseudomonas fluorescens* strain CHA0 and a derivative with improved biocontrol activity on the culturable resident bacterial community on cucumber roots. *FEMS-Microbiology-Ecology.* 27(4): 365-380.

Nowak-Thompson, B., N. Chaney, J. S. Wing, S. J. Gould, and J. E. Loper. 1999. Characterization of the pyoluteorin biosynthetic gene cluster of *Pseudomonas fluorescens* Pf-5. *J. Bacteriol.* 181:2166-2174.

Nowak-Thompson, B., S. J. Gould, and J. E. Loper. 1997. Identification and sequence analysis of the genes encoding a polyketide synthase required for pyoluteorin biosynthesis in *Pseudomonas fluorescens* Pf-56. *Gene,* 204:17-24.

Nowak-Thompson, B., S. J. Gould, J. Kraus, and J. E. Loper. 1994. Production of 2,4-diacetylphloroglucinol by the biocontrol agent *Pseudomonas fluorescens* Pf-5. *Can. J. Microbiol.* 40:1064-1066.

Pfender, W. F., J. Kraus, and J. E. Loper. 1993. A genomic region from *Pseudomonas fluorescens* Pf-5 required for pyrrolnitrin production and inhibition of *Pyrenophora tritici*-repentis in wheat straw. *Phytopath.* 83:1223-1228.

Raaijmakers, J. M. and Bonsall, R. F., Weller, D. M.1999. Effect of population density of *Pseudomonas fluorescens* on production of 2, 4-diacetylphloroglucinol in the rhizosphere of wheat. *Phytopath.* 89(6): 470-475.

Raaijmakers, J. M. and Weller, D. M. 1998. Natural plant protection by 2, 4-diacetylphloroglucinol-producing *Pseudomonas* spp. in take-all decline soils. *Molec. Pl. Micro Inter.* 11(2): 144-152.

Raaijmakers, J. M., Weller, D. M., and Thomashow, L. S. 1997. Frequency of antibiotic producing *Pseudomonas* spp. in natural environments. *Appl. Environmental Microbiol.* 63:881-887.

Raaijmakers J.M., Vlami M. and de Souza J.T.2000. Antibiotic production by bacterial biocontrol agents. Antonie van Leeuwenhoek vol. 81, no. 1-4. pp: 537-547(11).

Ramanathan, A., Shanmugam, V., Raguchander, T. and Samiyappan, R. 2002. Induction of systemic resistance in ragi against blast disease by *Pseudomonas fluorescens*. *Annals of Plant Protection Sciences*. 10(2): 313-318

Rangaraj, Nandakumar, Subramanian-Babu, Ramalingam-Radjacommare, Raguchander,T., Samiyappan, R Nandakumar, R., Babu, S., Radjacommare, R., Raguchander, T. and Samiyappan, R. 2002. *Pseudomonas fluorescens* mediated antifungal activity against *Rhizoctonia solani* causing sheath blight in rice. *Phytopathologia Mediterranea*. 41(2): 109-119.

Rodriguez, F. and W. F. Pfender. 1997. Antibiosis and antagonism of *Sclerotinia homoeocarpa* and *Drechslera poae* by *Pseudomonas fluorescens* Pf-5 *in vitro* and in planta. *Phytopath.* 87:614-621.

Sarniguet, A., J. Kraus, M. D. Henkels, A. M. Muehlchen, and J. E. Loper. 1995. The sigma factor RpoS affects antibiotic production and biological control activity of *Pseudomonas fluorescens* Pf-5. *Proc. Natl. Acad. Sci.* U. S. A. 92:12255-12259.

Schnider, U., C. Keel, C. Blumer, J. Troxler, G. Défago, and D. Haas. 1994. Amplification of the housekeeping sigma factor in *Pseudomonas* fluorescens CHA0 enhances antibiotic production and improves biocontrol abilities. *J. Bacteriol.* 177:5387-5392.

Schnider, U.; Keel, C.; Voisard, C.; Défago, G., and Haas, D. 1995. Tn5-Directed cloning of pqq genes from *Pseudomonas fluorescens* CHA0: mutational inactivation of the genes results in overproduction in the antibiotic pyoluteorin. *Appl. Environ. Microbiol.* 61:3856-3864.

Schnider-Keel, U., Seematter, A., Maurhofer, M., Blumer, C., Duffy, B., Gigot-Bonnefoy, C., Reimmann, C., Notz, R., Défago, G., Haas, D. and Keel, C. 2000. Autoinduction of 2, 4-diacetylphloroglucinol biosynthesis in the biocontrol agent *Pseudomonas fluorescens* CHA0 and repression by the bacterial metabolites salicylate and pyoluteorin. *J.Bacteriol.* 182:1215-1225.

Schmidli-Sacherer, P., Keel, C. and Defago, G.1997. The global regulator GacA of *Pseudomonas fluorescens* CHA0 is required for suppression of root diseases in dicotyledons but not in Gramineae. *Plant Pathol.* 46(1): 80-90.

Sharifi-Tehrani, A., M. Zala, A. Natsch, Y. Moenne-Loccoz, and G. Defago. 1998. Biocontrol of soil-borne fungal plant diseases by 2, 4-diacetylphloroglucinol-producing fluorescent *Pseudomonads* with different restriction profiles of amplified 16S rDNA. *Eur. J. Plant Pathol.* 104:631-643.

Shternshis, M. V., Ovchinnikova, L. A., Duzhak, A. B. and Tomilova, O. G .2002. The efficiency of viral and bacterial entomopathogens formulated with chitinase for biocontrol of lepidopteran cabbage pests. *Archives of Phytopathology and Plant Protection*. 35(2): 161-169.

Shternshis, M. V., Beljaev, A. A., Shpatova, T. V., Bokova, J. V. and Duzhak, A. B. 2002. Field testing of BACTICIDE(R), PHYTOVERM(R) and CHITINASE for control of the raspberry midge blight in Siberia. *Bio Control*. 47(6): 697-706.

Stutz, E. W., G. Defago, and H. Kern. 1986. Naturally occurring fluorescent pseudomonads involved in suppression of black root rot of tobacco. *Phytopath.* 76:181-185.

Takikawa,Y., Mori, H., Otsu, Y., Matsuda, Y., Nonomura, T., Kakutani, K., Tosa, Y., Mayama, S. and Toyoda, H .2002. Rapid detection of phylloplane bacterium *Enterobacter cloacae* based on chitinase gene transformation and lytic infection by specific bacteriophages. *J.Applied Microbiol.* 93 (6): 1042-1050.

Thangavelu, R., Palaniswami, A., Doraiswamy, S. and Velazhahan, R. 2003 . The effect of *Pseudomonas fluorescens* and *Fusarium oxysporum* f.sp. *cubense* on induction of defense enzymes and phenolics in banana. *Biologia-Plantarum.* 46(1): 107-112.

Tikhonov, V.E., Lopez-Llorca, L.V., Jansson, H.B., Monfort, E., Salinas, J. and Elad, Y. (ed.); Freeman, S. (ed.);Monte, E. 2001. Chitinases produced by the biocontrol agents *Verticillium suchlasporium* and *Verticillium chlamydosporium*. IOBC-WPRS Working Group "Biological Control of Fungal and Bacterial Plant Pathogens". *Proceedings of the sixth meeting, Biocontrol Agents: Mode of Action and Interaction with Other Means of Control*, Sevilla, Spain, November 30 - December 3, 2000. *Bulletin-OILB-SROP.* 24 (3) 329-332.

Viswanathan, R. and Samiyappan, R. 2001. Antifungal activity of chitinases produced by some fluorescent Pseudomonads against *Colletotrichum falcatum* Went causing red rot in Sugarcane. *Microbiol.Res.*155:309-314.

Xu, G. W., and D. C. Gross. 1986 Selection of fluorescent pseudomonads antagonistic to *Erwinia carotovora* and suppressive of potato seed piece decay. *Phytopath.* 414-422.

Wei, G., Kloepper, J. W. and Tuzun, S. 1996. Induced systemic resistance to cucumber diseases and increased plant growth by plant growth promoting rhizobacteria under field conditions. *Phytopath.* 86:221-224.

Whistler, C. A., N. A. Corbell, A. Sarniguet, W. Ream, and J. E. Loper. 1998. The two-component regulators GacS and GacA influence accumulation of the stationary-phase sigma factor RpoS and the stress response in *Pseudomonas fluorescens* Pf-5. *J. Bacteriol.* 180:6635-6641.

Whistler, C. A.; Stockwell, V. O., and Loper, J. E. 2000. Lon protease influences antibiotic production and ultraviolet tolerance of *Pseudomonas fluorescens* Pf-5. Appl. *Environ. Microbiol.* 66:2718-2725.

• • •

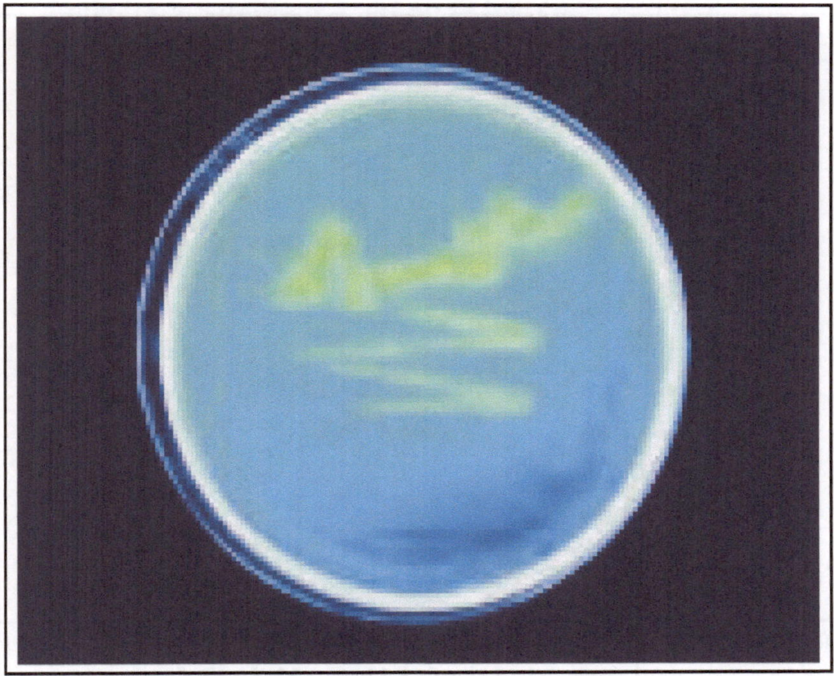

PLATE 6.1. Pure culture *Pseudomonas fluorescens*

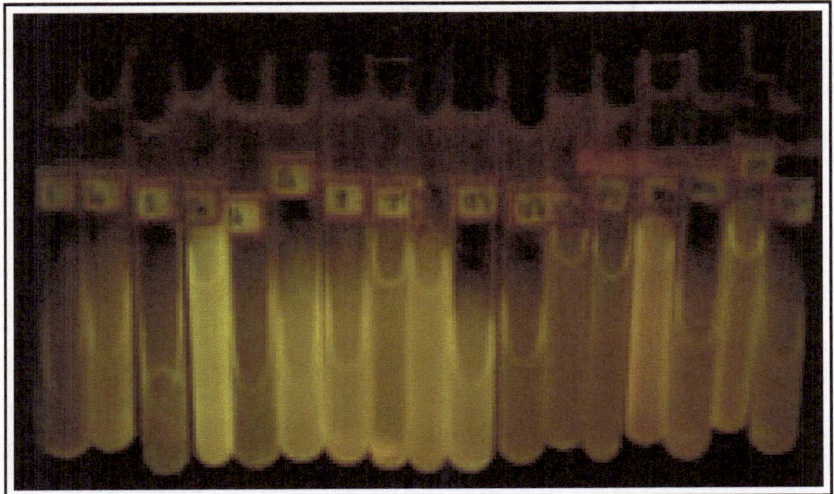

PLATE 6.2. Strainal variation of *Pseudomonas fluorescens* with respect to fluorescent pigmentation.

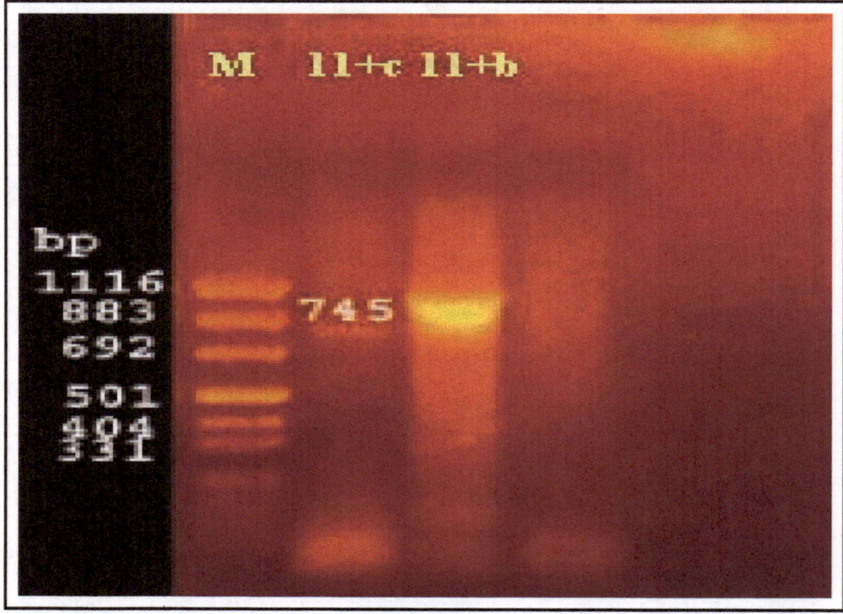

PLATE 6.3. Amplification of Phl-D gene by Phl 2a & 2b primers in *P. fluorescens*. M=$_{P}$UC mix 8 11+C & 11+b = *P. fluorescens*

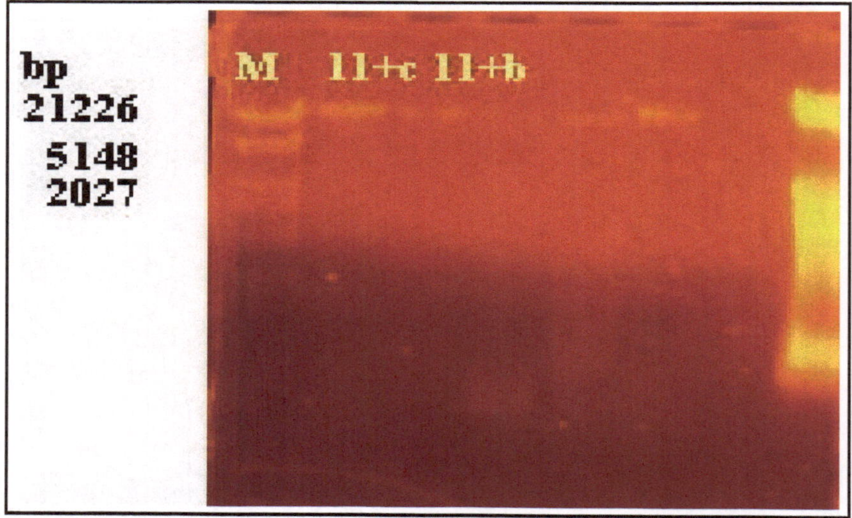

PLATE 6.4. No amplification of Phl-D gene by B 2BF & 2BPR4 primers in *P. fluorescens*. 1+C & 11+b= *P. fluorescens* M-EcoRI+ Hind III.

Advances in Soil Borne Plant Diseases, 2008
Editors: Naik and Devika Rani, pp. 165-217
New India Publishing Agency, New Delhi

7

Soil Borne Diseases of Wheat and their Management

Rashmi Aggarwal, P. Srinivas, K.D. Srivastava and D.V. Singh

INTRODUCTION

Wheat has served as a source of staple food to the mankind since times immemorial. The production of wheat in India got a fillip after green revolution. Wheat production in India has increased by over ten times in the past five decades and India has become the second largest wheat producer in the world. Today, wheat plays an increasingly important role in the management of India's food economy (Gandhi, *et al.*, 2001). Wheat, primarily the winter cereal of Indo-Gangetic plain, is a major contributor to the agricultural economy of India, occupying nearly 25.9 m ha area with 71.8-mt production (Anon., 2003).

India is also second most populous country with one billion people. Demographers indicate that by 2012 India's population will reach 1.2 billion. With this ever increasing population, there is also an increase in the demand for food grains. Depending on the rate and nature of economic growth and given the population growth rate about 4 or more per cent, annual rate of growth in the demand for wheat is likely to increase in the near future. But,

Division of Plant Pathology, Indian Agricultural Research Institute, New Delhi - 110 012 wheat production is expected to grow at about 2 to 3 per cent per year considering the various constraints unless new breakthroughs occur. Thus, ignoring fluctuations, a demand-supply gap may open at a rate of about 1 to 2 per cent per year, which is equivalent to 0.7 to 1.4 million tonnes of wheat, growing larger over the years (Gandhi *et al.*, 2001). But this demand-supply gap faces further serious impediments from many diseases and pests. Misuse of modern agricultural practices like intensive cultivation of genetically uniform and high yielding varieties, excessive use of fertilizers, close planting and frequent irrigation pose a threat of aggravating disease incidence in wheat. In order to sustain the high level of production, it is necessary to keep constant vigil on production and protection of the crop.

Wheat is known to be prone to many diseases like smuts, rusts, bunts, rots and foliar blights. For the last three decades, there has been a remarkable progress in breeding for rust resistant varieties which in turn prevent the calamity of rust epidemics by planned usage of vertical genes in different epidemiological sub-zones in the country (Nagarajan *et al.*, 1984; Bahadur *et al.*, 1994). These factors, therefore, have encouraged population build up of minor diseases (Joshi *et al.*, 1983; Singh *et al.*, 1986), among them soil borne diseases have emerged as major concern, which has attracted worldwide attentions. Karnal bunt, foliar blights, head scab, foot and root diseases, are some of the major soil borne diseases of wheat causing concerns (Joshi *et al.*, 1978; Kulkarni and Naragund, 1986, Saari, 1986). Among all soil borne pathogens, fungal pathogens hold more economic importance than bacterial and nematode pathogens (Joshi *et al.*, 1986, Kulkarni and Naragund, 1986, Mathur and Cunfer, 1993, Sarri, 1986, Sharma *et al.*, 1998). Some of the major soil borne pathogens causing symptoms like foot rots, root rots, bunt, seedling blights etc. in wheat have been reviewed below.

KARNAL BUNT

Karnal bunt has assumed great importance world over in recent times. This disease has been cited as an example of a minor disease turning to a major one within a short span of few years (Joshi, *et al.*, 1980, Joshi, *et al.*, 1983, Joshi, *et al.*, 1984, Singh *et al.*, 1985). Though the disease had been endemic to the Indian sub-continent, lately it was also reported from many parts of the world.

The widespread occurrence of Karnal bunt of wheat caused by *Neovossia indica* (Mitra) Mundkur was first detected in the experimental seed material at Karnal (Haryana, India) and was named as Karnal bunt (Mitra, 1931). The disease was considered endemic until 1969-70 crop seasons, when most of the Mexican wheat cultivars in India and Pakistan were affected by Karnal bunt (Hassan, 1973, Swaminathan *et al.*, 1971). Besides India and Pakistan, it has also been reported from Afghanistan (Loke and Watson, 1955), Nepal (Singh *et al.*, 1989), Iraq (CMI, 1989), Mexico (Prescott, 1984) and USA (Ykeima *et al.*, 1996). The disease reached the Western Hemisphere in 1972 when it was discovered in the Mexican state of Sonora. The first report of Karnal bunt in the United States came from Arizona in March of 1996 (Bonde, 1997).

In India, the disease has been established in the states of Punjab, Haryana, Jammu and Kashmir, Himachal Pradesh, Uttar Pradesh, Delhi, Rajasthan, Bihar, Gujarat, West Bengal and Madhya Pradesh (Aggarwal *et al.*, 1991; Singh, 1986).

This disease does not cause significant yield losses but is responsible for quality deterioration. Moreover, it does not pose any health hazards to human beings upon consumption (Bhat *et al.*, 1981). Karnal bunt has become a major impediment in exports to several countries where this disease is not reported till now. Efforts in recent times have yielded some success in demarking various disease free wheat growing areas in India which may contribute to exports of Karnal bunt free produce.

Pathogen

The Karnal bunt pathogen was first described by Mitra (1931) as *Tilletia indica* Mitra. However due to the fact that the fungus produces numerous sporidia which do not fuse and other taxonomic details, it was later designated as *N. indica* (Mitra) Mundkur (Mundkur, 1944, Munjal, 1969, Krishna and Singh 1982). Teliospores of the fungus are globose to subglobose in shape and are surrounded by a thick covering or sheath (2-4 mm). The spores are darker than those of *T. caries* and *T. foetida.* There is a wide variation in teliospore size (22-49 mm), average being 35 mm in diameter (Bansal, *et al.*, 1984a). Scanning and transmission electron microscopic studies on teliospore morphology have shown existence of three distinct layers, viz., the perisporium (sheath), the episporium and the endosporium (Khanna *et al.*, 1968). The teliospores of Karnal bunt require a period of dormancy for their germination. Soaking of teliospores in peptone or wheat-straw extract, benzaldehyde, furfuraldehyde, butyric acid, natural extracts etc. were used to break the dormancy and improve germination percentage (Singh, 1986). Treatment of teliospores in liquid nitrogen (-196°C) for 15 min. enhanced the germination percentage (Bansal, *et al.*, 1984b). On germination, teliospore produces promycelium of variable length, bearing large number of sporidia. The promycelium is a stout structure measuring 10-190 mm in length and 6-13 mm in breadth but at times it becomes unusually long or so much suppressed that the cluster of sporidia appears to be coming out of the teliospore itself

(Bansal, *et al.*, 1983). The branching of the promycelium was also observed but only one branch bears 60-185 primary sporidia at the tip. On potato-dextrose medium, the primary sporidia produce two types of secondary sporidia, falcate (allantoid), which are released forcibly and filiform like that of primary sporidia (Dhaliwal and Singh, 1988; Warham, 1987).

Variability is an essential requirement for survival and co-evolution of the organisms. Mitra (1935) reported two collections of *N. indica* as different physiologic forms differening in their teliosporic size. But Bansal *et al.*, (1984 a) noticed that teliospore size is influenced by environment and should not be taken as a criterion for differentiating strains of *N. indica*. Efforts have also been made to differentiate isolates of the pathogen on a set of standard differential hosts. Aujla *et al.*, (1987) identified four pathotypes from Punjab and Himachal Pradesh and designated as K1, K2, K3 and K4. Pathotype K1 and K2 were reported to be the most virulent forms and predominantly present. Dhiman and Bedi (1989) also identified three distinct pathotypes in Punjab. Singh *et al.*, (1991) emphasised that in *N. indica* true sense of race concept does not exist. This pathogen during annual cycle has the benefit of sexuality, as the + and - strains combine resulting in nuclear fusion. Therefore at each cycle, there is an assortment of virulence genes and due to that pathogenic isolates is a population with continuous variability. In the absence of clear-cut host-pathogen interaction, the aggressiveness as determined on the basis of difference of biomass (teliospore) production, was taken as the basis of variability by Singh *et al.*,(1991) and five pathotypes designated as KBAg-1, KBAg-2, KBAg-3, KBAg-4 and KBAg-5 were identified.

Symptoms

Karnal bunt usually is noticed first when broken and/or partially smutted kernels are seen in threshed grain. However, symptoms are present at the soft-dough stage in the form of blackened areas surrounding the base of the grain, which extend upward along the suture to varying degrees. Unless there is severe disease incidence, only a few florets per spike are affected and diseased spikes are not conspicuous. The glumes are not distorted by infected kernels and, the endosperm and the dorsal side of the seed except in very severe infection remain unaffected (Munjal and Chatrath, 1976; Holmes *et al.*, 1997).

Karnal bunt pathogen infects wheat at the flowering stage prior to seed formation and hence the symptoms are evident when grains have developed. All the ear-heads are not affected in a stool and also not all the grains in a spike. The grains are partially infected and therefore, the disease is also referred as 'Partial bunt' (Bedi *et al.*, 1949). Cashion and Luttrell (1988) have demonstrated that the causal fungus does not invade the seed and its growth is limited to the pericarp where it is entirely intercellular. The endosperm is shrunken to varying degrees and become cartilaginous in appearance. Aggarwal *et al.*, (1994) through light and Scanning Electron Microscopy also revealed that the middle parenchymatous layer between inner and outer pericarp of infected wheat seed is dissolved and results in shrunken endosperm.

In a standing wheat crop, the infected spike can be detected by the shinning silvery black spike lets, with glumes spread apart and swollen ovaries. The infected ears emit a foul smell of a volatile compound, trimethylamine.

Losses

Karnal bunt normally occurs sporadically but becomes serious in the years of epidemic and causes substantial losses to wheat crop. In 1976, the disease reduced annual yield of wheat by 0.2 per cent in Punjab and Jammu & Kashmir (Munjal, 1976). However, some individual samples may show 10-30 per cent loss (Joshi *et al.*, 1983). Gill *et al.* (1993) reported that in the foothills of Himachal Pradesh, Jammu and Kashmir, the percentage of wheat samples containing seeds infected by *T. indica* increased from 25% in 1977–78 to 86% for the 1982–83 crop years. Singh *et al.*, (1994) reported 0.3- 0.5 percent loss in India during the most severe years between 1982-89. Contrary to these estimates, Shukla *et al.*, (1982) estimated 5-20 per cent losses in some wheat cultivars in Uttar Pradesh during 1982. Results based on the extent of grain portion converted into bunt sorus showed that loss in case of small sorus was 5.2% by weight whereas the estimates were 19.9% and 51.5% in medium and large sori respectively. This amounted to an average of 25.5 per cent reduction in the weight of a grain lot having equal proportion of all the three grades (Aggarwal *et al.*, 1989a). The weight of infected grains is directly related to the severity of infection; as the degree of infection increases, the weight of the grains correspondingly decreases (Holmes *et al.*, 1997). The disease adversely affects seed germination. Seedlings raised from heavily infected grains are small and weak and their leaves contain less sugar and chlorophyll (Bansal *et al.*, 1984c).

Karnal bunt is a disease that generally causes little direct loss in either quantity or quality of wheat seed or grain. However, it can cause considerable economic loss due to international regulation of the pathogen. A grain or seed shipment can be prohibited by even the presence of a single teliospore of the fungus (Bonde *et al.*, 1997). The partial damage to the grain in terms of quantity caused by Karnal bunt is lesser than that of Hill bunts but main concern is again about the quality of grain, as even 3% infection could impart a fishy odour to the flour (Mehdi *et al.*, 1973) due to production of trimethylamine (Singh *et al.*, 1983). Flour prepared from grains having 3% infection emits a fishy odour and when the infection is 10% the flour becomes black and *chapatties* prepared from it are unpalatable (Mehdi *et al.*, 1973). Studies have conclusively proved that the bunted grains do not pose any health hazards (Aujla and Sharma, 1990) and the quality losses in case of bunts are generally due to the bad smell and blackish colour flour. Brennan *et al.*, (1990) estimated direct and indirect economic losses from Karnal bunt of wheat in Mexico. The direct loss, which resulted from yield, quality and seed export losses, was estimated to be about 0.12 per cent per year in North-Western Mexico. Indirect losses were those associated with the measures aimed to prevent the spread of Karnal bunt and to reduce the severity, including quarantine restrictions on planting in Karnal bunt infected areas, grain fumigation and restrictions on the use of Karnal bunt infected seed.

Disease cycle

Sources of inoculum are diseased seed and soil contaminated with spores from previous crops. Teliospores of *N. indica* in the infected seed get into the soil at the time of harvesting, threshing or winnowing. The delicate pericarp surrounding each sorus of teliospores is easily broken during harvest, dispersing the spores to contaminate the soil (Mitra, 1937, Smilanick *et al.*, 1985). Karnal bunt is thus considered to be soil, seed and air-borne disease. The spores remain viable for a number of years in the soil (Singh, 1986; Singh *et al.*, 1983). Teliospores lying 15 and 25 cm deep in soil retained viability for 36 and 26 months respectively (Dhiman and Bedi, 1989). Fresh teliospores require a dormancy period of up to 6 months, although a low level (6.7%) of germination of teliospores from freshly harvested kernels has been observed after 1 week of storage. They survive longer when buried in dry soil than in moist, cropped soils or on the soil surface (Holmes *et al.*, 1997). Teliospores under irrigated soil showed significantly higher germination than unirrigated soils (Aggarwal *et al.*, 1989b).

The teliospores germinate from the middle of February to middle of March depending upon suitable meteorological conditions (Bedi *et al.*, 1949). The disease gets established very well if the maximum temperature is in the range of 19-23°C and the minimum 8-10°C, followed by intermittent rains (Joshi *et al.*, 1981). Aujla *et al.*, (1977) noted that during anthesis period when the temperature was 18- 20°C and RH > 70%, severe infection occurred in many varieties. Poor incidence of Karnal bunt is recorded in the years of high temperature and poor rainfall at anthesis.

Analysis of Karnal bunt incidence and meteorological data during flowering stage of the crop in the month of March from 1990 to 2001 revealed that occurrence of disease was governed by the prevalence of minimum temperature (10.1-10.2°C), maximum temperature (25.9-26.0°C) and maximum relative humidity (90% and above). Some rainfall (40.9-46 mm) was also necessary for the development of Karnal bunt (Singh *et al.,* 2003). Normally the infection takes place at the anthesis stage; however, awn emergence and boot II stage (Dhiman and Bedi, 1981) are also reported to be susceptible for the establishment of Karnal bunt. Nagarajan (1991) hypothesised that boot leaf serves as the catchment area for the allantoid sporidia and boot leaf sheath as the reservoir.

Two types of sporidia- allantoid and filiform play important role in the disease cycle of the fungus. The filiform sporidia serve as the reproductive bodies to raise allantoid sporidia in successive generations (Dhaliwal and Singh, 1989). Singh *et al.,* (1990) analyzed the soil of Gurdaspur, Dhaulakuaon, Saharanpur, Dehradun and Hisar. The spore density at Gurdaspur and Dhaulakuaon was high where the disease appeared in severe form suggesting that high inoculum load in the soil is positively correlated with severe infection in nature. This suggests that the original inoculum from soil-borne teliospores play only a starting role in Karnal bunt epidemics.

Teliospores are also reported to spread by wind (air currents). Boratynski *et al.,* (1985) demonstrated the movement of *T. indica* teliospores from Mexico into California by railcars. Bonde *et al.,* (1987) found viable teliospores of *T. indica* at least 3000 m above fields being burned after harvest in Mexico (Bonde *et al.,* 1987). The updraft of air from burning fields most likely carries teliospores to great heights and acts as a mechanism of dispersion. Animals and birds also can transport viable teliospores on their bodies and in their digestive systems however, it is not known whether these spores could result in disease (Smilanick, *et al.,* 1986).

Management

Karnal bunt spreads through the seed, soil and air and it is, therefore, difficult to manage. Such interactions reveal that an integrated approach for tackling Karnal bunt is the most ideal. Some of the approaches which influence the disease are discussed as follows.

Cultural practices : Changing the crop environment and cultivation practices to directly or indirectly influence pathogen can be a viable strategy to combat Karnal bunt. Mitra (1937) advocated crop rotation, while Padwick (1939) recommended non-cultivation of wheat for two consecutive years to minimize the Karnal bunt incidence. Bedi *et al.,* (1949) noted that adjustment in water and fertilizer balance affected the disease incidence under field conditions. Higher dose of nitrogen increases the susceptibility of plants (Singh and Singh, 1985).

Mulching has been reported to be suited against over summering soil inoculum of *N. indica* in the form of teliospores in North Western India during March (Dhaliwal and Singh, 1988). Singh *et al.,* (1992) observed that mulching the infested soil with plastic sheet or by growing chickpea in between wheat rows could minimize the incidence up to 79.8% and 57.2% respectively. The Karnal bunt incidence has been reported to be influenced by sowing dates (Goel *et al.,* 1977). Singh and Prasad (1978) also observed that the incidence of Karnal bunt increased with delay in sowing. Therefore, sowing preferably in the third week of November has been recommended to reduce the incidence of Karnal bunt (Singh and Singh, 1985). Kalha *et al.,* (1987) showed the lowest incidence of Karnal bunt in late sown crop planted on 5th January.

Chemical application : Seed dressing fungicides, such as aureofungin, ethyl mercury chloride, ethyl mercury phosphate, fentin acetate, triphenyltin chloride, fentin hydroxide, oxycarboxin, indar, benomyl, carbendazim, triadimenol, methfuroxam, tetramethyl thiuram disulphide, carboxin etc. were used for seed dressing and were found them effective (Singh and Rai, 1986; Singh, 1986).

Foliar sprays with fungicides have also been tried by Singh *et al.,* (1985) using mancozeb, carbendazim or fentin hydroxide at early heading stage before flowering of the crop to control Karnal bunt. Bitertanol and diferpconazole sprayed at heading stage were found to be effective. Propiconazole (Tilt 250 EC) also showed promise in complete control of Karnal bunt when tested under Indian conditions (Bains and Dhaliwal, 1988). Singh *et al.,* (1989) could also reduce Karnal bunt significantly by single spray of Tilt @ 250 and 500 ml/ha. Single foliar spray at boot leaf stage with propiconazole and cyproconazole (SAN 619 F) @ 0.2 per cent concentration significantly reduced the disease (Singh and Srivastava, 1992).

Resistance: Exploitation of genetic resistance could be the most powerful management strategy against Karnal bunt. Unfortunately, immunity to Karnal bunt in *Triticum aestivum* varieties is lacking. Majority of the wheat cultivars grown in India are susceptible to the disease, but with the standardization of techniques to screen wheat germplasm, some sources of resistance could be detected (Aujla *et al.,* 1985; Singh *et al.,* 1988, Warham, 1990) and concerted efforts have been made to identify sources of resistances to Karnal bunt both in India and Mexico (Gill, 1985; Warham *et al.,* 1986).

Sawhney *et al.*, (1992) screened advance wheat strains and reported multiple resistance to Karnal bunt in DL 802-3, DL 790-1, DL 330-1, DL 484-2, DL 790-2, DL 770, DL 760-1 and DL 377 -8. Efforts have been made to identify morphological and physiological resistance in *bread* wheat, *durum* wheat and triticale. In nature, the resistance to Karnal bunt is rated highest in triticale followed by *T. durum* and *T. aestivum*, however, Warham (1988) observed same degree of susceptibility in both the wheats and triticale with boot as well as spray inoculation, indicating that resistance to natural infection in *durum* wheat and triticale may be morphological rather than physiological.

Aujla *et al.*, (1990) listed a number of wild species of wheat which remained disease free under epiphytotic conditions. Some diploid, tetraploid, hexaploid and amphidiploid wheat species, *viz.*, *Triticum tauschii, T. dicoccoides, T. spelta album, T. spelta, Agropyron elongatum,* Chinese spring *A. elongatum,* Chinese spring *A. junceum* were found to be immune to Karnal bunt (Tomar *et al.*, 1991). These genetic sources could be manipulated for breeding for disease resistance.

Certain factors which account for the morphological resistance such as compactness of spike lets on a spike, hairs present on the glume, peduncle length, plant height, ear length, boot leaf sheath length, awn characters etc. hold promise for their incorporation from varieties showing considerable resistance (Aujla *et al.*, 1990; Sidhartha, 1992). Genetic studies revealed that Karnal bunt resistance in bread wheat is polygenic and partially dominant but the additive components of variation are significant and predominantly higher than other components (Gill *et al.*, 1993). Bag *et al.*, (1999) have shown that resistance to Karnal bunt in HD 29, HP1531 and W485 is conditioned by a single recessive gene and all the three accessions carry a non allelic gene for resistance. However, presence of some modifiers was found affecting the degree of susceptibility in segregating generations. The gene for resistance was located on chromosome 2D of HD 29.

Biocontrol approach : Antagonistic potential of some biocontrol agents has been noticed against Karnal bunt. Aggarwal *et al.*, (1996) observed inhibitory effect of *Trichoderma viride, T. harzianum* and *Gliocladium deliquescens* on growth of *N. indica*. Amer *et al.*, (1998) used some fungal and bacterial microorganisms against *N. indica* and noticed that *T. pseudokoningii, T. lignorum, T. koningii, G. deliquescens* and *G. virens* antagonized teliospores and secondary sporidia of the pathogen, affecting their germination adversely. Scanning electron microscopic observations indicating overgrowth of *T. lignorum* on mycelium of *N. indcia* and resulting in disintegration of the host mycelium was reported by Aggarwal *et al.*, (1996).

Decon and Berry (1993) expressed that the future of biocontrol of soil borne plant pathogens lies in integrated control systems that combines the development of resistant crop cultivars, fungicides and cropping practices in a way that is intended to exploit both natural and inoculant biocontrol agents not alone but in concert. Amer (1995) also found that *T. lignorum* when applied as soil inoculant under field and glasshouse conditions showed significant decline in teliospore germination and also stimulated the growth of wheat plants. Increase in 1000- grain weight was also recorded.

FOLIAR BLIGHTS

Foliar blights have emerged as important disease of wheat in recent times due to successful suppression of leaf rusts. Foliar blights, basically the serious problem in the North-eastern region have now assumed significance far and wide in the country.

The disease is important not only in Indian subcontinent, but also in Africa, South America, Australia, Canada and Asia having warm and humid environments (Singh and Srivastava, 1997). In Indian subcontinent, such environments are encountered in the warmer parts of eastern India, most of Bangladesh and *tarai* region of Nepal. Rice and wheat rotations have occupied large area in Punjab, Haryana and western U.P. and late sowing of wheat is a common practice which experiences a warm and humid weather in February, which is favorable to spot blotch incidence (Singh and Srivastava, 1997). Therefore, spot blotch has emerged as a major disease in North-western as well as in the peninsular India and has acquired the status of national importance instead of regional one (Sharma *et al.*, 1998). Saari (1997) also reported that expansion of rice-wheat cropping system has added to the importance of leaf blight diseases. The disease recorded severe losses in Bihar, Bengal, eastern UP and some parts of Orissa and coastal regions of India (Joshi *et al.*, 1986).

Pathogens

Foliar blight is a complex disease and a lot of confusion prevails about the sympatomatology of the pathogens involved. Mohy in 1914 collected *Helminthosporium* for the first time from wheat (HCIO No. 12508), whereas *Alternaria triticina* as a pathogenic fungus was authentically identified after release of Kenphad wheats (Prasad and Prabhu, 1962).

Drechslera sorokiniana (Sacc.) Subram. & Jain [Syn. *Helminthosporium sativum*; perfect stage *Cochliobolus sativus* (Ito & Kuribay)] appears to be the major pathogen alongwith *D. tritici repentis* (Died.) Shoemaker, *D. tetramera*

(McKinney) Subram. & Jain, *Alternaria triticina* Prasada and Prabhu and *A. alternata* (Fr.) Keissler (Joshi *et al.*, 1974). Foliar blight is thus aptly regarded as a complex, since a number of pathogens causing blight; blotch and spot are associated with foliar blights of wheat in India (Misra, 1973; Joshi *et al.*, 1978).

The leaf blights/blotches caused by *Helminthosporium* sp., *Alternaria* sp. and *Septoria* sp., are restricted to areas under high humidity with high temperature during plant growth. Majority of infected samples over the years indicated that *D. sorokiniana* was the chief fungus causing blight or spot blotch. *D. sorokiniana* has been the predominant species in South East Asia and in the Indian subcontinent and appears in epidemic proportions (Dubin and Ginkel, 1991; Saari, 1986). Singh *et al.*, (1995) examined pathogenic behaviour of several fungi isolated from 1104 foliar blight samples and confirmed that *D. sorokiniana* was the most predominant pathogen involved in the foliar blight complex in the country.

Alternaria blight *(A. triticina)* also causes substantial losses ranging from 35-60 per cent in yield to wheat crop (Prabhu and Singh, 1974; Sokhi and Joshi, 1974). Earlier, Chenulu and Singh (1964) had also recorded maximum loss in yield (99 per cent) in highly susceptible varieties like Kenphad 21, Kenphad 32 and NP 830 when infection took place at the boot leaf stage. Unlike *D. sorokiniana,* which is widely distributed, the Alternaria blight is basically common in southern and eastern areas of the South Asian subcontinent. It has also been reported in South Italy, Mexico, Nigeria and Middle East countries. (Frisullo, 1982; Mathur and Cunfer, 1993).

D. sorokiniana isolate differs in their parasitic capabilities, irrespective of the host or the geographical distribution (Ashworth *et al.*, 1960; Gayed, 1962; Hamilton *et al*, 1960). Fungus is extremely variable; its progenies from a single conidium may differ in pathogenicity.

Losses

The importance of disease may be reflected by the yield losses but the estimates widely vary. Losses caused by leaf blight though variable in different areas averaging around 16-17%. De Milliano and Zadoks (1985) found a 30 per cent yield loss in growth chamber studies in the Netherlands. Losses of 85 per cent were assessed in Zambia (Raemaekers, 1988) and 40 per cent in Philippines (Lapis, 1985). Though, severe losses have been reported due to *Helminthosporium* sp. involved in foliar blights in eastern region of India but scientific details relating to the actual losses have not been worked out. Similarly *Alternaria* sp. has been reported to appear in severe from in experimental plots under artificial epiphytotics causing severe losses (Chenulu and Singh, 1964).

Estimated losses revealed that there could be almost total loss of yield in a highly susceptible variety like NP 830 (Chenulu and Singh, 1964). Prabhu and Singh (1974) using the same cultivar estimated approximately 60 per cent yield reduction due to combined effect of *A. triticina* and *Helminthosporium sativum*. According to Sokhi and Joshi (1974), extent of damage may be up to 35 per cent if top most second and third leaves are badly blighted at heading stage. Ram and Joshi (1979) under artificial conditions of infection observed that when disease appears at flag leaf stage the loss in number of grains per ear was up to 26.4 per cent. When the infection appears at the tillering stage the ear-head length is considerably reduced which in turn affects the number of grains, though, grains are not shriveled in such cases and 1000 grain weight is not much affected.

During favorable years losses from spot blotch alone were between 30 and 86 per cent in Brazil and in some fields loss reached 100 per cent (Mehta, 1985; Hetzler, *et al.*, 1991). In India, Nema and Joshi (1971a) reported that yield loss is variety dependent. They observed that at flag leaf stage in Sonora 64, the loss incurred was hardly 3 per cent, while S-227 showing maximum susceptibility suffered a loss of about 20 per cent.

Disease cycle

The disease first noticed in 1914 from Pusa, Bihar, is now one of the most widely distributed and serious concerns in the warm and humid environments. The disease has been frequently reported from Bihar, West Bengal, eastern Uttar Pradesh and Orissa (Chattopadhyay and Chakrabarti, 1968; Narain *et al.*, 1973).

Though recently it has also made its appearance in the North-western region of the Indo-Gangetic plains (Singh *et al.*, 1993). In 1990-91, the disease appeared in an epidemic form in some western districts of Uttar Pradesh on wheat varieties HD 2329 and HD 2285, mainly due to late seeding and warm-humid environment in March (Singh *et al.*, 1993). *D. sorokiniana* is seed and soil borne pathogen and in nature it can be transmitted by seed also (Mehta, 1993). Seed and soil borne inoculums are important in the establishment of leaf blight. Seed infection can adversely affect germination and root system developmentans and can kill seedlings with in a few days (Mehta, 1993). Rain during later stages of crop cycle, especially during grain filling may lead to staggering losses and complete loss of crop (Saunders, 1988). Even high RH in the absence of rains contributes for disease development. When mean temperatures average 25-28°C and RH above 70% spot blotch continues to develop (Lapis, 1985a, Raymundo 1992). Maity (2003) studied the epidemiology of foliar blight disease of wheat involving *H. sativum* and *A. triticina* and reported highest viability of conidia during February and early March at maximum temperature around 28 °C.

Management

Cultural practices : Dense vegetative growth of wheat varieties provides a distinct microclimate congenial for the build up of the disease through profuse tillering and thus wider row spacing has been reported to reduce disease severity (Mehta and Gaudenico, 1991). Nitrogen application may also be restricted in high disease incidence areas in order to avoid excessive vegetative growth by which humidity is increased within the crop (Saunders, 1988). However Maity, *et al.*, (2002) reported that leaf blight disease severity of wheat could be reduced by application of high N_2 and low K_2O in field. As foliar blights attained higher intensities in the late sown crop especially in the eastern part of Uttar Pradesh (Singh and Singh, 1994), it may be useful to avoid late sowings and excessive use of nitrogenous fertilisers in disease endemic areas. Heat therapy of wheat seed by hot water has been reported to be effective in checking the seed infection by leaf blight pathogen, *A. triticina* (Prabhu and Prasad, 1966).

Resistance : Considering the high yield losses, breeding for resistance commands high priority. Resistance to *D. sorokiniana* remains low among commercial wheat cultivars. Many varieties like HD 2329, HD 2285, Sonalika, HUW 234, K 8804, HS 284, UP 2338, UP 2003, C 306, WH 147, A-9-30-1 etc. gave susceptible reaction across several states in India (Singh *et al.*, 1995).

Intensive efforts in many countries are now underway to identify sources of resistance against *D. sorokiniana*. Among the Brazillian, Chinese and American varieties, *viz.* Horizon, Hadden, BH 1146, CNT 8, Suzhoe 8, Yang Mai 6 and PAT 7219 were reported to have the best resistance (Mehta, 1985; Kohli *et al.*, 1991). Resistance to *A. triticina* has been identified in some wheat strains (Dalela *et al.*, 1971; Garg, *et al.*, 1972; Katiyar and Singh, 1968; Singh *et al.*, 1977), but many varieties and genotypes including the current dwarfs are susceptible (Anonymous, 1995). Das *et al.*, (1999) identified wheat genotypes T 3336, Parula and Bob white, showing resistance and moderate resistance to *D. sorokiniana.* The resistance was attributed to the waxy leaf surface. Molecular charecterization of resistant and susceptible genotypes to identify polymerase chain reaction (PCR) based markers and isozyme based markers was elucidated by Das *et al.*, (2001; 2002).

Genes for resistance from several sources other than wheat can be exploited to achieve sufficient resistance. *Triticum aegilopoides* and *T.*

monococcum (diploid wheat) and *T. timopheevi* (tetraploids), which are highly resistant to spot blotch pathogen, hold good prospects (Harding, 1972). At CIMMYT, spot blotch resistance from *Agropyron distichum, A. curvifolium* and *Elymus gigantens* is being transferred to wheat (Kohli *et al.*, 1991).

Chemical management : Tests have indicated that contact fungicides like zineb (Dithane Z-78) or maneb (Dithane M-45) and systemic fungicides such as Propiconozole (Tilt 250 EC) and Triadimenfe control the foliage phase of the disease (Anon, 1968; Raemaekers, 1979, Mehta and Igarshi, 1985; Singh and Rai, 1986). However, their use under environmental conditions highly conducive to high disease pressure is not economical (Raemaekers, 1985).

Fungicides such as Triadimefon and Fentinacetate (Raemaekers, 1985) and Propiconazole (Lapis, 1985) provide effective control of spot blotch and Fytolan to Alternaria blight (Ram and Joshi, 1978). Seed treatments (prochloraz + mancozeb (as Prelude Fe), guazatine (as Panoctine L), tebuconazole + thiram (as Raxil TM) and carboxin + thiram (as Vitavax) were tested against soilborne pathogens of durum wheat by Balmas *et al.*, (1995) and they found that all these fungicides reduced infections of *B. sorokiniana.*

Biocontrol : Dickinson (1981) emphasised that the leaf surface antagonists hold some promise for control of foliar diseases. Biological control is becoming more possible in wheat diseases (Fokkema, 1973; Narain, 1972; Prasad *et al.*, 1978; Dandurand and Knudsen, 1993). *Epicoccum nigrum* [*E. purpurascens*] and *Trichoderma harzianum* were effective against *Drechslera sorokiniana [Cochliobolus sativus]* (Wu, 1976). A recent study on biological control of spot blotch of wheat has shown that culture filtrate of *Trichoderma reesei, T. pseudokoningii* and *Chaetomium globosum* antagonise the conidia of *D. sorokiniana* and inhibit their germination significantly. Soil application of *T. reesei* was able to enhance the germinability of infected grains and also subsided browning of collar region of the seedlings. Foliar spray with culture filtrate of this antagonist also reduced foliar infection substantially (Mandal *et al.*, 1995). *C. globosum* has been identified as potential antagonist of *B. sorokiniana.* The antagonist produces secondary metabolites/ antifungal compounds when fractioned on Thin Layer Chromatography plate (Biswas *et al.*, 2000). Several antagonistic mechanisms may play a role in disease suppression by *C. globosum. In vitro* hyphal coiling and penetration was observed in dual culture with *B. sorokiniana* and *C. globosum* through Scanning Electron Microscopy (Aggarwal *et al.*, 1996). The interaction between antagonists and the host mycelium by scanning electron microscopy has shown that the mechanism of parasitism is through penetration pegs, coiling and creation of holes. Thompson and Wildermuth (1989) have shown that vascular- arbuscular mycorrhizae (VAM) can reduce root infection by *D. sorokiniana.*

FLAG SMUT

Flag smut (*Urocystis agropyri* (Preuss) Schroet) is of major importance and is widely distributed in many wheat-growing parts of the world. The disease was first reported in Australia in 1868 (Mc Alpine, 1910). Though this disease was observed on *Agropyron* sp. in Europe in 1848 and the causal organism described as *Uredo agropyri* Preuss, the presence of flag smut disease on grass in Europe was observed 20 years before it was first observed on wheat in Australia. It is suspected that spores of flag smut occurring on grass might have contaminated the wheat seed imported into Australia where the fungus adopted to wheat (Purdy, 1965). The disease has been reported from Japan, China, India, Pakistan, South Africa, Egypt, Italy, Spain and United States (Ram and Arora, 1986). Recently it has been also reported from United Kingdom (Sansford *et al.*, 1998).

First report of disease in India was by Butler (1918) from Lyallpur in Punjab (now in Pakistan). The introduction of this disease in Punjab is believed to be through wheat grains imported from Australia (Bedi, 1957). Its occurrence was subsequently observed from Punjab, Haryana, Himachal Pradesh, Madhya Pradesh, Uttar Pradesh, Delhi, Bihar and Rajasthan (Ram and Arora, 1986). Recent surveys of fields in Uttar Pradesh have revealed an increase in incidence of flag smut of wheat (Pant *et al.*, 1996).

Although occurrence of flag smut is not widespread in the country, its importance cannot be underestimated as unlike other smuts, it is long persisting and if sowing method and environmental conditions are favourable and susceptible varieties are grown, the disease may assume serious form (Singh, 1978). The disease was observed in severe form in 1964-65 in some wheat fields near Jobner, (Rajasthan) resulting in heavy yield losses (Kothari and Dange, 1968).

Losses

Flag smut results in complete loss of productivity of infected plants leading to considerable reduction in overall crop yield. There is a long history of wheat losses due to flag smut in Australia and losses ranging from 10 per cent to 70 per cent have been reported (McAlpine, 1910; Brittlebank 1920). Kothari and Dange (1968) observed the disease during 1967-68 in moderate to severe form in many fields in Chittorgarh district of Rajasthan in India. Surveys of 37 localities in 8 provinces of China indicated the incidence of flag smut ranging from traces to 94 per cent (Yu *et al.*, 1936). In India yield of nine commercial wheat cultivars C 306, Sonalika, WH 147, WH 157, WH 416, HD 2009, HD 2285, Raj 3077 and PBW 226 was reported to be reduced by 23-65.4%, maximum in WH 147 and minimum in Sonalika. Yield components of different cultivars were reduced by various amounts: tillering was reduced by 15.7-45%; plant height by 37-62%; ear head length by 28-46%; and 1000 grain weight by 19-37% (Beniwal, 1992).

Earlier, several wheat disease surveys conducted by the Indian Agricultural Research Institute, New Delhi have revealed the presence of this disease in few isolated pockets of Rajasthan and Punjab since 1974-75 (Ram and Arora, 1986). Bhatnagar *et al.*, (1975; 1978) reported noticeable reduction in the number of ears per infected plant, 1000-grain weight and height of infected plants. Sethi and Singh (1971) observed a very heavy attack of flag smut on Kalyansona in Kulu valley of Himachal Pradesh. In some fields, they could observe 30-35 per cent plants affected and among affected plants about 80 per cent plants produced only diseased tillers. Widespread incidence varying up to 50% has also been observed in Himachal Pradesh and highest incidence was observed in the mid-hills sub-humid zone, at elevations of 651-1800 m (Bansandrai *et al.*, 1993).

Symptoms

The disease is observed from late seedling stage till maturity of crop and as a rule every shoot of the plant is infected. The sori are formed in the mesophyll tissue between the veins, and are covered by epidermis during early stages of development. Later the epidermis ruptures, releasing a black powdery mass of spores; leaves become twisted and assume drooping habit followed by withering. The leaves are shed and often the plant dies. The affected culms remain sterile bearing no grains. Even if some grains are formed, they are shriveled, valueless and show poor germination. In highly susceptible varieties the plants are stunted due to the reduction in elongation of internodes. Often spike development stops prior to its emergence from the leaf whorl. The infected plants besides giving distorted appearance often produce more tillers than the healthy plants. The root development is also inhibited (Ram and Arora, 1986). According to Angell (1934) flag smut can be detected in the very early seedling stage of plant development by a characteristic twisting and bending of the coleoptile associated with the formation of bleached spots on the coleoptile.

Pathogen

The fungus causing flag smut of wheat was identified first by Wolff (1873) as *Urocystis occulta* (Wallr.) Rab. *Urocystis tritici* Koern. was later accepted as valid name for the flag smut of wheat fungus till 1942 when Fischer (1943) suggested the combination of the morphologically similar species *Urocystis tritici, U. occulta* and *U. agropyri* (Preuss.) Schroet. on wheat, rye, and grasses respectively, under *U. agropyri*, recognizing the specialised varieties of the fungus on the different cereals and grasses. Fischer (1953) later included *U. tritici* under *U. agropyri* and retained *U. occulta* on rye largely on the basis of smaller spore balls and the location of sori in culm and leaf sheath rather than in the leaves as in *U. agropyri*.

The spores are borne in tiny balls consisting of one to six, mostly three, bright brown cells surrounded by a layer of flattened sterile peripheral cells. According to Fischer and Holton (1957) the fertile interior cells are spores and should be called as teliospores. Individual spores are globose to subglobose and reddish to olivaceous, smooth walled and measure 8-18 microns in diameter. The spore balls measure 18-52 microns in diameter. The sterile peripheral cells are smaller and more elliptical. Aggarwal *et al.* (1998) have studied the spore ball morphology under Scanning Electron Microscope and have reported that the surface of spore balls was ornamented having sterile cells all around the teliospores. The spores germinate *in situ* to produce a short promycelium, with or without septa. At the apex of promycelium, 3-4 hyaline cylindrical sporidia are borne. The sporidia usually remain attached to the promycelium and germinate by a slender germ tube which may be of considerable length and acts as an infection hypha. It has been observed by Noble (1924) that the nucleus of the spores of *U. tritici* (*U. agropri*) divides at the time of spore germination producing the haploid nuclei which move out through the promycelium into the sporidia, one nucleus to each sporidium.

Noble (1924) observed two distinct types of sporidia. In one case, the sporidium germinates and the nucleus divides and the germ tube becomes binucleate. Occasionally, this situation remains for a considerable length of time. In the second case, sporidial fusion sometimes occurs, resulting in production of a germ tube containing the two nuclei, which migrate from different sporidia. Pre-soaking of the spores in water for three days and then adding small pieces of wheat tissue could support good germination (Singh, 1978).

Disease cycle

The disease perpetuates through seed and soil. It can persist in the soil as spore ball for many years without access to the host (Walker, 1969). The fungus causes infection at seedling stage. The point of entry during infection is restricted to the young coleoptile. After infection, the fungus grows both inter- and intra-cellularly. Infection in the seedling stage is influenced to a considerable extent by environmental factors including soil moisture, soil temperature, soil pH as well as cultural practices such as planting date, sowing depth, host variety and its stage of development. The flag smut spores germinate best at pH 5.1 to 5.7 and temperature of 18 to 24°C (Yadava *et al.*, 1969). Minimum temperature for germination was reported to be below 15°C and maximum between 27° and 30°C. Noble (1934) found a direct relationship between the spore viability and relative humidity conditions under which the spores of flag smut fungus were stored. Spores can remain viable for as long as 10 years in the laboratory when kept at low humidity (Verwoerd, 1929). Mean disease incidence was higher (27.1% plant and 15.3% tiller infection) in sandy soil than in sandy loam (12.67% and 6.64%, respectively) and loamy soil (9.31% and 4.2%). Three off-season irrigations in addition to normal rainfall resulted in highest mean incidence (26.88% plant and 14.98% tiller infection). Infection was lowest (3.19% and 1.06%) under completely dry conditions (Sanjay Arya *et al.*, 1995). Sowing in relatively dry soil favors infection whereas, sowing in moist soil is detrimental to disease development. Although actual moisture requirements in terms of percentages are not known, it would appear that around 40 per cent would be optimum for most of the soils (Faris, 1933). The cardinal minimum, optimum and maximum soil temperatures for infection are in general 5, 20 and 28°C respectively.

Flag smut incidence is reported to increase with an increase in inoculum load applied to the seeds (Goel and Jhooty, 1989). Soil moisture content of 9% was optimum for infection. The failure of *U. agropyri* to infect wheat at relatively higher soil moisture contents was attributed to the anaerobic conditions in the soil, which prevented the stimulation of teliospore germination. The previous crop did not affect infection in wheat (Goel and Jhooty, 1989). It was reported by Clayton (1925) that the spores of *U. tritici* do not lose their germinability even after passage through the alimentary tract of domestic animals. In Australia, it has been found that the disease expression was more in soil pH range of 5.5 to 8.7 and disease was less common in calcium deficient soils (Millikan, 1939). Deep sowing leads to more infection probably due to the fact that coleoptile remains below the soil for a longer period and thus is exposed to infection for greater length of time (El-Helaly, 1948.).

Disease Management

Resistance : Sources of resistance to flag smut have been reported by several workers in India. Some wheat varieties, which were widely grown in northern parts of country, earlier such as Kalyansona, PV 18, Lerma Rojo and Sonora 64 were found susceptible to flag smut (Gothwal and Pathak, 1977). Singh and Sethi (1975) tested a number of wheat varieties for their resistance to flag smut and reported seven varieties viz., VL 426, HD 2117, HW 161, HB 121,

HB 113, Norteno and 112-P4 as resistant having less than 2 per cent infection and 4 varieties viz., VL 422, VL 431, VL 529 and HB 110 as moderately resistant with infection ranging from 2.13 to 4.40 per cent.

Goel and Gupta (1990) evaluated 163 wheat genotypes for resistance after artificial inoculation with *U. agropyri* and found 32 immune genotypes, 16 highly resistant and 28 moderately resistant genotypes. All 12 *T. durum* genotypes and the single *T. dicoccum* genotype tested were immune. Comparisons of the reaction of these genotypes at other locations indicated the presence of different virulences in the pathogen. Later, Goel (1991) derived information on genetic inheritance of resistance to flag smut from the data on percentage infection, measured as incidence of flag smut/plant at milk-soft dough stage, in 5 highly resistant wheat lines (CI14394, CPAN1425, E6840, HS86 and WL410), a highly susceptible variety (Agra Local) and their F_2 and F_3 progenies, grown during 1986-89 in Ludhiana, India. The results indicated the presence of 9-10 major genes for resistance in E6840, 3-4 in WL410 and 2-3 in the other 3 resistant lines, as well as minor genes and modifiers in some cases.

Among 148 entries, comprising *Triticum aestivum, T. durum* and triticale evaluated for resistance to *U. agropyri* in an infested village field at Aryanagar, Hisar during 1988-91, 23 entries of *T. durum* and triticale were immune to the disease, and of the 125 *T. aestivum* entries, 23 were immune and 7 highly resistant (Beniwal *et al.*, 1999). In a separate study 126 strains of *Triticum aestivum* and *T. durum* screened for reaction to flag smut and nineteen strains of *T. aestivum* and two of *T. durum* were resistant (Bhatnagar *et al.*, 1991).

Among several wheat and triticale genotypes, and wheat X triticale crosses, evaluated for resistance to flag smut under artificial inoculation conditions at Hisar during 1992-93 and 1993-94, all durum wheat and triticale genotypes included in the study (8 and 13, respectively) were completely free from flag smut incidence. Bread wheat genotypes gave variable disease reactions from no disease incidence to high susceptibility. Similarly, wheat X triticale crosses showed differential flag smut incidence ranging from no disease incidence to moderately resistant disease response (Singh *et al.*, 1997). Among nineteen wheat varieties commercially grown in the North-Western Plain Zone of India evaluated against flag smut during the 1995-96 and 1996-97 crop seasons at Hisar and Karnal, *bread* wheat varieties, namely, WH 283, WH 291, WH 542, Raj. 3765, PBW 343, PBW 373 and a *durum* variety WH 896 exhibited multiple disease resistance (Beniwal *et al.*, 1999).

Allan and Pritchett (1976) attempted to show the relationship between the type of spike and flag smut incidence. They reported that incidence of *U.*

agropyri was indirectly related to the club type of spikes. Paired comparisons between isolines showed that club types averaged 3 and 12.2 per cent more flag smut than lax type. Further they observed that the relationship between the C allele and increased severity of flag smut appeared to be indirect, since no linkage or pleiotropism was detected between the two characteristics. The C allele reduced rates of seedling growth and emergence, and delayed emergence may increase the chance for wheat seedlings to become infected.

The relationship between phenols and carbohydrates and disease resistance in wheat and triticale has been studied by Sindhan *et al.* (1996) using several varieties of triticale (TL 2656, TL2597), durum wheat (PBW-34, Raj. 1555) and bread wheat (HD 2329, WH 283, WH 147, C306). The triticale varieties, TL2656 and TL2597 were immune to *U. agropyri*, durum wheats, PBW 34 and Raj 1555 were highly resistant and bread wheats WH 283 and HD 2329 were resistant. Bread wheats, WH 147 and C 306 were highly susceptible. Total phenols, orthodihydric phenols, total sugars and reducing sugars were higher in healthy leaves of resistant than susceptible varieties and their concentration was reduced in infected leaves in all the varieties. This depletion was more pronounced in susceptible varieties.

Biocontrol approach : Seed treatment with bioagents-*Azotobacter chroococcum* (Strain No. MAC-27), *Trichoderma viride* and *Gliocladium virens* (alone and in combination with carboxin and carbendazim) and with solar heat and water flotation treatments, were investigated for control of flag smut of wheat by Sanjay Arya *et al.* (1998). They found that solar heat and water flotation treatments showed 2.3 and 3.1% plant, and 1.8 and 1.6% tiller infection, respectively, in pot and field experiments. Seed treatment with *A. chroococcum* showed a significant reduction in disease incidence compared with the controls. Disease control with the antagonists was relatively higher in field conditions compared with pot experiments.

Chemical management : Besides the use of resistant varieties, the disease can also be controlled by crop rotation, cultural practices, exclusion by quarantine and seed treatment. Seed treatment has been considered partly effective under conditions of soil infestation. The areas where resistant varieties are not under cultivation, the disease usually persists perennially in spite of careful seed treatment. However, where resistant varieties are available, flag smut has almost disappeared (Purdy, 1965).

Seed treatment with copper carbonate at the rate of approximately 2 ounces per 80 pounds of seed was found very effective (Dawson, 1934). Steeping the seed in formaldehyde solution (one pound formaldehyde in 40 gallons of water) has also been recommended. Tetrachloronitroanisole (TCNA) was found by Purdy (1962) as an excellent fungicide for seed and soil disinfection against this disease. Bedi (1957) reported that the flag smut infection could be suppressed by seed treatment with Agrosan GN and Ceresan at the rate of 1: 400. Tyagi (1968) found that the seed dressing with some organomercurials could bring about complete control of the smut. He carried out experiments to control soil-borne flag smut and reported seed treatment with Vitavax and TCNA as highly effective. According to Tariq *et al.*, (1992) seed treatment with Vitavax (carboxin), Bayletan (triadimefon), Sunlate, Vincit (flutriafol + thiabendazole) and Raxil (tebuconazole) gave 100% control of flag smut tested on a susceptible cultivar. Isoproturon and 2,4-D as foliar sprays reduced disease incidence. Disease incidence was lower in plants sprayed with 2,4-D than with isoproturon. Of the insecticides tested, phorate controlled *U. agropyri* best in field and greenhouse trials (Sanjay, Arya *et al.*, 1997).

Bedi *et al.*, (1975) evaluated Vitavax and Plantvax with other five common fungicides as seed treatment and found that Vitavax and Plantvax completely controlled the disease. Brassicol, Trametan and Thiram were significantly superior to Bisdithane and Dithane Z-78. Bhatnagar *et al.* (1975; 1978) found that seed dressing at the rate of 2 g/kg seed with Vitavax, Benlate and Campogran-M gave almost complete control of the disease. Raina *et al.* (1980) based on 4 years field tests reported carboxin, benomyl, carbendazim, captafol, Brassicol (quintozene), Captan and mancozeb as quite effective in reducing the flag smut incidence. Four fungicides, Vitavax 75 WP (carboxin), Bavistin 50 WP (carbendazim), G696 20 WP and Raxil 2DS (tebuconazole) at 1.5, 2.0, 2.5 and 3.0 g/kg were evaluated for the control of flag smut by Sindhan *et al.* (1996) before sowing in the field. Disease incidence and severity were recorded 90 days after planting. All the fungicides significantly reduced flag smut at all dose levels and, at their higher doses (2.5, 3.0 g/kg), they all increased the amount of total phenols, orthodihydric phenols, total sugar and reducing sugars at both stages of plant growth.

Cultural practices : Besides the use of resistant varieties and seed treatment, the incidence of flag smut can also be controlled adopting certain cultural practices such as crop rotation, early sowing, burning of stubbles, exclusion by quarantine and hot weather cultivation (Bedi, 1957; Mafiz, 1948). In Egypt, sowing by 'herati' method in which seeds are broadcast on moist land, results in more flag smut than in 'afir' method in which seeds are broadcast on dry land, harrowed, and then irrigated (El-Helaly, 1948). It has been shown that since the spore in soil is dry, it does not become infective while the seedling is in susceptible stage of development in the 'afir' method of planting. On the other hand, in the 'herati' method of planting spores receive the necessary pre-soaking and are infective at the proper time, resulting in high percentage of infection.

Hagiwara *et al.*, (1996) reported control of wheat flag smut by soil sterilization with hot water injection. Depth of sowing is also known to influence flag smut incidence. Four sowing depths (2, 4, 5 and 6 cm) were assessed by Sanjay Arya *et al.*, (1998) in pot and field experiments for their effect on the incidence of *U. agropyri* on wheat and found that sowing at a depth of 2 cm from the soil surface recorded a significantly low incidence of disease. The highest disease incidence was observed in pots at a sowing depth of 6 cm. Delaying of wheat sowing from 10 Nov. to 30 Nov. increased infection, but a further delay had an opposite effect, irrespective of the wheat cultivars tested (Goel and Jhooty, 1989). Disease symptoms appeared 4-17 days earlier on late-sown wheat, while the percentage disease incidence on the plants was less by 43.2-73.8% compared with the normal sown crop (Beniwal *et al.*, 1992).

FUSARIUM HEAD SCAB

Head blight or scab is an important disease of wheat, oats, barley and spelt in years when warm and wet weather persists during the heading and blossoming period of these cereal crops. Scab is not only a disease of small grains, but of many grasses including wheat, foxtail, quackgrass, crabgrass and bluegrass (Lipps, 1996). The disease is widely prevalent in all regions of world where humid conditions persist during anthesis and grain filling stage (Snijders, 1990). Crop intensification with rice-wheat cropping system favours Fusarium scab of wheat (Saari, 1986).

Pathogen

The fungi causing scab are all members of the genus *Fusarium*, the asexual spore forming stage. Scab, also known as head blight, is incited by several species of *Fusarium*. The principal pathogen is *Fusarium graminearum* (the sexual stage is *Gibberella zeae*), but *F. avenaceum* (*G. avenaceum*) and *F. culmorum* (no known sexual state) have also been reported (Lipps, 1996).

Losses

Wheat head scab epidemics can cause great, direct yield losses. In severe epidemics, the percentage of scabbed spikes (disease incidence) is usually 30 to 50%, but can exceed 60 to 70% in the most susceptible cultivars in severe epidemics (Zhang, 1998; Shi and Wang, 1991). Scab has been one of the most serious production problems for wheat in the 1990s, and the fungal disease has proven devastating to U.S. malting barley (Rick Ward, 1996). In India, the outbreak of scab was reported from Arunachal Pradesh and Assam in 1973 (Roy, 1973). It also appeared in some districts of Punjab adjoining hilly region and posed serious problem, causing 20-40 per cent loss in wheat yields in 1989-90 (Singh *et al.*, 1993). Although, it is still a disease of minor importance, it can be a potential disease in future, especially under maize-wheat and rice-wheat cropping systems (Liu, 1985; Saari, 1986).

Scab is important, not only because it reduces yield, but also because it reduces the quality and feeding value of the grain. The disease in addition to direct yield losses, also causes further indirect economic losses. Disease also results in mycotoxin contamination of affected grain. Diseased kernels are often shriveled, lower in weight, and frequently lost with the chaff during combining. The quality of scab infected wheat kernels is reduced due to changes in protein content and to contamination with mycotoxins (Xu and Chen, 1993; Wu *et al.*, 1997). The fungi causing scab may produce chemicals in the infected grain which are toxic to livestock and humans.

Wheat infected by foliar head blight contain deoxynivalenol (DON) a mycotoxin, also known as vomitoxin. This toxin is produced primarily by *F. graminearum* and *F. culmorum*. Non-ruminants are the most sensitive to the presence of DON (Annoymous, 2003a). The deoxynivalenol (DON) [vomitoxin] content of harvested grain was significantly correlated with Fusarium head blight incidence and severity. DON is considered to be a virulence factor in *F. graminearum* (Xu and Chen, 1993; Desjardins *et al.*, 1996). When isolates of *F. graminearum* with various degrees of virulence were compared for DON production, a high positive correlation was observed between virulence and levels of DON produced in cultures on autoclaved wheat grain (Xu *et al.*, 1990; Wei *et al.*, 1990).

Disease cycle

The earliest and most conspicuous symptom of scab occurs soon after flowering. *F. graminearum* infects wheat spike tissues primarily through anthers that protrude from the glumes and then spreads to the epidermis of the glumes and to the ovaries (Xu and Hideki, 1989). The fungus can also directly invade the external and internal epidermis of glumes, and can penetrate the external epidermis of anthers through natural openings (Xu and Hideki, 1989). Generally, spread to the ovary is much faster than spread throughout the glume, which can result in partial infection of individual spikelets. Diseased spikelets turn light-straw colored and have a bleached appearance due to premature death of tissues. When the fungus infects the stem immediately below the head, the entire head may die (Lipps, 1996).

Infected kernels are generally shrunken, wrinkled, and light in weight, with a rough, scabby appearance. These kernels range in color from light-brown to pink to grayish white. If the fungus invades and kills the rachis or main axis of the spike, the spikelets above that point die. Heads with diseased spikelets may become speckled with dark purplish-black fruiting bodies (perithecia) of the fungus if the weather remains cool and moist until harvest. These perithecia are a sign of the sexual stage, the *Gibberella* stage of the fungus (Lipps, 1996). The scab fungi overwinter and survive between crops in infected grain and grass stubble, chaff, and crop residue on the soil surface. They survive as asexual spores (conidia), mycelium, and perithecia within which are borne the sexual spores (ascospores). The fungi are also common contaminants on seed. These fungi continue to grow and produce spores from harvest until the residues decompose in the soil (Lipps, 1996).

The conidia are produced profusely during warm, moist weather on crop residues. Ascospores produced within perithecia are discharged into the air as the fruiting bodies wet and dry during fluctuations in moisture levels. Free moisture and warm temperature (18°C+) favour foliar and spike infection (Saari, 1986). The perithecia and ascospores produced on crop stubble may survive from one season to another and can be the incitant of scab infection (Lipps, 1996). Ascospores inside the perithecia require water as a stimulus to forcibly eject them into the air (Tschanz *et. al.,* 1975). The spread of this type of inoculum is by rain splashing over a short distance under favourable conditions (Tandon, 1985). These factors may be attributed to the high incidence of Fusarium scab during 1989-90 in some districts of Punjab.

Management

As no single measure can provide adequate protection to wheat from head scab, several strategies have to be formulated for successful management of this disease (Bai *et al.,* 1999).

Cultural practices: Selection of cultural practices aimed to reduce inoculum-borne residues will assist in the control of Fusarium head blight to a great extent. Changes in regional tillage practices, principally the move towards conservation tillage and reduced-till systems, contribute to the epidemics (Dill Macky and Jones, 2000). Cultural practices like draining wheat fields to lower the water table, ploughing fields to bury or remove crop residues before sowing to reduce the primary inoculum and changing the sowing date to avoid climate conditions that are favorable for disease at the flowering stage, have been applied in China to reduce disease severity (Liu, 1990; Xu and Chen, 1993). Cultivation of wheat may be avoided near old maize fields if stalk residues are left on the soil surface, in those areas. No-till wheat seeded in old maize residues greatly increases the chance of scab. Field sanitation with deep ploughing of all infested stubble and straw of cereals and weed grasses, maize stalks and rotted ears may reduce the disease incidence.

Crop rotation having wheat after a legume crop (soybean), while maintaining a rotation with 2 to 3 years between two wheat crops has been recommended. Fusarium head blight incidence and severity were greatest when wheat followed maize and least when wheat followed soybeans. Yields of wheat were approximately 15% lower in plots where wheat followed maize or wheat than in wheat following soybeans and were 10% greater in mould board ploughed plots than in either chisel ploughed or no-till treatments (Dill Macky and Jones, 2000). Complete burial of infested crop residues from previous suspect crop such as wheat or maize reduces inoculum. Early planting or planting of cultivars that flower early is an effective scab avoidance strategy in some parts of the world (Khonga and Sutton, 1988). Harvested grain may be dried to 13.5 percent moisture as soon as possible after harvesting to avoid mycotoxin contamination. The Fusarium fungi cannot grow or produce mycotoxins in grain at this moisture level (Lipps, 1996).

Resistance: The strategy for the integrated control of scab should involve the utilization of resistant varieties. Wheat resistance to head scab is an inheritable trait, but the number of genes controlling resistance is still controversial. Some investigators have hypothesized that resistance to wheat head scab is controlled by many minor genes, but others have provided evidence for the existence of genes with major effects (Bai *et al.*, 1990). There are few varieties of wheat highly resistant to scab. Differences in resistance/ susceptibility may be due to physical barriers to infection of spikelets. A significant difference among wheat cultivars with respect to sensitivity to DON produced by *F. graminearum* has been observed, and has been associated with scab resistance in wheat cultivars. Thus, adding DON to media during tissue culture could select variants tolerant to DON and resistant to wheat head scab (Wang *et al.*, 1989; Xu *et al.*, 1990).

In breeding programs, progress has also been made in improving scab resistance of moderately susceptible cultivars by using irradiation treatment (Bai *et al.*, 1999). In China, where wheat scab is a major problem, the cultivars belonging to 'Ning' system -Ning 7840 and others are resistant (Zong, 1988). Most wheat head scab resistant materials developed in China are related to Sumai 3 (Bai *et al.*, 1999).

Chemical management : Effective chemical control of this disease has been made possible with some of the new systemic fungicides. Significant reduction in scabs by application of foliar fungicides such as Triadimefon, Benomyl, Propiconazole or MCB has been recorded (Bergstrom, 1993). Xu and Chen (1993) have stressed that the timing of fungicide application is critical for effective control. Fungicides should be applied once or twice from the heading stage to the flowering stage, depending on the environmental conditions and the susceptibility of wheat cultivars. Spraying carbendazim at the heading or flowering stages of wheat generally gives 80 to 90% control of visible symptoms (Xu and Chen, 1993). Tebuconazole + thiram were effective in controlling *Fusarium culmorum*, followed by prochloraz + mancozeb (Balmas *et al.*, 1995). The interaction of cyclic hydroxamic acids, and their related benzoxazolinones have been reported to reduce the mycelial growth of *Fusarium culmorum* (Vilich, 1998).

Biocontrol approach : Several workers have reported potential candidates that can be developed as effective biocontrol agents against Fusarium head blight. *Epicoccum nigrum* (*E. purpurascens*) and *Trichoderma harzianum* have been reported to be effective against *Fusarium* apart from improving emergence and vigour of wheat seedlings. Pelleting seed with *Aspergillus clavatus, T. aureoviride or T. harzianum* conidia and binder improved emergence and vigour in soil infested with these three fungi (Wu, 1976). Toxic substances produced by *Aspergillus clavatus* reduced the growth of *Fusarium culmorum in vitro*. Hyphae of *Trichoderma aureoviride, T. harzianum, Epicoccum nigrum* penetrated those of *F. culmorum* and *R. solani* (Wu, 1977).

Application of *B. subtilis* strain A014, and *Bacillus* spp. isolate-2 and isolate 8-2 as spray over wheat spikes decreased visible symptoms of head scab by approximately 50% in the greenhouse and in field plots (Shi and Wang, 1991; Wang *et al.*, 1992). Among around 700 isolates from field soil in Finland for their activity against *F. culmorum* on spring wheat in sand and peat cultures the highest number of effective isolates belonged to *Trichoderma* sp., but isolates of *Gliocladium* and *Fusarium* spp. were also antagonistic (Tahvonen and Sarri, 1992).

FOOT AND ROOT ROT DISEASES

Among the major and most notorious soil borne diseases of wheat, the foot and root rot disease warrant most attention as they are most difficult group of diseases to manage due to their diversity and sustained presence in soil by nature. The terms root rot, foot rot, crown rot, pink rot, culm rot and stalk rot are used to describe a group of wheat diseases caused by one or more organism alone or in association (Burgess *et al.*, 1975). Seedling blights and root rots are usually present every year in most wheat fields. Root decay, seedling blight, stunting and premature death of older plants are characteristic of root rot infections. These diseases are not as easily observed as rusts and smuts, but yields are often seriously reduced. Diseased seeds and several soil-borne organisms are sources of root infections and seedling blight. Root rots are caused by many species of fungi.

Though root and foot rots are important diseases worldwide, the occurrence and frequency of causal agents vary from one region to another (Kulkarni and Nargund, 1986; Fernandez *et al.*, 1985; Lyamani, 1988). Several factors like amount of inoculum present, cultural practices apart form soil and climatic factors determine the actual damage caused by these diseases (Statler and Darlington, 1972, Tinline 1986, Windermuth and McNamara, 1991).

Common Symptoms

The foot and root rot disease generally reduce seedling vigour, impair the functioning of roots and crowns resulting in crop stand loss, reduced yield and lower grain quality (Fenster *et al.*, 1972 and Mergoum and Quick, 1990) apart from affecting the yield components like number of tillers per plant and number of kernels per spike (Ismail and Michalik 1980). Discolouration, necrosis or rolling of roots, sub crown internodes, crown stem browning constitutes some of the typical symptoms alone or in combination. Under severe disease pressure, symptoms such as stunting, later known as white heads (WH) or dead heads may occur (Scardaci and Webster, 1982; Stack, 1982; Lyamnai, 1988). In India root rot diseases have been reported in majority from "Barani" areas of the country (Joshi, *et. al.*, 1986).

Pathogens

Several pathogens are involved in root rot and foot rots of wheat. Some of the important pathogens reported frequently like *Sclerotium rolfsii, Helminthosporium* spp., *H. sativum, H. tetramera, Pythium graminicolum, Rhizoctonia solani, Fusarium* spp., *F. moniliforme, F. dimerum var. subglutinans, F. solani, F. equiseti, F. graminearum, F. avenacerum, Fusarium pseudograminearum, Ophiobolus graminis, Sclerotinia sclerotiorum, Curvularia verriceformis, Neovossia indica* (Subramaniam (1928), McRae (1923), Chattopadhyay (1953), Agarwal and Sahni, 1963, Padwick (1940), Misra and Singh (1967), Mishra and Munankami (1970), Kari, (2001).

Foliar and root samples were collected from different locations to monitor foliar and root diseases caused by soil-borne fungi at the booting stage of wheat crop in rice-wheat cropping areas of four districts of Punjab, Pakistan. The highest mean incidence of root rot was 100% in Narowal and Sialkot whereas the lowest of 95% was in Gujranwala. The highest mean severity was 2.2% in Narowal while the lowest 1.5% was in Sialkot. *Fusarium* spp., *Helminthosporium* spp., *Curvularia* spp., *Alternaria alternata, Macrophomona phaseolina, Coniothyrium* spp., *Mucor* spp., *Trichocladium* spp., *Rhizoctonia solani, Humicola* spp. and *Aspergillus* spp. were isolated from the plants showing root rot symptoms (Shazia, *et al.,* 2003).

Some of the major fungal pathogen associated with various foot and root rot diseases are reviewed below.

Rhizoctonia solani

Rhizoctonia solani has been reported to be associated with root rot of wheat in northern and central India (Sharma and Jain, 1967). The diseased seedlings of wheat remain stunted with yellowing of leaves which in turn dry up and curl. The injury to root ranges from necrosis of root tips to rotting of primary and secondary root system. Patches showing no growth in fields under intensive cultivation has also been observed because of successional killing of seedlings. Dry soil conditions and temperatures around 30°C favour the infection.

Disease appears in patches in which plants are stunted or killed (bare patch). Plants are weakened and killed, or more generally recover with delayed maturity and low yield. Mycelium of fungus is present in rotted tissues and root stubs near crown. Disease development is more at low temperatures, although some strains of the fungus incite disease at high temperatures and is favored by reduced or no-till practices (Palm, 1993).

It has been noted that the conditions favourable to *Sclerotium,* also encouraged the incidence of root rot and pre and post emergence damping off caused by *R. solani* in wheat. Infection of young plants results in wilting and death. Infected tissue turns discoloured and eye spot patches appear on the stem and sheath and occasionally on leaves (Saari, 1986). Mazzola *et al.* (1997) observed stunted and spindly seedlings having roots girdled and severed by lesions typical of *Rhizoctonia* root rot.

Sclerotium rolfsii

Sclerotium rolfsii is one of the major seedling root and stem rot pathogen frequently encountered in tropical environments. It causes pre and post emergence damping off, seedling blight and death of plant at any stage of development. Presence of 'white heads' in the standing crop indicates premature death of adult plants (Saunders, 1988). *S. rolfsii* was isolated from the root of wheat plants from India and Burma (Subramanian, 1928) and several other workers have reported the association of this organism with foot rot of wheat in India (Kulkarni and Nargund, 1986). Dubin and von Ginkel (1991) have considered *Sclerotium rolfsii* as most serious soil borne pathogen over *Fusarium* and *B. sorokiniana* in a survey conducted in Pakistan.

Fusarium spp.

The soil and stubble borne fungal diseases of wheat also include Fusarium crown rot (*Fusarium pseudograminearum, Fusarium avenaceum, Fusarium* spp.). Foot rot generally begins as an invasion of seedling roots. Infected plants remain stunted and later turn yellow and die. Roots of diseased seedlings are rotted, reddish-brown in color and may be covered with a mass of grayish or pink mold. Root rots or crown rots, which occur as plants approach maturity, cause premature ripening and death of plants. Dead plants may have a bleached appearance. The occurrence of the disease complex was high under low rainfall. Diseased tillers were either unproductive or with shrivelled seeds, thus both grain yield and kernel weight of infected tillers were conspicuously lower compared with non-infected tillers (Kari, 2001).

Foot rots are favored by cool temperatures. The fungus is more a secondary invader of roots initially infected by the common root rot fungus. The fungus surviving on crop residues and rotation may reduce disease severity. Losses are due to reduction in the number of heads, reduced seed size and fewer seeds per head.

Bipolaris sorokiniana

Common dry land root rot is caused by *Cochliobolus sativus* (Ito & Kurib.) Drechs. ex Dastur (teleomorph) and *Bipolaris sorokiniana* (Sacc.) Shoem. (anamorph) (= *Helminthosporium sativum*). The fungus infects the sub-crown internode of young wheat seedlings, widely distributed on wheat seedlings, causing root and shoot rotting. Post-emergence death occurs due to rotted stems.

Symptoms of common root rot include brown spots or discoloration on the roots, sub-crown internodes, lower leaf sheaths and crown. Affected plants tend to be scattered over a field, and they may be slightly stunted. Initial symptoms of common root rot originate on young seedlings from inoculum carried on the seed or from infections originating from soil borne conidia near the seedling (Piening, 1997, Wiese, 1987). Dark brown lesions appear on the outer coleoptile tissue and/or on the leaf base. Lesions may coalesce into long areas of necrotic brown tissue. In extreme cases, the entire seedling may die. Even if the seedling survives, the growth of developing plant may be stunted and plant may die prior to heading (Mathre *et al.*, 2003). Common root rot may also result in decreased number of tillers on an infected plant as compared to the number on a healthy plant (Duczek, 1989).

Disease cycle

Most infections are initiated by soil-borne inoculum; however, infection can also be carried on or in the seed. New spores are produced on infected plants or on crop debris. Spores are spread by wind, water, infected seeds and cultivation practices, and can remain viable in the soil for several years. (Mathre *et al.*, 2003). Root rot occurs in both wet and dry soil, however, the disease tends to be worse under dry conditions.

Harris and Moen (1985) reported that tillering predisposes wheat to a succession of secondary infections of common root rot and minor pathogens including *Fusarium equiseti, F. avenaceum, Cochliobolus sativus, Microdochium bolleyi* and *Alternaria alternata.*

Management

The sustainable disease management in wheat has to be focused on a systems approach against these soil borne diseases by suppressing the pathogens at economic threshold level. Such a system approach should incorporate various components like selection of variety and resistant cultivars, cultural practices viz, crop rotation, sanitation, time of planting, fertility, water management and time of harvest and chemical and biological approaches.

Cultural practices : Chemical control of this disease complex is not effective due to its soil-borne character; therefore, emphasis should be placed on tolerant cultivars to root and foot rot, and to proper cultural practices (Kari, 2001). Dastur (1932) reported that late sowing (middle to end of October and early November) reduces the incidence of foot rot disease. Sharma and Jain (1967) also observed that there was a change in the incidence of *F. graminicolum* due to shift in sowing dates.

Nargund *et al.*, (1982) reported that Safflower oil cake and groundnut oil cakes were effective in reducing the mortality of wheat seedlings in artificially infested soil to control foot rot of wheat. *Fusarium* crown and foot rot in cereals could be controlled by reduction or delay in plant water stress by preventing evaporation, delaying sowing and applying less N, which resulted in increased yields in diseased fields by 10-20% (Cook, *et al.*, 1981).

The potential of Indian mustard (*Brassica juncea*) seed meal as an in-furrow treatment for suppressing soil-borne fungal pathogens of wheat was assessed by Kierkegaard *et al.,* (1996). But wheat seed germination and growth were reduced when 10kg/ha mustard seed meal was applied in the drill row with the seed, but up to 1000 kg/ha incorporated into the top 5 cm or placed as a band 4 cm below the seed had no adverse effect on seedling growth (Kierkegaard *et al.,* 1996). Mazzola *et al.,* (1997) suggested that burning done routinely to improve the yield of bluegrass seed can also reduce the potential for damage caused by *Rhizoctonia* and *Pythium* root rots of wheat in the field. After tillering, *R. solani* could rarely be recovered from inoculated wheat plants in a pot experiment and the *Rhizoctonia* attack was confined to seminal roots and only rarely to nodal roots (Harris and Moen, 1985). Studies of changes in severity of Rhizoctonia root rot showed that the levels of disease suppression could be modified over time to the point where complete control of the disease can be achieved. The development of disease suppression occurred at a faster rate under no-till however tillage practice does not appear to change the final level of suppression of Rhizoctonia root rot (Gupta, *et al.,* 2003).

Biocontrol approach : Soil-borne diseases result from a reduction of biodiversity of soil organisms. Restoring beneficial organisms which counter or antagonize disease-causing pathogens will render a soil disease-suppressive. Plants growing in such a disease-suppressive soil resist diseases much better than in soils low in biological diversity. Beneficial organisms can be added directly, or the soil environment made more favorable for them by the use of compost and other organic amendments.

Manjappa (1979) reported that *T. viride* and *Penicillium* sp. were found to be antagonistic to *S. rolfsii.* Jone (2001) reported a commercial formulation with some bacterial antagonists named "Harpin", which he proclaimed promising biological pesticide against soil borne pathogen, *Fusarium* spp., which may be used as a component of integrated pest management programmes also. Among the organisms isolated from amended field, *Trichoderma viride, Penicillium* sp.; *Bacillus subtilis* and *Streptomyces* sp. were antagonistic to *S. roljsii* (Nargund *et al.,* 1982). Hegde *et al.,* (1980) reported effective control of *S. rolfsii* with *B. subtilis in vitro.*

Epicoccum nigrum (*E. purpurascens*) and *Trichoderma harzianum* placed in *soil* infested with *Rhizoctonia solani* improved emergence and vigour of wheat (Wu, 1976). *Aspergillus clavatus* produced toxic substances which reduced the growth of *Drechslera sorokiniana (Cochliobolus sativus)* and *Rhizoctonia solani in vitro.* Hyphae of *Trichoderma aureoviride* and *T. harzianum* penetrated hyphae of *R. solani* (Wu, 1977). Seed treamtnet with *Bacillus subtilis* strain A-B could reduce population of *R. solani, Pythium* sp. and *Fusarium* sp. apart from increasing tillering and yield of wheat crop. Biological control of *Sclerotium rolfsii* with *B. subtilis*, *Trichoderma* spp. and *Penicillium* sp. has also been reported (Hegde, *et al.*, 1980; Luthra and Sattar, 1934). Some effective bacteria *Pseudomonas alcaligenes* PsA15, *Bacillus polymyxa* (*Paenibacillus polymyxa*) BcP26, *Mycoplana bullata* MpB46 and *Mycobacterium phlei* MbP18 produced the phytohormone, sp., auxin and reacted antagonistically to soil-borne plant pathogen *Fusarium oxysporum* apart from increasing yield of wheat (Egamberdiyeva *et al.*, 2001). Biological control of the root pathogen has also been reported to be effected by *Pseudomonas fluorescence* (Pandian and Singh, 2000).

Chemical management : Several workers reported the effect of fungicides on foot-rot organisms (Agrawal, and Singh, 1969, Joshi and Vyas, 1974, Pan and Sen, 1977, Singh and Mehrotra, 1978, Kulkarni, 1980, Nargund *et al.*, 1982). Seed treatment can be utilized effectivelly to combat seed and soil borne diseases. Pre-sowing seed treatment with carboxin, carbendazim or Benomyl @ 2.0 g /kg seed provides protection to many diseases such as seedling blight, root rot, foot rot (Singh and Rai, 1986). When used in combination with other active ingredients, triticonazole has given excellent control of the major seed- and soil borne diseases of wheat (Beal *et al.*, 2001).

Agrawal and Singh (1969) studied the efficiency of ten seed dressing fungicides against *S. rolfsii*. Arasan at 0.2 per cent was the best followed by Captan, Thiram and Bisdithane. Bahadur *et al.*, (1974) found BAS 3191F, Vitavax, Rhizoctol and Brassicol as effective against S. *rolfsii*. Plantvax (oxycarboxin) and Vitavax (carboxin) as seed dressing @ 2g/kg gave effective control of *S. rolfsii* upto 35 days after seeding (Reddy, *et al.*, 1971). Singh and Mehrotra (1978) studied the effect of 20 fungicides at 200 ppm on *H. sativum*, the root rot and foot rot pathogens of wheat. They found that, fungicides reducing mycelial growth by 70 per pent or more were generally found to inhibit the respiration rate significantly. Das and Srivastava (1971) found that Parzate and Arasan were effective against this pathogen. Joshi and Vyas (1974) suggested the use of Demosan and Thiram for the control of foot and root rot of wheat caused by *F. oxysporum*. Some promising systemic fungicides as seed treatment against *Sclerotrium rolfsii* are reported to have activity against *R. solani* also (Lapis, 1980b).

Resistance : Much emphasis should be given on growing resistant varieties. Rao and Joshi (1974) stressed the importance of breeding for disease resistance to different wheat diseases. Reddy *et al.*, (1971) found in their studies that Lerma Rojo 64 A, was the most resistant among the lines tested against *S. rolfsii*. Similarly, when Nargund *et al.*, (1982) screened twenty-five wheat genotypes including varieties and multilines, Lerma Rojo showed maximum disease resistance while Agra local showed high susceptibility to *S. rolfsii.* Maximum resistance to the foot rot disease was shown by CC 464, UP 301 and HD 2189 among thirty- five wheat cultivars screened against *S. rolfsii* as reported by Kulkarni *et al.,* (1978). All the 15 wheat varieties evaluated were highly susceptible to the root-rot complex against four root- rot pathogens, viz., *P. graminicolum, H. sativum, S. rolfsii* and *Fusarium* sp. under glasshouse conditions in autoclaved soil. In un-autoclaved soil, however, RR-21 wheat variety was moderately susceptible to S. *rolfsii* while all others showed susceptible or highly susceptible reactions (Kauraw and Singh, 1975). Fifty varieties of wheat were evaluated for their resistance against *R. solani* at Powerkheda. Varieties, Kalyansona, HY 38, HY 172, HY 277, HY 278, HY 5, S 331, PKD 9, MP 14, HY 34, MPO 212, HY 173, HY 633, MP 4, NP 404, NP 839, HI 316, HD 1739, and HD 1977 were found to be resistant (Chaurasia, *et al.,* 1973). Saksena and Kumar (1967) also recorded least susceptibility of Lal Bahadur and UP 301 against *R. solani.*

CONCLUSION

Soil borne diseases are ever present and pose a serious threat to all crops. These diseases are important because of losses they cause to the cultivated crops. In wheat, soil borne diseases have become a matter of great concern to the research workers due to variability in the causal fungi from place to place and the various kinds of symptoms i.e. foot rots, root rots and seedling blight make them still complicated. The most important disease management strategies aim either to prevent infection or to restrict the spread of the pathogen. In fact, curative measures are generally rare and single method does not offers complete control to soil borne fungal pathogen of wheat. Therefore, a rational combination of management strategy elements, like selection of variety, resistant cultivars, cultural practices and time of harvest must be taken into account. Emphasis for crop feeding strategy that should involve the use of legume in rotation, organic recycling and balanced fertilizer application in accordance with the requirements for macro and micronutrients in different soil types.

Rotation with non-susceptible crops like chickpea or cluster bean is supposed to minimize the risk of flag smut, Karnal bunt and root and foot rots. Together with culture practices amendment of the soil with suitable bioformuations and need based chemical applications help in the containment of the diseases. Thus, by manipulation of different IDM approaches, it should be possible to minimize the risks of diseases. However, the success would depend largely on an effective disease monitoring system, frequent communication among the various disciplines involved in the programme and active link between research scientist and farmers.

REFERENCES

Adlakha, K. L., Wilcoxson, R. D. and Raychaudhuri, S.P. 1984. Resistance of wheat to leaf spot caused by *Bipolaris sorokiniana*. *Plant Dis.* 68: 320-321.

Aggarwal, R., Joshi, L.M. and Singh, D. V. 1989 a. Inoculum production in relation to varieties in Karnal bunt disease of wheat. *Indian Phytophath.* 42: 569-571.

Aggarwal, R., Singh, D.V. and Joshi, L.M. 1989 b. Survival of Karnal bunt teliospores in soil. *Indian Phytopath.* 42: 273.

Aggarwal, R., Singh, D.V., Srivastava, K.D. and Bahadur, P. 1991. Karnal bunt of wheat: what it does and how it spread. *Indian Farmg.* 40: 9-10.

Aggarwal, R., Singh, D.V., Srivastava, K.D. and Bahadur, P. 1996. The potential of antagonists for biocontrol of *Neovossia indica* causing Karnal bunt of wheat. *Indian J. Bioi. Control.* 9: 69-70.

Aggarwal, R., Srivastava, K.D. and Singh, D.V. 1996. Mechanism of combat between plant pathogens and antagonists. Presented in *Annual Meeting of Indian Phytopathological Society* held at PAU, Ludhiana, 14-16 Feb. 1996. pp.7.

Agrawal, S. C. and Singh, S. P. 1969a. A varietal reaction of wheat against *Sclerotium rolfsii*. *Indian Phytopath.* 22: 511-13.

Agrawal, S. C. and Singh, S. P. 1969b. Efficacy of some fungicides in controlling foot rot of wheat. *Indian Phvtopath.* 22: 476-77.

Allan, R. E. and Pritchett, J. A. 1976. Relation of flag smut incidence to spike type in wheat isolines. *Crop Sci.* 16: 596-597.

Amer, G. A. M. 1995. Biocontrol of Karnal Bunt of Wheat. *Ph.D. Thesis*, Division of Mycology and Plant Pathology, Indian Agricultural Research Institute, New Delhi.

Angell, H. R. 1934. A preliminary note on the recognition of flag smut or bunt, infection on the deformation of seedlings. *Jour. Austr. Counc. Sci. and Indu Res.* 7: 110.12.

Anonymous, 1968. Fly on fungicides, new way to fight leaf diseases in cereals. *World Farming.* 10: 14-15.

Ashworth, L.J., Lahr, K.A. and Collin, R.J. 1960. Pathogenic specialization of *Helminthosporium sativum*. *Phytopath.* 50: 627 (Abstr.).

Aujla, S.S. and Sharma, L. 1990. Compatibility system in *Neovossia indica*. *Indian Phytopath.* 43: 222- 223.

Aujla, S.S., Grewal, A.S., Nanda, G.S. and Sharma, I. 1990. Identification of stable resistance in wheat to loose smut. *Indian Phytopath.* 43 : 90-91.

Aujla, S.S., Sharma, I. and Gill, K.5. 1990. Sources of resistance in wheat to Karnal bunt. *Indian Phytopath.* 43: 91-93.

Aujla, S.S., Sharma, I. and Singh, S.S. 1987. Physiological specialization of Karnal bunt of wheat. *Indian Phytopath.* 40: 333-336.

Aujla, S.S., Sharma, I., Gill, K.S. and Grewal, A.S. 1985. Variable resistance in wheat germplasm to *Neovossia indica. Third Nat. Seminar on Genet. and wheat Improv.*, Shimla, May 8-10, pp. 8-10 (Abstr.)

Aujla, S.S., Sharma, Y.R., Karamchand and Sawhney, S.S.1977. Influence of weather factors on the incidence and epidemiology of Karnal bunt disease of wheat in Punjab. *Indian J. Ecol.* 4: 71-74.

Bag, T.K., Singh, D.V. and Tomar, S.M.S. 1999. Inheritance of karnal Bunt (*Neovossia indica*) resistance in some Indian bread wheat (*Tritcum aestivum*) lines and cultivars. *J. Genet. & Breed.* 52: 67-72.

Bahadur, P., Singh, D.V. and Srivastava, K.D. 1994. Management of wheat rusts -a revised strategy for gene deployment. *Indian Phytopath.* 47: 41-47.

Bahadur, P., Sinha, V. C. and Upadhyaya Y. M. 1974. Evaluation of some systemic and non-systemic fungicides as seed dressing against foot rot of wheat caused by *Sclerotium rolfsii* Sacc. *Pesticides.* 8: 42-44.

Bai G.H, Chen L F, Shaner G.E. 1999. Breeding for resistance to head blight of wheat in China. *(In:)* Kurt Leonard (Ed.). *Wheat Fusarium Head Blight.* APS Press .

Balmas, V., Basili, O. and Corazza, L. *1995.* Comparison between seed dressing products of *wheat* for the control of *soil*-borne fungal pathogens. *Sementi-Elette.* 41: 6, 3-7.

Bansal, R., Singh, D. V. and Joshi, L. M. 1984 b. Effect of liquid nitrogen treatment on germination of teliospores of Karnal bunt. *Indian Phytopath.* 37: 368-369.

Bansal, R., Singh, D. V. and Joshi, L. M. 1984 c. Effect of Karnal bunt pathogen *[Neovossia indica* (Mitra) Mundkur] on weight and viability of wheat seed. *Indian J. Agric. Sci.* 54 : 663-666.

Bansal, R., Singh, D. V. and Joshi, l.M. 1983. Germination of teliospore of Karnal bunt of wheat. *Seed Res.* 11: 258-261.

Bansal, R., Singh, D.V. and Joshi, l.M. 1984 a. Comparative morphological studies in teliospores of *Neovossia indica. Indian Phytopath.* 37: 355-357.

Bansal, R., Singh, D.V. and Joshi, l.M. 1984 b. Effect of liquid nitrogen treatment on germination of teliospores of Karnal bunt. *Indian Phytopath.* 37: 368-369.

Bansandrai, A. K; Sood-AK; Sud-AK 1993 Prevalence of flag smut of wheat (*Urocystis agropyri*) in Himachal Pradesh. *Plant Dis. Res.* 8: 1, 85.

Beal, S. J., Cavell, P; Holt, T. and Biddle, A. J. 2001.Triticonazole based cereal seed treatments for the control of seed- and *soil*-borne diseases. *Seed-treatment: challenges and opportunities. Proceedings of an International Symposium,* Wishaw, North Warwickshire, UK, 26-27-February-2001., 99-104.

Bedi, K. S. 1957. Fighting the flag smut. *Indian Farmg.* 7: 25-27.

Bedi, K. S., Sikka, M. R. and Mundkur, B. B. 1949. Transmission of wheat bunt due to *Neovossia indica* (Mitra) Mundkur. *Indian Phytopath.* 4: 20-26.

Bedi, P. S., Singh, G. and Singh, H. 1975. Comparative efficacy of some fungicides in controlling smut of wheat under field conditions. *J. Res.* (Ludhiana) 11 : 56-58.

Beniwal, M. S. 1992. Effect of flag smut on yield and yield components of wheat varieties. *Crop-Research-Hisar*. 5: 2, 348-351.

Beniwal, M. S., Virk,K. S., Rajender, Singh and Singh, R. 1999. Relative resistance of commercially grown wheat varieties of North-Western Plain Zone of India against important diseases. *Crop-Research-Hisar* 18: 2, 296-298.

Beniwal, M.S., Karwasra, S. S., Anil, Gupta and Gupta, A. 1992. Sources of resistance to flag smut of wheat. *Annals-of-Biology-Ludhiana*. 8: 2, 213-215.

Bergstrom, G. C. 1993. Scab (head blight), *(In:)* Mathur, S.B. and Cunfer, B.M. (Eds.), *Seed-borne Diseases and Seed Health Testing of Wheat* pp 83-94. Danish Government Institute of Seed Pathology for Developing Countries, Copenhegen, Denmark. pp:68.

Bhat, R. V., Roy, B., Roy, D. N., Vijayaraghavan, M. and Tulpule, P.G. 1981. Toxicological evaluation of Karnal bunt in monkeys -A Report. *gull. Food and Drug Toxicological Research Centre,* Jamaiosmania Hyderabad, India. Aujla, S.S. and Sharma, 1.1990. Compatibility system in *Neovossia indica. Indian Phytopath.* 43: 222- 223.

Bhatnagar, G. C., Gupta, R. B. L. and Mishra, V. L. 1978. Relative efficacy of different fungicides in controlling flag smut *(Urocystis agropyri)* of wheat in field conditions. *Pesticides.* 12: 27-28,

Bhatnagar, G. C., Gupta, R. B. L., Mishra, V. L. and Singh, G. 1975. Assessment of losses and chemical controlof flag smut of wheat caused by *Urocystis tritici. Indian J. Mycol. PI. Pathol.* 5: 35-36 (Abstr.).

Bhatnagar, G. C., Mali, S. N. and Bhatnagar, V. K. 1991. Identification of sources of resistance against flag smut (Urocystis agropyri (Preuss) Schroet) of wheat. *Wheat-Information-Service.* 73, 5-7.

Biswas, S. K., Srivastava, K. D., Aggarwal, R., Prem Dureja and Singh, D. V. 2000. Antagonism of *Chaetomium globosum* to *Drechslera sorokiniana,* the spot blotch pathogen of wheat. *Indian Phytopath.* 53: 436-440.

Bonde, M. R., Peterson, G. L., Schaad, N. W. and Smilanick, J. L. 1997. Karnal bunt of wheat. *Plant Dis.* 81:1370-1377.

Bonde, M. R., Prescott, J. M., Matsumoto, T. T., and Peterson, G. L., 1987. Possible dissemination of teliospores of *Tilletia indica* by the practice of burning wheat stubble. *Phytopath.* 77:639. (Abstr.)

Boratynski, T. N., Matsumoto, T. T., and Bonde, M. R., 1985. Interceptions of *Tilletia indica* at the California-Mexican border in Mexican railroad boxcars. (Abstr.) *Phytopath.* 75:1339.

Brennan, J. P., Warham, E. J., Hernandez, J., Byerlee, D. and Coronel, F., 1990. *Economic losses from Karnal Bunt of wheat In Mexico. CIMMYT Economics working Paper* 90/02. Mexico.

Brittlebank, C. C. 1920. Flag smut. *Jour. Dept. Agric. Victoria.* 18: 240-43.

Burgress, L. W., Wearing, A. H. and J. A. Joussoun, 1975. Survey of Fusaria associated with crown rot or wheat in eastern Australia. *Aust. J. Agric. Res.* 26: 791-799.

Butler, E. J. 1918. *Fungi and Diseases in Plants.* Thacker Spinck & Co. Calcutta, India. pp:547.

Carne, W. M. 1924. Flag smut of wheat. *J. Dept. Agrtc. West Austr.* I: 142-47.

Cashion, N. L. and Luttrell, E.S. 1988. Host parasite relationships in Karnal bunt of wheat. *Phytopath.* 78: 75-84.

Chattopadhyay, S. B. and Chakrabarti, K. B. 1968. Role of different pathogens in inciting the *Helminthosporium* disease of wheat in West Bengal. *Indian J. Agric. Sci.* 38: 937-940.

Chaurasin, R. K., Mishra. A. N. and Singh, S. P. 1973. Screening of wheat varieties against *Rhizoctonia solani. JNKVV Res. J.* 7: 178-79.

Chenulu, V. V. and Singh, A. 1964. A note on estimation of losses due to leaf blight of wheat caused by *Alternaria triticina. Indian Phytopath.* 17: 254-256.

Clayton, E. S. 1925. A common source of infection with flag smut. *Agric. Gaz. N. S. Wales.* 36:860.

Clayton, E. S. 1925. A common source of infection with flag smut. *Agric. Gaz. N.* S. *Wales.* 36: 860.

CMI Distribution Maps of Plant Diseases, 1989. *Tilletia indica* Map No. 173. Commonweath Agricultural Bureaux International, Wallingford, Oxon, U.K.

Cook, R. J., Reis, E., Jenkyn, J. F. and Plumb, R. T. 1981. Cultural control of soil borne pathogens of *wheat* in the Pacific North West of the U.S.A. *(In:) Jenkyn, J.F. and Plumb, R.T. (Eds). Strategies for the control of cereal disease.* 167-177. Blackwell Sci. Pub. Oxford.

Dai, F. M., Zhou, S. M. 1995. Isolation and utilization of microbial strains to control wheat scab. *Acta Phytophylacica Sinica.* 22(2):112-116.

Dalela, G. G., Sankhla, H. C., Sankhla, B. and Mathur, L.1971. Reaction of wheat varieties to Rajasthan isolates of *Alternaria triticina,* incitant of leaf blight of wheat. *Indian J. Mycol Pl. Pathol.* I: 121-125.

Dandurand, R.M. and Knudsen, G.R.1993. Influence of *Pseudomonas fluorescence* on hyphal growth and biocontrol activity of *Trichoderma harzianum* in the spermosphere and rhizosphere of pea. *Phytopath.* 83: 265-270.

Das, A. M. and Srivastava, D. N. 1971. Evaluation of some seed dressing fungicides against foot rot of wheat incited by *Helminthosporium sativum. Indian J. Agric. Sci.* 41: 387-89,

Dastur, J. F. 1932. Foot rot and black point diseases of wheat in the Central Provinces. *Agriculture Livestk. India.* 2: 275-82.

Dawson, G. T. 1934. Seed borne flag smut infection effectively controlled by copper carbonate treatment. *Agric. Gaz. N. S. Wales* 45: 431-32.

Duran, R., and Cromarty, R. 1977. Tilletia indica : A heterothallic wheat bunt fungus with multiple alles controlling incompatibility. *Phytopath* 67: 812-15.

De Milliano, W. A .J. and Zadoks, I. C. 1985. The effect of early foliar infection by *Helminthosporium sativum* on some yield components of two African wheats. *(In:)* Villareal, R.L. and Klatt, A.R. (Eds.) *Wheats for More Tropical Environments* pp: 154-157. *Proceedings of the International Symposium,* CIMMYT , Mexico, 354.

Desjardins A E, Proctor R H, Bai G. H, McCormick S. P., Shaner G. E., Buechley G. and Hohn TM. 1996. Reduced virulence of trichothecene antibiotic-nonproducing mutants of *Gibberella zeae* in wheat field tests. *Molec. Plant Microb. Inter.* 9:775-781

Dhaliwal, H. S. and Singh, D. V. 1988. Up-to-date life cycle of *Neovossia indica* (Mitra) Mundkur. *Curr. Sci.* 57: 675-677.

Dhaliwal, H. S. and Singh, D. V. 1989. Production and interrelationship of two types of secondry sporidia of *Neovossia indica. Curr. Sci.* 58: 614-618.

Dhiman, J. S. and Bedi, P. S. 1981. A method for inoculation and detection of Karnal bunt of wheat under field conditons. *Proc. 3rd Intern. Symp. Plant Path.,* IARI, New Delhi, Dec. 14-18, pp:230 (Abstr.).

Dhiman, J. S. and Bedi, P. S. 1989. Studies on the mode of perpetuation and pathogenic variability in *Neovossia indica* causing Karnal bunt of wheat. *Indian Phytopath.* 42:323.

Dickinson, C. H. 1981. Leaf surface microorganisms as pathogen antagonists and as minor pathogens. *(In:)* Jenkyn, J.F. and Plumb, R.T. (Eds.). *Strategies for the control of cereal diseases* pp:109-121. Blackwell Sci. Pub., Oxford.

Dill Macky, R. and Jones, R. K. 2000. The effect of previous crop residues and tillage on Fusarium head blight of *wheat. Plant Dis.* 84: 1, 71-76.

Directorate of Wheat Research 1995. *Report of the Coordinated Experiments,* 1994-95, Vol. I, Crop Protection (pathology) (Eds. Prasher, M., Nagarajan, S., Goel, L.B. and Kumar, *I.)* pp: 43-54, AICWIP ,Directorace of Wheat Research, Karnal. pp:206.

Dubin, H. J. and M. Van Ginkel 1991. The status of what diseases and disease research in warmer area. (*In*:) D.A. Saunders (Ed), Wheat for the Non-traditional warm areas. *The Proceedings of the International Conference,* pp: 125-145. Mexico, D. F.: CIMMYT.

Dubin, J. and Ginkel, M. 1991. The Status of wheat diseases and disease research in warmer areas, *(In:)* Saunders, D. A. (Eds.). *Wheat for the Non-traditional Warm Areas* pp: 125-145, CIMMYT , Mexico. pp: 549.

Egamberdiyeva, D. Mamiev, M., Hoflich, G., Garcia, T. L., Benites, J. and Martinez, V. A. 2001. Emergence promoting bacteria for improvement of plant growth in semi arid region. *Conservation-agriculture,-a-worldwide-challenge.*-First-World Congress on conservation agriculture, Madrid, Spain, 1-5-October, 2001.Vol-2:-offered-contributions. 2001. 491-495.

El-Helaly, A. F. 1948. The influence of cultural conditions on the flag smut of wheat. *Phytopath.* 38: 688-97.

Faris, I. A. 1933. Influence of soil moisture and temperature on infection of wheat by *Urocystis tritici. Phytopath.* 23: 10-11.

Fenster, C.R., Booslis, M.G. and J.L. Weihing. 1972. Date of planting studies of winter barley in relation to root rot an crown rot, grain yields and quality. Nebr. Agric Exp Res. Bull. 250.

Fernandez, J. A., Wofford, D. S. and J. L. Horton, 1985. Interactive effects of freezing and common root rot fungi on winter wheat. *Phytopath.* 75 : 845-847.

Fischer, G. W. 1943. Some evident synonymous relationships in certain graminicolous smut fungi. *Mycologia.* 35: 610.

Fischer, G. W. 1953. *Manual of North American smut fungi.* Ronald Press Co. New York.

Fischer, G. W. and Holton, C. S. 1957. *Biology and control of the smut fungi.* The Ronald Press Co., New York.

Fokkema, N. J. 1973. The role of saprophytic fungi in antagonism against *Drechslera sorokininana (Helminthosporium sativum)* on agar plates and on rye leaves with pollen. *Physiol. Plant Pathol.* 3 : 195-205.

Frisullo, S. 1982. Fungal parasites of plants in southern Italy. I. *Alternaria triticina* Prasada and Prabhu on durum wheat. *Phytopathol. Mediter.* 21: 113-115.

Gandhi, V. P., Zhang-Yue Zhou and John Mullen. 2001. Wheat Economy of India: Development, Nature and Trade Prospects October 10th, 2001. *Agribusiness Perspectives –Paper.* pp:46.

Garg, V. R., Arya, H. C. and Kant, U. 1972. Varietal reaction and physiologic specialization in *Alternaria triticina* causing leaf blight of wheat in Rajasthan. *Indian J. Mycol. Pl. Pathol.* 2: 24-31.

Gayed, S. K. 1962. The pathogenicity of six strains of *Helminthosporium sativum* to three cereals with special reference to barley. *Mycopath. Mycol. Appl.* 18: 271-279.

Gill, K. S. 1985. *Sources* of resistance of Karnal bunt and loose smut and their use in breeding resistant wheat varieties. *(In:)* A.K. Gupta, S. Nagarajan and R.S. Ranci, (Eds.) *Genetic and Wheat Improvement* ICAR, New Delhi. pp:134-153.

Gill, K. S., Sharma, I. and Aujla, S. S. 1993. *Karnal bunt and wheat production.* Punjab Agricultural University, Ludhiana.

Goel, L. B., Singh, D. V., Srivastava, K. D., Joshi, L.M. and Nagarajan, S. 1977. *Smuts and bunts of wheat in India.* Indian Agric. Res. Insti., New Delhi. pp:38.

Goel, R. K. 1991. Inheritance of resistance to flag smut (*Urocystis agropyri*) in wheat (*Triticum aestivum*) lines. *Indian J.Agricul.Sciens.* 61: 11, 844-849.

Goel, R. K. and Gupta, A. K. 1990. Reaction of some wheat cultivars and lines to flag smut. *Indian J. Gen. Plant Breed.* 50: 2, 185-188.

Goel, R. K. and Gupta, A. K. 1990. Reaction of some wheat cultivars and lines to flag smut. *Indian J. Gen.Plant Breed.* 50: 2, 185-188.

Goel, R.K. and Jhooty, J.S. .1989. Some factors affecting the incidence of flag smut of wheat. *Indian J.Agricul Res.* 23: 4, 200-204.

Gothwal, B. D. and Pathak, V. N. 1977. Pathogenic races of *Ustilago tritici* in India. *Indian Phytopath.* 30 : 311-14.

Gupta, V. V. S. R., Roget, D. K. and Bill Davoren 2003 Managing the soil biota - Beneficial functions of soil microflora and microfauna in agricultural systems Research Update for Advisers, CSIRO Land and Water, Adelaide. http://www.grdc.com.au/growers/res_upd/south/03S/soilbiota.htm.

Hagiwara, H., Sato, T. and Saitoh, H. 1996. Control of wheat flag smut by soil sterilization with hot water injection. [*Urocystis agropyri*]. *Proceedings of-the-Kanto-Tosan-Plant-Protection-Society.* 43, 57-58.

Hagiwara, H., Sato, T. and Saitoh, H. 1996. Control of wheat flag smut by soil sterilization with hot water injection. [*Urocystis agropyri*]. *Proceedings-of-the-Kanto-Tosan-Plant-Protection-Society.* 43, 57-58.

Hamilton, D. G., Clark, R.V., Itannah, A. E. and Loiselle, R. 1960. Reaction of barley varieties and selections to root rot and seedling blight incited by *Helminthosporium sativum* P .K. & B. *Canad. J. Plant Sci.* 40 : 713-720.

Harding, H. 1972. Reaction to common root rot of 14 *Triticum* species and its incidence of *Bipolaris sorokiniana* and *Fusarium* spp. in subcrown internode tissue. *Canad. J. Bot.* 50 : 1805-1810.

Harris, J. S. and Moen, R. 1985. Replacement of *Rhizoctonia solani* on *wheat* seedlings by a succession of root-rot fungi. *Trans.Bri. Mycol.Soc.* 84(1):11-20.

Hassan, S. F. 1973. Wheat disease and their relevance to improvement and production in Pakistan. *Proc. Fourth FAO/Rockfeller Found. Wheat Seminal.* pp:84-86.

Hegde, R. K., Kulkarni, S., Siddaramaiah, A. L. and Krishna Prasad, K. S. 1980. Biological control of *Sclerotium rolfsii* Sacc. causal agent of foot rot of wheat. *Curr. Res.* 9: 67-69.

Hetzler, J., Eyal, Z., Mehta, Y. R., Fehrmann, H., Kvshnir, U. and Ckaria, J. 1991. Interactions between spot blotch *(Cochliobolus sativus)* and wheat cultivars, *(In:)* Saunders, D. A., (Ed.), *Wheat For The Nontraditional Warm Areas* pp:146-164. CIMMYT, Mexico. pp:549.

Holmes, G. J., Jackson, L. F. and Perring, T. M. 1997. Imperial Valley conditions limit Karnal bunt in wheat. *California Agri.* 51(3):29-33.

Huang C. H, Li J. H., Zhang Y. Q., Wang M. and Liu S. 1993. A study of the control effect of *Bacillus*-HCH on scab of wheat. *J. Huazhong Agril. Univ.* 12(6): 566-569.

Jone, J. 2001. Harpin. *Pesticide Outlook* 12(4): 134-135.

Joshi, L. K. and Vyas, S. C. 1974. Screening of systemic and non-fungicides against *Fusarium oxysporum* causing foot and root-rot of wheat. *Indian Phytopath.* 27: 650-51.

Joshi, L. M., Gera, S. D., Adlakha, K. L, Srivastava, K. D., Basant Ram and Palmer, L. T. 1974. Some foliar diseases of wheat during 1969-70 crop season. *Indian Phytopath.* 27 : 178- 181.

Joshi, L.M., Singh, D.V., Sirvastva, K.D. and Wilcoxson, K. 1980. Karnal bunt: A minor disease that is now threat to wheat. *Bot. Rev.* 49: 309-30.

Joshi, L.M., Singh. D.V. and Srivastava, K.D. 1984. Flactuations in incidence of rusts and other wheat diseases during past decade and strategies for their containment. *Proc. 23rd All Indian Wheat Res. Worker's Workshop* heldat at GBPUAT, Pantnagar, August 1985.

Joshi, L. M., Singh, D. V., Srivastava, K. D. and Wilcoxson, R. D. 1983. Kanal bunt a minor disease that is now a threat to wheat. *Bot. Rev.* 43: 309-338.

Joshi, L. M., Srivastava, K. D., Singh, D. V., Goel, L .B. and Nagarajan, S. 1978. *Annotated Compendium of Wheat Diseases in India,* ICAR, New Delhi, pp: 332.

Joshi, L.M., Singh, D.V. and Srivastava, K.D. 1986. Wheat and wheat diseases in India. *In: Problems and Progress of Wheat Pathology in South Asia.* pp. 11-19. Malhotra Publishing House, New Delhi.

Joshi, L.M., Singh, D:V. and Srivastava, K.D.1981. Meteorological conditions incretion to incidence of Karnal bunt of wheat In India. *Proc. 3rd Intern, Symp. Plant Pathol.* IARI, New Delhi, Dec. 14-18. pp:11 (Abstr.).

Kalha, C. S., Anand, S. N., Gupta, R. S. and Gupta, R. 1987. Incidence of Karnal bunt of wheat in relation to different dates of sowing. *Res. and Develop. Reptr.* 4:100-101.

Kari, A. G.2001. Disease incidence of common root and foot-rot on durum wheat in semi-arid conditions in Cyprus. *Technical Bulletin Cyprus Agricultural Research Institute.* 208: 9.

Katiyar , R. L. and Singh, D. V. 1968. Resistance of wheat varieties to Alternaria blight. *Indian J. Microbiol.* pp: 55-56.

Kauraw, L. P. and Singh, R. S. 1975. Studies on the root rot complex of wheat in India-I varietal reaction. *Indian J. Mycol. Pl. Path.* 5: 96.97.

Khanna, A., Payak, M. M. and Mehta, S. C. 1968. Teliosporemorphology of smut fungi I. Electron microscopy. *Mycologia.* 58: 562-569.

Khetrapal, R. K. and Agarwal. V. K. 1979. Studies on some aspects of black-point and Karnal bunt diseases of triticale. *Indian Phytopath.* 32 : 292-294.

Khonga, E. B. and Sutton, J. C. 1988.Inoculum production and survival of *Gibberella zeae* in maize and wheat residues. *Cana. J. Plant Pathol* 10 : 232-239.

Kirkegaard, J. A., Wong, P. T. W., Desmarchelier, J.M. and Sarwar, M.1996. Suppression of *soil*-borne cereal pathogens and inhibition of *wheat* germination by mustard seed meal. *Proceedings of the 8th Australian Agronomy Conference*, Toowoomba, Queensland, Australia, 30 January-2 February, 1996, 353-356.

Kohli, M. M. Mann, C. E. and Rajaram, S. 1991. Global status and recent progress in breeding for the warmer areas, *(In:)* Saunders, D.A. (Ed.), *Wheat for Nontraditional Warm Areas* pp 96-112., CIMMYT, Mexico, 549 pp.

Kothari, K. L. and Dange, S. R. S. 1968. Flag smut, a potential threat to wheat cultivation in India. *Farmer& Parliament*, June, 1968.

Krishna, A. and Singh, R. A. 1982. Taxonomy of Karnal bunt fungus: evidence in support of genus *Neovossia. Indian Phytopath.* 35: 544-545.

Kulkarni, G. S. 1924. Report of the work done in Plant Pathology Section during the year 1922- 23. *Rep. Dept. Agric. Bombay Presidency* for the year 1922-23. pp:167-171.

Kulkarni, S, Siddaramaiah, A. L. and Krishna Prasad, K S. 1978. Varietal resistance of promising wheat cultures to foot rot disease of wheat. *Curr. Res.* 7: 708-709.

Kulkarni, S. 1980. Chemical control of foot rot of wheat in Karnataka.*Pesticides* 14:29-30.

Kulkarni, S. and Nargund, V. B. 1986. Soil borne diseases of wheat. *(In:)* Joshi, L.M., Singh, D.V. and Srivastava, K. D. (Eds.) *Problems and progress of wheat pathology in South Asia.* pp: 216-229. Malhotra Publishing House, New Delhi.

Lapis, D. B. 1986b. Chemical control of wheat diseases in Philippines. In : larula, P .N. and Srivastava, O. P. 1971. Genetics of *Alternaria* resistance in wheat. *Indian J. Genet.* 31: 105-107.

Lapis, D.B. 1985. Insect pests and diseases of wheat in the Philippines. *(In:)* Villareal, R.L. and Klatt, A. R. (Eds.). *Wheats for More Tropical Environments.* pp:354. *Proceedings of the International Symposium.* CIMMYT, Mexico.

Lipps, P. E., 1996. Head blight or of small scab. Extension fact sheet-Plant pathology, Ohio State University. http-ohioline.osu.edu-ac-fact-pdf-0004.pdf. AC-04-96: 1-3.

Liu T. R. 1990. Integrated management of wheat head scab. Pages 211-226. (*In*:) Li G-B, Zheng S-M, Li Z-Q. (Eds.). *Integrated management of wheat pests.* China Agricultural Science and Technology Publishing House, Beijing.

Liu, Z. Z. 1985. Recent advances in research on wheat scab in China, *(In:)* Villareal, R.L. and Klatt, A.R. (Eds.) *Wheats for More Tropical Environments*, pp 174-181. *A proceeding of the International Symposium*, CIMMYT, Mexico, 354 pp.

Loke, C. M. and Watson, A. J. 1955. Foreign plant diseases intercepted in quarantine inspection. *Plant Dis. Reptr.* 39 : 518.

Lyamani, A. 1988. Wheat root rot in west central morocco and effects of *Fusarium culmorum* and *Helminthosporium sativum* seed and soil-borne inoculum and root rot development, plant emergence, and crop field. *Ph.D. Thesis*, Iowa State University, Ames, USA.

Mafiz, A. 1948. Smut diseases and their control by cultural methods. *Sci. J. R. Coll. Sci.* 18 : 77-80.

Maity, S. S. 2003. Effect of metrological factors on production of conidia in *Helminthosporium* spp. and *Alternaria* spp., the incitant of foliar blight of wheat. *Ann Pl. Protec. Sci.* 11 (2): 315-318.

Maity, S. S., Sanyal, R. P. and Das, S. 2002. Effect of inorganic nutrients on leaf blight severity in wheat caused by *Helminthosporium sativum. Ann Pl. Protec. Sci.* 10: 106-110.

Mandal, S. 1995. *Effect Of Some Antagonists On Drechslera sorokiniana, Causal Agent Of Spot Blotch Of Wheat, M.Sc. Thesis,* Indain Agricultural Research Institue, New Delhi, pp: 74.

Manjappa, B. H. 1979. Studies on survival and variation in Sclerotium rolfsii Sacc. *M. Sc. (Agri) Thesis*. University of Agricultural Sciences, Bangalore. pp: 86.

Mathre, D. E., Johnston, R. H., and Grey, W. E. 2003. Diagnosis of common root rot of wheat and barley. Online. *Plant Health Progress doi: 10.1094/PHP-2003-0819-01-DG.*

Mathur, S. B. and Cunfer, B. M. 1993. *Seed-Borne Diseases and Seed Health Testing of Wheat,* Danish Government Institute of Seed Pathology for Developing Countries, Copenhegen, Denmark, pp:168.

Mazzola, M., Johnson, T. E. and Cook, R. J. 1997.Influence of field burning and *soil* treatments on growth of *wheat* after Kentucky bluegrass, and effect of Rhizoctonia cerealis on bluegrass emergence and growth. *Plant Pathol.* 46: 5, 708-715.

Mc Rae, W. 1924. Economic Botany, Part ill, Mycology. *Rep. of Board of Scientific Advice, 1ndia.* 1922-23. pp: 31-35.

McAlpine, D. 1910. The smuts of Australia; their structure, life history treatment and classification. *J. Dept. Agric. Victoria.* 7: 288.

Mehdi, V., Joshi, L. M and Abrol, Y. P. 1973. Studies on chapatti quality VI. Effect of wheat grains with bunts on the quality of chapatties. *Bull. Grain Technol.* 11:195-197.

Mehta, Y.R. 1981 b. Indentification of races of *Helminthosporium sativum* of wheat in Brazil. *Pesquisa Agropecuaria Brasiliera* (Brasilla) 16: 331-336.

Mehta, Y.R. 1981a. Conidial production, sporulation period and extension of lesion of *Helminthosporium sativum* on flag leaves of wheat. *Pesquisa Agropecuaria Brasiliera* (Brasilia) 16: 79-99.

Mehta, Y.R. 1985. Breeding wheat for resistance to spot blotch. (*In:*) Villareal, R.L. and Klatt, A.R. (Eds.). *Wheats for More Tropical Environments.* pp: 135-144. *Proceedings of the International Symposium*, CIMMYT, Mexico. pp: 354.

Mehta, Y.R. and Gaudenico, CA. 1991. The effect of tillage practice and crop rotation on the epidemiology of some wheat diseases. *(In:)* Saunders, D.A. (Ed.) *Wheat for the Non-traditional Warm Areas*, pp 266-283, CIMMYT, Mexico, 549 pp.

Mehta, Y.R. and Igarashi, S., 1985. Chemical control measures for the major diseases with special attention on spot blotch. *(In:)* Villareal, R. L. and Klatt, A.R (Eds.) *Wheats for More Tropical Environments*. pp 196-200. *Proceedings of the International Symposium*, CIMMYT, Mexico, 354 pp.

Mergoum, M. and J.S. Quick. 1990. Implication of root rot inoculation and nitrogen fertilization of wheat cultivars under varying moisture stress in west central of Morocco. Agron Abstr. 101, ASA, San Antanio, TX.

Millikan, C. R. 1939 The influence of nutrition on the reaction of wheat to *Urocystis tritici* Koern. *J. Dept. Agric. Victoria*. 37: 302-08.

Milus, E. A. and Parsons, C. E. 1994. Evaluation of foliar fungicides for controlling Fusarium head blight of wheat. *Plant Dis*. 78: 697-699.

Misra, A. P. 1973. *Helminthosporium* species occurring on cereal and other Graminae. Final report on PL 480 Project, Tirhut College of Agriculture, Dholi, Muzaffarpur (Bihar), India. pp:289.

Mitra, M. 1931. A new bunt of wheat in India. *Ann. Bioi*. 18: 178-179.

Mitra, M. 1935. Stinking smut (bunt) of wheat with special reference to *Til/etia indica* Mitra. *Indian J. Agric. Sci.5* : 51-74.

Mitra, M. 1937. Studies on the stinking smut or bunt of wheat in India. *Indian J. Agric. Sci.7* : 469-478.

Mundkur, B. B. 1943. Karnal bunt, an air-borne diseae. *Curr. Sci*. 12: 230-231.

Mundkur, B.B. 1936. Studies in cereal smuts V. Mode of transmission of the Karnal bunt of wheat. *Indian J. Agric. Sci*. 13: 54-58.

Mundkur, B.B. 1944. A second contribution towards a knowledge of Indian ustilaginales. *Trans.Brit. Mycol.Soc*.24:32-34.

Munjal, R. L. 1969. Application of serology and spectroscopy to the taxonomy of phytopathogenic fungi. *Indian Phytopath*. 22 .154.

Munjal, R.L. 1976. Studies of Karnal bunt *(Neovossia indica)* of wheat in northern India during 1968-69 and 1969-70. *Indian J. Mycol. Pl. Pathol*. 5: 185-187.

Munjal, R.L. and Chatrath, M.S. 1976. Studies on mode of infection of *Neovossia indica* incitant of Karnal bunt of wheat. *J. Nuclear Agric. Biol*. 5: 40-41.

Nagarajan, S. 1991. *Epidemiology of Karnal bunt of what incited by Neovossia indica and an attempt to develop a disease predication system -CIMMTY Technical Report*. International Maize and Wheat Improvement Centre (CIMMYT), Mexico.

Nagarajan, S., Bahadur, P. and Nayar , S.1984. Contemplating the brown rust resistance I *(Lr)* genes to mItIgate the spread of *Puccinia recondita* f. sp. *tritici*. *Indian Phytopath.* 30 : 490- 497.

Narain, A., Sinha, S.K., Mahapatra, B.K., Scchler, D.T. and Poehlman, J.M. 1973. Incidence of *Helminthosporium* blight of wheat in Orissa State, India. *Plant Dis. Reptr.* 57: 278-280.

Narain, N. 1972. Control of some fungal diseases by antibiotics and crude plant extracts. *Indian Nat. Sci. Acad.* 37B : 332-340.

Nargund, V. B., Kulkarni, S. and Hegde, R. K. 1982. Studies on foot rot of wheat caused by *Sclerotium rolfsii* Sacc. in Karnataka. *Proc. 21st All India Wheat Workers Workshop held at UAS Bangalore, from 27th to 31st August* 1982.

Nema, K.G. and Joshi, L.M. 1971a. The spot blotch disease of wheat caused by *Helminthosporium sativum. Proc. 2nd Int. Symp. Plant Path.*, IARI, New Delhi, pp:42.

Nema, K.G. and Joshi, L.M. 1971b. Symptoms and diagnosis of the spot blotch and leaf blight diseases of wheat. *Indian Phytopath.* 24: 418-419.

Noble, R. I. 1924. Studies on the parasitism of *Urocystis tritici* Koern. the organism causing flag smut of wheat. *J. Agric. Res.* 27: 451-89.

Noble, R. I. 1934. Note on the longevity of spores of the fungus *Urocystis tritici* Koern. *J. and Proc. Roy. Soc. New S. Wales* 567: 403-10.

Padwick, G.W. 1939. Report of the Imperial Mycologist. *Sci. Report, Agric. Instt.*, New Delhi 1937-38, pp:105-112.

Palm, E. W. 1993. Wheat Diseases in Missouri Agricultural publication G04319. http://muextension.missouri.edu/explore/agguides/crops/index.htm

Pan, S. and Sen, C. 1977. Chemical control of foot rot of wheat caused by *Sclerotium rolfsii* Sacc. *Proc. Indian Sci. Acad.* B. 43: 79-85.

Pant, S. K; Mittal, R. K; Bhatt, J. C; Sharma, A. K. 1996. Occurrence of flag smut of wheat in the hills of Uttar Pradesh. *J. Hill Res.* 9(1): 156.

Prabhu, A. S. and Prasada, R. 1966. Plant pathological and epidemiological studies on leaf blight of wheat caused by *Alternaria triticina. Indian Phytopath.* 19 : 95-112.

Prabhu, A. S. and Singh, Amar, 1974. Appraisal of yield losses in wheat due to foliage diseases caused by *Alaternaria triticina* and *Helminthosporium sativum*. *Indian Phytopath.* 27: 632-634.

Prabhu, A.S. and Singh, A. 1974. Appraisal of yield loss in wheat due to foliage diseases caused by *Alternaria triticina* and *Helmionthosporium sativum. Indian Phytopath.* 27 : 632-634.

Prasad, K.S.K., Kulkarni, S. and Siddararnaiah, A.L. 1978. Antagonistic action of *Trichoderma* sp. and *Streptomyces* sp. on *D. sorokiniana [Helminthosporium sativum* (pam. King and Bakke) Subr. & Jain] *Curr. Res.* 12 : 202-203.

Prasada, R. and Prabhu, A. S. 1962. Leaf blight wheat caused by a new species of *Alternaria. Indian Phytopath.* 15 :292-293.

Prescott, J.M. 1984. Overview of CIMMYT's Karnal bunt programme. *Proc. Conf. on Karnal bunt,* Ciudad Obregon, Sonora, Mexico from April, 16-18. pp:2-4.

Purdy L. H. 1962. Tetrachloronitroanisole a new wheat smut fungicide. *Phytopath.* 52: 25-26.

Purdy, L. H. and Hot ton, C. S. 1964. Reactions of wheat to flag smut in the Pacific North West. *Plant Dis. Reptr.* 48: 895-99.

Raemaekers, R.H. 1979. The efficacy of fungicides in a severe spot blotch epidemic of wheat in Zambia. IX *Inter. Cong. Plant Prot.,* Aug. 5-11, Washington, D.C. (Abstr.).

Raemaekers, R.H. 1985. Breeding wheats with resistance to *Helminthosporium sativum* in Zambia, *(In:)* Villareal, R.L. and Klatt, A.R. (Eds.), *Wheat for More Tropical Environments* pp 145-148. *A Proceeding of the International Symposium,* CIMMYT. Mexico. pp:345.

Raemaekers, R.H. 1988. *Helminthosporium* : Disease complex on wheat and sources of resis- tance in Zambia, *(In:)* Klatt, A.R. (Ed.), *Wheat Production Constraints in Tropical Enviroments* pp 175-186. CIMMYT, Mexico pp: 410.

Raina, G. L, Aujla, S. S. and Chohan, I. S. 1980. Evaluation of systemic and non-systemic fungicides in controlling flag smut of wheat. *Pesticides.* 14: 36-37.

Ram B. and Arora, P.C. 1986. Smuts of wheat in India. (*In*:) Joshi, L.M., Singh, D.V. and Srivastava, K.D. (Eds.) *Problems and Progress of Wheat Pathology in South Asia.* Malhotra Publishing House, New Delhi. pp.92-123.

Ram, B. and Joshi, L.M. 1978. Spray schedule of Fytolan for leaf blight of wheat and its effect on yield components. *Indian Phytopath.* 31 : 348-351

Ram, B. and Joshi, L.M. 1979. Effect of artificial inoculation of *Alternaria triticina* on yield components of wheat. *J. Plant. Dis. Prot.* 86: 741-744.

Rao, M. V. and Joshi, L. M. 1974. Strategy for control of diseases of wheat in India. *Indian J. Genetics.* 34A: 481-85.

Raymundo, A.D. 1992. Disease management in wheat. In: *Wheat Review and Planning. Proceedings of the workshop,* 1989, pp. 97-102. Los Banos, Laguna, Philippines : PCARRD and CIMMYT.

Reddy, H. R., Fazalnoor, K. and Hegde, R. K. 1971. Screening of commercial varieties and genetic stocks of wheat against foot rot caused by *Sclerotium rolfsii* Sacc. *Mysore J. Agri Sci.* 5: 252.56.

Rick, Ward. 1996. Defining the U.S. Wheat & Barley Scab Initiative. *Scab News.* 2:1. 2.

Roy, A.K. 1973. Ear blights and scab of wheat in Arunachal Pradesh. *Curr. Sci.* 43 : 162.
Saari, E.E. 1985. Distribution and importance of root rot diseases of wheat, barley

and triticale in South and South-East Asia, *(In:)* Villareal, R.L. and Klatt, A.R. (Eds.) *Wheats for More Tropical Environments* pp 189-195. *Proceedings of the International Symposium,* CIMMYT, Mexico. pp:354.

Saari, E.E. 1996. Wheat Disease problems in South-East Asia, *(In:)* Joshi, L.M., Singh, D.V. and Srivastava, K.D. (Eds.). *Problems and Progress of Wheat Pathology in South Asia* pp: 31- 40. Malhotra Publishing House, New Delhi, India, pp:401.

Saari, E.E. 1986. Wheat diseases in South East Asia. *(In:)* Joshi, L.M., Singh, D.V. and Srivastava, K.D. (Eds.) *Problems and Progress of Wheat Pathology in South Asia.* Malhotra Publishing House, New Delhi. pp:31-40.

Sakesena, H; K. and Kumar, K. 1967. Rhizoctonia root rot of cereals in India. *Proc. Second lnt. Symp. Pl. Path. I. A. R.I., New Delhi.* pp:72.

Sanjay, Arya, Sher, Singh, Rakesh, Kumar, Arya, S., Singh, S. and Kumar, R. 1997. Non-targeted effects of weedicides and insecticides on flag smut of wheat. *Plant Dis. Res.* 12: 1, 97-98.

Sanjay-Arya; Sher, Singh, Surender, Kumar, Rangpal-Singh, Arya, S., Singh, S., Kumar, S. and Singh, R. 1998. Role of bioagents and physical methods of seed treatment in controlling flag smut of wheat. *Crop Research Hisar.* 15: 2-3, 275-278.

Sanjay-Arya; Sher-Singh; Surender-Kumar; Arya-S; Singh-S; Kumar-S. 1995. Effect of soil type and off-season irrigation on the incidence of flag smut of wheat caused by *Urocystis agropyri. Plant Dis.Res.* 10: 2, 122-125.

Sanjay-Arya; Sher-Singh; Surender-Kumar; Rakesh-Kumar; Arya-S; Singh-S; Kumar-S. and Kumar-R 1995. Effect of depth of sowing on the incidence of flag smut of wheat incited by *Urocystis agropyri* (Preuss) Schraet. *Annals Biol. Ludhiana.* 11: 1-2, 290-292.

Sanjay-Arya; Sher-Singh; Surender-Kumar; Rakesh-Kumar; Arya-S; Singh-S; Kumar-S; Kumar-R 1995. Effect of depth of sowing on the incidence of flag smut of wheat incited by *Urocystis agropyri* (Preuss) Schraet. *Annals. Biol. Ludhiana.* 11: 1-2, 290-292.

Sansford, C.E. Beales, P.A. and Clarkson, J.D.S. 1998. First report of flag smut of wheat caused by Urocystis agropyri in the United Kingdom. *Plant Dis.* 82: 12, 1403.

Saunders, D.A. 1988a. Characterization of tropical wheat environments: Identification of production constraints and progress achieved -South and South East Asia, *(In:)* Klatt. A.R. (Ed.), *Wheat Production Constraints in Tropical Environments Proc. Intr. Conf.* pp: 12-26, CIMMYT, Mexico. pp:410.

Saunders, D.A. 1988b. Characterization of tropical wheat environments: Identification of production constrains and progress achieved. South and South east Asia. In: A. R. Klat (Ed.) *Wheat Production Constraints in Tropical Environments. Proc. Intr. Conf,* 1987. pp:12-26.

Sawhney, R.N., Sharma, J.B., Sharma, D., Chowdhury, H.B., Singh D.V. and Srivastava, K.D. 1992. High yielding Karnal bunt and rust resistance sources of bread wheat. *Annual Wheat Newsltr (USA)* 38: 125-126.

Sethi, G. S. and Singh, H. B. 1971. Flag smut threatens Kalyansona. *Intensive Agric.9* : 13.

Sharma, A.K., Kumar, J. and Nagarajan, S. 1998. Disease management strategy in wheat. *Indian Farmg.* 18: 52-53.

Sharma, O. P. and Jain, A. C. 1967. Root rot of wheat in Madhya Pradesh. *Indian Phytopath.*20: 267-69.

Shazia Iram, M. Ashraf and Iftikhar Ahmad Prevalence. 2003 Incidence and Severity of Soil-borne Diseases and Fungi of Wheat in Rice-wheat Cropping System of Punjab Province of Pakistan During the Cropping Season 1999 .*Pak. J. Biol. Sci.* 6 (1): 36-40.

Shi, Z.X, Wang, D.B. 1991. Inhibitory effect of *Bacillus* spp. as biocontrol agent for *Gibberella zeae* (Schw.) *Petch. J.Huazhong Agril.l Univ.* 10(2): 151-154.

Sidhartha, S. V. 1992. *Epidemiological Studies on Karnal Bunt of Wheat. Ph.D. Thesis,* Indian Agricultural Research Institute, New Delhi.

Sindhan, G.S., Parashar, R.D., Indra, Hooda and Hooda, I. 1996. Relationship between biochemical parameters and flag smut resistance in wheat. *Indian.J.Mycol.Pl.Pathol.* 26: 3, 291-293.

Sindhan-GS; Parashar-RD; Indra-Hooda; Hooda-I. 1996. Effects of chemical seed treatment on incidence of flag smut and biochemical constituents of wheat plants. *Indian.J.Mycol.Pl.Pathol.* 26 (3): 289-291.

Singh, A. and Prasad, R. 1978. Date of sowing and metrological factors in relation to occurance of Karnal Bunt of wheat in U.P. Tarai, *Indian J. Mycol. Pl. Path.* 8: 2.

Singh, A. and Rai, R.C. 1986. Chemical control of wheat diseases in India, *(In:)* Joshi, L. M.. Singh, D.V. and Srivastava, K.D. (Eds.) *Problems and Progress of Wheat Pathology in South Asia,* pp 374-401. Malhotra Publishing House. New Delhi. 401pp.

Singh, A. and Rai, R.C. 1986. Chemical control of wheat diseases in India. *(In:)* L.M.Joshi, D. V. Singh and K.D. Srivastava, (Eds.) *Problems and Progress of Wheat Pathology in South Asia* Malhotra Publising House, New Delhi, pp. 374-394.

Singh, D. Sharma, B. K. Singh, A., Gartan, S. L. Ravinder Kumar and Amit Ghanaksh. 2003. Incidence of Karnal bunt (*Tilletia indica*) in relation to meteorological factors. *Res. on Crops.* 4 (1): 144-145.

Singh, D. V., Joshi, L. M., and Srivastava, K. D. 1986. Varietal susceptibility and spread of Karnal bunt of wheat in India. *Rachis Barley Wheat Newsltr.* ICARDA 4:10–16.

Singh, D. V., Joshi, L.M. and Srivastava, K.D. 1983. Karnal bunt new threat to wheat in India. *(In:)* A. Hussain, Kishan Singh, B.P. Singh and V.P. Agnihotri, (Eds.) *Recent Advances in Plant Pathology Prof: H. K. Saksena Festschrift,* Print House (India), Lucknow, pp. 121-135.

Singh, D. V., Joshi, L.M. and Srivastava, K.D. 1986. Foliar blight and spot of wheat in India. *Indian J Genet.* 46 : 217-245.

Singh, D.V. 1986. Bunts of wheat in India. *(In:)* L.M. Joshi, D.V. Singh and K.D. Srivastava, (Eds.) *Problems and Progress of Wheat Pathology in South Asia* Malhotra Publishing House, New Delhi, pp.124-161.

Singh, D.V. and Srivastava, K.D. 1992. *Investigations on Karnal bunt (Neovossia indica) of Wheat - Final Technical Report of PL-480 Project,* Division of Mycology and Plant Pathology, IARI, New Delhi.

Singh, D.V. and Srivastava, K.D. 1992. *Investigations on Karnal bunt (Neovossia indica) of Wheat - Final Technical Report of PL-480 Project,* Division of Mycology and Plant Pathology, IARI, New Delhi.

Singh, D.V. and Srivastava, K.D. 1997. Foliar blights and Fusarium scab of wheat: present status and strategies for management. *(In:)* Agnihotri, A.P., Sarbhoy, A.K. and Singh, D.V. (Eds.) *Management of Threatening Plant Disease of National Importance.* Malhotra Publishing House. 1-16.

Singh, D.V., Aggarwal, R. and Nagarajan, S. 1991. Variability in *Neovossia indica* in the context of host resistance and disease management. *Golden Jubilee Sym. Genet. Res and Edu,* Indian Society of Genetics and Plant Breeding, IARI, New Delhi from Feb. 12-15, 1991.

Singh, D.V., Aggarwal, R., Shreshtha, J.K., Thapa, B.R. and Dubin, H.J. 1989. First report of *Tilletia indica* on wheat in Nepal. *Plant Dis.* 73: 273.

Singh, D.V., Siddique, K.A., Aggarwal, R., Srivastava, K.D., Maurya, A.K. and Arora, P. 1990. Extraction of teliospores of *Neovossia indica* from soil mixed with teliospores. *Indian Phytopath.* 43: 500-503.

Singh, D.V., Srivastava, K.D., Aggarwal. R. and Bahadur, P. 1993. Wheat Disease problems: The Changing Scenario. *(In:)* Sharma, R.C. and Rao, V. (Eds.), *Pests and Pest Management in India -The Changing Scenario.* pp 116-120 Plant Protection Association of India, Hyderabad. pp:312.

Singh, D.V., Srivastava, K.D., Joshi, L.M. and Nagarajan, S. 1980. Note on the occurrence of foliar pathogens of wheat in India. *Indian J. Agric. Sci.* 50: 633-634.

Singh, H. B. and Sethi, G. S. 1975. Reactions of wheats to flag smut. *Indian J. Genet.* 35: 450-53.

Singh, L. and Sharma, M. 1978. Pesticides in relation to germination and growth of *Helminthosporium sativum* and some Indian wheat varieties *in vitro. Acta Botanica Indica.* 6: 139-47.

Singh, P. J., Dhaliwal, H.S. and Gill, K.S. 1989. Chemical control of Karnal bunt *(Neovossia indica)* of wheat *(Triticum aestivum)* by s!ngle spray of fungicides at heading. *Indian J. Agric. Sci.* 59: 131-133.

Singh, P.P., Sohi, H.S. and Bedi, P.S. 1983. Effect of Karnal bunt of wheat on seed viability, embryo vigour and metabolities of plants. *J. Res. (PAU)* 20: 481-484.

Singh, R. S. 1978. *Plant Diseases.* Oxford and I B H Publication, New Delhi. pp: 564.

Singh, R.A. and Krishna, A. 1982. Susceptible stage for inoculation and effect of Karnal bunt on viability of wheat seed. *Indian Phytopath.* 35 : 54-56.

Singh, R.P., Singh, A.K. and Singh, R.N. 2003. Effect of Neem products on growth of *Alternaria triticina. Annals of Plant Protection Sciences.* 11(2): 384-3845.

Singh, R.V. and Singh, A.K. 1994. Studies on the epidemiology of foliar blight pathogens. *(In:)* Goel, L.B., Kwnar, J. and Nagarajan, S. (Eds.) *Results of the Coordinated Experiments* 1993-94, *Crop Protection (Pathology),* pp 213-214, AICWIP, Directorate of Wheat Research, Karnal, 253 pp.

Singh, R.V., Singh, A.K. and Singh, S.P. 1995. Progress of work under National Centre on Foliar Diseases during 1994-95. Paper presented in the 34th *All India Wheat Research Workers' Workshop*, AICWIP, ICAR, at UAS, Dharwad, August 1995.

Singh, S., Ram, B. and Srivastava, M.P. 1997. Flag smut resistance in wheat varieties/ lines and triticales. *Crop-Research-Hisar* 13: 3, 649-652.

Singh, S.I. and Singh, P.P. 1985. Cultural practices and the Karnal bunt *(Neovossia indica)* incidence. *Indian Phytopath.* 38 : 594.

Singh, V.K., Khanna, B.M. and Verma, P.C. 1977. Reaction of wheat varieties to Alternaria blight. *Indian J Mycol. Pl. Pathol.* 7 : 190-191.

Singh. A., Tiwari. A.N. and Rai, R.C. 1985. Control of Karnal bunt of wheat by spray of fungicides. *Indian Phytopath.* 38: 104-108.

Singh. K. B. and Mehrotra, R. S. 1978. Effects of some fungicides on mycelial growth and respiration of *Helminthosporium sativum. J. Indian Bot. Soc.* 57: 1-5.

Singh. K. B. and Mehrotra. R. S. 1978. Effects of some fungicides on mycelial growth and respiration of *Helminthosporium sativum. J. Indian Bot. Soc.* 57: 1-5.

Smilanick, J. L., Dupler, M., Goates, B.J., Hoffmann, J. A., Clark, D., and Dobson, D., 1986. Germination of teliospores of Karnal, dwarf, and common bunt fungi after ingestion by animals. *Plant Dis.* 70:242–244.

Smilanick, J.L, Hoffman, J.A, Royer, M.H. 1985. Effect of temperature, pH, light and desiccation on teliospore germination of *Tilletia indica. Phytopath.* 75:1428-31.

Snijders, C.R.A. 1990). Fusariwn head blight and mycotoxin contamination of wheat, a review. *Neth. J. Plant Pathol.* 96: 187-198.

Sokhi, S.S. and Joshi, L.K. 1974. Estimation of losses in yield due to leaf blight disease of wheat caused by *Alternaria triticina. Indian J. Mycol. Pl. Pathol.* 4 : 29-33.

Srivastavaa D. N. and Durgapal, J. C. 1963. Foot rot of Wheat. *Agric. Res.* 3: 170.

Stack, R.W. 1982. Root and Crown rots of small garins. Cooperative Extension Service, Ext. Circular, pp. 785. North Dakota State University, USA.

Statler, G.D. and L.C. Darlington. 1972. Resisitance of hard reel spring wheat and durum wheat to seedling blight and crown rot. *Plant Dis. Reptr.* 56 (9) : 788-792.

Swaminathan, M.S., Raychaudhuri, S.P., Joshi, L.M. and Gera, S.D. 1971. Incidence of black point and Karnal bunt in North-western plains zone during 1969- 70. *Wheat Disease Newsletter* (New Delhi) 4: 1-4.

Tahvonen, R. and Sorri, P. 1992. Studies on the biological control of *Fusarium culmorum* in Finland. *Bulletin OILB SROP.* 15: 1, 24-26.

Tandon, J.P. 1985. Wheat improvement programs for the hotter parts of India, *(In:)* Villareal, R.L. and Klatt, A.R. (Eds.). *Wheats for More Tropical Environments. A Proceedings of the International Symposium,* pp 63-67, CIMMYT, Mexico, 354 pp.

Tariq, A. H., Khan, S. H. and Saleem, A. 1992. Control of flag smut (*Urocystis agropyri*) through screening of varieties and systemic seed-dressing fungicides. *Pakistan.J.Phytopath.* 4: 1-2, 32-36.

Tinline, R.D. 1986. Agronomic practives and common root rot in spring wheat: Effect of depth and density of seedling disease. *Can. J. Plant Pathol.* 8 : 429-435.

Thompson, J.P. and Wildermuth, G.B. 1989. Colonization of crop and pasture species with vesicular-arbuscular mycorrhizal fungi and a negative correlation with root infection by *Bipolaris sorokiniana. Can. J Bot.* 69 : 687-693.

Tomar, S.M.S., Srivastava, K.D. and Singh, D. V.1991. Screening of wheat species and few amphidiploides of Karnal bunt (KB) resistance. *Annual Wheat Newsltr (USA)* 37: 63-64.

Tschanz, A.T., Rorst, R.K. and Nelson, P.E. 1975. Ecological aspects of ascospore discharge in *Gibberella zeae, Phytopath.* 55: 597-599.

Tyagi, P. D. 1968. Control of loose smut and flag smug of wheat. *Ist Summer* v' *Instt. in Plant Disease Control* IARI. New Delhi. pp:80.

Verwoerd, L. 1929. The biology, parasitism and control of *Urocystis tritici* Koem., the causal organism of flag smut of wheat as the cause of "Stem smut, in rye. S. *African Dspt. Agric. Sci. Bull.* 76: 52.

Walker, J. C. 1969. *Plant Pathology.* McGraw-Hill Book Company, New York.

Wolff, R. 1873. Beitrag Zur Kenntnis der Ustilagineen. *Bot Ztg* 31: 657-61. 195.

Wu, Y. S. 1949. Temperature and cultural studies on *Urocystis* tritici. *Can. J. Res.* 27: 66-72.

Yadava, H. R., Mathur, R. S. and Tandon, I. N. 1969. Results of recent resistance tests of wheat and barely against loose smut in Uttar Pradesh. *Labdev. J. Sci. Technol. 7-B:* 237-39.

Yu, T. F., Wang, H. R. and Tsiang, C. T. 1936. Varietal resistance and susceptibility of wheats to flag smust *(Urocystis tritici* Koem) Ill. Physiologic specialization in *Urocystis tritici* Koem. (Chinese summary). *Bull. Chin. Bot Soc.* 2: 111-13.

Wilddermuth, G.B. and McNamara R.B/ 1991. Effect of cropping history on soil population of Bipolaris sorokiniana and common root rot of wheat. *Aust. J. Agric. Res.* 41: 1-10.

Vilich, V.1998. Effect of benzoxazolinone allelochemicals from *wheat* on selected *soil*-borne pathogens and antagonists. *Proc. 50th Intr. Symp.Crop Prot*, Gent, 5 May 1998. Part IV. 63: 3b, 971-976.

Wang, Y.P., Liu, Y.Q., Pan, N.S. and Chen, Z.L. 1992. Preliminary studies of the control of wheat scab with *Bacillus subtilis* A014 strain. *Chinese J. Biol. Contr.* 8(2): 54-57.

Wang, Y.Z, Chen, H.G, Yang, X.N, Lu, M. and Wu Z.F. 1989. Bioactivity of crude toxin produced by *Fusarium graminearum* and its application in identification of *Fusarium* head blight resistance of wheat cultivars. *Scientia Agricultura Sinica* 22(4): 54-57

Warham, E.J. 1984. A comparison of inoculation methods for Karnal bunt *Neovossia indica. Phytopath,* 74: 856-857.

Warham, E.J. 1987. *Studies on Karnal bunt Disease of Wheat. Ph.D. Thesis,* University of Wales, U.K.

Warham, E.J. 1988. Screening for Karnal bunt *(Tilletia indica)* resistance in wheat, triticale, rye and barley. *Cand. J. Pl. Pathol.* 10: 57-60.

Warham, E.J. 1990. A comparison of inoculation techniques for assessment of germplasm susceptibility to Karnal bunt *(Tilletia indica)* disease of wheat. *Ann. Appl. Bioi.* 116 : 43-60.

Warham, E.J., Kazi, A.M. and Rosas, V. 1986. Karnal bunt *(Tilletia indica)* resistance screening of *Aegiolops* species and their practical utilization for *Triticum aestivum* improvement. *Cand. J. Pl. Pathol.* 8: 65-70.

Wei, C. M, Lu S. H, Xue W. L, Xu S. Z. and Zhang W-T. 1990. The distribution along the middle and lower reaches of Yangtze River of isolates of *Fusarium graminearum* producing deoxynivalenol. *Shanghai Agricultural Science and Technology* (6):19-20

Wu, W. S.1976. Biological control of seed- and *soil*-borne fungi associated with wheat and oats. Increasing the effectiveness of seed-treated fungicides by the pelleting method. *Botanical Bulletin of Academia Sinica*.17: 2, 161-168.

Wu, W. S.1977. Antibiotic and mycoparasitic effects of several fungi against seed- and *soil*-borne pathogens associated with *wheat* and oats. *Bot. Bull. Acad. Sinica.* 18: 1, 25-31.

Wu, Y.L., Wu G. Q. and Ye H Z. 1997. Changes in protein components of wheat grain infected with *F. graminearum. J.Sichuan Agril. Univ.* 15(3):329-334.

Xu, Y.G, Yao C.L. and Fang Z D. 1990. Toxic effect of DON from scabby kernels on wheat. *J. Nanjing Agril. Univ.* 13(2):58-63.

Xu, Y.G. and Chen, L.F. 1993. *Wheat Scab: Theory and Practice on Control.* Jiangsu Science and Technology Publishing House, Nanjing.

Xu, Y.G. and Hideki N. 1989. The infection process of the wheat scab pathogen. *J.Nanjing Agril. Univ.* 12(3): 33-38.

Zhang, Q. C. 1998. A severe epidemic of wheat scab in Yanggu County. *Plant Protection* 24(6):52.

Zong, W.Y. 1988. Screening techniques and sources of resistance to Fusarium head blight, *(In:)* Klatt, A.R. (Ed.). *Wheat Production Constraints in Tropical Environments,* pp:239-250 CIMMYT, Mexico. pp: 410.

• • •

Advances in Soil Borne Plant Diseases, 2008
Editors: Naik and Devika Rani, pp. 219-231
New India Publishing Agency, New Delhi

8

Soil Borne Diseases of Temperate Fruits and their Management

J.P. Sharma and R.C. Sharma

INTRODUCTION

In India, temperate fruits are grown successfully at locations above 3000 feet mean sea level in hilly areas of Himachal Pradesh, Jammu and Kashmir, Uttranchal, Sikkim, parts of North Eastern states like Assam, Arunachal Pradesh, Nagaland, Meghalaya etc. Mount Abu hills of Rajsthan and Nilgiris of Tamil Nadu also provide suitable agro-climatic conditions for growing some temperate fruits. Apart from apple; pear, peach, plum, apricot, cherry, almond and walnut are major temperate fruits being grown over 3.38 lakh hectares of area and play pivotal role in the economy of hilly farmers whose land is not otherwise considered suitable for traditional agriculture. They contribute about 11.5 percent of total area under fruits in the country, yet production is only 5 percent of the total. Incidence of many fungal, bacterial and viral diseases in these fruit crops is a major impediment in their successful cultivation and soil borne pathogens are of foremost importance and are responsible for premature death of the trees in the orchards. Amongst these, white root rot, collar rot, seedling blight, crown gall, hairy root are the major constraints which can cause as high as 40 percent damage in certain situations (Sharma and Sharma, 2001). Threat from the replant problem is causing concern due to its negative effect on the establishment of replanted sites due to soil sickness.

Department of Mycology and Plant Pathology
Dr. Y.S. Parmar University of Horticulture and Forestry Nauni, Solan -173230 (H.P.)

A detailed account on some of the major soil borne disease problems in temperate fruits and their management is presented in this chapter.

WHITE ROOT ROT

It is a destructive disease of many fruit trees including apple and pear, as well as about 170 other species of plants in 63 genera. It is widely distributed in temperate regions of the world including western Himalayas. In India, the disease was first observed in the hills of erstwhile Uttar Pradesh in 1939. The annual losses estimated in India due to this disease are about Rs. 1.3 million (Agarwala and Sharma, 1966).

Symptoms

The characteristic underground symptom is the rotting of small roots by white mycelium, which then invades larger roots. The invasion extends through the cortex and cambium to the trunk and occasionally the fungus ruptures the bark. Infected roots and adjoining soil is covered with strands of white cottony mycelium. Later, the mycelium turns greenish grey or black, and the pathogen forms indefinite plaques within the bark and loosely aggregated external hyphal strands. Under suitable moisture and temperature conditions, white mycelium appears at the ground level on the crown region of the tree. Affected trees are easily uprooted from the soil. Aerial symptoms are sparse growth and premature discolouring of foliage and absence of new shoot growth. Leaves are smaller than normal and premature defoliation may occur. Fruits cease growth and may shrivel.

Causal organism

The disease is caused by *Rosellinia necatrix* Prill. (Ana.: *Dematophora necatrix* Hartig). The perfect stage of the pathogen is not known to occur in India. The hyphae of the fungus are septate and have peculiar pear shaped swellings near each septum, a distinctive diagnostic character that separates this fungus from other root rot pathogens. These swellings occur most frequently in older hyphae. The fungus produces rough, irregular black micro sclerotia (98 x 130 µm). The teleomorphic stage produces perithecia, which are rarely found in nature and require a long period to develop. Asci are filiform, cylindrical (250 -380 x 8-12 µm), long stalked, unitunicate and eight spored. Ascospores (30-50 x 5-8 µm) are one celled, straight or curved, uniseriate, and dark brown, with a longitudinal slit running parallel to the long axis for about one third of the length of the spore (Jones and Aldwinckle, 1990). Conidiophores are produced on brown, rigid synnemata which are up to 1.5 mm high. Conidia (3-4.5 x 2-2.5 µm) are one celled, solitary, ellipsoidal or ovoid, hyaline to pale brown and smooth.

Epidemiology

The pathogen can survive on rotted roots in the ground and occasionally as mycelium or loose micro-sclerotia in the soil or adhering to root debris. The pathogen infects fine roots entering by direct penetration at the point of contact. The hyphae branch freely in the cortex, destroying it as they grow towards larger roots. The fungus comes out from the rotten cortical region to form the characteristic white mycelium on the roots. Spores do not play any significant role in the dissemination of the pathogen. The pathogen spreads through root contacts and cultural operations. The disease is favored by high soil moisture (near to field capacity) and temperature of 20 to 25°C. The disease is common in heavy soils (Gupta and Sharma, 1999).

Management

Infected trees should be uprooted and destroyed. All the roots should be removed from soil and destroyed. Infected trees or the' orchards should be isolated by digging trenches around them and cultivation through the basins of such trees may be avoided. The pathogen can be controlled before planting by either soil fumigation with methyl bromide or soil solarization (Gupta and Sharma, 1999; Sharma *et al.*, 1999). Application of carbendazim (0.1%) through deep holes (15-20 cm) made in the basin of the tree at 30 cm from each other help in checking the root rot in the orchards., Almost all the apple root stocks have shown susceptibility to this pathogen, however, some degree of resistance occurs in MM 109, MM 104 and M16. Some bio-control agents have also been tested with little success.

COLLAR ROT

Collar rot and crown rot, although similar, are distinct diseases. Unfortunately, the two terms are often used interchangeably, which has resulted in confusion. Collar rot is a disease of the scion portion of the tree, affecting bark tissues of the lower trunk at or above the soil line, a portion of the trunk infrequently referred to as the collar. Crown rot is a disease of the root stock portion of the tree affecting bark tissues of the root crown region (i.e., the point where the roots join the stem) and the proximal ends of the primary roots (Jones and Aldwinckle, 1990). In India, the disease is serious in the entire apple growing areas.

Symptoms

The symptoms are often not visible unless the outer layers of bark (the periderm) of crown or collar region are removed to expose the inner phloem tissues down to the cambium. Affected tissues are necrotic and orange to red brown in colour and eventually becoming dark brown as they decay. In crown rot, the necrosis often extends some distance along the primary roots and occasionally up to the graft union, but rarely above it. In collar rot, the necrosis may extend one meter or more up the trunk but does not extend much below the graft union. Foliar symptoms are not diagnostic but are merely general indicators of root or vascular problems. Affected trees are unthrifty, having poor terminal growth and become stunted. Foliage is sparse, chlorotic and may develop an early purple discoloration in autumn. Fruits tend to be small and colour prematurely.

Causal organism

Collar rot is mainly caused by *Phytophthora cactorum* (Lebert & Cohn) Schroter though other Pythiaceous fungi are also capable of causing similar symptoms in apple and pear trees. *P. cactorum* is semi aquatic, soil borne homothallic fungus which produces distinctive papillate sporangia on sympodially branched sporangiophores. The antheridia are paragynous. Both sporangia and oogonia form readily on many solid media.

Epidemiology

The fungus is known to survive in the orchard soils and a soil temperature of 12 to 20°C and soil pH 5 to 6 are optimum for its spread. The fungus survives as chlamydospores in plant debris or soil and may colonize fallen apple and pear fruits on the orchard floor. The most important propagule of infection is probably zoospores produced within sporangia. Maximum production of sporangia occurs when the soil is saturated or nearly saturated with water. The disease is severe in poorly drained soils and sites where standing water collects for longer periods. The active movement of zoospores to tree crowns and roots is probably a chemostatic response, operating only over several centimeters. The epidemiological importance of directly germinating sporangia, oospores, and chlamydospores as infective propagules is unknown. The susceptibility of scion bark tissues to infection by *P. cactorum* is apparently influenced by the rootstock, being greater in trees on the vigorous rootstock clones. A similar influence of the scion cultivar on rootstock susceptibility has also been demonstrated. Physiological changes associated with transmission from a vegetative to a reproductive state also affects susceptibility, as significant disease occurrence is noted during the early bearing years of an orchard (Gupta and Sharma, 1999).

Management

An integrated approach to disease management, employing cultural practices, resistant plant material, and chemical treatments is recommended.

Plantation should be done only on soils with adequate drainage. As the pathogen also enters through graft union, so grafting may be practiced about 30 cm above the soil level. The disease is also suppressed by localized superficial heat treatment of infected region with blow lamp flame. Apple rootstocks M23, M4, MMII0, MM114 and MM115 and crab apple possess high degree of resistance against this disease. Scion cultivars also vary in resistance to collar rot. They are generally much less susceptible than most common clonal rootstocks. The fungicides are applied around the affected trees or by painting the wounds with fungicidal paints. Affected portion is scarified up to healthy tissue and then Bordeaux paint or Metalaxyl paint or Chaubbatia paste is applied. Ridomil MZ and Aliette applied as soil drench and paint on affected portions effectively control the disease. Fosetyl-Al applied as foliar spray also provides effective control of the disease (Sharma *et al.*, 1999). Recent investigation suggests that biological control of this disease with *Bacillus subtilis* and *Enterobacter aerogenes* is feasible and require commercial exploitation (Sharma *et al.*, 1999).

SEEDLING BLIGHT

It occurs in nurseries and orchards on trees of about 3 years of age. In India, the disease was first reported in 1937 from Mysore. The pathogen has been reported on over 200 species of plants.

Symptoms

A web of white mycelium is found on soil at the base of an infected seedling. The mycelium may progress up the stem for 7 to 17 cm. During period of high humidity, masses of tan to brown sclerotia develop. They may be found at the base of the seedlings and on roots up to 14 cm into the soil. The fungus gradually girdles the crown, and the dry papery, brown leaves usually hang on the seedlings which ultimately die (Gupta and Sharma, 1999).

Causal organism

Seedling blight of apple and pear is caused by *Sclerotium rolfsii* Sacco. It does not produce asexual spores. The basidial stage, *Athelia rolfsii* (Curzi) Tu & Kimbrough is rarely seen in the field, and strains of the pathogen differ in morphology and pathogenicity. The sclerotia have a long but variable survival period.

Epidemiology

Sclerotia are the principal means of survival and serve as a source of primary inoculum. They germinate better in light and well drained soils.

Infection usually occurs directly by the penetration of young, uninjured bark but wounds facilitate the entry of the pathogen. The pathogen produces both pectic and cellullolytic enzymes. Temperature of 25-35°C, high soil moisture, good aeration and abundance of plant debris favour disease epidemics. Disease incidence in the nursery is directly proportional to the number of sclerotia in the soil. Maximum infection occurs when the sclerotia are adjacent to the seedling stem. Sclerotia more than 3 cm away from stem rarely causes infection. Young trees of 1 to 3 years age are more susceptible than the older trees. The number of viable sclerotia associated with a dead tree decreases with time.

Management

Weed control in the nursery is essential, since this pathogen grows profusely in humid conditions. Sites with 5 or more sclerotia per kilogram of soil should be avoided for raising plants. The nursery site must be changed every 4 to 5 years. Growing maize continuously for 4 to 5 years also helps in reducing pathogen population in the soil (Sharma *et al.,* 1999). Fertilizing with ammonium type fertilizers, following crop rotation and soil treatment with various contact fungicides also keeps the disease under control (Sharma and Sharma, 2000). Rootstock M9 shows some resistance to S. *rolfsii.*

CROWN GALL AND HAIRY ROOT

The disease occurs on a number of dicotyledonous, herbaceous and woody plants belonging to 140 genera of more than 93 families and out of these host plants, it is destructive on stone and nut fruits. Affected nursery plants are unsolvable and fruit plants with galls or tumors grow poorly with reduced yields. Hairy root is primarily of concern in apple and develops only infrequently on stone and nut fruits.

Symptoms

Galls develop on roots, crowns and occasionally on branches. They vary in size from microscopic to more than 10 cm in diameter. The galls are composed of disorganized vascular and parenchyma tissues, rapidly growing, white, fleshy in the beginning and becoming tan to brown in the dormant season. New galls frequently develop adjacent to those from the previous year. Hairy root symptoms on young apple trees are characterized by an extensive proliferation of adventitious roots at the base of the stem, crown and roots (Sharma, 2000).

Causal organism

The crown gall is caused by a gram negative soil bacterium *Agrobacterium tumefaciens* (Smith & Townsend) Conn. This bacterium is rod shaped and aerobic with a few peritrichous flagella and all strains reported from Himachal Pradesh fall within biovar (Jindal and Sharma, 1988). Virulent isolates of the pathogen contain one to several large plasmids composed of circular double stranded DNA. One of these plasmids carried the genes for tumor induction and is called the tumor-inducing plasmid (Ti-plasmid). Isolates devoid of Ti plasmid are a virulent. Genes for host range of the bacterium are also located on the Ti plasmid. When the bacterium infects, a segment of plasmid becomes incorporated into the plant chromosome. In response, the host cell over produces plant hormones, thereby initiating tumour formation. The T -DNA also directs the cells to produce unusual amino acids (opines) which are utilized by the pathogen. These opine impart a nutritional advantage to the pathogen over other microorganisms in the galls. *Agrobacteirum rhizogenes* Conn. is the causal organism of hairy root. Nearly all isolates of this species fall within biovar 2. Pathogenic strains have large root inducing (Ri) plasmid which also possesses T -DNA and the latter is transferred to the host genome upon infection.

Epidemiology

These pathogens are soil inhabitants and may be isolated from soil, galls and apparently healthy *roots.* The pathogen is disseminated with plants or by movement of soil or water. Crown gall infections are initiated at wounds. The most susceptible wounds on stone and nut fruits are at the crown or on the roots and rarely on trunks and limbs. Infections are favored by moist, alkaline, poorly drained soils and can be stimulated by the feeding damage due to nematodes. Previous cropping history as it relates to inoculum levels in soil affects disease development.

Management

Sanitation and cultural practices are important for reducing the amount of disease. These include using certified disease free propagative material, planting disease free plants choosing a rootstock with low slanting in well drained areas, adopting cultural practices which reduced wounding, and rotating contaminated sites with non-host plants such as monocots (Sharma *et al.,* 1999). Chemical control for crown gall and hairy root has been either ineffective or impractical for use at commercial scale (Ogawa *et al.,* 1995). Satisfactory and nearly cent per cent control of crown gall is achieved by dipping seeds or root system of nursery plants or the rootstocks up to crown region in *Agrobacterium radiobacter* strain 84 suspension (Kerr, 1980). This biocontrol agent occupies the infection sites and produces a bacteriocin agrocin 84 which is inhibitory to the pathogen. A derivative strain of K 84, designated as K 1026, is superior to its original K 84 strain (Jones and Kett, 1989).

REPLANT PROBLEM

Replant problems in agriculture and horticulture are far from rare. These problems are really man-made and occur due to replanting, narrow crop rotation and monoculture. In temperate fruit trees, poor growth of the replanted orchards has been observed when apple after apple and cherry after cherry were grown (Thompson, 1959).

It has also been observed that apple after cherry and cherry after apple grow normally. The situation resulting in this poor growth is generally known as apple replant disease (ARD) where biotic causes are involved and apple replant problem (ARP) when a biotic factors are considered as its causative factors. This problem can be specific and shows poor growth of apple specifically after apple and therefore sometimes referred to as specific apple replant disease (SARD). The peach replant disease affects only peaches planted on the site of uprooted peach trees. A few related species, as apricots planted after peaches are sometimes also affected (Gur and Cohen, 1988).

Causes

There are many causes of poor growth of young trees of apple and other temperate fruit trees, which vary from region to region. They include weak, damaged or diseased nursery stock, poor planting and management techniques, water deficiency or excesses, spray toxicity, weed competition or damage from insect pests, diseases and rodents. But major concern here is only with soil related causes which include biotic (harmful microorganisms) and abiotic (nutritional deficiencies or excesses, soil pH, phytotoxins) factors in the soil. As in Kotkhai area of Himachal Pradesh where dead or poor quality plants or the ones which have cultivated their commercial productive life are being replaced by new plants even in the orchards where most of the plants are still standing. The poor growth of the newly replanted trees has been observed to be due to the attack of Sanjose scale, wooly aphid and root rot disease (J.N.Sharma, unpublished), which may not be considered as a case of replant disease or problem and should be treated accordingly.

However, several workers have reported fungi, bacteria, actinomycetes and nematodes as potential biotic agents responsible for apple replant disease. Many fungi belonging to oomycetes, hyphomycetes and basidiomycetes were reported as causal agents of replant problems, causing root rots and progressive decline of fruit trees (Yadava and Doud, 1980). *Fusarium (F. equiseti, F.oxysporum, Fsolani)* and *Rhizoctonia* species have been associated with replant disease of peach and cherries in USA, *Theilaviopsis basicola* in England and Holland (Hoestra, 1965) and *Rosellinia necatrix* in Hungary. Utkhede and Li

(1988) reported *Penicillium janthinellum* and *Cortentinella terrestris* as the cause of replant disease. *Peniophora sacrata* was responsible for decline and replant problems in apple and stone fruits in New Zealand. Among the oomycetous fungi, *Pythium sylveticum* and *P. irregulare* were reported as its potential fungal pathogens in UK and Canada (Sewell, 1981). Two strains of *Bacillus subtilis* were associated with stunted growth of apple seedlings in British Columbia (Utkhede and Li, 1988) and *Pseudomonas putida* in Czechoslovakia (Catska *et al.*, 1982). Direct evidence for the involvement of actinomycetes in ARD was provided from Germany. *Pratylenchus penetrans,* a root knot nematode was also reported to be an important agent in causing replant disease. Synergistic interactions of fungi, bacteria and nematodes were found responsible for causing apple replant disease.

Phytotoxins, unbalanced nutrition, poor soil moisture, soil pH are some of the common abiotic factors that cause replant problem (Sharma and Bhardwaj, 1999). Toxic chemical prunasin produced by the decomposing roots of peach plants have been established to be the primary cause of replant problems of peaches alongwith hydrocynamic acid and benzaldehyde (Mizutani *et al.*, 1988). Deficiency of major nutrients such as N, P, and K in the soil can cause physiological stresses that might lead to replant problem (Mai and Abawi, 1981). In some soils, growth of apple is strongly mycorrhiza-dependent thus indicating the role of phosphorus solublization and its availability. However, in a diagnostic trial conducted in Kotkhai apple belt of Himachal Pradesh, there was no indication of any inhibited growth of apple seedlings in any of the 12 test soils collected from the orchards ranging from 0 to 60 years of age. There was no symptom of replant problem in trees planted as gap filling in 40 year old standing orchard at Jubbal (J.N.Sharma, unpublished). Absence of ARD in our conditions might be due to the comparatively more tolerant seedling rootstock and more planting distance as compared to the developed world where high density plantings on clonal rootstocks are popular.

Management

As the causes of replant problems vary from region to region and even orchard to orchard, therefore different control measures are practiced depending on their conditions. Many cultural practices involving site selection, sub soiling to remove hard pan, liming, intercropping, clean cultivation, improving drainage etc. have been recognized to lower this problem. Crop rotation is not an option as the temperate fruit crops are perennial and can not be rotated frequently and apple trees can still be affected even after 20 years after removal of an orchard (Slykhuis, 1990). Treatment with liquid fertilizer and potassium sulphate increased apple growth two fold. Application of FYM and sulphur containing potassium fertilizer also reduced the replant disease effects. Similarly, application of nitrogen alone and in combination with phosphorus in soil infested with fungi and bacteria causing ARD significantly increased seedling height, whereas phosphorus alone had no effect. Replacing the soil at planting site with 10-20 kg of potting soil mixture comprising peat, sand or soil, dolomite lime, NPK fertilizer and minor elements followed by fertigation could overcome the replant problem in the Netherlands. Slykhuis and Li (1985) demonstrated the increased growth of apple trees by preplanting treatment with mono ammonium phosphate. Intercropping with white mustard, radish and some other plants as catch crops before planting greatly improve the apple seedling growth. Antagonistic plants such as marigold *(Tagetes petula)* substantially reduce the population of nematode and *Pythium* species responsible for replant disease. Steam heating of replant sites at 90°C increased the apple plant growth by 120 percent as compared to control. Soil mulched with black polythene film after planting supported the greatest plant girth increment, total extension growth and highest yields. In a similar field trial in Kotkhai, apple trees planted as gap filling in the standing orchard in Chaithla village where the plant had been killed by soil borne insect pests and diseases, showed almost equal girth and incremental shoot growth by following normal package of practices recommended for planting new orchards in 1992 (J.N. Sharma, unpublished).

Although pre-planting fumigation with methyl bromide, formalin or chloropicrin is expensive, hazardous and difficult, yet these chemicals have been proved quite effective in reducing the effect of replant problems in apple and other temperate fruits. Weekly and biweekly application of metalaxyl after planting was found as effective as steam sterilization controlling ARD (Jensen and Buszard, 1988). Biocides such as dazomet, formalin, mancozeb, methem, methyl bromide and 1, 3 dichloropropane, methyl isocyanate and chloropicrin also improved the apple tree growth by 20 to 50 percent.

Kumar *et al.,* (1996) reported an increased height of Chaubattia Princess and Early Shanburry apple varieties by treating old orchard sites with carbendazim and zineb.

Inoculation of apple seedlings with *Agrobacterium radiobacter* reduced the amount of phytotoxic micromycetes in the rhizosphere that inhibit the growth of apple seedlings in soil with apple replant problem. Application of B8 strain of *Enterobacter aerogenes* alone was found effective and was the best treatment under orchard conditions in combination with mono ammonium phosphate (Utkhede and Li, 1989). Application of Bact-I, EBW4, *Bacillus subtilis* and B8 soil drenches also increased tree growth in ARD sick soils. VAM fungus *Glomus etunicatum* was found effective in first six months of growth when apple seedlings were grown for first 3 weeks in a sterile substrate. Also, inoculation of apple seedlings with *Glomus mossae* in ARD sick soil that was neither pasteurized nor fertilized with mono ammonium phosphate significantly increased tree growth. In Kullu valley of Himachal Pradesh, growth improvement up to 29.4 to 64.1 and 85.4 percent was observed in case of pre-planting with additional phosphorus treatment, formalin drench and their combination respectively in old apple sites (Bhardwaj and Sharma, 2001). All said and done, the development of resistant/tolerant rootstocks for apple seems to be the best control measure to provide eco-friendly, economic and sustainable management of replant problem of apple and other fruit crops and future efforts should concentrate on this aspect.

REFERENCES

Agarwala, R.K. and Sharma, V.C. 1966. White root rot disease of apple in Himachal Pradesh. *Indian Phytopath.* 19: 82-86.

Bhardwaj, S.S. and Sharma, I.M. 2001. Replant problem in apple-causes and management. (*In:*) Jindal, K.K. and Gautam, D.R. (Eds.). *Productivity of Temperate Fruits, Issues and Strategies.* Dr. Y. S. Parmar University of Horticulture and Forestry, Nauni, Solan (H. P.). pp:417-421.

Catska, V, Vancura, V., Hudska, G. and Prikryl, Z. 1982. Rhizosphere microorganisms in relation to the apple replant problem. *Plant Soil.* 69: 187-197.

Gupta, V.K. and Sharma, R.C. 1999. Concepts and practices in the management of soil borne diseases. (*In:*) Bagyaraj, D.J., Verma Ajit, Khanna, K.K. and Kehri, H.K. (Eds.). *Modern Approaches and Innovations in Soil Management,* Rastogi Publications. pp:333-344

Gur, A. and Cohen, Y. 1988. Causes of soil sickness in replanted peaches: The role of cyanogenesis in peach soil sickness. *Acta Hortic.* 233: 25-31.

Hoestra, H. 1965. *Thielaviopsis basicola,* a factor in the cherry replant problem in the Netherlands. *Neth. J Plant Pathol.* 71: 180-182.

Jensen, P. and Buszard, D. 1988. The effects of chemical fumigants, nitrogen, plastic mulch and metalaxyl on the establishment of young apple trees in apple replant disease soil. *Can. J Plant Sci.* 68: 255-260.

Jindal, K.K. and Sharma, R.C. 1988. Crown gall and hairy root in fruit plant nurseries in Himachal Pradesh. *Proc. 5th Int. Congo Plant Pathology,* Kyoto Japan. pp:88 (Abst).

Jone, D. A. and Kerr, A. 1989. *Agrobacterium radiobacter* strain KI026, a genetically engineered derivative of strain K-84, for biological control of crown gall. *Plant Dis.* 73:15-18.

Jones, A.L. and Aldwinckle, H.S. 1990. *Compendium of Apple and Pear Diseases*. APS Press, St Paul, Minnesota USA. pp:P101.

Kerr, A. 1980. Biological control of crown gall through production of agrocin 84. *Plant Dis.* 64: 25-30.

Kumar, R., Pandey, J.C., Verma, B.L. and Kumar, R. 1996. Chemical control of replant disease.ofapple. *Recent Hortic.* 3: 134-138.

Mai, W.F. and Abawi, G. W. 1981. Controlling replant diseases of some and stone fruits in north eastern United States by pre plant fumigation. *Plant Dis.* 65: 859-864.

Mizutani, F., Hirota, R. and Kadoya, K. 1988. Growth inhibiting substances from peach roots and their possible involvement in peach replant problems. *Acta Hortic.* 233: 37-43.

Ogawa, J.M., Zehr, E.l., Bird, G.W., Ritrchie, D.F., Uriyu, K. and Uyemoto, J.K. 1995. *Compendium of Stone Fruit Diseases.* APS Press, Minnesota, USA.

Sewell, G. W .F. 1981. Effects of *Pythium* species on the growth of apple and their possible causal role in apple replant disease. *Ann. Appl. Bioi.* 97: 31-:42.

Sharma, J.N. and Sharma, R.C. 2001. Diseases of apple and their management -A review. pp. 93-129. *(In:)* Trivedi, P.C. (Ed.). *Plant Pathology*, Pointer Publishers, Jodhpur.

Sharma, R.C. 2000. Bacterial diseases of pome and stone fruits. *(In:)* Gupta, V.K. and Sharma, S.K. (Eds.), *Diseases of Fruit Crops*. Kalyani Publisher, Ludhiana. pp:133-153.

Sharma, R.C. and Sharma, S. 2000. Concepts and methods in temperate fruit disease management. -An overview. *(In:)* Sharma, P. C. and Naveen, B. S. (Ed.), *Pest Control Administration and Environment* 4-H Club Jammu and Eco- development Project, Deot Sidh, Hamirpur. pp:187-197.

Sharma, R.C., Sharma, S. and Sharma, J.N. 1999. Present concepts and strategies of disease management in pome and stone fruits. *Int. J.Trop. Pl. Dis.* 17: 69-80.

Sharma, S.K. and Bhardwaj, L.N. 1999. Apple replant problem. pp. 105-119. *(In:)* Verma, L.R. and Sharma, R.C., (Eds.). *Diseases of Horticultural Crops-Fruits* Indus Publishing Company, New Delhi.

Slykhuis, J.T. 1990. Replant disease. (*In*:) Jones, A.L. and Aldwinckle, H.S. (Eds.) *Compendium of Apple and Pear Diseases,* APS Press St. Paul, USA. pp:47-48.

Slykhuis, J.T. and Li, T.S.C. 1985. Responses of apple seedlings to biocides and phosphate fertilizers in orchard soils in British Columbia. *Can. J. Plant Pathol.* 7: 294- 301.

Thompson, J.A. 1959.-The occurrence of areas of poor growth in a fruit tree nursery. *Rep. E. Mailing Res. Stn.* 58: 80-82.

Utkhede, R.S. and Li, T .S.C. 1988. The role of fungi, bacteria and their interactions in apple replant disease complex in soils of British Columbia. *Acta Hortic.* 233: 75-80.

Utkhede, R.S. and Li, T.S.C. 1989. Chemical and biological treatments for control of t apple replant disease in British Columbia. *Can. J. Plant Pathol.* 11: 143-147.

Yadava, V.L. and Doud, S.L. 1980. The short life and replant problems of deciduous fruit trees. *Hortic. Rev.* 2: 1-116.

• • •

Advances in Soil Borne Plant Diseases, 2008
Editors: Naik and Devika Rani, pp. 233-243
New India Publishing Agency, New Delhi

9

Wilt Diseases of Cotton

A. Kannan[1], and C.D. Mayee[2]

INTRODUCTION

'Wilting' is a broad terminology used by Plant Protection Scientists. Any disturbance or damage to the normal physiology of the plant leads to 'wilting' of plants. Wilting in cotton is attributed to many factors and it is necessary to single out the actual cause for wilting before suggesting any remedial measure (Kannan, 1991).

Wilting in cotton could be due to any one or more factors as indicated below:

- Attack by stem weevil at the collar region or above
- Attack by ash weevil grab at the root tip level
- Attack by root mealy bug
- Damage to roots during field operations
- Placement of fertilizer very close to the plants
- Drought
- Water stagnation
- High concentration of pesticidal spray

[1] Central Institute for Cotton Research, Regional Station, Coimbatore, India
[2] Agricultural Scientist Recruitment Board, Krishi Anusandhan Bhavan, New Delhi

- Incompatible mixtures of pesticidal spray
- Soil borne pathogens *viz., Fusarium* or *Verticillium*
- Soil borne Nematodes
- Unknown etiology

Wilting due to the fungal pathogens *Fusarium* and *Verticillium* and unknown etiology are discussed in this chapter.

VERTICILLIUM WILT

Verticillium wilt, a vascular wilt disease affecting cotton is caused by *Verticillium dahliae* kleb and *Verticillium albo atrum*. Reinke & Berthold these fungi are soil-invading pathogens with narrow host ranges and have the ability to survive in the soil for a long period even in the absence of a susceptible host. Upon infection, the host plant is killed by the secretion of toxic fungal metabolites leading to vascular occlusion and moisture deficit.

Veriticillium wilt of cotton was first reported in 1914 in Virginia in the USA. It is now known to occur in most countries growing cotton expect perhaps Egypt where the highly resistant species *Gossypium barbadense* is grown (Schnathorst, 1981). The disease was reported to occur in central Asia, Bulgaria, and Greece (Butler 1933; Miles 1934). Subsequently the disease was also observed in Peru, Brazil, Uganda, China, Soviet Union, Australia, Zimbabwe and Spain. The disease has become a serious threat to cotton in Peru, Syria, and parts of erstwhile USSR. Verticillium wilt has been catalogued as most serious disease in South Africa (Swanepoel 2003).

The disease was first noticed in India on a crop of MCU – 5 at Coimbatore in 1967 (Natarajan *et al.*, 1968). The disease had probably been prevalent in this area for over a decade and was noticed primarily as a disease of brinjal (Egg plant). It appears to have passed on to *hirsutum* cottons from brinjal. The farmers of this area are in the habit of cultivating brinjal after cotton or cotton after brinjal. About 6000 hectares of MCU – 5 grown during the winter season were affected in Coimbatore district alone. Subsequently the disease was observed in summer grown cotton in Salem (Srinivasan *et al.*, 1976). The disease also occurs in Karnataka (Srinivasan and Santhanam 1974).

The disease is severe in low lying, poorly drained areas and may be seen affecting circular patches sporadically, mostly in winter grown crop. Over the years, the disease spreads to the entire area through irrigation water or through ploughing operations. After a few seasons the disease spreads to the entire field and the affected area may become unsuitable for cotton cultivation.

The disease is commonly seen from November to January during which time both the soil and ambient temp is low (around 15 to 18^0 C). The disease has also been observed as early as in August in years when there is an early and appreciable fall in temperature. The appearance of symptoms most often coincides with the formation of squares and flowers.

Symptoms

Plants infected early are severely stunted. The typical symptoms are bronzing of the veinlets followed by interveinal chlorosis and yellowing. The leaves start drying from the margins and become cup shaped. The interveinal yellowing and drying take on a characteristic appearance of "Tiger stripes". Affected leaves dry and ultimately drop off leaving bare branches. Plucking of an affected leaf, one can observe at the base of the petiole, a curved line of brown spots, the points where the vascular elements enter the petioles from the stem. In a healthy stem these points are pale green, but in a wilt-affected plant they are dark brown. Splitting open the stem, a dark brown discoloration of the vasculature may be seen in wilt-affected plant. Affected plants may bear few bolls and even these are small and have immature fiber. The defoliated plants may rejuvenate producing fresh short branches and diminutive, chlorotic leaves. However, these branches do not produce any bolls. The fiber from affected plants has been found highly immature.

Losses

Verticillium wilt is the most important disease causing economic losses to the cotton crop in China, erstwhile USSR, USA, Australia, Greece, Syria, Zimbabwe, Peru, South Africa and Spain. 25 - 30 % annual loss has been reported from former Soviet Union. In China a loss of 100 million Kg has been estimated (Hillocks 1992).

Causal organism

The genus Verticillium was described by T. F. L. Nees von Esenbeck in 1860. Two species of these genus have been mentioned as causing the wilt of cotton, namely *Verticillium albo-atrum* and *V. Dahliae*. Verticillium has predominantly live vegetative phase and colonies with moderate growth rates. Conidiophores produced on solid media are erect bearing whorls of slender flask shaped divergent phialides. Conidia are hyaline, unicellular, and ellipsoidal (Srinivasan 1994).

V. albo- atrum causes severe wilt of cotton in the green houses at low temperature (20 -24^0 C). *V. dahliae* produces discreet black microsclerotia as resting structures and only *V. dahliae* has been found to affect the cotton under field conditions (Hillocks, 1992).

Host Range

V. dahliae has a wide host range of over 400 species of angiosperms. In India this pathogen has been found to affect brinjal (Egg plant), bhendi (Okra), chilli, pepper and agathi (*Sesbania grandiflora*).

Variation in the Pathogen

Five strains are known to attack the upland cotton in USA. The most common are relatively mild, non-defoliating strain called SS.4 and more aggressive defoliating strain T.1 (Schnathorst, 1981). Russian scientists have distinguished six physiologic races (Kas'yanenko *et al.*, 1978). Work at CICR, Regional Station, Coimbatore has revealed the presence of atleast three races of the pathogen. Race IV.1 (Indian Verticillium) affects a large number of *hirsutum* cottons but most of the *G. barbadense* show resistant or tolerant reactions. Race IV.2 isolated from summer cotton is able to infect cotton grown up to a temperature of 30^0 C. Race IV.3 isolated from *G. barbadense* Suvin (Kannan and Srinivasan 1984) causes severe and rapid defoliation and death of plants. This race produces abundant microsclerotia and sparse conidia in culture (Kannan, 2002).

Screening Techniques

A comparative study involving different techniques of inoculation of cotton plants with *V. dahliae* was carried out to assess the reliability and quicker simulation of field symptoms. A "cool - bed" technique was also devised for lowering the soil temperature to a desired level. Among the different methods of inoculation studied, placing a drop of conidial suspension on the base of the stem and pricking through it twice or thrice with entomological pins and then incubating the inoculated plants in "moist cool sand beds" have produced the typical symptoms of the diseases in 7-10 days time. Stem puncture also ensures direct introduction of conidia into vasculature (Srinivasan and Kannan 1986) and the pathogen reached the top leaves within 48 hours after inoculation.

As the temperature are critical for infection and disease development the lowering of desired temperature could be brought out by frequently moistening the sand bed in which potted inoculated plants are kept buried and incubated. With this method it was possible to screen a large number of populations of breeding materials and germplasm accessions in the seedling stage itself through out the year (Srinivasan and Kannan 1986).

Development of wilt resistant cultivars/hybrids

In tests at Coimbatore over 1800 varieties/lines belonging to four cultivated species of Gossypium were screened in wilt sick plots. Only four out of 1800 *G. hirsutum* proved resistant. They are Hopicala (USA) Aleppo 40 (Syria) Nazili 82 (Egypt) and a semi wild *G. hirsutum* spp *mexicanum* var. *nervosum.*

As the disease is soil borne, development of resistant lines is the most suited and assured way of Management. Hence, a three-pronged approach was adopted for development of resistance lines to Verticillium wilt in India.

Intra - Varietal selection for wilt resistance from existing lines

Tolerance in *G. hirsutum* has been developed in Verticillium infested plots (Barrow 1973). In similar way, a wilt tolerant selection was made in 1982 from MCU 5, a highly wilt susceptible *hirsutum* cultivar. MCU 5 an extra long staple *G. hirsutum* cotton released by Agricultural College and Research Institute, Coimbatore in 1968 made a revolutionary spread in peninsular India and the lint of this quality fetches a premium price in the cotton market even today. However, this variety is highly susceptible to Verticillium wilt and the disease also spread to a larger area along with the spread of the Variety, causing considerable yield loss in areas of occurrence of the disease.

Hence, through a series of isolation of resistant individual population of MCU 5 over the years from 'hotspot' endemic area and screening populations through 'stem prick' inoculation and incubation by 'cool-sand bed', a wilt tolerant selection with the same fibre standards and yield of susceptible MCU 5 was developed and released for commercial cultivation as MCU 5 VT in 1982. This variety is currently in great demand and occupies a large area.

One such resistant selection of MCU5 VT was used as a female parent in heterosis breeding. An *intra hirsutum* hybrid Savita involving the resistant selection of MCU5 VT possessed very high tolerance to Verticillium wilt and released for commercial cultivation in 1987.

Biochemical mechanism of resistance

Biochemical aspects of resistance to wilt in cotton assume great significance in the control of development of resistant type. Pre-infection and post-infection phases of pathogenesis following stem prick inoculation were studied using susceptible and resistant genotypes. Resistant genotypes *viz.,* MCU 5 VT, Surabhi (VRS 7), VRS 9, AVTH, VLV selections had higher constitutive and inducible flavanols as compared to susceptible genotypes viz., MCU 9 and LRA 5166. Inoculation with the pathogen brought about 18-40% increase in tannin and 7-22% rise in gossypol content in tolerant genotypes 5-8% decrease in tannin and 7-10% in gossypol content was observed in susceptible genotypes. The spread of the disease exhibited inverse relationship with the inducible profile of enzymes like IAA oxidase and peroxidase in susceptible line. However, characteristic rise to an extent of 20 -50 % was recorded in the activities of these enzymes up on inoculation with the pathogen in tolerant genotype, depicting the capacity of enhanced protein synthesis and efficient turn over of defensive enzymes. Concomitant rise in protein content was also evident in case of incompatible host pathogen interaction. The genetic potential of the resistant selection manifested in the maintenance of integrity of sub cellular organelles by way of moderate rise (10%) in hydrolytic enzymes like acid and alkaline Phosphatase and Ribonuclease as compared to 20 - 45% rise in compatible interaction, which resulted in loss of membrane integrity, discolouration of leaves, drying and defoliation.

Introgressive Breeding

The semi-wild *G. hirsutum sp. mexicanum,* var. *nervosum* of Leningrad strain has been found to be resistant to Verticillium wilt (Wilhelm *et al.,* 1974; Srinivasan *et al.,* 1976).

The resistance available in this semi wild species has been transferred to elite cultivated variety MCU5 VT through a series of six backcrosses to the recurrent parent and further by pedigree selection method a wilt resistant selection VRS 7 was identified and released as Surabhi for commercial cultivation (Kannan and Gururajan, 1992). The wild characters associated with the semi wild species such as brown lint, wild growth; poorer fiber strength has been removed by frequently back crossing to the recurrent parent. The variety Surabhi with a yield potential of 30-32 qtls / ha of seed cotton and spinning potential of 60_s count is under cultivation in about 20,000 ha in Tamil Nadu, Andhra Pradesh and Maharashtra.

Several resistant selections of the same cross *viz.,* VRS 9, VRS 15, VRS 16, VRS 19, VLV 3 have been found promising and attempts are under way to register them as germplasm accessions with National Bureau of Plant Genetic Resources.

G. barbadense cultures viz., Sujata, Suvin developed by this centre have shown high resistance to *Verticillium* when cultivated in wilt infested soils. However, a new race affecting *G. barbadense* cotton 'Suvin' has been identified. This new race appears similar to sudden defoliating strain T1 of USA. This has been identified in Coimbatore District (Kannan and Srinivasan, 1984).

Non Genetic Control measures

a. Crop Rotation

Any disruption of continuous growing of susceptible varieties will naturally reduce the incidence and severity of the disease compared to the monoculture. Disease Resistant varieties or rotation of non-host crops prevent the spread and build-up of inoculum potential. Monoculture of Acala SJ-2 in California has been found to increase the propagule densities of Verticillium in the soil over the years and had a devastating effect (Devay and Pullman 1984). At Coimbatore, rotation with a puddle crop of paddy reduced the wilt incidence in the following cotton to 3.3. to 15.6 percent against 38 to 54 percent after a crop of garden land Sorghum.

A two or three year rotation involving Lucerne (*Medicago sativa*) or Chrysanthemum (Marigold) reduced wilt incidence by 55 to 70%.

Soil amendments

Application of 125 tonnes per hectare of compost or refuse containing a large proportion of undecomposed cellulosic material gave 85 percent reduction of the disease.

Solarization

Solarization treatments have been very effective in eradicating most *V. dahliae* propagules from soils (Pullman *et al.*, 1981a). Solarization normally increases soil temperatures and most of the propagules of the fungus get killed in about a fortnight (Pullman 1981 b).

Chemical Control

Chemicals that eradicate the fungus from soil or prevent its development in cotton plant are not usually preferred because the cost is prohibitive and highly toxic to natural flora of the soil. Soil fumigation like Vapam, Nemagon applied to the soil 3 weeks prior to sowing has brought out the good control of wilt disease at the wilt sick village Sandegoundenpalayam. However, the usage has been stopped due to its deleterious effect in the soil and prohibitive cost. Fumigation with Chloropicrin; methyl bromide eradicates the pathogens from soil but has stunted the cotton plants leading to severe yield loss (Wilhelm *et al.*, 1966).

Benzimidazole fungicides gave complete control in plots but in the field the control is incomplete (Buchenauer and Erwin 1981). Near complete control has been obtained by use of benomyl or Uzgen @ 59-100 kg / ha (Yunnsou *et al.*, 1973).

Biological control

Trichoderma viride has been used in Russia to control wilt. *Trichodermin* a dried preparation grown on peat or lignin supplemented with nutrients applied as seed coat, pre-plant broadcast or side dressing with fertilizer have reduced the wilt incidence.

Gliocladium roseum added as an antagonist in Vermiculite: bran (3:1) preparation has been found to kill the microsclerotica in soil (Keinath *et al.*, 1990).

FUSARIUM WILT

Fusarium Wilt caused by *Fusarium oxysporum* f.sp. *vasinfectrum (Atk)* Suyol and Hars were first diagnosed in Alabama region of USA in 1892. The disease is worldwide in distribution. This disease is wide spread in black soils of Central, Western and Southern India on diploid cotton viz., *Gossypium arboreum* and *G. herbaceum*. In India, the disease was first observed at Nagpur.

The disease usually occurs in patches in the field and progressively increases in size. The disease causes severe losses through the death of seedling and growing plants. The disease may be seen at any stage of the growth. In the seedling stage cotyledons may appear yellow and chlorotic. A little later stage the foliage may lose turgidity, droop, shrivel and finally defoliation occurs. The symptoms start with the oldest leaves at the base and gradually proceed upwards. Wilted plants exhibits brown discoloration of the vasculature. The pathogen enters the roots and invades the xylem vessels and spread mainly by mycelial growth. In USA, certain nematodes *Meloidogyne incognita, Belanolaimu gracilis* and *Pratylenchus brachyurous* were found to be associated with the wilt syndrome and increase the severity of the disease. Monumental and classic research work has been carried out by the Madras School of Plant Pathology of the Department of Botany, University of Madras and has been well documented (Kalyanasundaram, 1971).

Six Physiologic races have been identified in *Fusarium*. Race 4 is known only in India and affects Asiatic species *viz., Gossypium arboreum* and *G. herbaceum* but not on *G. hirsutum* and *G. barbadense*. Races of Fusarium and response to different *Gossypium* are furnished in Table 9.1.

It is significant that the race 4 is confined to India, attacks only Asiatic cottons, which have been grown for thousands of years but has not adapted itself to the tetraploid hirsutum cottons.

With the release and cultivation of Fusarium wilt resistant varieties of Asiatic cotton on the one hand and simultaneous replacement of resistant hirsutum cottons, Fusarium wilt is fast disappearing (Srinivasan, 1999).

Table 9.1. Races of *Fusarium oxysporum* sp. *vasinfectum* and response to *Gossypium* species

Race No.	Origin	*G.arboreum*	*G.barbadense*	*G. barbadense*	*G.hirsutum*
Race 1	USA	R	S	S	S
Race 2	USA	R	S	S	S
Race 3	Egypt	S	R	S	R
Race 4	India	S	R	R	R
Race 5	Sudan	S	S	S	R
Race 6	Brazil	R	S	S	R

Source: Armstrong and Armstrong (1978)

ADILABAD WILT

A disorder resembling wilt, which had assumed serious proportions in early eighties, was first noticed in 1978 on intra hirsutum hybrid JK Hy: 1 at Adilabad in Andhra Pradesh. This malady has been called variously as Adilabad wilt (Srinivasan 1984a), New wilt (Raj *et al.*, 1987), Para wilt (Mayee, 1991). This malady affects almost all commercial hybrids *viz.*, JK Hy.1, DCH 32, MECH 1.

Symptoms

This malady appears scattered in the field at any stage of crop growth. The drooping of lamina and reddening of leaves, petioles, stem and branches are observed. No discoloration of the stem or root is observed. At certain places lint development inside the boll does not occur. Bolls may look normal in appearance, but when split open, nil or meager development of fiber has been observed. The causal agent of this malady has not yet been determined and the etiology still remains unknown.

REFERENCES

Barrow, J.R. 1973. Genetics of *Verticillium* tolerance in *Verticillium* wilt of cotton *(In:)* Verticillium wilt of cotton - publication ARS - 5 -19, USDA, Washington D.C. pp 89 -97.

Buchenaner, H, and Erwin D. C, (1981) control of *Verticillium* wilt of cotton by spraying foliage with benomyl and thio bendozole. *Phytopath.* 61: 433 - 434.

Butler E.J. 1933, Emp cotton grows. *Rev.* 10: 91 - 99.

Devay J. E. & Pulman G.S. 1984. Epidemiology and ecology of diseases caused by *Verticillium* species with emphasis on Verticillium wilt of cotton. *Phytopathologia Medit.* 23: 95 - 108.

Hillocks R. J. 1992 *Verticillium* wilt - cotton diseases, CAB International, pp. 87 – 126.

Kannan A. and Srinivasan K. V. 1984. A new race of *Verticillium dahliae* affecting cotton. *Curr. Sci.* 53:104.

Kannan, A. 1991. Soil borne diseases of cotton. *(In:)* G. Krishnamohan and R. Jayarajan (Eds.) *Manual on Crop Diseases,* Pubd by Sandoz (India) Ltd. pp: 61-64.

Kannan, A. and Gururajan K.N. 1992 Development of *Verticillium* resistant varieties in cotton - (*G. hirsutum*) *AICCIP Silver Jubilee Symp*, Coimbatore (Abstr.28) Publication CICR, Nagpur.

Kannan A. 2002 Verticillium wilt of cotton - Twenty-five years Achievements in cotton Pathology at CICR, (1976 - 2001) (Ed.) Sheoraj, Central Institute of Cotton Research, Nagpur. pp. 46 – 48.

Kas'yanenko, A. G., Gor'kovtseva, E.A., and Ryabora I. M. 1978. *Plant Breed. Abstr.* 49 (9): 8212.

Keinath, A. P., Fravel D.R. and Papvizas G. C. 1990. Evaluation of potential antagonism for bio control of *Verticillium dahliae* in *Proc. Fifth Int. Verticillium Symp,* Leningrad, USSR, p. 82.

Mayee, C.D. Chanlwar, A.B. and Quadri, S.M. 1991. Etiology and epidemiology of parawilt of cotton, Final Rep. *Marathwada Agric. Univ. Parbhani.* P.60

Miles L.E., 1934. Verticillium wilt of cotton in Greece, *Phytopath.* 24: 558 – 559.

Natarajan, M. K., Sivaprakasan, K. and Ramakrishnan, K. 1968 *Madras Agric. J.* 55: 455-457.

Pullman, G.S. Devay J. E., and Garber R. H. (1981a) Soil solarization, effects on Verticillium wilt of cotton and soil population of *Verticillium dahliae. Phytopath.* 71: 954 –959.

Pullman, G.S. Devay J. E., and Garber R. H. (1981b) Soil solarization, and thermal death: a logarithmic relationship between time and temperature for four soil borne plant pathogens. *Phytopath.* 71: 959 - 964.

Raj, S. Meshram M.K and Mukewar P.M. 1987. Proc. GMP. Discussion on new wilt in cotton. Central Institute of Cotton Research, Nagpur. pp: 33-34.

Schnathorst. W. C 1981. In G. M. Watkins (ed) compendium of cotton diseases, Amor. *Phtyopathol. Soc.* pp. 41 – 44.

Srinivasan K.V., R. Krishnamoorthy and A. Kannan. 1976. Progress report on Verticillium wilt control. *J. Indian Soc. Cott. Improvement,* 1: 77 – 83.

Srinivasan, K.V. 1984 a. Proc Group Discussion on new wilt in cotton. Cent. Instt. Cott. Research, Nagpur, pp: 9-13.

Srinivasan K. V. and A. Kannan 1986. Verticillium wilt of cotton. An appraisal of Techniques of testing resistance. *J. Indian Soc. Cott. Imp.* 1986 (No.1).

Srinivasan K.V. 1994. Cotton diseases. Indian soc. for cott. Imp, Bombay, pp. 41 – 59.

Srinivasan K.V. 1999. Cotton diseases and their control. (*In*:). V.Sundaram, A.K. Basu, K.R. Krishna Iyer, S.S. Narayanan & T.P. Rajendran. (Eds). *Handbook of Cotton in India* Indian Soc. Cotton Improvement, Mumbai.

Swanepoel, A.2003. Assessment of the impact and main dynamics of cotton diseases in Southern and Eastern Africa: Paper presented in *World Cotton Research Conference* – 3, Cape Town. South Africa.

Yunnosov, Ibragimov F.V, Popov V.I and Tkachenko M.P., 1973. Effectiveness of new preparation, *Khlopkovodslvo* 4: 24 – 26.

Advances in Soil Borne Plant Diseases, 2008
Editors: Naik and Devika Rani, pp. 245-259
New India Publishing Agency, New Delhi

10

Mycorrhizae as Biological Deterrent of Plant Pathogens

M.N. Sreenivasa

INTRODUCTION

The fine feeder roots of many vascular plants are invaded by specific nonpathogenic fungi forming a distinct morphological unit known as mycorrhiza (In Greek, Mykes mean mushroom or fungus, rhiza mean root). Physiologically mycorrhiza represents a case of symbiosis.

The term mycorrhizae were coined by a German botanist Albert Bernard Frank in 1885 which literally mean "fungus-roots". Mycorrhizal fungi form a symbiotic association with agricultural and horticultural crops in addition to certain tropical tree species.

Mycorrhizal fungi are broadly classified in to two groups:

1) Ectomycorrhizae

2) Endomycorrhizae.

Ectomycorrhizae

This is otherwise referred to as ectotrophic mycorrihizae. They are formed by the invasion of actively growing absorbing roots of Gymnospermous and Angiospermous plants by Basidiomycetes or Ascomycetes fungi. A compact

Department of Agricultural Microbiology, University of Agricultural Sciences
College of Agriculture, Raichur-584 101, Karnataka, India

mantle of mycelium is formed on the surface of roots. The infested roots become morphologically different from normal roots. Ectomycorrhizal roots are short, fleshy, dichotomously branched and devoid of root hairs.

The ectomycorrhizae are characterized by the presence of an intercellular fungal network in the root cortex termed as "Hartig net". The Basidiomycetous genera, *Amanita, Boletus, Russula, Laccaria, and Pisolithus* etc form ectomycorrhizal association in free species belonging to Pinaceae, Fagaceae, and Juglandaceae etc. Among Ascomycetes, Truffles (Tuberaceae) are known to form ectomycorrhizal association with trees like oak and beech. Each genus show a wide host range, inturn, a single tree species may harbour different mycorrhizal fungi.

The ectomycorrhizal fungi can be easily isolated on a laboratory medium (Melin Norkran's medium) using a carbon source and certain growth factors such as thiamine, aminoacids etc. They play a key role in afforestation programmes.

Endomycorrhizae

The endomycorrhizae includes smaller groups like Arbutoid, Ericoid and Orchid mycorrhizae in addition a predominant group, vesicular-arbuscular mycorrhizae (VAM). VAM fungi are formed by zygomycetous fungi belonging to the genera *Glomus, Gigaspora, Acaulospora, Sclerocystis, Entrophospora* and *Scutellospora* in the family Glomaceae of the order Glomales. All these genera produce azygospores as they are produced with out sexual reproduction. VAM fungi occur in soil universally and are very well established globally as plant growth promoting fungi.

The AM fungi penetrate the epidermis of the root and produce intercellular non septate hyphae. The hyphae produce minutely branched structures that appear intracellularly soon after the entry of the fungus to root cortex. These structures are called as "arbuscules" which are involved in bidirectional transfer of nutrients. The hyphae also produce thin walled spherical to oval bladder like structures called as "vesicle" which act as storage organ of extra nutrients and release them under nutrient deficient conditions.

Of late, VAM associations are called as Arbuscular mycorrhizal (AM) associations as vesicle formation is not seen in few genera.

In most AM associations, roots do not under go visible morphological alterations as found in ectomycorrhizae. The AM roots will be normal and bear root hairs in maize, onion and other Liliaceae members. AM roots can also be recognized by their yellow colour in few cases.

Real interest in AM was stimulated when the scientists demonstrated P uptake by AM fungi. Since then scientists are trying to use AM for better utilization of P in poorly endowed soils. An AM fungus is also known to improve plant nutrition through synergistic interactions with N_2 fixers and P-solublizers (Sreenivasa and Krishnaraj, 1992). AM fungi also act as biological deterrents of soil borne root infecting pathogens (Azcon-Aguilar and Barea, 1996) and help in drought tolerance (Thippeswamy and Sreenivasa, 1999), in addition to production of plant harmones. Thus AM fungi are gaining importance in afforestation and reclamation of deserts.

BIOLOGICAL DETERRENT ACTIVITY OF AM FUNGI

It is very well established that AM fungi help plants to withstand different stresses and to ward off soil borne root infecting pathogens. Thus AM fungi safeguard plant health and in turn promote crop yield (Bethlenfalvay and Linderman 1992; Barea and Jeffries, 1995). Though the literature indicates the first interaction study between a plant pathogenic fungus and AM fungus was carried out during 1968 by Safir and a great deal of information has been published on the subject (Dehne 1982; Bagyaraj, 1984; Schenck 1987. Paulitz and Linderman, 1991; Hooker *et al.*, 1994; Azcon Aguilar and Barea 1996), a little is known about underlying mechanisms.

In the present chapter, a brief account of possible role played by AM fungi in suppressing root pathogens along with mechanisms involved are discussed.

Biological deterrent activities of AM fungi on fungal root pathogens

A perusal of literature on this aspect suggests that the inoculation of AM fungi decrease the disease severity caused by plant pathogenic fungi. However, at the same time, few reports indicate no effect or increase in disease severity due to AM inoculation (Ross 1972, Davis *et al.*, 1979, Menge *et al.*, 1979).

Though a plenty of literature appeared on the possible use of AM fungi as biological deterrent of root pathogens like *Phytophthora, Pythium, Sclerotium Thielaviopsis, Rhizoctonia, Verticillium, Gaeumannomyces, Cylindrocladium, Fusarium* (Liu, 1995; Chand and Jalali, 1998; Carling *et al.*, 1989), only few scientists have worked out the mechanisms of suppression of pathogens. All these studies, however, clearly revealed that AM fungi can mitigate or alleviate the severity of the disease caused by root infecting pathogens. Few scientists reported better tolerance to stress caused by pathogenic fungi (Schonbeck and Dehne, 1977) or decrease in production of chlamydospores by pathogenic fungi (Baltruschat and Schoenbeck 1975). Pre-colonisation of AM fungi was found to be effective in reducing damage caused by pathogen in citrus root stocks (Schenck *et al.*, 1977) and tomato (Dehne and Schonbeck, 1975). AM fungi have also been reported to reduce the disease severity caused by *Rhizoctonia* sp (Iqbal *et al.*, 1977; Prashanthi *et al.*, 1997), *Sclerotium rolfsii* (Krishna and Bagyaraj, 1983; Sreenivasa *et al.*, 1992; Sreenivasa 1994; Sreenivasa, 1997).

Biological deterrent activities of AM fungi on bacterial root pathogens

As compared to interaction studies between AM fungi and root infecting pathogenic fungi, the studies on the possibility of controlling plant pathogenic bacteria using AM fungal inoculation are restricted to bacterial canker of peach caused by *Pseudomonas syringae* (Weaver and Wehunt, 1995) and bacterial wilt of tomato (Halos and Zorillia, 1979) caused by *Pseudomonas solanacearum* (Suresh and Rai, 1991).

Biological deterrent activities of AM fungi on plant parasitic nematodes

Generally nematodes especially sedentary endoparasitic nematodes which penetrate the plant roots as vermiform javeniles cause higher damage to the crops. The literature indicates much work on interactions between sedentary endoparasitic nematode especially belonging to genus *Meloidogyne* and AM fungi.

The suppression of 25 per cent of larvae of root knot nematode *Meloidogyne incognita* was recorded in tobacco plants with AM inoculation (Baltruschat *et al.*, 1973). The population of *M. incognita* which developed into adults also reduced in oat, tomato and carrot plants with AM inoculation up to 4 months (Sikora and Schonbeck, 1975). Working on interaction between root knot nematode and AM fungi, Jaizme-vega *et al.*, (1997) observed reduction in galling and nematode reproduction in banana. Habte *et al.*, (1999) reported effectiveness of *Glomus* species in protecting white clover against nematode damage. A similar observation was made by Labeena *et al.*, (2003) in tomato infected by *M. incognita.*

Few scientists have observed that pre-inoculation of tomato with mycorrhizal fungi suppressed nematode population as well as damage as compared to simultaneous or subsequent inoculation of AM fungi (Suresh and Bagyaraj 1984; Suresh *et al.*, 1985).

Few reports on interactions between migrating nematodes and AM fungi have appeared in the literature. In cotton plant roots inoculated with AM fungi, the population of *Pratylenchus* sp reduced significantly. Similar results of reduction of *Radopholus similis* in banana (Umesh *et al.*, 1988) and rough lemon (O'Bannon and Nemec, 1979) with AM inoculation has been reported.

Earlier Schenck *et al.*, (1975) opined that inoculum level of nematode, varietal resistance etc. will reflect on biological deterrent activity of AM fungi. In most of the studies it was observed that mycorrhizal stimulation of plant growth was not affected suggesting the beneficial influence of AM fungi in making host plant tolerant to the parasite. This aspect has been clearly demonstrated by Roncadori and Hussey (1977) using nematode susceptible and nematode resistant cotton varieties in which AM inoculation offset the nematode damage to the susceptible cultivar. This indicates that host vigor played a key role in offsetting nematode damage to cotton cultivars. Further Bagyaraj *et al.*, (1979) observed reduction in number and size of primary galls produced by nematode in tomato inoculated with AM fungi. Similar observations were made by Cooper (1981) in lucerne, clover and tomato in which AM inoculation mitigate the damage caused by nematodes. However it has been observed that nematode exert very little influence on mycorrhizal root colonization though it has variable effect on sporulation by AM fungi.

Mechanisms of biological deterrent activity of AM fungi

Several mechanisms have been attributed in the literature for crop protection against root infecting pathogens due to AM inoculation. They include improvement in plant nutrition, competition for sites for root colonization, photosynthates, enhancement of plant defense mechanisms through antagonists in rhizosphere, production of morphological, anatomical and biochemical changes in plant roots etc. These mechanisms are broadly grouped under physical (morphological), chemical (Physiological) and biological alterations in the host plant.

Physical (Morphological) Mechanisms

Several scientists have reported thickening of the host plant cell walls through lignifications and polysaccharides by AM fungi which prevent penetration and growth of root infecting pathogens like *Sclerotium, Fusarium, Phoma, Meloidogyne* etc. The reduction in the number of giant cells or syncytia of nematodes in mycorrhizal plants has been reported (Dehne and Schonbeck, 1979).

Inoculation of AM fungi also suppressed the size of nematodes and adult formation. Mycorrhizal inoculations also resulted in stronger vascular tissues which reduces the effect of vascular pathogens (Bagyaraj, 1996). AM fungi also were found to enhance branching of root systems and modify the root architecture. Also localized competition for space between AM fungi and soil borne root pathogens have been reported (Cordier *et al.*, 1996).

AM fungi and root pathogens also compete for infection sites (Linderman, 1994). It has been observed that AM fungi and pathogenic fungi while colonizing the same host usually developed in separate cortical cells of root which clearly indicate competition for space (Smith 1987; Jalali and Jalali 1991). Further evidence for this mechanism was reported by Cordier *et al.*, (1996) who found no colonization of *Phytophthora* in host cells having arbuscule of AM fungi.

Several compounds studied in plant defense mechanisms of mycorrhizal plants are phenolics, phytoalexins, enzymes (Chitinase, Beta 1-3 gluconase, peroxidase), callose, pathogenesis related proteins (PR proteins) etc (Bowles, 1990, Gianinazzi Pearson *et al.*, 1994).

Higher concentration of phytoalexin in soybean colonized by AM fungi was reported by Wyss *et al.*, (1991) which perhaps inturn offer protection to roots against *Rhizoctonia solani.* Phytoalexins generally accumulate at the infection site of pathogens which were not detected during AM colonization but were found at later phase of symbiosis. (Harrison and Dixon, 1993: Morandi *et al.*, 1984; Duchesne *et al.*, 1987).

Higher levels of pathogenesis related proteins (PR proteins) and hydroxyproline rich glycoproteins (HRGP) (Boller 1993; Gianinazzi pearson *et al.*, 1994), chitinase (Blee and Anderson, 1996), peroxidase (Gianinazzi and Gianinazzi Pearson, 1992), chalcone synthase (Volpin *et al.*, 1995) phenylalanine ammonium lyase (PAL), phenyl propanoid related enzymes (Volpin *et al.*, 1995) phenolic compounds (Krishna and Bagyaraj 1984), Sulphur containing amino acids (Suresh and Bagyraj, 1984) have been reported in mycorrhizal plants which are known to reduce disease severity caused by soil borne root pathogens (Huynh *et al.*, 1992). Sharma *et al.*, (1992) found higher levels of arginine in the root exudates of mycorrhizal plants that reduced the chlamydospore production of *Thielaviopsis brasicola*. Similar results were observed by Bagyaraj (1996) in which higher content of serine and phenylalanine in tomato roots inoculated with AM fungi inhibited root knot nematode.

Higher content of arginine is involved in enzyme activities of ornithine cycle (Dehne 1982) which is stimulated in mycorrhizal roots that results in the formation of arbuscules.

Biochemical and physiological mechanisms

Nutrient uptake

The AM fungi improve the nutrient status of plants by increased nutrient uptake which inturn enhance plant vigor and tolerance to pathogenic damage through compensation for loss of root biomass (Cordier *et al.*, 1996, Bagyaraj 1996: Pinochet *et al.*, 1996). The fungal hyphae increase the surface area of nutrient absorption and thus compensate for loss of root biomass. Bodker *et al.*, (1998) suggested that non-p- mediated mechanisms are also involved in plant tolerance to pathogen. They have used non mycorrhizal plants with similar tissue P concentration as that of mycorrhizal plants to draw this conclusion. Further they opined that an induced systemic factor may also be responsible for enhanced resistance in plants inoculated with fungi. Maria *et al.* (2002) worked on the localized versus systemic effect of AM fungi a defense responses to *Phytophthora* infection in tomato using split root experimental system. AM fungi reduced disease symptoms produced by pathogen perhaps due to combined influence of local and systemic mechanisms. The biochemical analysis of different plant defense related enzymes showed a local induction of mycorrhiza related new isoforms of the hydrolytic enzymes chitinase, chitosanase and β-1,3-glucanase as well as superoxide dimutase, an enzyme involved in cell protection against oxidative stress. Studies on the lytic activity against *Phytophthora* cell wall of root protein extracts also corroborated a systemic effect of mycorrhizal symbiosis on tomato resistance to pathogen.

Morandi *et al.*, (1984) observed accumulation of isoflavonoids (which exhibit antimicrobial properties) which may increase resistance to plant pathogenic fungi and nematodes in mycorrhizal plants.

Tangming *et al.*, (2003) reported certain biochemical changes in plants induced by AM fungi. These changes were found to be positively correlated with AM formations. They include relative turgidity, peroxidase activity and total phenolic content. An AM fungus was found to enhance tolerance to pathogen by promoting the uptake of water and nutrients. Both AM fungi and pathogen compete for the host photosynthates (Linderman, 1994). Precolonization of plants with AM fungi will result in utilization of photosynthates (carbon) by AM, thus, inhibiting pathogenic growth. However there is insufficient evidence in this regard in biocontrol of pathogens by AM fungi.

AM colonization of plants alters host physiology especially with reference to root exudation. Higher phospholipids content in cell membrane of plants colonized with AM fungi may reduce root exudation and thus helps in reducing infection by soil borne root pathogens. (Mada and Bagyaraj 1993; Smith *et al.*, 1994).

Graham and Menge (1982) reported that high phosphate status of mycorrhizal roots was found to decrease root exudation in wheat plants which inturn affected root colonization process by pathogen. Further evidence for decrease in root exudation at higher host P status was given by Ratnayake *et al.,* (1978) who showed that high P status in host plants increased the phospholipids content which in turn decreased the root membrane permeability that led to less root exudation.

Biological Mechanisms

It is well documented that mycorrhizosphere microflora is altered quantitatively and qualitatively by colonization of plant roots by AM fungi as it changes the host root exudation pattern (Pozo *et al.,* 1999, Balestrini *et al.,* 1999). Certain microflora may be antagonistic to soil borne root pathogens (Secilia and Bagyaraj, 1987). Many a times it has been observed that the prophylactic ability of AM fungi could be enhanced in association with other rhizosphere microflora like *Trichoderma, Pseudomonas* etc.

Sreenivasa (1997) reported significant reduction in sclerotial bodies produced by *Sclerotium rolfsii* when chilli crop was co-inoculated with AM fungi and *Trichoderma harzianum.* Similar results were observed by Prashanthi *et al.,* (1997) in sunflower root disease management with co-inoculation of AM fungi and *T. viride.* Certain plant growth promoting rhizobacteria (PGPR) are known to encourage the development of AM fungi (Barea *et al.,* 1996) in addition to their ability to reduce population of pathogenic flora (Linderman, 1994; Bagyaraj, 1996; Bude *et al.,* 1999). This is due to the production of antibiotics siderophores, HCN or competition for colonization sites and carbon compounds (Bagyaraj 1996, Fester *et al.,* 1999, Leeman *et al.,* 1995; Azcon Aguilar and Barea, 1996).

Of late, the recent literature also indicates AM stimulation by certain bacteria known as mycorrhiza helper bacteria (MHB). Occurrence of certain bacterial associates on spores of AM fungi have been reported (Ho and Trape, 1979; Mosse, 1970) which helped AM fungi to colonise crop roots due to production of enzymes or growth promoting substances (Edwards *et al.,* 1999).

Williams (1981) observed certain "companion fungi" living in close association with AM fungi. Hence the current focus is on the integrated approach in impeding the pathogen attack in addition to improvement of plant growth and health which reflects on yield.

CONCLUSION

Of late, biological control of plant pathogens has gained importance as an alternative approach since it preserves quality of environment. The use of pesticides is hazardous to both human health as well soil health. In many places, pesticides are being dumped which led to imbalance in soil ecosystem and inturn reduction in crop yield. Many researchers have successfully tried AM fungi as biocontrol agent. However, literature clearly indicates that mycorrhizae have not suppressed the pathogen completely.

AM fungi can usually (though not always) deter the disease severity caused by soil borne root pathogen. On the contrary, few reports indicate increase in host susceptibility for fungal and viral diseases (Schoenbeck, 1979; Schonbeck and Spengler, 1979). Hence drawing general conclusions on AM influence on soil borne root pathogens is difficult. Moreover interactions between AM-pathogen-Host Plants are influenced by environment. Perhaps for this reasons, the mechanisms of biocontrol of root pathogens by AM fungi are not completely understood.

Though mycorrhiza is known to offer an inexpensive and alternative approach for biological control of soil borne root pathogens, more studies are required to be carried out in this direction using molecular techniques to understand mechanism of suppression of pathogens by AM fungi. The major bottleneck in commercial utilization of AM fungi is its obligate symbiotic nature which makes it difficult to produce large quantities of inoculum. However, few techniques have been developed recently to produce clean AM fungi (Sreenivasa and Bagyaraj 1989 a & b).

It is always better to use AM fungi in conjunction with other beneficial microorganisms like PGPR, mycorrhiza helper bacteria, companion fungi etc to suppress root infecting pathogens.

Certainly use of mycorrhizal technology to defer root pathogens looks promising and has long lasting implications in biocontrol of soil borne root pathogen of agriculture crops as well as forest trees.

REFERENCES

Azcon - Aguilar, C. and Barea, J.M. 1996. AM and biological control of soil borne plant pathogens - An over view of the mechanisms involved. *Mycorrhiza.* 6: 457-464.

Bagyaraj, D.J. 1984. Biological interactions with VA mycorrhizal fungi (*In:*) Bagyaraj, D.J. C.L. Powell, (Eds.) *VA Mycorrhiza,* CRC press, Boca Raton pp. 131-153.

Bagyaraj, D.J. 1996. Mycorrhizal symbiosis in tropical trees in relation to control of root pathogens. (*In:*) Impact of diseases and insect pests in tropical forests. Kerala Forest Research Institute.

Bagyaraj, D.J., Manjunath, A. and Reddy, D.D.R. 1979. Interaction of VAM with root knot nematodes in tomato. *Plant Soil.* 51: 397.

Balestrini, R., Perotto, S., Gasuerde, E. Dahiya, P., Guldman, L.L., Brewin, M.J. and Bonfante, P., 1999. Transcription of gene encoding a lectin like Glycoprotein is induced in root cells harbouring AM fungi in *Pisum sativum*. *Mol. Plant Microbe Interactions.* 12: 785-791.

Baltruschat, H and Schonbeck, F. 1975. The influence of endotrophic mycorrhiza on the infestation of tobacco by *Thielaviopsis basicola*. *Phytopathol* Z. 84:172-.

Baltruschat, H, Sikora, R.A. and Schonbeck, F. 1973. Effect of VA mycorrhizae (*Endogone mosseae*) on the establishment of *Theilaviopsis basicola* and *Meloidogyne incognita* in tobacco. In: *Proc. 2nd Int. Cong.Plant Pathol.* Minneapolis.

Barea, J.M., Azcon - Aguilar, C and Azcon. R. 1996. Interaction between mycorrhizal fungi and rhizosphere microorganisms with in the context of sustainable soil - plant system. *(In:)* Ganga, A.C. and Brown, V.K. (Eds) *Multitrophic Interactions in Terrestrial systems*. Blackwell, Oxford.

Barea, J.M. and Jeffries, P. 1995. Arbuscular mycorrhizas in sustainable soil plant systems. (*In*:) *Mycorrhiza Structure, function, molecular biology and biotechnology*, Springer, Heldelberg pp. 521-529.

Bethlenfalvay, G. J. and Linderman, R.G. 1992. Mycorrhizae in sustainable agriculture. ASA special publication No 54 Madison. Wisconsin P. 124.

Blee, K.A. and Anderson, A.J. 1996. Defence related transcript accumulation in *Phaseolus vulgaris* L. colonised by AM fungus, *Glomus intraradices. Plant Physiol.* 100: 675-688.

Bodker, L., Kjoller, R. and Rosendahl. S., 1998. Effect of phosphate and the AM fungus *Glomus intraradices* on disease severity of root rot of peas (*Pisum sativum*) caused by *Aphanomyces enteiches*. *Mycorrhiza.* 8: 169-174.

Boller, T. 1993. Antimicrobial functions of the plant hydrolases, chitinase and b-13-glucanase. (*In*:) Fritig, B. and Legrand M. (Eds.). *Mechanisms of Plant Defense Responses.* Kluwar Academic Pulbishers, Dordrecht. pp. 391-400.

Bowles, D.J. 1990. Defence related proteins in higher plants. *Ann. Rev. Biochem.* 59:873-907.

Bude, S.W., Van Tuinen, D., Martinotti, G. and Gianinazzi, S. 1999. Isolation from the *Sorghum bicolor* mycorrhizosphere of a bacterium compatible with AM development and antagonistic towards soil borne fungal pathogens. *Appl. Env. Microbiol.* 65:5148-5150.

Carling, D.E., Roncadori, R.W. and Hussey, R.S. 1989. Interaction of VAM fungi, root knot nematode and P-fertilization on soybean. *Plant Dis.* 73:730-733.

Chand, H, and Jalali, B.L. 1998. Dynamics of plant pathogen interactions in legume crops. *(In:)* Rai B., Upadhya R.S. and Dube N.K. (Eds). *Trends in Microbial Exploitation,* Int. Soc. for conservation of National Resources Publication, Varanasi.

Cooper, K.M. 1981. Mycorrhizal fungi act as deterrents to soil borne root pathogens. In: Proc. *N.Z. Microbiol. Soc. Annu. Conf. Auckland.*

Cordier C., Gianinazzi S. and Gianinazzi pearson. V. 1996. Colonization patterns of root tissues by *Phytophthora nicotianae var. parasitica* related to reduced disease in mycorrhizal tomato. *Plant Soil.* 185: 223-232.

Davis, R.M., Menge, J.A. and Erwin, D.C., 1979, Influence of *Glomus fasciculatus* and soil P on *Verticillum* Wilt of Cotton. *Phytopath.* 69:453.

Dehne, H.W. 1982. Interaction between VAM fungi and plant pathogens. *Phytopath,* 72: 1115-1119.

Dehne, H.W. and Schonbeck, F. 1975. The influence of endotrophic mycorrhiza on the fusarial wilt of tomato. *Z. Pflanzenkr. Pflanzenschutz,* 82: 630-634.

Dehne H.W. and Schonbeck, F. 1979. Investigation on the influence of endophytic mycorrhiza on plant diseases II phenol metabolism and lignification. *Phytopath,* 95: 105-110.

Duchesne, L.C., pererson, R.C. and Ellis, B.E. 1987. The accumulation of plant produced antimicrobial compounds in response to ectomycorrhizal fungi: A review. *Phytoprotection.* 68: 17-27.

Edwards, S.G., Young, J. P. W. and Fitter, A.H. 1999. Interaction between *Pseudomonas fluorescens* biocontrol agents and *Glomus mosseae,* an AM fungus with in the rhizosphere. FEMS *Microbiol Ecol.* 28:93-99.

Fester, T., Maier, W. and Strack, D. 1999. Accumulation of secondary compounds in barley and wheat roots in response to inoculation with an AM fungus and co-inoculation with rhizosphere bacteria. *Mycorrhiza.* 8:241-246.

Gianinazzi, S. and Gianinazzi- pearson, V. 1992. Cytology, histochemistry and Immunochemistry as tools for studying structure and function in Endomycorrhiza: Techniques for the study of mycorrhiza. (*In*:) Norris, J.R., Read, B.J. and Varma, A.K. (Eds.). *Methods in Microbiology* Vol 24 Academic press, London. pp. 109-139.

Gianinazzi-Person, V, Gollote, A., Dumas - Guadot, E., Franken, P and Gianinazzi S. 1994. Gene expression and molecular modifications associated with plant responses to infection by AM fungi *(In:)* Daniels, M., Downic, J.A. and Osbourn, A.E. (Eds.) *Advances in Molecular Genetics of Plant - Microbe Interactions.* Kluver Dordecht. pp 179-186.

Graham, J.H. and Menge, J.A. 1982. Influence of VAM and soil P on take all disease of wheat. *Phytopath.* 72: 95-100

Habte, M., Zhang, Y.C. and Schmit, D.P. 1999. Effectiveness of Glomus species in protecting white clover against nematode damage. *Can. J. Bot.* 7: 135-139.

Halos, P.. and Zorilla, R.A. 1979. VA mycorrhiza increase growth and yield of tomatoes and reduce infection by *Pseudomonas solanacearum. Phillippines Agri.* 62: 309-315.

Harrison, M.J. and Dixon, R.A. 1993. Isoflavonoid accumulation and expression of defense gene transcripts during the establishment of VA mycorrhizal associations in roots of *Medicago truncatula. Molecular Plant Microbe Interactions.* 6:643-654.

Ho, I, and Trappe, J.M. 1979. Interaction of a VAM fungus and a freeliving N_2 fixing bacterium on the growth of tall fescue. In: *Abstr 4th North American Conf. Mycorrhizae.* Fort collins, Colorado, USA.

Hooker, J.E., Jaizme - vega, M. and A. ktinson, D. 1994. Biocontrol of plant pathogens using arbuscular mycorrhizal fungi. In*: Impact of Arbuscular Mycorrhizas on Sustainable Agriculture and Natural Ecosystems.* Birkhauser, Basel, pp. 191-200.

Huynh, Q.K., Hironara, C.M., Levine, E.B., Smith, C.E. Borgmeyer, J.R. and Shah., D.M. 1992. Antifungal proteins from plants. Purification, molecular cloning and antifungal properties of chitinases in maize seeds. *J. Biol. Chem.* 26: 6635-6640.

Iqbal, S.H. Qureshi, K.S. and Ahmad, J.S. 1977. Influence of VAM on damping off caused by *Rhizoctonia solani in Brassica napus. Biologia.* (Lahore) 23: 197-199.

Jaizme-vega, M.C., Tenoury, P., Pinochet, J and Jouemot, M. 1997. Interaction between the root knot nematode *Meloidogyne incognita and Glomus mosseae* in banana. *Plant Soil.* 196: 27-35.

Jalali, B.L. and Jalali, I. 1991. Mycorrhiza in plant disease control. (*In*:) Arora, K. Rai, B. Mukerji, K.G. and Knudsen, G.R. (Eds.) *Hand Book of Applied Mycology*, Marcel Dekker, New York, pp. 131-154.

Krishna, K.R. and Bagyaraj, D.J. 1983. Interaction between *Glomus fasciculatus* and *Sclerotium rolfsii* in peanut. *Can. J. Bot.* 61: 2349-2351.

Krishna, K.R. and Bagyaraj, D.J. 1984. Phenols in mycorrhizal roots of *Arachis hypogea Experientia.* 40:85-86.

Labeena, P. Sreenivasa, M.N. and Lingaraju, S. 2003. Interaction effect between AM fungi and root knot nematode in tomato. *Indian J. Nematol.* 32:118-120.

Leeman, M. Vanpelt, J.A., Denouden, F.M. Hunsbroek, M., Barker, P. and Schippers, . 1995. Induction of systemic resistance by *Pseudomonas flourescens* in radish cultivars differing in suscptibility to *Fusarium* wilt, using a novel bioassay. *European. J. Plant Pathol.* 101: 655-664.

Linderman, R.G. 1994. Role of VAM fungi in biocontrol. (*In*:) *Mycorrhizae and Plant Health*, ASP, St. paul, pp. 1-26.

Liu, R.J. 1995. Effect of VAM on *Verticillium* wilt of cotton. *Mycorrhiza.* 5: 293-297.

Mada, R.J. and Bagyaraj, D.J. 1993. Root exudation from *Leucaena leucocephala* in relation to mycorrhizal colonization. *World. J. Microbiol. Biotechnol.* 9: 342-344.

Maria J.P., cordier, C., Ehane, D.G., Gianinazzi, S. Barea, J. M. and Azcon - Aguilar, C. 2002 Localized versus systemic effect of AM fungi on defence responses to *Phytophthora* infection in tomato plants. *J. Exptl. Bot.* 53: 525-534.

Menge, J.A. Nemec, S. Davis, R.M. and Minassian, V. 1979. Mycorrhizal fungi associated with citrus and their possible interactions with pathogens. *In Proc. Int. Soc. Citric.* Vol.3. William G. (Ed.) Univ. Florida press. Lake Alfred. pp 872.

Morandi, D., Bailey, J.A. and Gianinazzi-pearson V. 1984. Isoflavonoid accumulation in soybean roots infected with VAM fungi. *Physiol. Plant. Pathol.* 24: 357-364.

Mosse, B. 1970. Honey coloured sessile *Endogone* spores. I changes in fine structure during spore development. *Arch. Microbiol.* 74:129-132.

O Bannon and Nemec, S. 1979. The response of citrus lemon seedlings to a symbiont, *Glomus etunicatus* and pathogen, *Rhodopholus similis. J. Nematol.* 11: 270-275.

Paulitz, T.C. and Linderman, R.G. 1991. Mycorrhizal interactions with soil organisms. (*In*:) Arora, D.K., Mukerji, K.G. and Knudsen, G., Marcel Dekker (Eds.), *Hand Book of Applied Mycology*, New York P 728.

Pinochet, J., Calvet, C., Camprubi, A. and Fernandez, C. 1996. Interaction between migratory endoparasitic nematodes and arbuscular fungi in perennial crops: A review. *Plant Soil.* 185:183-190.

Pozo, M.J., Azcon- Aguilar, C., Dumas-Gaudot, E. and Barea, J.M. 1999. b 1-3 glucanase activities in tomato roots inoculated with AM fungi and / or *Phytophthora parasitica* and their possible involvement in bioprotection. *Plant Sci.* 141: 149-157.

Prashanthi, S.K., Srikant Kulkarni and Sreenivasa, M.N. 1997. Integrated Management of root rot disease of safflower caused by *Rhizoctonia bataticola. Environ. Ecol.* 15: 800-802.

Ratnayake, M. Leonard, R.T. and Menge, J.A. 1978. Root exudation in relation to supply of phosphorus and its possible relevance to mycorrhizal formation. *New Phytol.* 81:543-548.

Roncadori, R.W. and Hussey, R.S. 1977. Interaction of the endomycorrhizal fungus, *Gigaspora margarita* and root knot nematode on cotton. *Phytopath.* 67:1507-1510.

Ross, J.P. 1972. Influence *of Endogone* mycorrhiza on *Phytophthora* rot of soybean. *Phytopath.* 62:896-899.

Schenck, N.C. 1987. Vesicular arbscular mycorrhizal fungi and control of fungal root diseases. In: *Innovative Approaches to Plant Disease Control.* Wiley, New York PP 179-191.

Schenck, N.C., Kinlock, R.A. and Dickson, D.W. 1975. Interaction of endomycorrhizal fungi and root knot nematode on soybean. *(In:)* Sanders, F.E., Mosse, B. and Tinker, P.B. (Eds) *Endomycorrhizas*, Academic press, London pp 605.

Schenck, N.C., Ridings, W.H. and Cornell, J.A. 1977. Interaction of two VAM fungi and *Phytophthora parasitica* on two citrus root stocks. In: *Abstr. 3rd North Amer. Conf. Mycorrhizae*, Athens, pp 9.

Schonbeck, F. 1979. Endomycorrhizas is relation to plant diseases. In: *Soil borne plant Pathogens. Acad. Press.* New York. pp. 271-280.

Schonbeck, F and Dehne, H.W. 1977. Damage to mycorrhizal and nonmycorrhizal cotton seedlings by *Thielaviopsis basicola. Plant Dis. Rept.* 61: 266-270.

Schonbeck, F. and Spengler, G. 1979. Detection of TMV in mycorrhizal cells of tomato by immunofluorescence. *Phytopath.* 94:84-89.

Secilia, J. and Bagyaraj, D.J. 1987. Bacteria and actinomycetes associated with pot cultures of VAM. *Can. J. Microbiol.* 33: 1069-1073.

Sharma, A.K., Johri, B.N. and Gianinazzi, S. 1992. VAM in relation to plant disease. *World J. Microbiol. Biotechnol.* 8: 550-563.

Sikora, R.A. and Schonbeck, F. 1975. Effect of VA mycorrhiza *(Endogone mosseae)* on the population dynamics of the root knot nematodes *Meloidogyne incognita* and *M. hapla*. In: *Proc. 8th Int. congress plant prot. Mascow.*

Smith, C.S. 1987. Interaction of nematodes with mycorrhizal fungi, (*In:*) Vech. J.A. and Dickson, D.W. (Eds). *Vistas on Nematology Society of Nematol., Hyattsille,* M.D.

Smith, S.E., Gaininazzi pearson, V., koide, R. and Cainey, J.W.G. 1994. Nutrient transport in mycorrhizas. Structure, physiology and consequences for efficiency of the symbiosis. (*In:*) *Management of mycorrhiza in agriculture, horticulture and forestry.* Kluwer, Dordecht, pp. 103-113.

Sreenivasa, M.N. 1994. Biological deterrent activities of VAM and *Trichoderma harzianum* on *Sclerotium rolfsii* at different p-levels in chilli. *Environ. Ecol.* 12: 319-321.

Sreenivasa, M.N. 1997. Influence of VAM and *Trichoderma harzianum* on *Sclerotium* in three varieties of chill. *Environ. Ecol.* 15: 343-345.

Sreenivasa, M.N. and Bagyaraj D.J. 1989a. Mass producing plant inoculum. *Approp. Technol.* (UK):16:4.

Sreenivasa, M.N. and Bagyaraj D.J. 1989b. Use of pesticides for mass production of VAM inoculum. *Plant Soil.* 119: 127-132.

Sreenivasa, M.N. and Krishnaraj, P.U. 1992. Synergistic interactions between VAM and a phosphate solublizing bacterium in chilli. *Zentralbl. Mikrobiol.* 147: 126-130.

Suresh, C.K. and Bagyaraj, D.J. 1984. Interaction between a VAM and a root knot nematode and its effect on growth and chemical composition of tomato. *Curr. Res.* 12: 31-39.

Suresh, C.K., Bagyaraj, D.J. and Reddy, D.D.R. 1985. Effect of VAM on penetration and development of root knot nematode in tomato *Plant Soil.* 87: 305-308.

Suresh, C. K. and Rai, P.V. 1991. Interaction of *Pseudomonas solanacearum* with astagonistic bacteria and VA mycorrhiza. *Curr. Res.* 20: 36-37.

Tangming, Chen Hui and Sheng S.H. 2003 Mechanisms of VAM fungi enhanced resistance of plants to canker, *Scientia Silvae Sinicae.* 36: 87-92.

Thippeswamy, R. and Sreenivasa, M. N., 1999. Physiological, nutritional and yield responses of tomato to the inoculuation of *Glomus fasciculatum* under water stress condition. *J. Maharastra Agric. Univ.* 23: 265-268.

Umesh, K.C., Krishnappa, K. and Bagyaraj, D.J. 1988. Interaction of burrowing nematode, *Rhodopholus* similis and VA mycorrhiza, *Glomus fasciulatum* in banana. *Indian J. Nematol.* 18: 6-11.

Volpin, H., Phillips. D.A., Okon, Y. and Kapulnic, Y. 1995. Suppression of an Isoflavanoid Phytoalexin defence response in mycorrhizal alfalfa roots. *Plant Physiol.* 108: 1449-1454.

Weaver, D. J. and Wehunt, E.J. 1995. Effect of soil pH on susceptibility of peach to *Pseudomonas syringae. Soil Biol. Biochem.* 65: 984-998.

Williams, P.G. 1981. The prospects of using VAM in broad scale agriculture to improve the efficiency of phosphate fertilizer use. *Rep. Australian Meat Res. Council. Workshop.* Cambridge publication.

Wyss, P., Boller, T. and Wiemken, A. 1991. Phytoalexin response is elicited by a pathogen (*Rhizoctonia solani*) but not by a mycorrhizal fungus (*Glomus mosseae*) is soybean roots. *Experientia* 47: 495-499.

• • •

Advances in Soil Borne Plant Diseases, 2008
Editors: Naik and Devika Rani, pp. 261-283
New India Publishing Agency, New Delhi

11

Management of Endoparasitic Phytonematodes Particularly *Meloidogyne* spp. and *Heterodera* spp. through Non-chemical Methods

S.K. Patel, N.B. Patel and D.J. Patel

INTRODUCTION

Burgeoning human population and its elevated living standard have greatly demanded more agricultural production. But unfortunately agricultural crops suffer deadly due to many biotic and abiotic stresses. Of biotic stresses, phytonematodes, the hidden enemies of crops are real threat to successful and profitable cultivation of agricultural crops (Plate. 11.1). These, endoparasitic nematodes, *Meloidogyne* spp. and *Heterodera* spp., are mainly dreadful causing wrecking havoc to agricultural scenario. Chemical control of these, tiny parasites created residue, resistance, resurgence, hazardous and pollution problems to the growers as well as soil and atmospheric environment. To overcome these hazards, efforts are being made to go for non-chemicals means of nematodes management strategies in crops. Hence, a research work carried out on management of these endoparasitic nemas particularly *Meloidogyne* spp. and *Heterodera* spp., through non-chemical methods is reviewed in subsequent status.

Department of Nematology, B.A. College of Agriculture,
Gujarat Agricultural University, Anand campus, Anand-388 110

ROOT-KNOT NEMATODES (*MELOIDOGYNE* SPP.)

The root-knot nematodes being sedentary endoparasites, a common name collectively give to *Meloidogyne* spp., produce knots/galls of various sizes and shapes on the roots of wide varieties of plants. They are considered the number one farmers' enemy of agricultural crops in most developing nations. The crop disease produced by root-knots was first reported on the greenhouse grown cucumbers in England by Berkley (1985). Whereas in India, it was first reported on tea from, southern state of Kerala. Till 1991, Nickle recognized 75 species of *Meloidogyne* genus. The four most common species in 66 of 75 countries in the world are *M. incognita, M. javanica, M. arenaria* and *M. hapla.* At present 99 *Meloidogyne* species have been reported to encounter world over. Among these at least 17 species have been reported from India (Table 11.1). Globally, a frequency estimate of samples from 75 countries showed that *M. incognita* is most widely distributed with 53 % occurrence, *M. javanica* 50 %, *M. arenaria* 8 %, *M. hapla* 8 % and other species 2 %.

In India, over 350 plants species have been recorded as mild or severe hosts of *Meloidogyne* spp. (Sen and Dasgupta, 1982), *M. incognita* alone infecting about 250 and *M. javanica* infecting 150 genera of plants (Krishnappa, 1985). Frequency of *Meloidogyne* spp. range between 17-44 % in different surveys of the eastern Indian subcontinent. Physiological races/biotypes/ pathotypes are known to be prevalent in few species of root-knot nematodes in India. The International *Meloidogyne* Project (IMP) has identified four races of *M. incognita* and two of *M. arenaria.* Races 1, 2 and 3 (out of four races) of *M. incognita* and race 2 (out of two races) of *M. arenaria* are reported to be widely distributed in continent and subcontinents of the world. (Sasser, 1982). The distribution of different *Meloidogyne* races present in India is also worked out (Table 11.2). Patel *et al.,* (1988) first time reported *M. javanica* pathotype 2 on peanut in Kapdwanj area of kheda district of Gujarat State, India (Plate. 11.2).

Host range

Meloidogyne spp. is noticed to attack over 2500 hosts. The more preferred hosts are dicot vegetables, pulses, fibers, fruit crops (Papaya, grapes), tobacco, peanut, betel vine, tea, coffee, mulberry, flowers, spices, ornaments and some monocots like sugarcane, sorghum, banana, rice as well as large numbers of weeds. Wheat previously being non-host now faces serious threat when grown after tobacco in Gujarat, India. Presently, *M. graminicola* is emerging as serious threat to rice cultivation in several states of India.

Symptoms

The root-knot nematodes are basically parasitizing plant roots or underground stem/pods/rhizomes/corms etc., the above ground symptoms, hence are those of slow ability of root in its function of nutrient and water uptake and translocation. The nematode infected plants exhibit yellowing, dwarfing, stunting, leaf margin, drying and such other symptoms as if it is suffering from nutritional deficiencies (Plate. 11.3). The very broad leave plants like tobacco or cucurbits may virtually show daytime wilting. Poor emergence and death of young seedlings may occur in heavily infested soil but death of grown up plants is rarely observed unless some bacteria or fungus infection takes place. On uprooting of such infected plants, severe root galling/knotting of various sizes along with egg sacs is observed (Plate. 11.3). The general appearance of the crop in a field is of patchy growth due to uneven distribution of the nematodes. The poorly growing patches grow in size each year and spread more in direction of plough and irrigation water.

Losses

Yield losses invariably depend on initial nematode population present in the soil. The avoidable losses estimated on various crop around the country due to *Meloidogyne* spp. are summarized in (Table 11.3).

CYST NEMATODES (*HETERODERA* SPP.)

The second most threat to farmers next to root-knot nematodes is cyst nematodes because of its wide distribution and protection to eggs by cyst formulation. More than 81 *Heterodera* spp., 12 *Globodera* spp. and 9 *Cactodera* spp. are recorded in the world with few species of *Sarisodera, Punctodera* and *Dolichodera*. From India, 16 species of *Heterodera*, 2 species of *Globodera*, 2 species of *Cactodera* from Bangalore have been reported. The first report of *Heterodera* sp. from India was that of *H. avenae* which was then reported to be associated with molya disease of wheat and barley crops in Rajasthan. Taking into consideration of crops and distribution of various cyst forming species in the country, *Heterodera avenae*, the causal organism of molya disease of wheat, stands out as the number one enemy. *H. avenae, H. cajani* and *H. zeae* were key cyst forming nematodes having wide distribution in various states of the country (Table 11.4). *Globodera rostochiensis* and *G. pallida* are still major constraints to potato production in Tamil Nadu, being prevalent in about 3050 ha of Nilgiri hills and about 200 ha of Kodaikanal hills. Survey has indicated that cyst nematodes are prevalent in all 22 localities surveyed in Nilgiri hills with population (below 25/100 g soil) at Berhahatti and Masanagudi, which represent low elevation levels of 950 to 1250 meters above sea level heavy populations (100-200 cysts/100 g soil) at 13 localities situated at 2000 meters above (Krishna Prasad, 1986).

Mathur and his coworkers in 1975 reported the existence of five biotypes of *H. avenae* from Rajasthan and Haryana states. Two host races of *H. cajani* have been encountered. Race A which cannot multiply on clusterbean and sunhemp and race B which could multiply on these hosts (Walia and Bajaj, 1988). Different species and their races are recorded in various states of India. (Table 11.5).

Host range

Heterodera spp. is known to attack over 1500 hosts. The most preferred hosts are cereals, pulses, tobacco, sugarcane, potato, sugar beet, rye, some vegetables as well as some weed species.

Losses

The avoidable yield losses due to *Heterodera* spp. may vary from crop to crop depending upon the initial soil nematode population. Information available on yield losses on various crops is depicted in Table 11.6.

Symptoms

Nematode attacked plants initially exhibit yellowing of foliage with stunted and patchy crop growth. On uprooting such plants, several whitish to brown females are seen on the roots. As a result, reduced root size and quantity have been observed (Plate. 11.3).

Management

A number of methods like physical, cultural, biological, legal and development of crop cultivars for disease resistance (DHDR) are being experimented for effective and economic management of such a devastating nematodes.

Physical method

This includes oven heating, steam treatment, robbing with slow burning materials, hot water treatment, hot water drenching & soil solarization, etc.

a) Oven heating

Placing small containers of soils in an oven and heating them at about 121° C for one hour is quite enough to kill nematodes present in soil. Of course, this method, though gave good nematode control, could not be practiced on large scale adoption.

b) Steam treatment

Making soil free from these parasites through use of stem is a long established practice fulfilling small requirements of home growers, nursery and many other types of plant production (small scale) where large acreages are not involved. Steam is usually introduced through perforated pipes buried in the soil and heating takes place by condensation of soil particles. It is comparatively better than dry heating because of short treatment, easy to apply and non-toxic to man.

However, it is limited to small quantity of soil only perhaps because of difficulties in generating large quantities of steam. It is generally recommended that the temperature at the coolest point of the soil must reach to at least 82° C for minimum of about half an hour. The contents of potato cysts nematode in dry soil are affected at 70° C and die at temperature of 90° C, while the contents of cysts from moist soil generally die at a temperature of 60° C, with an exposure of 10 hrs.

c) Rabbing:

The most primitive way of heating the soil on large scale is by putting a fire over it. Dry heat of this type is generally used by many homer owners to sterilize potting soil and for seed germination. Burning a layer of dry leaves 10 cm thick on the soil killed root-knot nematodes at a depth of 9 cm only. Choudhary and Hoque (1982) reported 77 % and 63 % kill of *Meloidogyne* spp. by soil burning with 5 cm thick layers of rice husk and saw dust, respectively. Complete elimination of *M. incognita* from tobacco seedbeds by burning 20 kg rice husk/ m^2 seedbed (Krishnamurthy and Elias, 1969). Burning materials outside and incorporating ash into the soil surface satisfaction nematode kill. Hence burning materials on soil samples is advocated. Rabbing of bajra (pearl millet) husk, paddy husk and wheat straw plus tobacco stalk each @ 6 kg/ m^2 in tobacco nursery yielded 147.78, 170.31 and 174.40 % higher production of transplantable seedlings with 43.48, 39.61 and 45.49 % reduction in root-knot index on seedling realizing Rs. 46,670, 32,190 and 44,680 net realization/ha, respectively over non rabbed soils, (Patel *et al,* 1987). Even in tomato nursery, burning of bajra husk, @ 6 kg/ m^2 significantly enhanced plant height, fresh shoot and root weights and transplantable seedling with benefit to cost ratio of 1:5.12 (Patel *et al.,* 1989).

d) Hot water drenching

Soil can also be disinfested by putting (pouring) boiling water into the infested soil. Hot water drenching of soil is considered impracticable due to the requirement of enormous quantity of water and creation of water logged condition of soil.

e) Hot water treatment

This is a method followed for killing nematodes within plant structures by submerging them for short periods in hot water. It is best adapted to denematizing, rhizomes, bulbs, corms, tubers and fleshy roots of plantation and planting materials. It has been proved successful in several crops. But the caution, the growers should take into consideration, is the constant maintenance of temperature and exposure time failing which will result in toxicity on establishment. Treatment of infected papaya seedlings roots at 50° C for 10 minutes (Edward *et al.*, 1963) and incubation of potato tubers at 45° C for 48 hrs (Nirula and Bassi, 1965) had yielded satisfactory control of root-knot nematodes. Hot water treatment to banana corms at 55 ° C for 10 minutes followed by application of neemcake @ 1 kg/plant enhanced 31 % higher banana yield by reducing root-knot index by 35.5 % over untreated ones (Anon., 2001-03) (Table 11.7).

f) Soil solarization / Slow soil pasteurization (ss)

Another approach of killing nematodes through heat is soil solarization (SS). The technique consists of covering the moist soil with a good quality clear transparent film (LLDPE 25m) for 2-6 weeks during the period of intense sunshine and increasing substantial soil temperature over non-solarized soil. Increased soil temperature coupled with soil moisture invariably resulted in a significant reduction in population densities of species of *Meloidogyne* and *Heterodera*. Solarization by covering the soil with clear plastic film (100 m) for 6 weeks during hot summer period increased soil temperature (8° C) compared to unsolarized soil. Increased soil temperature together with soil moisture resulted in a significant reduction in population densities of *M. incognita* (58.1 %) (Rao and Krishnappa, 1995). Tobacco beds tarped with LDPE (clear) films yielded 215 and 152 % higher production of transplants and total seedlings with reduction in root-knot disease, weed count and weight to the tune of 58, 77 and 70 % respectively (Patel *et al.*, 1991 a) (Plate 11.4). The cyst nematode (*Heterodera avenae*) of wheat was completely controlled by solarization at 90 cm or more inward from the edge of mulched pots. SS by covering soil with transparent polyethylene (100mm) sheets during summer months significantly reduced population densities of *Heterodera cajani* (93 % eggs and juveniles) (Sharma and Nene, 1990). Covering tomato nursery beds with LDPE-UV and LDPE clear films, for 15 days in summer significantly reduced root-knot disease by 50.0 and 49.1 % weed population by 87.5 and 87.0 % and thereby yielded 41.5 and 63.8 % higher tomato transplants (Fig. 11.1 Patel *et al.*, 1991 b). In tobacco nursery SS through 100 gauge LLDPE clear film for 15 days in summer increased production of healthy seedlings by 412 % with 75 and 81

% reduction in RKI and weeds respectively ultimately giving net profit of Rs. 8,432=00 from 1000 m^2 area (Plate. 11.4). Similarly in tomato nursery also 615 % more production of seedlings with 66 and 92 % reduction in RKI and weeds respectively were recorded ultimately giving net profit of RS. 7,661=00 from 1000 m^2 area. The population of all plant parasitic nematodes of pigeonpea and chickpea fields was markedly reduced by SS. SS with irrigation was most effective in reducing population of cysts and nematode eggs and larvae.

Prevention of dissemination/spread through selection of healthy planting materials

Parasitic nematodes move only a few centimeters by virtue of their own locomotion. They are spread through soils or infected plant parts, water, wind, farming implements etc. Selection of nematode free healthy planting material is quite useful and important when the area is free from important parasitic nematodes and also when the nematode spread is usually through seeds, nursery stocks or seedlings. This can be accomplished by growing clean plants in nematode free soils. Preventive measures are most effective and economic way for control of RKN on banana and vegetable seedlings. Therefore banana corms and vegetable seedlings should be selected from uninfected land.

Cultural method

Practicing cultural practices attempts to adopt crop husbandry systems so as to minimize crop losses due to nematodes. Cultural control of nematodes can be achieved by crop rotation, growing antagonistic and trap crops, water flooding, fallowing, deep summer ploughing removal of infected plant debris prevention of spread and multiplication by selection of nematode free healthy planting materials, application of organic manures and fertilizers etc.

Crop rotation (Biological decontamination)

Control of plant parasitic nematodes by crop rotation/ sequences is based on the fact that some species of nematodes are able to feed and multiply on certain crop plants only but not on others. Hosts on which they cannot feed and abundantly reproduced are referred to as non-hosts. The crops to be grown in between the susceptible host crops should be either resistant or immune to nematodes or at least non-host plants. Rotation with graminaceous poor hosts such as wheat, barley, ragi, maize, sorghum, etc. or dicot like mustard, sesame, *Crotalaria spectabilis*, asparagus and so on has some antagonistic effects to suppress root-knot nematodes (Sundaresh and Setty, 1977; Roy, 1978; Patel *et al.*, 1979; Sharma *et al.*, 1980; Gaur and Mishra,

1983). Two years intensive cropping systems including sweet corn-soybean-wheat-soybean-spinach and turnip-peanut-cucumber-turnip-cucumber-soybean had effective suppressing of *M. incognita* race 1 in USA (Johnson *et al.*, 1976; Johnson, 1985). Incorporation of mustard, sesame, maize, wheat etc. also suppressed nematode population in India (Haque and Gaur, 1985). Crop rotation with non-hosts and keeping the field fallow during rabi decreased nematode population by 50 % at the end of one year and up to 75% by the end of second year, whereas cultivation of host crops like wheat and barley for a single season only can increase population by 250 to 335% in initial population of *H. avenae*. Two years crop rotation of tobacco cv. Anand 119 (*kharif* and *rabi*)- Pearl millet cv. CJ 104 (summer)- castor cv. Anand 39-1 (*kharif* and *rabi*)- fallow (summer) reduced 74.2% root-knot disease and consequently increased tobacco yield by 28.5% giving net realization Rs. 18059/- over conventional practice of tobacco every year (Patel *et al.*, 1990). Tomato cv. SL 120, Castor cv. A 39-1 and cotton cv. Hy-6 cultivated in *kharif* - *rabi* followed by summer green gram in summer effectively checked root-knot disease (13-17%) and enhanced grain yield of summer green gram cv. Gujarat 2 (53-76%) (Patel *et al.*, 1997).

Growing traps/Antagonistic crops

The basic concept with trap crops is that when they are planted/grown in nematode infested soils, the second stage juveniles penetrate and host plants are destroyed completely carefully before parasites lay mature eggs.

A more realistic approach is to plant crops that are highly susceptible (super susceptibility) to invasion by nematodes but simultaneously do not allow further development of juveniles to mature. Several species of Crotalaria (*C. spectabilis*) have been reported to reduce population of *Meloidogyne* spp. Castor (*Ricinus communis* L.) is also reported to trap *M. incognita* and *M. javanica* and thereby deplete the *Meloidogyne* population (Table 11.8) (Patel *et al.*, 1985). Other antagonistic plants such as *Tagetes erecta, T. patula* and Chrysanthemum also proved to be effectively reduced *Meloidogyne* spp. These crops are included in normal farming schedule in nematode prone field (Khan *et al.*, 1971).

Flooding

When water is available in plenty and nematode infested area is uniformly leveled, flooding can be employed as a routine practice to reduce parasitic nematodes. Period from 7 to 40 days is quite effective to reduce root-knot infestation. It produces anaerobic condition to nematodes by releasing CO_2 and H_2S gases. It is more feasible in black soil rather than loamy/sandy soils, whose water holding capacity is low.

Fallowing and deep summer ploughing

Fallowing is a common practice of keeping the land free from all vegetations for specific period by occasional ploughing during hot and dry weather. Complete fallow without allowing any plant or weed to grow is the best method. This method invariably ensures that parasitic nematodes will have no host to feed and thereby will not be reproduced. Alternatively the infested fields can be ploughed during hot summer months when the temperature is bit high. This practice exposes nematode eggs and other stages to desiccation. Under field conditions, root-knot nematodes may be killed by drying soil for two days in 2.5 cm depth, 4 days in 5 cm depth and 32 days in 15 cm depth. The nematodes present in deeper soil can also be taken care off giving frequent turning/ raking of soil (Krishnamurthy *et al.*, 1969). The field trial was conducted to study effect of 1 to 5 deep summer ploughings at weekly intervals during May and June. The reduction in population of *H. avenae* and increase in the wheat yield had direct relation to the number of ploughings. The reduction ranged from 10.6 to 42. 4 % and wheat yield increased from 4.5 to 97 %. Bhatti and Dabur (1982) studied effect of 1 to 3 summer ploughings at 10 days intervals during May and June on the population of *H. avenae* and subsequent growth yield. After 133 days of last ploughing, soil nematode population was 5, 4 and 3 eggs and larvae/ g soil in plots ploughed 1, 2 and 3 times respectively whereas initial nematode population was 7 eggs and larvae/g soil in unploughed plots. Yield increased was to 2.7, 5.6 and 11.2 % for three treatments.

Application of organic manures

The organic substances include organic manure, green crop residues, green leaf manures, soil amendment, oil cakes and oils. Mishra and Prasad (1978) tested the nematicidal bioefficacy of neem, mahuva, groundnut, sesamum and karnaj cakes on different levels of *M. incognita* infesting tomato and observed that all the treatments significantly affected nematode population. No penetration in chickpea roots was observed in case of juveniles hatched from egg masses treated with S/10 and higher concentration. Water soluble portion of oil seed cakes of neem, mustard, karnaj and groundnut were reported to have nematicidal effect against *M. incognita* (Mojumder and Mishra, 1991). Again in 1992 they further reported that neem seed coat could be used for the management of *M. incognita* infecting chickpea. Soil application @ 2 % w/w combined with seed coating @ 20 % w/w yielded 66.5 % reduction in root-knot galls and found to be most effective in respect of plant growth characters as well. Application of neem and castor cake each @ 1.0 t/ha significantly enhanced grain yield by 30.3 and 24.7 with 22.3 and 21.8 %

reduction in RKI realizing Rs. 1806 and Rs. 1266/ha respectively (Patel *et al.*, 1993 a). Piludi, neem and mustard cakes each applied @ 1.0 t/ha a week prior to tomato seeding increased number of transplants by 37.7, 44.9 and 22.5 % and decreased RKI by 27.3, 31.7 and 30.8 %. Whereas their application at 2.0 t/ha enhanced production of transplants by 62.3, 12.3 and 76.8 % and decreased RKI by 33.7, 25.0 and 23.4 % respectively (Patel *et al.*, 1993 b). Application of aak (*Calotropis procera*), water hyacinth (*Eichhornia crassipes*), naffati/besharmi (*Ipomea fistulosa*) and congress grass (*Parthenium hysterophorus*) each @ 3.0 kg/m^2 tremendously pushed seedling height, fresh shoot weight and number of transplantable seedlings by 64.3, 74.3 & 85.6, 59.3, 30.4 & 94.7, 65.7, 47.9 & 107.4 and 55.0, 41.3 & 78.2 % respectively (Patel *et al.*, 1993 b) (Plate. 11.5) Patel and his coworker in 1995b observed poultry manure application @ 2-3 t/ha in tomato nursery to increase 33.3 to 51.4 % production of transplantable seedlings with reduction in RKI by 27.8 to 32.3 % reduction in RKI.

Biological method

For biological control of RKN and cyst nematodes, more than 100 bioagents belonging to fungi, bacteria, viruses, nematodes protozoans, etc. are reported world over in last 5 decades. Of which, few promising bacterial and fungal agents were identified and now commercialized and marketed for farmers use in last decade. The Royal 300 and Royal 350 products based on *Arthrobotrys* spp. in UK., *Paecilomyces lilacinus* based product Bicon in Philippines (Asia) and Peru (South America) are in use. Recently, a *Bacillus* antagonist of RKN is isolated and developed by Ecogen Inc., which will be shortly available in the market of Israel.

Nematophilic bacteria

Gokte and Swarup (1988) tested five bacterial isolates against *M. incognita* and reported *Bacillus subtilis* and *P. pumilis* to give 100 % larval mortality at 2.0 O.D. concentrations. Similarly Somasekhar and Gill (1990) studied *Pasteuria penetrans* (3.5 g spore/kg soil in pots) against *M. incognita* J_2 in laboratory as well as on tomato plants and recorded 92 % spore attachment on larvae with 65 and 51 % reduction in nematode penetration in host root and gall formation respectively. Reduction of 94 % *M. incognita* population by *P. penetrans* at 1000 mg/kg soil on tomato in pots was obtained by Jayraj and Mani (1988). Somasekhar and Gill (1991) further studied the effect of *P. penetrans* and carbofuran alone as well as in combination and reported 69 % reduction in gall formation in *P. penetrans* alone treatment. Application of *Pseudomonas fluorescens* @ 2.5 kg/ha at seedling significantly decreased *Heterodera cajani* population in roots at 45 DAS by 18.32 % and final nematode population by

29.55 % and thereby gave 35.93 & 32.46 % higher pod and grain yield respectively.

Nematophilic fungi

Walia and Swarup (1985) studied effect of culture filtrate of *Arthrobotrys, Dactylaria* and *Curvularia* on hatching of *M. incognita* eggs and reported 100 % concentrated filtrate of *Arthrobotrys* allowed only 3.8 larval hatching compared to 45 in water alone. Sharma and Trivedi (1987) tried different *Paecilomyces lilacinus* doses against *M. incognita* on brinjal in pots and reported 75 g dose to be most effective giving 91 % nematode control. *P. lilacinus* followed by *M. incognita* in sequential application realized maximum yield and minimum nematode infection in okra (Saikia and Roy, 1994). Mortality of *M. incognita* eggs at 5 x 10^7 spore dose of *P. lilacinus* in laboratory bioassay was obtained by Vyas *et al.,* (1995). Patel *et al.,* (1995 a) studied bioefficacy of *P. lilacinus* against *M. javanica* pt.2 on groundnut in pots and recorded 3 to 5 % (w/w) fungus application to be most effective for nematode management but seed dressing through fungal spores proved ineffective. Application of *P. lilacinus* spore dust @ 50 kg/ha gave significantly higher groundnut dry pod yield by 25.56 % and fodder yields by 21.61 % with reduction of RKI by 50.92 % [Patel *et al.,* (1998 b)]. Similarly application of *P. lilacinus* @ 25 kg / ha in cotton gave 10.68 % higher seed cotton yield with 28.71 % in RKI (Vyas *et al.,* 1998).

Development of hybrids for disease resistance

Replacing nematode susceptible varieties with genetically resistant ones appears to be the most possible approach to nematode management. In India, several attempts have been made to screen the available indigenous and exotic crop germplasm and hybrids to locate resistance sources with little efforts to breed varieties specifically for resistance to nematodes. Khanna and Nirula (1964) reported a potato selection HC 294 resistant to *M. incognita.* Similarly, a tomato selection Pusa 120 exhibited resistance to this nematode (Prasad and Dasgupta, 1964; Singh and Choudhary, 1973). Krishnappa (1985) listed some of the varieties of potato, tomato, egg plant, cucurbits, sugar beet, capsicum, cowpea, mungbean, clusterbean, tobacco and rice found resistant to *M. incognita, M. javanica* or *M. graminicola.* O'Brien and Fisher (1979) reported two resistant wheat varieties viz. AUS 10894 and AUS 90248 from Australia. Bhatti *et al.,* (1976) identified barley BP 263, BP 264 and C 164 to be resistant to *H. avenae.* Cowpea variety C 152 a fodder type has been reported to be resistant to *M. incognita* and *M. javanica* (Patel *et al.,* 1977). Sethi and Sharma (1976) reported Barsati Mutant to be resistant to *M. incognita.* Cowpea

TABLE 11.1. Occurrence of root-knot nematodes in India

S. No.	*Meloidogyne species*	Host & locality	Reported by
1	*M. incognita*	Jute, West Bengal	Chhatopadhyay & Sengupta (1955)
2	*M. javanica*	Vegetables, West Bengal	-
3	*M. arenaria*	Tomato & Bean, Bihar	Lall & Ansari (1960)
4	*M. hapla*	Tea, Assam	-
5	*M. africana*	Citurs, Delhi	Chitwood & Toung (1960)
6	*M. brevicauda*	Tea, Tamil Nadu	Seshadri & Kumarswami (1963)
7	*M. thamesi*	Okra,Delhi	Sethi *et al.* (1964)
8	*M. exigua*	Coffee, South India	-
9	*M. graminicola*	Rice, Orissa	Patnaik (1969)
10	*M. indica*	Citrus	-
11	*M. lucknowica*	Spongegourd, U.P.	-

Other root-knot species reported are *M. chitwoodi, M. graminis, M.* artiellia, *M. nassi, M. triticoyzeae* and *M. piperi.*

TABLE 11.2. Important root-knot species distributed in India.

S. No.	Species	State
1	*M. incognita*	Delhi, Karnataka, Uttar Pradesh (U.P.)., Kerala, Tamil Nadu(T.N.)., Assam, Rajasthan, Haryana, Gujarat, Orissa, Bihar, Himachal Pradesh (H.P)., West Bengal(W.B.), Maharashtra. Madhya Pradesh (M.P.).
2	*M. javanica*	Delhi, Karnataka, U.P., Kerala, Rajasthan, Haryana,Gujarat, Bihar, H.P., W.B., M.P.
3	*M. arenaria*	Haryana, Gujarat, Bihar, H.P., U.P., Punjab, Delhi, T.N., Maharastra.
4	*M. hapla*	H.P., T.N.
5	*M. graminicola*	Assam, Haryana, Orison, U.P., Punjab, H.P., Karnataka, Kerala, M.P.
6	*M. triticoryzae*	Delhi, U. P.
7	*M. piperi*	Kerala
8	*M. indica*	Gujarat

TABLE 11.3. Yield losses in different crops due to root-knot nematodes in India.

Crop	Nematodes	% avoidable yield loss
1. Bidi tobacco	Root-knot nematodes *Meloidogyne incognita* & *M. javanica*	51.0 (Nursery) 50.0 (Field)
2. Brinjal	Root-knot nematodes, *M. incognita*	26.2 - 50.0
3. Chickpea	Root-knot nematodes, *Meloidogyne* spp.	84.0
4. Chilli	Root-knot nematodes, *M. incognita*	28.0
5. Cowpea	Root-knot nematodes, *Meloidogyne* spp.	30.3 - 71.3
6. Fennel	Root-knot nematodes, *M. javanica* pt.2	28.1
7. Ginger	*M. incognita* & Pin nematodes (*Paratylenchus* sp.)	51.2
8. Groundnut	Root-knot nematodes, *M. arenaria*, *M. javanica* pt.2	59.2 33.54 - 36.64
9. Jute	Root-knot nematodes, *M. incognita*	81.0
10. Pigeonpea	Root-knot nematode, *Meloidogyne* spp.	25.0 - 31.0
11. Sunflower	Root-knot nematodes, *M. incognita*)	38.0-84.0
12. Tomato	Root-knot nematodes, *M. incognita*	28.0 - 47.0
13. Turmeric	Root-knot nematodes (*M. incognita*)	31.3
14. Wheat	Root-knot nematodes (*M. incognita* & *M. javanica*) pt.1	32.92
15. Cotton	*M. incognita*	19.9

TABLE 11.4. Occurrence of cyst nematodes in India.

Name	Host & Locality	Reported by
HETERODERA		
H. avenae	*Triticum aestivum*, Sikar,Rajasthan	-
H. zeae	*Zea mays*, Chapli, Rajasthan	Koshy *et al*. (1971)
H. cajani	*Cajanus cajan*, New Delhi	Koshy (1967)
H. oryzicola	*Oryza sativa*, Kerala	Rao & Jay Prakash (1978)
H. sorghi	*Sorghum vulgare*, Gaziabad, UP.	-
H. chaubattia	Hills of U. P.	Gupta and Edward (1973)
H. vigni	Hills of U. P.	-
H. carotae	Field soil, Shimla Hills	Swarup *et al*. (1964)
H. galeopsidis	Field soil, Shimla Hills	Swarup *et al*. (1964)
H. cyperi	*Cynodon dactylon*, Karnataka	-
H. delvii	*Eleusine coracana*, Bangalore	-

contd....

Table contd....

H. graminis	*Cynodon dactylon*, New Delhi	Sharma & Swarup (1984)
H. mothi	*Cyperus rotundus* , Aligarh	
H. sacchari	*Saccharum spontaneum*, Delhi	Swarup *et al.* (1964)
		-
H. trifolii	Meadow grass, Gulmarg, J&K	
H. chaubattia	Hills of U.P.	Gupta and Edward (1973)
H. Vigni	Hills of U.P.	Edward and Mishra (1968)
GLOBODERA		
G. rostochiemis	*Solanum tuberosum*, Nilgiri hills	Jones (1961)
G. pallida	*Solanum tuberosum*, Nilgiri hills	Jones (1961)

Other cyst nematode reported in India are H. oryzae, H. cacti, H. gambiensis, H. cardiolata, H. raskii , H. spinicauda, H. goettingiana, H. indocyperi, , H. schachtii, H. skohensis, H. swarupi, H. filipjevi and H. irri.

TABLE 11.5. Important cyst nematode species distributed in India

Species	**State**
H. avenae	Delhi, H.P. (Leh), Rajasthan, Jammu and Kashmir (J & K), Haryana, Punjab, U.P.
H. zeae	Delhi, H.P., J & K, Haryana, U.P., Rajasthan, Gujarat, Maharastra, M.P., Punjab, A.P., Bihar.
H. cajani	Delhi, U.P., Haryana, Gujarat, Bihar, Rajasthan, W.B., T.N., Maharashtra, Karnataka, A.P., Punjab, H.P., M.P.
H. graminis	Delhi, Rajasthan, U.P., H.P.
H. oryzicola	Kerala, Orissa, Haryana, W.B., M.P.
H. gambiensis	Bangalore
H. sorghi	Haryana, U.P., M.P., Maharashtra
H. swarupi	U.P., Rajasthan
H. filipjevi	Haryana
G. rostochiensis and G. pallida	Kerala, Karnataka, T.N.

TABLE 11.6. Yield losses in different crops due to cyst nematodes in India.

S.No.	Crop	Nematode	% available yield loss	Reference
1	Wheat	*H. avenae*	50-100	Swarup *et al.*, 1976
2	Oat, barley	*H. avenae*	45-48	Duggan, 1963
3	Rice	*H. oryzae*	21-42	Kumari & Kuriyan, 1981
4	Banana	*H. oryzicola*	20.5-56.6	Charles, 1989
5	Pigeonpea	*H. cajani*	30.1	Saxena and Reddy, 1987
6	Mungbean	*H. cajani*	67.5	Saxena and Reddy, 1987

TABLE 11.7. Management of root-knot nematodes in banana.

Treatment	Plant stand (n=6) at		Plant ht., cm.	Stem girth cm/ plant	No. of leaves/ plant	No. of hands/ bunch	No. of fingers/ band	Yield kg/ha	RKI (1-5)*
	Initial	Harvest							
PHW	6	6	150.3	56.0	14	8	14	49062	3.09
NMC	6	6	151.6	51.8	14	9	15	50614	2.65
CAR	6	6	151.5	57.7	14	8	13	54697	2.27
PHW+ NMC	6	6	152.6	59.6	15	6	16	63074	2.76
PHW + CAR	6	6	160.9	57.9	14	9	14	64066	2.80
NMC + CAR	6	6	156.2	60.7	15	9	15	62513	2.56
PHW + NMC + CAR	6	5	158.9	62.0	15	9	15	65427	2.35
CON	6	6	148.3	50.0	13	7	14	48241	4.28
S.Em	0.11	0.33	3.2	1.7	0.55	0.4	0.38	671 1	0.16
C.D.0.05	NS	NS	NS	NS	NS	NS	1.01	3236	0.45
C.V.%	5.7	24.2	6.3	9.3	11.2	14.1	7.5	24.2	14.7

* 1= Free, 5= Maximum Disease Intensity.

TABLE 11.8. Effect of different crops on root-knot nematodes.

Crop	Root-knot index (0-5)* on	
	Different crops	Test crop Tomato
Castor	0.00	2.15
Castor + Pigeonpea		
Castor	0.00	3.23
Pigeonpea	3.69	-
Pigeonpea	3.91	4.10
Tomato (control)	4.78	4.88
Control	-	4.08
S.Em	-	0.30
C.D. 0.05	-	0.90
C.V.%	-	15.8

* 0 = Free; 5= Maximum disease intensity.

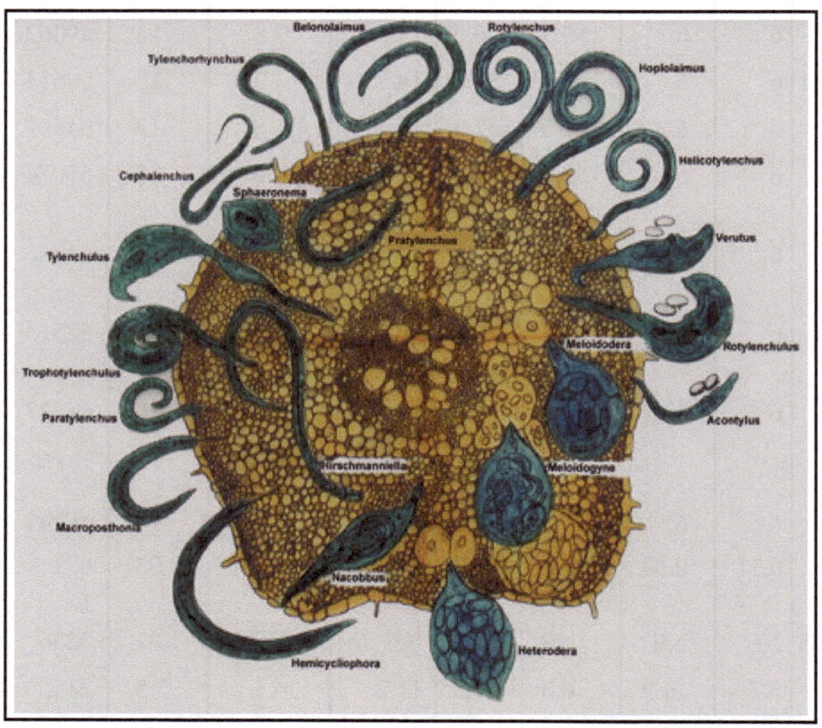

PLATE 11.1. Phytonematodes attaching crop

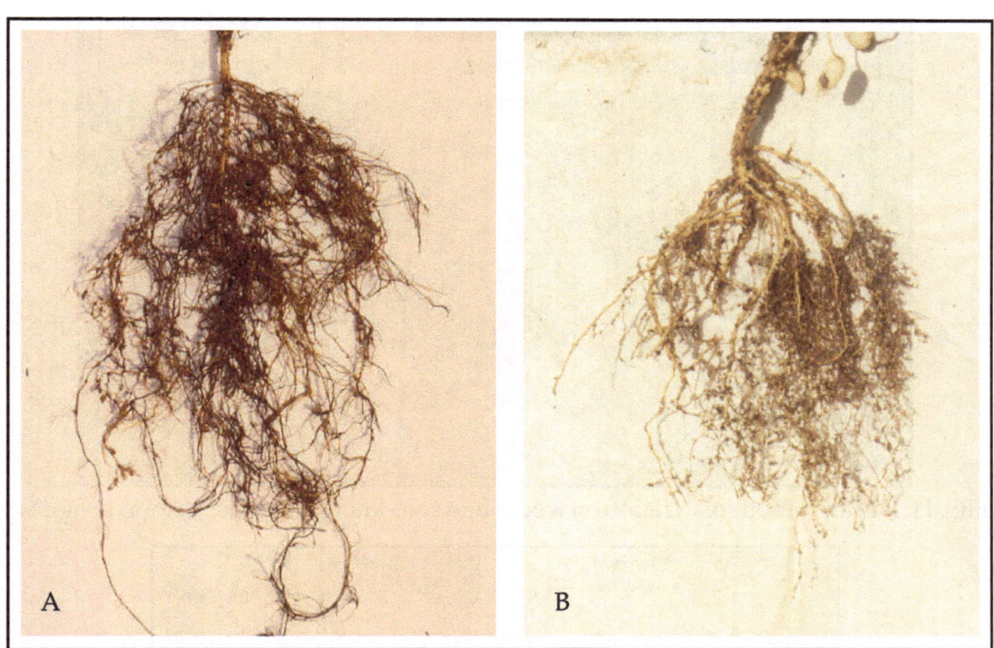

PLATE 11.2. Galling pattern of *Meloidogyne arenaria* (A) and *Meloidogyne javanica* pt. (B)

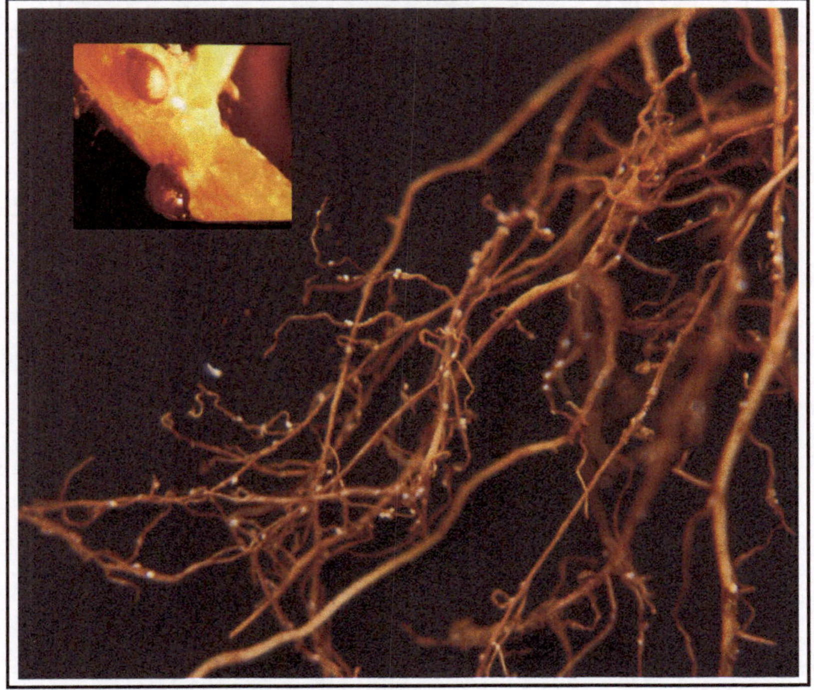

PLATE 11.3. Pigeonpea roots showing severe infection of cyst nematode (*Hetrodera cajani*)

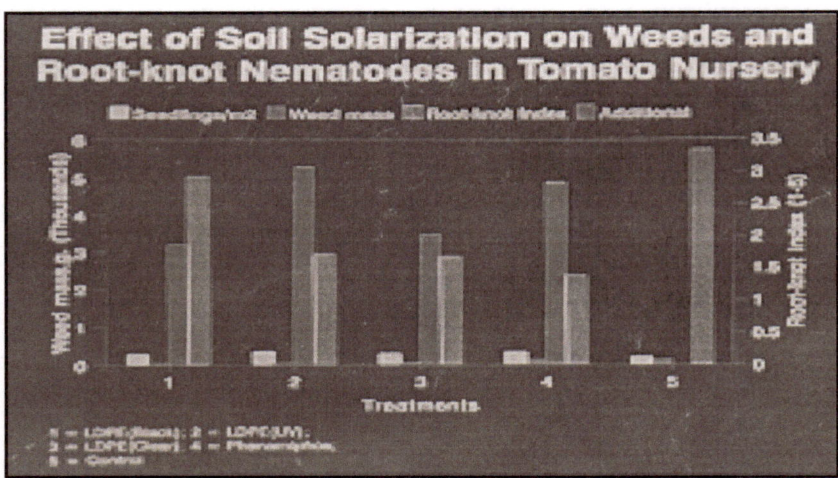

FIG. 11.1. Effect of soil solarization on weeds and root-knot nematodes in tomato nursery

PLATE 11.4. Impact of soil solarization on tobacco nursery

PLATE 11.5. Effect of different plant materials on tomato seedlings

GC 1 had resistant reaction with *M. incognita* (diseases index 1.00/5.00 and) *M. javanica* (disease index 0.80/5.00) (Patel *et al.*, 1984). (Plate 11.7). Tobacco variety McNair 135 later nematodes as PCT 4 had high degree of resistance to both the species, *M. incognita* and *M. javanica* (Patel *et al.*, 1986). Evaluation of selected tomato varieties against *Meloidogyne* spp. indicated that cultivars NT 8 and SL 120 were highly resistant whereas NT 3 and NT 12 were resistant to *M. incognita* whereas tomato NT 3 and NT 8 recorded highly resistant and NT 12 and SL 120 resistant reaction (Patel *et al.*, 1990).

REFERENCES

Anonymous. 2001-03. Management of nematode complex on banana. *AICRP Biennial Report.* Department of Nematology, B. A. College of Agriculture, GAU, Anand. pp. 44.

Berkley, M. J. 1985. Vibrios forming cysts on the roots of cucumbers. *Gdnrs. Chron.* 14: 220.

Bhatti, D. S. and Dabur, K. R. 1982. Effect of summer ploughing on *Heterodera avenae* population and wheat yield. *J. Nematol.* 14: 429.

Bhatti, D. S., Dahiya, R. S., Dalal, M. R. and Dhawan, S. C. 1976. Resistant Barley varieties for the control of *Heterodera arenaria,* Noteswaber, 1924 *Curr. Sci.* 45: 678.

Chattopadyay, S. B. and Sengupta, S. K. 1955. Root-knot disease of Jute in West Bengal. *Curr. Sci.*4: 276.

Chitwood, B. G. and Toung, Mo-chu. 1960. *Meloidogyne* from Taiwan and New Delhi. *Phytopath.* 50: 631-632.

Choudhary, B. C. and Hoque, M. O. 1982. Control of root-knot nematodes in brinjal seed beds by heat treatments. *Bangladesh J. Agric. Res.* 7: 55-57.

Edward, J. C., Singh, R. B. and Mishra, S. L. 1963. Survey of plant parasitic nematodes of farm soils of the Allahabad Agricultural Institute, *Allahabad Farmer.* 37: 1.

Gaur, H. S. and Mishra, S. D. 1983. Influence of irrigation and nitrogen on inter specific plant parasitic nematodes and performance of Tomato. *Ann. Agric. Res.* 4: 133-139.

Gokte, N. and Swarup, G. 1988. On the potential of some bacterial biocide against root-knot and cyst nematodes. *Indian J. Nematol.* 18: 152-153.

Gupta, Pramila and Edward, J. C. 1973. A new record of a cyst forming nematode (*Heterodera chaubatha* sp.) from the hills of Uttar pradesh. *Curr. Sci.* 42: 618-620.

Haque, M. M. and Gaur, H. S. 1985. Effect of multiple cropping sequences on the dynamics of nematode population and crop performance. *Indian. J. Nematol.* 15: 262-263.

Jayraj, S. and Mani, A. 1988. Bio control of *Meloidogyne javanica* with the bacterial spore parasite *Pasteuria penetransint. Nematol. Network. Newsltr.* 5: 3-4.

Johnson, A. W. 1985. Specific crop rotation effects combined with cultural practices and nematicides. *(In:)*, J. N. Sasser and C. C. Carter (Eds.), *An Advanced Treatise on Meloidogyne* Vol. I IMP, N.C. State Univ., pp. 283-301.

Johnson, A. W., Summer, D. R., Dowler, C.C. and Glaze, N.C. 1976. Influence of three cropping systems and four levels of pest management on population of root-knot and lesion nematodes. *J. Nematol.*8: 290-291.

Jones, F. G. W. 1961. The potato root eelworm, *Heterodera rostochiensis* wall in India *Curr. Sci.* 30:187.

Khan, A. M., Saxena, S. K. and Sidiqi, Z. A. 1971. Efficacy of *Tagetes* erecta in reducing root infesting nematodes of tomato and okra. *Indian Phytopath.* 24: 166-169.

Khanna, M.L. and Nirula, K.K. 1964. Breeding potatoes for resistant to root-knot nematodes, *Curr. Sci.* 23: 314-315.

Koshy, P.K. 1967. A new species of *Heterodera* from India. *Indian Phytopath.* 20: 272-274.

Koshy, P. K., Swarup, G. and Sethi, C. L. 1971. *Heterodera zeae* n.sp. a cyst forming nematode on zea mays. *Nematologica.* 16: 511-516.

Krishnamurthy, G. V.G. and Elias, N.A. 1969. Note on effects of nematicides and heat treatment on the control of root-knot nematode in tobacco nurseries. *Indian. J. Agric. Sci.* 39: 263-265.

Krishnamurthy, G. V. G., Elias, N.A., Chandwani, G.H. and Reddy, N.S. 1969. Studies on root-knot control on tobacco in Hunsur (Mysore State) and Rajahmundry (Andhra Pradesh) *First All India Nematology Symp.*, I.A.R.I., New Delhi. pp. 42-44 (Abstr.)

Krishnappa, K. 1985. Nematodes in developing countries India – IMP Region III. (*In:*) J.N. Sasser & C.C. Carter (Eds.), *An Advanced Treatise on Meloidogyne* vol. 1 IMP, N.C. State Uni., pp. 379-398.

Krishna Prasad, K. S. 1986. Nematode Problems of Potato. (*In:*) Plant Parasitic Nematodes of India. Problems and Progress. IARI, New Delhi. pp. 350-370.

Kumari, U. and Kuriyan, K.J. 1981. Cyst nematodes, *Heterodera oryzicola* on rice in Kerala. Estimation of loss in rice due to *H. oryzicola* in field condition. *Indian J. Nematol.* 11: 106.

Lall, B. S. and Ansari, M .N. A. 1960. Field studies on root-knot nematodes *Meloidoygne* spp., Nematoda, *Heteroderidae*. *Sci. & Cult.* 26: 279.

Mathur, B. N., Arya, H. C., Mathur, R. L. and Honda, D. K. 1975. The biology of cereal cyst nematodes *Heterodera avenae* in India III. Factors affected the emergence of larvae from cysts. *Nematologica.* 20:19-26.

Mishra, S. D. and Prasad, S. K. 1978. Effect of soil amendments on nematodes and crop yields II oil seed cakes, organic matter and inorganic fertilizers at different levels of *Meloidogyne incognita. Indian J. Ent.* 40: 42-53.

Mojumder, V. and Mishra, S. D. 1991. Nematicidal efficacy of some plant products and management of *Meloidogyne incognita* in pulse crops by soaking seeds in their aqueous extracts. *Curr. Nematol.* 2: 27-32.

Nirula, K. K. and Bassi, K. K. 1965. Thermotherapy for root-knot nematode, *Meloidogyne incognita* in potato tubers. *India Potato J.* 7: 9-11.

O'Brien, P. C. and Fisher, J. M. 1979. Reaction of cereals to populations of *Heterodera avenae* in South Australia. *Nematologica.* 25: 261-267.

Patel, D. J., Patel, B. A., Chavda, J. C. and Patel, H. V. 1988. Record of *Meloidogyne javanica* on groundnut in Gujarat, India. *Int. Arachis. Newsltr.* 3: 16-17.

Patel, D. J., Patel, H. R., Patel, S. K. and Makwana, M. G. 1991 a. Soil solarization through clear polyethylene trarping for management of root-knot disease in a tobacco nursery. *Tob. Res.* 35: 63- 64.

Patel, D. J., Patel, H. V. and Patel S. K. 1991 b. Comparative efficacy of LLDPE film for soil solarization in control of root-knot nematodes in tomato nursery. *Plasticulture.* 3: 33-66.

Patel, D. J., Patel, H. V., Patel, S. K. and Patel, B. A. 1993 b . Nematicidal properties of some plant materials for management of root-knot nematodes in tomato nursery. *Indian. J. Pl. Prot.* 21: 242-244.

Patel, D. J., Patel, H. V., Patel, S. K. and Patel, S. T. 1987. Effect of rabbing on root-knot control in bidi tobacco nursery.*Tob Res.* 13: 47-50.

Patel, D. J., Patel, S. K. and Patel, H. V. 1993 a. Evaluation of organic amendments for the effective management of root-knot nematodes in tomato nursery and in green gram. *World Neem Conference* held at Bangalore during 24-28 Feb., 1993 p. 68 (Abstra).

Patel, D. J., Valand, G. B. and Patel, G. J. 1986. PCT-4 a root-knot resistant cultivar. *Indian. J. Nematol.* 16: 115.

Patel, D. J., Vyas, R. V., Patel, B. A. and Patel R. S. 1995 a. Bio efficacy of *Paceilomyces lilacinus* in controlling *Meloidogyne javanica* (Pathotype 2) on groundnut. *Int. Arachis Newsltr.* 15: 46-47.

Patel, G. J., Patel, D. J., Patel, S. K., Patel, N.R. and Patel, H. D. 1990.Influence of crop sequences on yield, quality and root-knot disease control in tobacco. Paper presented in *Second International Nematology Congress,* during 11-17, August, 1990 at Veldhoeven. The Netherlands. p.122 (Abstra.)

Patel, G. J., Patel, D., Shah, H.M. and Valand. G. B. 1984. Cowpea 82 -1B, a root-knot resistant variety. *Indian J. Plant. Prot.* 12: 63-64.

Patel, G. J., Patel, S. H., Patel, N. M. and Patel, D. J. 1979. Feasibility of successful cultivation of bidi tobacco after summer bajra. *Tob. Res.* 5: 37-42.

Patel, G. J., Shah, H. M. and Patel, D. J. 1977. Screening of cowpea cultivars against root-knot nematodes. *Indian J. Nematol.* 7: 169-170.

Patel, G. J., Shah H. M., Patel D. J. and Patel, S. K. 1985. Use of castor in management of root-knot nematodes. *Tob. Res.* 11:78-80.

Patel, H. V., Patel, B. A., Patel, D. J. and Patel, S. K. 1990. Evaluation of selected tomato cultivars for reactions to *Meloidogyne incognita* and *M. javanica. Intl. Nematol. Network Newsltr.* 7: 8-9.

Patel, H. V., Patel, S. K. and Patel, D. J. 1989. Effect of rabbing on root-knot control in tomato nursery. *Indian. J. Plant Prot.* 17: 19-22.

Patel, H. V., Patel, S. K. and Patel, D. J. 1995 b. Poultry manure - An effective organic amendment for management of root-knot nematodes in tomato nursery. *Indian J. Mycol. & Pl. Pathol.* 25: 51.

Patel, N. B., Vyas, R.V. and Patel, D. J. 1998 b. Efficacy of biocontrol fungi, *Paecilomyces lilacinus* for management of *M. javanica* pt. 2 on groundnut. Proc. *National Symp. Rational Approaches in Nematode Management for Sustainable Agriculture* organized by NSI, IARI, New Delhi & BACA, GAU, Anand during 23-25, Nov. pp.23-24.

Prasad, S. K. and Dasgupta, D. R. 1964. Plant parasitic nematodes associated with rose in Delhi. *Indian. J. Ent.* 26: 235-238.

Rao, V. K. and Krishnappa, K. 1995. Soil solarization for the control of soil borne pathogens complex with special reference to the *Meloidogyne incognita* and *Fusarium oxysporum* f. sp. *cicier. Indian Phytopath.* 48: 300-303.

Rao, Y. S. and Jayprakash, A. 1978. *Heterodera oryzicola* n. sp. (Nematoda : *Heteroderidae*) a cyst nematode on rice (*Oriyza sativa* L.) from Kerala State, India. *Nematologica.* 24: 341-346.

Roy, A. K. 1978. Effectiveness of rotation with non host or fallow on reducing infestation of *Meloidogyne gramimcola. Indian J. Nematol.* 8:156-158.

Saikia, A. and Roy, A.K. 1994. Efficacy of *Paecilomyces lilacinus* on the reduction of attack of *Meloidogyne incognita* in okra. *Indian J. Nematol.* 24: 163-167.

Sasser, J. N. 1982. Plant Pest interaction with environmental stress and breeding for pest resistance - Nematodes. pp. 375-389. (In:) M.N. Shrishansen and C.E. Lewis, (Eds). Breeding plant for less favourable environments, New York Interscience.

Saxena, R. and Reddy, D.D.R. 1987. Crop losses in pigeonpea and mung bean by pigeonpea cyst nematodes (*Heterodera cajani*). *Indian J. Nematol.* 17: 91-94.

Sen, K. and Dasgupta, M. K. 1982. Control of root-knot nematodes in India. *Pesticides.* 16: 14-17.

Sethi. C. L. and Sharma, N. K. 1976. Nature of resistance of cowpea to the root-knot nematodes *Meloidogyne incognita* and the pigeonpea cyst nematode *Heterodera cajani. Indian. J. Nematol.* 6: 81-85.

Sethi, C L., Gill, J. S. and Swarup. G. 1964. A note on the prevalene of *Meloidogyne* spp. in vegetable crops . *Indian Phytopath.* 17: 69-70.

Sharma, A. and Trivedi, P. C. 1987. Stem nematode (*Ditylenchus angustus*) a potential pest of Rice in Assam and West Bengal, India. *Int. Nematol. Network. Newsltr.* 4: 24-25.

Sharma, S. B. and Nene, Y. L. 1990. Effect of soil solarization on nematodes parasitic to chickpea and pigeonpea. *J. Nematol.* 22: 658-664.

Sharma, S. B. and Swarup, G. 1984. Cyst forming nematodes of India, Cosmo Publ. New Delhi, pp. 152.

Sharma, S. K., Singh, I. and Sakhuja, P. K. 1980. Influence of different cropping sequences on the population of root-knot *Meloidogyne incognita* and the performance of the subsequence mung bean crop. *Indian J. Nematol.* 10: 53-58.

Singh, B. and Choudhary, B. 1973. The chemical characterisities of tomato cultivars resistant to root-knot nematodes (*Meloidogyne spp.*) *Nematologica* 19: 443-448.

Somasekhar, N. and Gill, J. S. 1990. Penetration and multiplication of *Melidogyne incognita* as influenced by *Pasteuria penetrans. Indian J. Nematol.* 20: 213-218.

Somasekhar, N. and Gill, J. S. 1991. Efficacy of *Pasturia penetrans* alone and in combination with carbofuran in controlling *Meloidogyne incognita Indian J. Nematol.* 21: 61-65.

Sundaresh, H. N. and Setty, K. G. H. 1977. Haustorial appendages of giant cells : Root-knot of Tobacco (*Meloidogyne incognita* chitwood). *Curr. Sci.* 46: 157-158.

Swarup, G., Mathur, R. L., Seshdri, A. R., Raski, D. J. and Mathur, B. N. 1976. Response of wheat and barley to soil fumigation by DD and DBCP against "Molya" disease caused by *Heterodera avenae. Indian. J. Nematol.* 6: 150-155.

Vyas, R. V., Patel, H. R., Patel, B. A. and Patel, D. J. 1995. Parasitism of muscardine fungi on *Meloidogyne incognita* eggs. *Curr. Sci.* 69: 635-636.

Vyas, R.V., Patel, N. B. and Patel, D. J. 1998. Integrated management of root-knot nematodes in cotton. *Proc. National Symp. Rational Approaches in nematode Management for Sustainable Agriculture* organized by NSI, IARI, New Delhi & BACA, GAU, Anand during 23-25, Nov., 1998 p. 51-52.

Walia, R. K. and Bajaj, H. K.1988. Further studies on the existence of races in pigeonpea cyst nematodes, *Heterodera cajani*. *Indian J. Nematol.* 18: 269-272.

Walia, R. K. and Swarup, G. 1985. Effect of some fungi on nematode hatching and larvae root penetration. *Indian J. Nematol.* 15: 174-176.

Whitehead, A. G. 1968. Taxonomy of *Meloidogyne* (Nematoda : *Heteroderidae*) with descriptions of four new species. *Trans. Zool. Soc. Lond.* 31: 263-401.

• • •

Advances in Soil Borne Plant Diseases, 2008
Editors: Naik and Devika Rani, pp. 285-310
New India Publishing Agency, New Delhi

12

Use of On-farm Wastes in Management of Soil Borne Diseases in Hot Arid Region

Satish Lodha, Sunil Israel and Ritu Mawar

INTRODUCTION

In arid and semi-arid regions of India, besides weather aberrations, diseases caused by certain soil-borne pathogens are responsible for low productivity of agricultural crops. In addition to losses caused in annual crops, occurrence of these diseases in forestry takes away a major share. Because of the requirement for free water, diseases of aerial plant parts are much less common than those involving roots when plants are grown in dry regions where humidities are low and rains are infrequent and/or of short durations. Under these conditions, pathogens like *Macrophomina phaseolina, Ganoderma lucidum, Cylindrocarpon lichenicola, Neocosmospora vasinfectum, Sclrotium rolfii* and species of *Fusarium (F. oxysporum, F. solani, F. equiseti,* etc.) cause root rots and wilts in many commercially valuable and perennial plants.

Certain agro-climatic factors are attributed for the development of few specific diseases caused by soil-borne plant pathogens. Low organic matter, small microbial population coupled with poor moisture retention capacity of soil favour survival and multiplication of soil-borne pathogens (Fig. 12.1). Inoculum of these pathogens increases in the agricultural lands with increased

---

Central Arid Zone Research Institute, Jodhpur -342 003

years of cultivation of susceptible crops and the inoculum density in the soil is directly proportional to disease intensity in the field.

For years, soil fumigation was the most common approach for the control of soil- borne pests. Chemical soil disinfestations eradicate most microbial residents in the soil. While plants benefit from such conditions for a short time, the microbial vacuum created often leads to a rebounding of parasitic agents, resulting in plant diseases that cause more damage than those originally targeted for control. Soils, especially those with low microbial populations are often more vulnerable to pathogen reinvasion following fumigation. One way of increasing microbial population and diversity is to introduce an adequate energy source. Food sources can be soil amendments, such as plant residues, animal manure or organic waste.

The incorporation of plant residues in soil as green manure or at the end of crop growth has been a common practice for years. Higher plants contain and release an enormous variety of biologically active compounds, some of which can be exploited as potential pesticides. Secondary metabolites are usually produced inside the plant tissue and can be released during plant growth as a defense mechanism, or when plant debris is incorporated into the soil and subjected to microbial degradation. Glucosinolates are present in various quantities in many dicotyledonous plants. Enzymatic hydrolysis of glucosinolates results in the production of various sulphur compounds, some of which possess antimicrobial activity. The cruciferae are among the plant families with high content of glucosinolates in their tissues. These are also characterized by a high content of other sulphur containing compounds (Walker *et al.*, 1937; Sang *et al.*, 1984; Mayton *et al.*, 1996). Antifungal volatile compounds such as allyl isothiocyanate have been found in leaf extracts of various *Brassica* species. The toxicity of isothiocyanates to various microorganisms is well documented (Brown and Morra, 1997).

Regular additions of on-farm wastes such as animal manures and crop residues are of utmost importance in maintaining the tilth, fertility and productivity of agricultural soils; protecting them from wind and water erosion; and preventing nutrient losses through runoff and leaching. These materials have predictable effects on soil physical properties such as increased water-holding capacity, soil aggregation, soil aeration and permeability and decreased soil crusting, compaction and bulk density.

In arid and semi-arid regions of India, agriculture is generally referred as "resource deficient farming" because agriculture and livelihood is largely dependent on the rain God. In the years of low rainfall it is difficult to sustain only from the meager farm produce, thus inhabitants have found alternative livelihood in animal husbandry. Sustenance of livestock is, in turn, dependent mainly on on-farm produce, which are not utilized for human consumption and thus are referred as fodder and feed. On the other hand, environmentally sound organic farming requires a system of recycling or returning to the soil what have been extracted out during the course of agriculture depleting soil fertility. However, resource deficient farmers will certainly be reluctant to amend agricultural fields with those residues, which are used for cattle/ livestock feeding.

Thus, many crop residues can not be considered, but a few can be termed as on farm wastes because of their unpalatability as cattle/livestock feed. Incidentally, amending soil with on-farm wastes has also not been preferred due to release of certain toxic volatiles during their decomposition that may cause phytotoxicity in subsequent agricultural crops. Finding appropriate time for their recycling as soil amendment after proper scientific experimentation will not only manage dreadly soil-borne pathogens but may also boast crop yields. There are certain prerequisites for the development and successful implementation of relevant technology by the use of organic amendments for the management of soil-borne pathogens. These include -

1. Existence-of the problem
2. Availability of efficient on-farm wastes
3. Appropriate time of application
4. Easy application
5. Promotion to resident antagonists
6. Synergism with other management strategies

An effort has been made for the past two decades to utilize certain on-farm wastes for the management of important soil-borne plant pathogens at the Central Arid Zone Research Institute, Jodhpur. Salient research findings and their implications are presented in this chapter.

PROBLEM

Clusterbean (*Cyamopsis tetragonoloba* (L.) Taub) is the most important legume cultivated in about 2.0 million hectares in hot arid region of India. The crop has assumed a great importance in recent years mainly due to the presence of gum (galoctomannan) in its endosperm that constitutes about 30-32% of the whole seed. Guar gum has diversified uses in various industries. India earns a foreign exchange of approximately 1000 crores from guar gum. Under concurrent heat and moisture stress this crop suffers heavily due to charcoal or dry root rot caused by *Macrophomina phaseolina* (Tassi) Goid. The sclerotial inoculum of *M. phaseolina* increases in the soil with the increased years of cultivation of susceptible crops and the inoculum density in the soil is directly proportional to disease incidence in the field.

In irrigated pockets, cumin *(Cuminum cyminum* L.) is grown in the same fields during winter season and has a great commercial value. It is grown in the states of Rajasthan and Gujarat in about 3 lakh hectares with an average yield of 5 q ha^{-1} but it may vary from 3 to 10 q ha^{-1}. Low yield per hectare has been attributed to absence of high yielding varieties and occurrence of diseases. Yield losses due to wilt *(Fusarium oxysporum* f sp. *cumini* Prasad and Patel) alone may often reach 40% (Lodha *et al.,* 1986). Frequently growers are left with no alternative except to abandon cumin cultivation after a few successive years of cropping.

Thus, occurrence of *M. phaseolina* under rainfed conditions and *F. oxysporum* in winter season in the same fields has become a serious limiting factor for the profitable production of clusterbean and other legumes and cumin, respectively.

SELECTION OF EFFECTIVE ON-FARM WASTES

In a laboratory study, population changes of *M. phaseolina* and *F. oxysporum* were followed in a moistened sandy soil amended with mustard and castor oil-cakes or nitrogen enriched pearl millet residue, singly or in combined form (Sharma *et.al.,* 1995). The pathogen infested soil was sub-divided into many lots for mixing with pearl millet residue (PMR, 0.9%) enriched with urea (40 kg N ha^{-1}), mustard oil cake (MC, 1 %) and castor cake (CC, 1 %) separately and a combination of above. The moisture level in all the treatments was maintained at 50% of MHC. The population densities of *M. phaseolina* were determined on Chloroneb-Mercury-Rose bengal-Agar (CMRA) medium (Meyer *et al.,* 1973) and that of *F. oxysporum* f sp. *cumini* on modified Peptone-PCNB medium (Papavizas, 1967). Total microbial populations in soil were enumerated at 15-day interval until a significant reduction was achieved for both the soil-borne pathogens. Antagonistic actinomycetes and bacteria were detected on Czapeck's dox agar (Ghaffar *et al.,* 1969).

A significant reduction in the population of *M. phaseolina* occurred in the cake and pearl-millet residue amended soil (Table 12.1). In the MC alone and N + PMR + MC amended soil, 100% reduction was achieved within a period of 30 and 45 days, respectively. In N+PMR amended soil, numbers of viable propagules of *M. phaseolina* were reduced by 94% in 45 days. Moist addition of urea decreased the C: N ratio and hastened the decomposition of crop residues. Low C:N ratio amendments have been found effective in reducing the population of *M. phaseolina* (Dhingra and Sinclair, 1974). A decline of 50% propagules in the non-amended soil maintained at 50% of MHC confirmed the significance of soil moisture alone in reducing the *M. phaseolina* population (Dhingra and Sinclair, 1975).

In MC-amended soil a 100% reduction in *Fusarium* population was achieved within 30 days (Table 12.1). Stimulatory and inhibitory effects of soil amendments with oil- cakes on *Fusarium* populations have also been reported by Singh and Singh (1970). Crucifer plant residue incorporation in the soil was shown to reduce the population of soil-borne pathogens effectively (Ramirez-Villapudua and Munnecke, 1987; 1988). The population of bacteria and actinomycetes increased considerably in amended soils. Over 90% of the total number of actinomycetes were antagonistic to *M. phaseolina,* with the highest numbers in the N+PMR+CC amended soil. However, populations of actinomycetes antagonistic to *Fusarium* propagules were less as compared to *M. phaseolina.* The number of bacterial antagonists did not differ significantly in amended soils. Thus, apart from the toxic effects of cakes, increased populations of actinomycetes might have also contributed in reducing the population of *M. phaseolina* and *Fusarium.* These results demonstrate the efficacy of mustard cake in controlling *M. phaseolina* and *Fusarium* within a period of 30 days.

UTILIZING SOLAR HEAT

During previous experiments on soil solarization for the management of these soil-borne pathogens (Lodha and Solanki, 1992; Lodha, 1995), considerable reduction in soil population densities of both the pathogens were recorded in the treatment where soil was irrigated once but was not covered with polyethylene sheet (wet non-mulched control). Possible reasons for this control were explored. During summer months in arid region, temperature of bare soil often reached lethal ranges (50-60°C), but still were ineffective in reducing the viability of sclerotia of *M. phaseolina* and chlamydospores of *Fusarium,* because the soil remained dry. Soil moisture (one summer irrigation) greatly affected the sensitivity of resting structures to a heat treatment (Kat an *et al.*, 1976). The mechanism of control induced by irrigation in heated soil could be by dilution of fungistatic behaviour of soil, stimulating germination of resting structures, which are more vulnerable to increased heat conduction and microbial activity (Dhingra and Sinclair, 1975). In different years of experimentation, soil temperatures in merely irrigated plots also reached 51-55°C. Thus, just by one summer irrigation, a reduction of 33.0-42.8% and 28.8-42.0% was achieved in the counts of *M. phaseolina* and *F. oxysporum,* respectively. This prompted us to intensify our efforts to improve the efficiency of summer irrigation by integrating readily available on-farm wastes as soil amendments as a substitute of soil solarization.

COMBINING SOLARIZATION / SUMMER IRRIGATION WITH CRUCIFEROUS RESIDUES

After ascertaining the efficiency of mustard oil-cake in simultaneous control of both the soil-borne pathogens, a field experiment was undertaken to evaluate the efficacy of cruciferous residues, mustard oil-cake (4 ton ha^{-1}) and cauliflower leaf residue (5 ton ha^{-1}) combined with summer irrigation and/or solarization on the population of *M. phaseolina* at 0-15 and 16-30 cm soil depth (Lodha *et al.,* 1997). Two kg portions of *M. phaseolina* infested soil were mixed uniformly to 30 cm depth in 28 sub-plots (3 x 1 M). Eight sub-plots were amended with air-dried ground-up mustard cake (4 ton ha^{-1}) and equal number with cauliflower residue (5 ton ha^{-1}). One irrigation was applied to field capacity in all sub-plots except four, which served as dry control. Four sub-plots of each amendment and merely irrigated plots were covered with polyethylene sheet. Soil temperatures were recorded at 7.5 and 22 cm depth. Polyethylene mulching was done for 15 days and then soil samples were collected from 0-15 and 16-30 cm depth to determine population of *M. phaseolina* and lytic bacteria (Greenbergar *et al.,* 1987).

One summer irrigation (SI) of the dry plots (without amendment) initially reduced the soil temperature to 37°C but a gradual increase to 51°C (7.5 cm) and 46°C (22 cm) followed within 6-7 days. This caused 40% reduction in *M. phaseolina* counts at 0-30 cm depth. In non-amended, irrigated solarized plots (SI+S), the temperature was 3- 7°C higher than in corresponding non-solarized plots. Amendment of soil with cruciferous residues augmented the efficacy of SI by eliminating a sizeable proportion of *M. phaseolina* propagules in non-solarized plots. Soil solarization of amended and irrigated plots raised the maximal soil temperature to 4-6°C above that of the corresponding non-solarized plots. The combined effect of moisture, amendments and temperature was almost complete elimination of viable propagules of *M. phaseolina,* irrespective of soil depth. Cruciferous residues increased the population of lytic bacteria slightly in irrigated non-solarized soil but their density increased significantly in amended solarized soil, particularly at 0-15 cm depth.

The results of this study revealed that combining cruciferous residues with polyethylene mulching as natural heating of moistened soil improved the reduction in populations of *M. phaseolina* in hot arid climates. Soil temperatures under natural heating or solarization in the present study were higher than those observed in other parts of the world with a similar climate

(Mihail and Alcorn, 1984; Abdel-Rahim *et al.*, 1988). The sharp decline in the viability of pathogenic propagules in the top soil layer could be attributed primarily to high soil temperatures (Sheikh and Ghaffar, 1984; Lodha and Solanki, 1992). This is also supported by the fact that temperatures at 22 cm remained relatively low (< 50°C) and mortality of the pathogen was also low at this depth.

In this study, pronounced reduction at the lower soil depth could be attributed to a combination of mechanisms. A weakening effect of sub-lethal soil temperature (47-50°C) may have facilitated the action of sulphur-containing toxic volatiles from cruciferous residues and microbial antagonism (Lewis and Papavizas, 1970; Lifshitz *et al.*, 1983). After incorporation of cruciferous residues in moist soils, a greater release of toxic volatiles such as methyl sulphide, mercaptan and isothiocyanates at high temperatures was reported by Gamliel and Stapleton (1993). However, isothiocyanates were not detected at low temperatures (Lewis and Papavizas, 1970; 1971) and these are mainly responsible for the inhibition or reduction of soil-borne pathogens (Angus *et al.*, 1994). A possible role of increased lytic bacterial density in reducing counts of *M. phaseolina* at lower soil depth in mustard cake amended soil can not be excluded as these bacteria are capable of lysing fungal mycelium of soil-borne pathogens (Mitchell and Alexander, 1961). Thus, combining amendments of cruciferous residues with one irrigation during hot summer days as a substitute for polyethylene mulching may be practical in the cultural control of *M. phaseolina*. This finding has a potential value and important implications for irrigated pockets of hot arid zone of India as well as for many countries in the appropriate climatic conditions.

DETERMINING COST-EFFECTIVE DOSE

After working out the potential value of mustard oil-cake, next step was to determine cost-effective field application dose. Concurrently our effort to search other readily available cheaper sources of cruciferous residues ended in the identification of mustard pod straw that was found effective in laboratory studies. Thus, efficacy of mustard oil-cake and pod straw was evaluated at concentrations ranging from 1 to 4 ton ha^{-1} combined with summer irrigation in small 7 cm pits. Soil temperatures were recorded for 15 days before collecting soil samples. Population of *M. phaseolina* was determined from air-dried samples.

Studies on evaluation of different dosages of mustard oil-cake and pod straw revealed that within 15 days, population of *M. phaseolina* was almost completely eliminated at 2 and 3 ton ha^{-1} concentrations of mustard cake and pod straw, respectively (Lodha *et al.*, 2002) (Table 12.2). Population of lytic bacteria also increased in naturally heated amended soil. Reduction in population of *M. phaseolina* was 11-17% lower under shade conditions. This study established that (i) mustard oil-cake and pod straw at lower dosages are effective in reducing population of *M. phaseolina* under field conditions and prevalent high soil temperatures accelerates the efficiency of *Brassica* amendments.

ENHANCEMENT OF RESIDUE EFFICIENCY BY SUB-LETHAL HEATING

In this region, during hot summer months (April-June), temperature of bare soil often reaches 50-60°C eliminating only a part of the inoculum (10-15% *M. phaseolina* or *Fusarium)* because soil remains dry. However, these sub-lethal dosages of temperature also may affect the surviving and possibly weakened propagules and often render them more vulnerable to other management strategies (Lifshitz *et al.,* 1983; Freeman and Katan, 1988). If any control measure is integrated after excerting a weakening effect, it is expected that relatively less energy, time or concentration of amendment would be required to achieve maximum control of viable pathogenic population. 'Hit the iron when it is hot' principle required scientific testing in this case.

Field experiments, therefore, were conducted in crop-free period to study combined effects of sub-lethal heating, cruciferous amendments and summer irrigation in controlling *Macrophomina* and *Fusarium* (Lodha and Mawar, 2000). The experiment related to *M. phaseolina* was arranged in four sets (A-D). Three sets (A-C) of pits (30x30 cm) were dug 1 cm apart in an open field while pits of set D were dug under the dense canopy of trees. The soil in pits of set A was separately amended with mustard oil-cake (0.09 and 18% w/w) and mustard pod straw (18 and 36% w/w) and then refilled in pits. Non-amended pits having only *M. phaseolina* infested soil served as dry and wet controls. On April 1, one irrigation was applied to field capacity in all the pits except dry controls. On the same day, infested soil without any amendment was filled in other three sets. Soil from sets B and D was withdrawn on April 30 while from set C on May 31 and amended with same dosages of amendments except in set D, where only one dose of amendments was separately mixed. Amended soil was re-filled in the pits same day and irrigation was applied on May 1 (set B and D) and June 1 (set C). Soil temperature was recorded at 15 cm depth in one representative pit of each treatment of all the sets from second day of irrigation. Soil samples were collected at 0-30 cm depth on July 1 from all the sets.

The studies related to *F. o.* f sp. *cumini* were conducted in 1998. Preparation of infested soil and digging of pits were similar to the manner described for

M.phaseolina except that April month treatment was not kept. Four amendment combinations viz., mustard residues (MR 0.09%), MR + Mustard oil-cake (0.09 + 0.04%, 4: 1), MR + Eroca (0.09 + 0.04%, 4:1), MR + Cauliflower (0.09 + 0.04%, 4:1) along with non-amended controls were arranged in different pits in two sets (A-B). These were exposed to summer heat from April 1. In set A irrigation was given on May 1 while in set B on June 1 so as to receive 30 and 60 days of summer exposure, respectively. All other details were similar to the manner described for *M. phaseolina.*

In 1995 the maximum soil temperature at 15 cm depth ranged from 35.5-41 °C (April), 41-45 °C (May) and 41-45.1 °C (June) in cruciferous amended pits. These temperatures are remained 0.5-1.5 °C (April), 0.5-3 °C (May) and 1-4.1 °C (June) higher than non-amended irrigated pits. Under shade conditions, soil temperature ranged from 30-37.5 °C in amended as well as non-amended pits. In 1998, maximum soil temperatures ranged from 38-45 °C (June) in residue amended pits, which were 1-4 °C higher than corresponding non-amended wet controls.

Amendment of soil with mustard oil-cake and pod straw at both the dose were significantly superior in reducing *M. phaseolina* population than control in April, Maysor June. The magnitude of reduction, however, increased with increasing days of summer exposure and was more at 0-15 cm compared to 16-30 cm soil depth. Mere exposure to dry heat for 90 days reduced only 10% sclerotia of *M. phaseolina.* Summer irrigation increased this reduction by 27-35% in April to 35-42% in June. Amendment of soil with mustard pod straw improved this reduction by 61-68% in April to 89-94% in June. Mustard oil cake was more effective in reducing viable propagules of *M. phaseolina* compared to pod straw. Thus, it further augmented the reduction and almost 94-96% reduction in *M. phaseolina* propagules occurred when soil was exposed to summer heat for 60 days before amendment.

Weakening effect exerted by summer heat for action of cruciferous residues on disinfestations of soil from *Fusarium* was also evident in second experiment. The reduction in viable propagules of *Fusarium* in all residue amended pits ranged from 69- 80.2% in set A where irrigation was given in May compared to 75-86.5% in set B which was irrigated in June. Thus merely by 30 days more exposure to dry summer heat improved efficiency amendments by increasing the reduction in *Fusarium* propagules by 6.8 to 8.7%.

More than one mechanism might have operated concurrently or in a sequence in eliminating viable propagules of *M. phaseolina* and *Fusarium* from soil. Sub-lethal heating in dry soil for 90 days (April 1-June 30) though reduced only a small number of viable propagules of both the soil-borne pathogens but exerted a weakening effect on the surviving propagules. The weakening effect depends on temperature level, exposure time and the environment into which the preheated propagules are introduced. However, a certain threshold of heating has to be reached to obtain a detectable weakening effect (Freeman and Katan, 1988).

Accelerated decomposition of cruciferous residues in moist soil at high temperature in our study subsequent to pronounced weakening effect facilitated action of sulphur containing toxic volatiles and microbial antagonism. Enhanced decomposition with increased temperature was tenable by greater difference in temperature of amended soil in May and June compared to merely irrigated pits. Further, presence of more residual soil moisture in amended compared to non-amended pits encouraged microbial antagonism against remaining weakened propagules. Increased colonization of heat treated sclerotia of *Sclerotium rolfsii* by bacteria has also been reported by Lifshitz *et al.,* (1983).

Our studies demonstrate that prior weakening of infective propagules of *Macrophomina* and *Fusarium* by dry summer heat can improve the efficiency of cruciferous residues in disinfesting soil in hot arid region.

RESIDENT ANTAGONISTS AGAINST *M. PHASEOLINA*

During frequent tests for antagonistic activity of bacteria and actinomycetes from cruciferous residue amended soil, a soil bacterium highly antagonistic to *M. phaseolina* was isolated. Dual culture tests were performed on Czapeck's dox agar to confirm its antagonistic activity. Subsequently, agar discs cut from pigmented zone (Without bacterium) were also tested for its activity against *M. phaseolina.* Separate dual culture tests were also performed with fungi found responsible for decomposition of residues in soil, *Trichoderma harzianum* and *F. o.* f sp. *cumini.*

The bacterium was identified as a species of *Bacillus.* Dual culture tests for confirming its antagonistic activity against *M. phaseolina* developed an inhibition zone between both the microbes. This zone had scarlet pigmentation. Agar discs cut from pigmented zone, when transferred to fresh *M. phaseolina* seeded plates again developed inhibition zone. This indicated that some antibiotics are released from *Bacillus* spp that inhibits the growth of *M. phaseolina.* In separate dual culture tests performed against *Aspergillus terreus, A. niger, A. fumigatus,* no inhibitory zone or scarlet pigmentation was observed in the presence of *Bacillus.* Similarly mycelial growth of *Trichoderma harzianum* was also not inhibited suggesting their possible use in combination against *M. phaseolina.*

Characteristic feature of growing well on a wide range of temperature (22-45 °C) is certainly beneficial for arid soil where temperature goes very high during different periods of the year. Broadbent *et al.*, (1971) also reported that the percentage of *Bacillus* isolates antagonistic to plant pathogens increased with the temperature of soil upto 60 °C for 10 min., and then declined. Presence of certain compounds in crucifers and release of volatiles during hydrolysis could be attributed for stimulating multiplication of *Bacillus*. Nutrients from decomposing residue and certain compounds such as ethylene, ammonia, acetone, ethanol, methanol, formaldehyde and acetaldehyde were reported to stimulate germination of microorganisms (Harman *et al.*, 1980).

INTEGRATION OF ON-FARM WASTES AND BIO-CONTROL AGENT

After working out potential value of cruciferous residues, sub-lethal heating on survival of both the pathogens and antagonistic effect of bacterium against *Macrophomina,* next step was to study effects of these strategies in an integrated manner on both the diseases. A field experiment was, therefore, conducted (1998-2000) to study combined effects of *Brassica* amendments and summer irrigation on survival of *M. phaseolina* and *Fusarium* in soil, and dry root rot intensity on clusterbean (July-Oct) and wilt of cumin (Nov-Mar) in the same field.

Two *Brassica* amendments, mustard oil-cake and pod straw (2.5 ton ha^{-1}) were amended in pathogenic infested field either in the end of May with one summer irrigation (MC+SI, MR+SI) or in July (MC+NS, MR+NS). In case of dry root rot of clusterbean, antagonistic bacterium was also integrated as seed treatment (MR+SI+ST). Summer irrigation, seed treatment and normal sowing (SI, ST, NS) were kept as corresponding controls.

Results of field study have shown that both the amendments combined with one summer irrigation were significantly superior in reducing incidence of dry root rot and wilt on clusterbean and cumin, respectively (Mawar and Lodha, 2002). The soil temperature of amended soil after one summer irrigation in June ranged from 38-44 °C at 15 cm depth. These temperatures were 0.5-5 °C higher than those recorded in unamended soil for the same period and 6-16 °C higher when amendments were incorporated in July. Merely one summer irrigation in unamended plots reduced dry root rot severity significantly by 34-39% compared with the u1namended control without irrigation (Table 12.3). *Bacillus* coated seeds also reduced mortality, which was significantly lower than SI alone. A single irrigation in summer further augmented the efficiency of cruciferous residues to control dry root rot in both the years. Of the residues, mustard oil-cake was 38% more effective than mustard residues in reducing disease severity. However, planting of *Bacillus* coated seeds in combination with MR+SI was significantly better than MR+NS.

Maximum wilt incidence (26.8%) was recorded in the unamended control plots (Table 12.4). A single summer irrigation, or soil amendment with residues either combined with SI or applied in July at the time of clusterbean planting significantly reduced wilt incidence. Amending the soil with the MR or MC in summer or with mustard oil-cake in July significantly reduced wilt incidence as compared with SI alone. Of the residues, mustard oil-cake was significantly more effective than mustard pod straw with 34% greater reduction in wilt incidence, but differences relating to time of application were not significant (Table 12.4).

In an earlier study, it was found that amending the soil with urea-N and farm-yard manure improved the effectiveness of summer irrigation in controlling *M. phaseolina* and *Fusarium oxysporum* (Lodha, 1995). Substituting *Brassica* residues for these amendments in the present study further reduced the pathogenic agents. This reduction may be a cumulative effect of bio-toxic volatile compounds released during the decomposition of the residues at prevalent high soil temperatures (38-12 °C) and subsequent microbial antagonism. Amendment of the soil with mustard oil-cake caused a significant reduction in populations of *M. phaseolina* and *Fusarium* (Sharma *et al.*, 1995; Lodha *et al.*, 1997). A lower dose of this amendment in the present field experiment was almost equally effective; probably because the prolonged exposure of infested soil to dry heat exerted a weakening effect on pathogenic propagules (Lodha and Mawar, 2000).

Statistically significant control of dry root rot merely by coating clusterbean seeds with the *Bacillus* sp. is also of practical value in low-input sustainable agriculture (LISA). Greater control of dry root rot with this *Bacillus* in the 1998 crop season was probably due to adequate soil moisture enhancing bacterial antagonism (Dhingra and Sinclair, 1975; Lodha, 1996). The bacterial activity decreased with decreasing soil moisture (Griffin and Quail, 1968). There was no significant improvement in disease control when *Brassica* residues were combined with the *Bacillus* coating of the seeds, which was most likely due to concurrent operation of two different management strategies. However, coating will help in establishing the *Bacillus* sp. in amended soil for inducing suppressiveness.

The ability of mustard pod residues to control pathogens and diseases, demonstrated in this study, is important in resource-deficient farming since this source of *Brassica* residue is cheaper and more readily available than oil-cake. Wheat-mustard- cumin rotation in the winter seasons ensures that *Brassica* residues are available in March.

ON-FARM WASTES FOR MANAGING *M. PHASEOLINA* IN RAINFED AGRICULTURE

a. Influence of moisture conservation techniques (MCTs)

Soil water stress to host plant has been considered to be an important predisposing factor for the development of *Macrophomina* induced diseases (Ghaffar and Erwin, 1969). Studies have indicated a relationship between soil moisture, plant density, soil fertility and disease incidence (Blanco-Lopez and Jimenez-Diaz, 1983; Ghaffar *et al.*, 1969; Patil *et al.*, 1982). A field experiment was carried out to study effect of moisture conservation techniques in influencing the population of *M. phaseolina* and dry root rot intensity on clusterbean (Lodha, 1996).

Studies revealed that mulching with a layer of pearl millet stover (3.5 ton ha^{-1}), farm yard manure (10 ton ha^{-1}) and low plant population (1.6 ton ha^{-1}), alone or in combination effectively conserved available soil moisture of varying levels during different stages of crop growth. This resulted in increased population of resident bacteria with a corresponding decrease in sclerotial counts of *M.phaseolina.* Correlations of soil moisture with total bacteria were positive but these factors were negatively correlated with *M. phaseolina.* In path coefficient analysis, the highest negative indirect effects of soil moisture modified the behaviour of antagonistic microorganisms. Enhanced soil moisture and a decrease in *M. phaseolina* population significantly reduced the dry root rot intensity and increased seed yield of clusterbean (Lodha, 1996).

b. Compost as soil amendment

The incorporation of on-farm wastes (crop and weed residues) changes the soil characteristics in many ways including enrichment of nutrients and improvements in microbial activity, water holding capacity and permeability. However, low moisture content, small microbial populations and high temperature prevailing in arid regions prolongs the process of decomposition of crop residues when incorporated in a nutrient deficient sandy soil. Another disadvantage of direct incorporation is that crop residues may carry plant pathogens (Hoitink and Fahy, 1986). Composting is a method to inactivate pathogens and decompose crop residues more rapidly. Most pathogens are inactivated during the heat phase of composting (Bollen *et al.*, 1989; Ylirnaki *et al.*, 1983). Since such information was not available for *M. phaseolina,* a heat-tolerant pathogen studies were initiated. Composts have been used with various levels of success for suppression of several soil borne pathogens

(Hoitink and Fahy, 1986; Ben Yephet and Nelson, 1999). Therefore, studies were also initiated on effect of composts on disease suppressive characteristics (Lodha *et al.*, 2002).

Fully dried residues of pearl millet, clusterbean, cauliflower and a mixture of off- season weeds were used for the preparation of composts. The process of composting was initiated under partially anaerobic conditions in separate pits according to the principles of the Indore type. A 5 cm cowdung layer was spread over the base of all the pits. In each pit, a 30 cm layer was filled with 40 kg of residues enriched with 400 g gypsum (1%) and 800 g urea (2%). Each layer was provided with 60% moisture (w/v) and than covered with a 10 cm thick layer containing a mixture of 60 kg cowdung and 100 kg field soil. Four such layers of residues with soil-dung mixture were piled in each pit. These were finally covered with a 5 cm layer of weeds. Loss of moisture was replenished by adding water at an interval of 10-15 days. C:N ratio was estimated of all the residues, cowdung and soil.

In clusterbean and pearl millet residue amended compost pits, several samples (20 g) each containing 2 g pieces (0.5-1.5 cm) of *M. phaseolina* infected roots and 18 g uninfected (apparently healthy) residues of clusterbean, enriched with 2 or 4% urea-N, were separately placed in small nylon pouches. These were tied with long nylon strings and buried at 30 and 60 cm depth. Temperatures at both depths were recorded every day at 14 and 16 h for 6 weeks. Pouches were retrieved after 2 and 4 months with a minimum disturbance from pits. These samples were air-dried before estimating viable propagules of *M. phaseolina.* After 4 months, all the composting material of each pit was given three turnings at an interval of 15 days.

Density of viable sclerotia of *M. phaseolina* was significantly lower in the partially decomposed clusterbean residue samples retrieved from 60 cm depth compared to those from 30 cm depth in both the pits. The counts of *M. phaseolina* were significantly lower in the residues enriched with 4% compared to 2% urea. No viable propagules of *M. phaseolina* was reduced drastically in all the samples retrieved after 4 months with similar trend for all the treatments except that differences were not significant in the samples enriched with urea-N at 30 cm depth from pearl millet pit.

The process of composting was completed in all the pits in a period of 6 months. Maximum temperatures observed were 48-51°C at 30 cm and 60-62°C at 60 cm depth after 9-13 days and 14-18 days, respectively, after filling the compost pits. The heat phase lasted 3 to 4 weeks. A reduction of 28 to 35% of the initial volume was observed in compost pits with maximum being in the cauliflower pit. In general, considerable decrease in C:N ratio was estimated in all the composts.

Composts retrieved from pits in May were separately spread over soil surface as a 10-12 cm thick layer in an open field during hot summer days of June and were moistened once at 10% (W/W) with water. Temperatures at 5 cm depth were recorded every day at 14 h. After seven days of exposure, three sub-samples (50 g) were collected from each compost for estimating the density of *M. phaseolina* propagules.

During a 7 day period of natural heating, maximum air temperature ranged from 39-47°C. Moistening the composts with water initially brought down the maximal soil temperature to 3°C but a gradual increase of 53°C at 5 cm depth followed within 2-3 days. As a result, 53-61% reduction in counts of *M.phaseolina* was estimated in different composts.

Suppressive effect of composts on dry root rot of clusterbean was studied during 1994 and 1995. After summer exposure 4 week old composts of pearl millet, clusterbean, weeds and cauliflower residues (each 4 t ha^{-1}) were incorporated to a depth of 0-30 cm. Data on plant mortality due to dry root rot and seed yield of clusterbean were recorded. Populations of *M.phaseolina,* total bacteria, total fungi, total actinomycetes and *Nitrosomonas* at 0-30 cm depth were estimated following standard procedures after the harvest of clusterbean crop. Lytic bacterial density and antagonistic actinomycetes were also estimated (Greenberger *et al.*, 1987; Ghaffar *et al.*, 1969).

During both years of field experiments, disease severity due to dry root rot was significantly reduced and seed yield of clusterbean was significantly enhanced by soil amendment with the different composts (Table 12.5). Among composts, the lowest disease severity was recorded with pearl millet compost except in 1995 where disease suppression was not significantly different from that obtained in the presence of weed compost. However, seed yield was the highest in plots amended with cauliflower compost. The increase in seed yield in amended treatments over control was more conspicuous in a normal rainfall year 1994 (28.1-52.3%) than in a low rainfall year such as 1995 (28.9-38.4%).

Amendment of soil with compost, in general, increased the population of antagonistic actinomycetes, lytic bacteria and *Nitrosomonas* and decreased the population of *M. phaseolina* in soil. Among composts, increase in antagonistic actinomycetes was significantly higher in cauliflower compost amended soil that also had maximum reduction of *M. phaseolina* density. In both years, the population of *Nitrosomonas* spp. was also significantly higher in cauliflower compost followed by pearl millet compost amended soil compared to the other two composts.

Release of *M. phaseolina* inoculum from infected plant residues was significantly reduced by organic composting. Many factors may be involved in reducing population of *M. phaseolina* during composting (i) heat generated in the first phase (Bollen *et al.*, 1989), (ii) toxicity of conversion products formed during or after the self-heating process (Berestetsky and Kravchenko, 1984), and (iii) microbial antagonism in presence of moisture (Dhingra and Sinclair, 1975). Amendment of soil with composts showing a reduced density of *M. phaseolina* lead to a significant decrease of dry root rot severity and a significant increase of seed yield of clusterbean both in normal and low rainfall conditions.

In the present study low survival of *M. phaseolina* propagules, particularly at 60 cm could be attributed to elevated temperatures during the heat phase that reached values reported to be lethal (60 °C for 3 seconds) for sclerotia of *M. phaseolina* (Bega and Smith, 1962). Further reduction in viable sclerotia of *M. phaseolina* after the heat phase could be a result of combined effects of fungitoxic compounds and microbial antagonism in the presence of moisture and nitrogen. Several workers have demonstrated production of fungitoxic volatile compound during decomposition of crop residues (Berestetsky and Kravchenko, 1984; Spring *et al.*, 1980).

The higher reduction in viable propagules of *M. phaseolina* in the decomposing tissues enriched with 4% urea-N than from 2% urea-N recorded in the present study could result from a higher concentration of nitrogen. Filho and Dhingra (1980) have also reported that higher nitrogen levels hasten the decomposition of plant residues accompanied by an increase in microbial populations that in turn caused reduction in *M. phaseolina* counts. Germination of sclerotia of *M. phaseolina* has been found to be stimulated by amendments (Smith, 1969). The resulting germ-tube and hyphae are known to be sensitive to bacteria and actinomycetes leading to lysis of fungal cell walls (Kovoor, 1954).

Our effort to expose moistened composts to prevailing high temperatures was found highly effective in further reducing the viable propagules of *M. phaseolina.* Sub- lethal temperatures (48-53 °C) exerted a weakening effect on remaining sclerotia of *M. phaseolina,* which might have accelerated microbial antagonism.

Organic amendment may increase, decrease or not affect diseases caused by soil- borne pathogens (Hoitink and Fahy, 1986; Huber and Watson, 1970). In the present study, disease suppression due to addition of composts into the soil could possibly be ascribed to the following factors: (i) composts support high levels of total microbial population including antagonists and (ii) composts improves the moisture holding of the soil, which in turn could reduce the pathogenic propagules and disease.

Increased densities of bacteria and actinomycetes along with antagonists by amendment of compost in the field study could be ascribed to its blending with field soil in each layer of residues during composting. Their population multiplied rapidly from initial 1.6 x 10^5 g^{-1} soil to 1.2 x 10^6 g^{-1} compost at high temperature attained during heat phase, more so, in the presence of nutrients like urea. No *Bacillus* spp. that can withstand peak heating may induce biological control in composts. One *Bacillus* strain was found to inhibit growth *of M. phaseolina* in separate *in vitro* tests. Phae *et al.,* (1990) also isolated a strain of *Bacillus subtilis* from compost that induced suppressive effect on phytopathogenic microorganisms. Amended soil held more water than the non-amended sandy soil, which in turn reduced *M. phaseolina* population or its infection on the host plant due to enhanced antagonism or competition for site (Filho and Dhingra, 1980: Sheikh and Ghaffar, 1979). Soil moisture is the most critical factor in determining *M. phaseolina* infection besides increasing antagonism particularly by bacteria (Dhingra and Sinclair, 1975; Lodha, 1996). Low disease levels in spite of substantial inoculum density (253-510 sclerotia g^{-1} soil) may be due to suppressive effects of the compost. Lewis *et al.* (1992) also observed induction of suppression without reduction in population of *Pythium ultimum* in compost amended soils. Mechanisms involved in the suppressiveness of amended soils described here could result from enhanced lytic bacterial activity and antagonism by actinomycetes, microbial competition and increased availability of soil moisture. These hypotheses, however, remains to be further evaluated.

These results showed greater potentials of cauliflower compost amendment to reduce *M. phaseolina* density compared to other composts, in which increased antagonist density brought proportionate reduction in dry root rot incidence. Beneficial effects of incorporating cruciferous residues on growth and yield of watermelon and wheat due to disease suppression and increase in microbes beneficial to plant growth has been well documented (Keinath, 1996; Kirkegaard *et al.,* 1994). Increased seed yield in all the compost-amended treatments may be a cumulative effect of reduced disease incidence, more retention of soil moisture, availability of nutrients and qualitative and quantitative improvements in microbiological properties. Improvement in the population of *Nitrosomonas* in amended soil in present study is an indication of mineralization of N from compost (Hadas *et al.,* 1996). *Nitrosomonas* oxidizes ammonium to nitrite, which in turn is converted into nitrate by *Nitrobacter.* The *Nitrosomonas* population is more stable in soil throughout the year than of *Nitrobacter* (Skujins and Fulgham, 1978), even in dry soil (Allison and Prosser, 1991). Increased microbial populations particularly those of antagonistic actinomycetes and lytic bacteria against *M. phaseolina* in compost amended soil probably lead to long-term beneficial effects. The high microbial activity and biomass caused by the "general soil microflora" in compost amended soil prevents germination of pathogenic propagules and infection of the host, presumably through microbiostasis (Hoitink and Boehm, 1999).

This study demonstrated that after sanitation of crop residues from *M. phaseolina,* amendment of compost in nutrient deficient sandy soil reduced dry root severity, improved microbial properties of soil and seed yield. In resource deficient farming of arid region, the beneficial effect of compost as an integral part of low-input sustainable agriculture can also be a practical way of managing soil-borne pathogens.

In a subsequent field study, efficacy of three composts prepared from residues of pearl millet, neem and weeds was ascertained on dry root rot caused by *M. phaseolina* nitrogen fixation and seed yield of clusterbean and cowpea (Lodha and Burman, 2000). Amendment of soil with pearl millet and weed composts for two consecutive years reduced resident population of *M. phaseolina* by 20-40% compared to non-amended plots. Associated increase in antagonistic actinomycetes (28-50%) and lytic bacteria (60- 70%) accounted for 20-60% reduction in plant mortality due to dry root rot in both the crops in amended plots. This resulted in higher seed yield of clusterbean (23-50%) and cowpea (13-17%). Further, activity of nitrogenase was maximum when both the crops were grown in amended plots. The number of nodules on crops grown did not corroborate with their nitrogenase activity. Moreover, neem compost did not benefit nitrogenase activity, which can be attributed to release of toxic compounds affecting the ability of *Rhizobium* to form effective nodules.

CONCLUSION

Results of these studies conclusively demonstrated that combining sub-lethal heating, cruciferous residues particularly mustard pod straw (2.5 ton ha^{-1}) and one summer irrigation were effective in reducing pathogens, *Macrophomina phaseolina* and *Fusarium oxysporum.* Concerted research efforts carried out at CAZRI has led to development of this eco-friendly, viable and cost-effective technology. Generally, farmers are adopting a wheat-mustard-cumin rotation. Mustard residues are, thus, available in March-April. This technology requires application of these residues in May end or early June during intense solar irradiations and high temperatures. This is a crop-free period. More so, sufficient time remains available for release of toxic volatiles before succeeding crop is grown in rainy season. Amendment of soil with these residues should be done after adequate heating of soil so as to get maximum benefit of weakening effect on chlamydospores/sclerotia of these

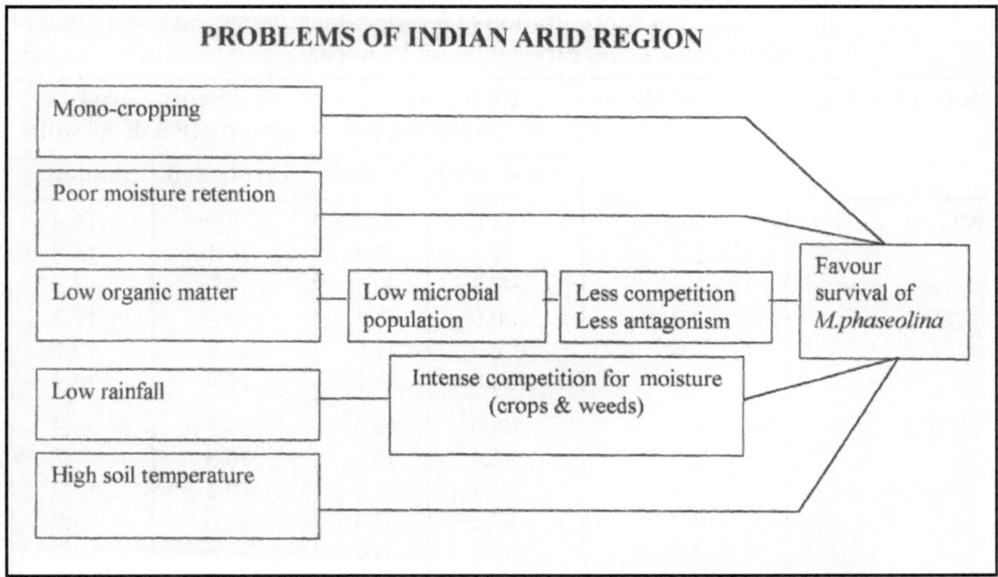

FIG. 12.1 Factors influencing survival of *M. Phaesolina* in arid soil

TABLE 12.1. Effect of soil amendments of population of *Macrophomina phaseolina* (M), *Fusarium oxysporum* f.sp. *cumini* (F) and the antagonistic actinomycetes population g^{-1} soil after 45 and 60 days.

Amendments[a]	***M.phasolina***	***Fusarium oxysporum***	**Antagonistic actinomycetes**	
			M(x 10⁶)	**F(x 10⁶)**
N+PMR	150	1.3	4.3	4.3
N+PMR+MC	0	8.3	23.3	10.6
N+PMR+CC	366	2.0	38.3	14.5
MC	0b	0.0b	27.3	11.0
CC	200	1.3	24.0	6.6
None	1132	1.7	2.3	0.3
Initial population	2753	0.3	0.9	0.3

[a]N - Nitrogen (40 kg ha^{-1} as urea; PMR-pearl millet; MC-mustard cake (1%); CC-castor cake (%) and [b]same as incubation for 30 days.

TABLE 12.2. Effect of different concentration of *brassica* amendments on population of *Macrophomina phaseolina* propagules under unshaded and shaded condition.

Amendments[a]	Concentration (ha^{-1})	Reduction in *M.phaseolina* (%)		Dehydrogenase activity (P kat[c] g^{-1} soil)	
		Un shaded	Shaded	Un shaded	Shaded
MC+SI	0.5	88.8	27.6	15.9	15.8
	1.0	91.6	38.3	16.1	16.5
	1.5	97.2	39.1	18.3	17.3
	2.0	100.0	43.1	19.5	17.7
	2.5	100.0	47.9	16.7	19.2
	3.0	100.0	59.7	16.9	16.9
MR+SI	1.0	88.0	40.4	14.3	16.2
	2.0	94.4	45.6	16.5	17.7
	3.0	94.4	52.3	18.1	17.9
	4.0	97.2	52.4	19.8	18.5
	5.0	100.0	59.7	20.6	17.7
	6.0	100.0	63.8	17.8	17.6
SI	-	52.7	15.2	13.5	15.0
DS	-	12.3	8.8	10.8	13.3
LSD (P-0.05)		15.5	6.3	1.3	1.2

[A]MC-Mustard oil-cake, MR-Mustard residue,
SI -Summer irrigation, DS-Dry soil
[b]Under dense canopy of trees

TABLE 12.3. Effect of cruciferous residue as soil amendments on population of *Macrophomina phaseolina* and clusterbean mortality due to dry root rot on clusterbean

Treatments	Mortality (%)		*M.phaseolina* population[b] (sclerotia g^{-1} soil)	
	1998	1999	1998	1999
MC+SI	4.7	5.1	180	120
MC+NS	9.8	8.7	220	168
MR+SI	7.2	6.5	206	148
MR+NS	10.3	10.7	248	188
MR+SI+ST	6.8	5.4	200	136
MR+NS+ST	7.2	10.6	240	170
SI	11.3	13.9	284	220

Contd...

Table 12.3 Contd...

ST	7.9	11.6	352	388
Control (unamended)	18.8	21.3	385	432

[a]MC – Mustard oil-cake (2.5 t/ha); MR-Mustard residues (2.5 t/ha); NS, amendment at the time of clusterbean sowing in July; SI, Summer irrigation in June; ST, Seed coating with *Bacillus spp.*
[b]Initial population was 340 sclerotia g^{-1} soil

TABLE 12.4. Effect of cruciferous residue as soil amendments on populations of *Fusarium oxysporum* f.sp. *cumini* and wilt incidence on cumin

Treatments	**Mortality (%)**		***M.phaseolina* population[b] (sclerotia g^{-1} soil)**	
	1998	**1999**	**1998**	**1999**
MC+SI	5.2	7.2	8.5	4.2
MC+NS	12.1	10.2	11.2	8.9
MR+SI	8.6	7.8	10.4	4.9
MR+NS	11.4	12.7	12.5	9.3
SI	17.8	15.2	13.2	13.8
Control (unamended)	23.3	26.8	18.6	21.4

[a]MC – Mustard oil-cake (2.5 t/ha); MR-Mustard residues (2.5 t/ha); NS, amendment at the time of clusterbean sowing in July; SI, Summer irrigation in June.
[b]Initial population was 17.5 x 10^{-3} cfu g^{-1}

TABLE 12.5. Effect of different compost on dry root rot mortality induced by *M.phaseolina* and seed yield of cluster bean

Compost amended soil*	**Mortality (%)**		**Seed yield (kg/ha)**	
	1994	**1995**	**1994**	**1995**
Pearl millet	2.0	4.4	508.3	296.8
Cluster bean	3.0	6.3	572.7	299.5
Weeds	4.2	6.8	585.7	311.4
Cauliflower leaves	2.2	6.0	604.1	318.7
Control	5.4	16.0	396.5	230.2
LSD (P<0.05)	1.2	5.5	54.5	45.5

* Soil was amended with 4 t compost per hectare

pathogens. Once these volatiles particularly isothiocyanates are released under prevailing temperatures, decomposed material will serve as an amendment to enrich nutrient deficient sandy soils.

The State of Rajasthan ranks first in the country with 2.2 million hectares under mustard cultivation, of which 6 lakh hectares is in hot arid region. Besides, other crucifers like *Eruca,* cauliflower and cabbage are also under cultivation in more than 0.5 Lac hectares. Since these residues are generally not fed to cattle and livestock, their proper disposal is a serious problem. Their use as fuel also is not acceptable because on burning these causes irritation in human eyes.

Utilization of these on-farm wastes in rainfed and irrigated farming for cultivation of cumin, in winter season and clusterbean and other legumes in rainy season may increase seed yield by 10-15%. Simultaneously in rainfed agriculture, composts prepared from different types of on-farm wastes can be used for suppression of soil-borne pathogens besides improving soil-fertility and moisture holding capacity of sandy soils. Enhancement in the population of antagonistic is additional advantage for inducing suppressiveness.

REFERENCES

Abdel Rahirn, M. F., Satour, M. M., Mickeail, K. Y. and Elbraki, S. A. 1988. Effectiveness of soil solarization in furrow-irrigated Egyptian soil. *Plant Dis.* 72: 143-146.

Allison, S. M. and Prosser, J. I. 1991. Survival of ammonia oxidising bacteria in air-dried soil. *FEMS- Microb. Ltr.* 79: 65-68.

Angus, J. F., Gardner, P. A., Kirkegaard, J. A. and Deshmarchelier, J. M. 1994. Biofumigation : iso-thiocyanates release from brassica root inhibit growth of the take all fungus. *Plant Soil.* 162: 107-112.

Bega, R. V. and Smith, R. S. 1962. Time-temperature relationships in thermal inactivation of sclerotia of *Macrophomina phaseolina. Phytopath.* 52: 632-635.

Ben Yephet Y and Nelson E. B. 1999. Differential suppression of damping-off caused by *Pythium aphanidermatum, P. irregulare* and *P. myriotylum* in composts at different temperatures. *Plant Dis.* 83: 356-360.

Berestetsky, O. A. and Kravchenko, L. V. 1984. Volatile products of plant residue decomposition and their effect on soil microflora. *(In:)* Jszegi (Ed.). *Soil biology and conservation of the biosphere* Vol. I, Academiai Kaido, Budapest. pp: 419-425.

Blanco-Lopez, M. A. and Jimenez-Diaz, R. M. 1983. Effect of irrigation on susceptibility of sunflower to *Macrophomina phaseolina. Plant Dis.* 67: 1214-1217.

Bollen, G. J., Volker, D. and Wijnen, A. P. 1989. Inactivation of soil-borne plant pathogens during small scale composting of crop residues. *Neth.J. Pl. Path.* 95, Supplement 1: 19-30.

Broadbent, P., Baker, K. F. and Waterworth, Y. 1971. Bacteria and actinomycetes antagonistic to fungal rot pathogens in Australian soils. *Australian J. Biol. Sci.* 24: 925-944.

Brown, P. D. and Morra, M. J. 1997. Control of soil borne plant pests using glucosinolates containing plants. *Adv. Agron.* 61: 167-231.

Dhingra, O. D. and Sinclair, J. B. 1974. Effect of soil moisture and carbon:nitrogen ratios on survival of *Macrophomina phaseolina* in soybean stems in soil. *Plant Dis. Reptr.* 58: 1034-1035.

Dhingra, O. D. and Sinclair, J. B. 1975. Survival of *Macrophomina phaseolina* sclerotia in soil: Effects of soil moisture, carbon:nitrogen ratios, carbon sources and nitrogen concentrations. *Phytopath.* 65: 236-240.

Filho, E. S. and Dhingra, O. D. 1980. Population changes of *Macrophomina phaseolina* in amended soils. *Tranc. British Mycol. Sco.* 74: 471-481.

Freeman, S. and Katan, J. 1988. Weakening effect on propagules of *Fusarium* by sub- lethal heating. *Phytopathol.* 78: 1656-1661.

Garnliel, A. and Stapleton, J. J. 1993. Characterization of antifungal volatile compounds evolved from solarized soil amended with cabbage residue. *Phytopath.* 83: 899-905.

Ghaffar, A. and Erwin, D. C. 1969. Effect of soil water stress on root rot of cotton caused by *Macrophomina phaseolina. Phytopath.* 59: 795-797.

Ghaffar, A. Zentmeyer, G. A. and Erwin, D. C. 1969. Effect of organic amendments on severity of *Macrophomina phaseolina* root rot of cotton. *Phytopath.* 59: 1267-1269.

Greenberger, A. Yogev, A. and Katan, J. 1987. Induced suppressiveness in solarized soils. *Phytopath.* 77: 1663-1667.

Griffin, D. M. and Quail, G. 1968. Movement of bacteria in moist particulate systems. *Australian J. Biol. Sci.* 21: 579-582.

Hadas, A. Kautsky, L. and Portnoy, R. 1996. Mineralization of composted manure and microbial dynamics in soil as affected by long- term nitrogen management. *Soil Biol. and Biochem.* 28: 733-738.

Harman, G. E., Mattick, L. R., Nash, G. and Nedrow, B. L. 1980. Stimulation of fungal spore germination and inhibition of sporulation in fungal vegetative thalli by fatty acids and their volatile per-oxidation products. *Can .J. Bot.* 58: 1541-1547.

Hointink, H. A. J. and Boehm, M. J. 1999. Biocontrol within the context of soil microbial communities: A substrate-dependent phenomenon. *Ann. Rev.Phytopath.* 37: 427-446.

Hoitink, H. A. J. and Fahy, P. C. 1986. Basis for the control of soil-borne plant pathogens with composts. *Ann. Rev. PlantPath.* 24: 93-114.

Huber, D. M. and Watson, R. D. 1970. Effect of organic amendments on soil-borne plant pathogens. *Phytopath.* 60: 22-26.

Katan, J., Greenberger, A. H. and Grinstein, A. 1976. Solar heating by polyethylene mulching for the control of diseases caused by soil-borne pathogens. *Phytopath.* 66: 683-688.

Keinath, A. P. 1996. Soil amendment with cabbage residue and crop rotation to reduce gummy stem-blight and increase growth and yield of watermelon. *Plant Dis.* 80: 564-570.

Kirkegaard, J. A., Gardner, P. A., Angus, J. F. and Koetz, E. 1994. Effect of bras sica crops on the growth and yield of wheat. *Australian J. Agril. Res.* 45: 529-545.

Kovoor, A. T.A. 1954. Some factors affecting the growth of *Rhizoctonia bataticola* in soil. *J. Madras Uni.* 24: 47-52.

Lewis, J. A. and Papavizas, G. C. 1970. Evolution of volatile sulfur containing compounds from decomposition of crucifers in soil. *Soil. Biol. Biochem.* 2: 239- 246.

Lewis, J. A. and Papavizas, G. C. 1971. Effect of sulfur containing volatile compounds and vapors from cabbage decomposition on *Aphanomyces euteiches. Phytopath.* 61: 208-214.

Lewis, J. A., Lumsden, R. D., Millner, P. D. and Keinath, A. P. 1992. Suppression of damping off peas and cotton in the field with composted sewage sludge. *Crop Protection.* 11: 260-266.

Lifshitz, R., Tabachnik, M., Katan, J. and Chet, I. 1983. The effects of sub-lethal heating on sclerotia of *Sclerotium rolfsii: Can. J. of Microb.* 29: 1607- 1610.

Lodha, S. 1995. Soil solarization, summer irrigation and amendments for the control of *Fusarium oxysporum* f sp. *cumini* and *Macrophomina phaseolina* in arid soils. *Crop Protection.* 14: 215-219.

Lodha. S, 1996. Influence of moisture conservation techniques on *Macrophomina phaseolina* population, dry root rot and yield of clusterbean. *Phytopath.* 49: 342-349.

Lodha, S. and Burman, U. 2000. Efficacy of composts on nitrogen fixation, dry root rot and yield of legumes. *Indian J. Agric. Sci.* 70: 846-849.

Lodha, S. and Mawar, R. 2000. Utilizing solar heat for enhancing efficiency of cruciferous residues for disinfesting soil-borne pathogens from arid soils. *Acta Horti.* 532: 49-52.

Lodha, S. and Solanki, K. R. 1992. Influence of solar heating on control of dry root rot *(Macrophomina phaseolina)* and weeds in arid environment. *Indian J. Agril. Sci.* 62: 838-843.

Lodha, S., Gupta, G. K. and Singh, S. 1986. Crop disease situation and some new records in Indian arid zone. *Ann. of Arid Zone,* 25: 311-320.

Lodha, S., Sharma, S. K. and Aggarwal, R. K. 2002. Inactivation of *Macrophomina phaseolina* propagules during composting and efficacy of composts on dry root rot and seed yield of cluster bean. *European J. Plant Path.* 108(3): 253-261.

Lodha, S., Sharma, S. K. and Aggarwal, R. K. 1997. Solarization and natural heating of irrigated soil amended with cruciferous residues for improved control of *Macrophomina phaseolina. Plant Pathol.* 46: 186-190.

Mawar, R. and Lodha, S. 2002. *Brassica* amendments and summer irrigation for the control of *Macrophomina phaseolina* and *Fusarium oxysporum* f sp. *cumini* in hot arid region. *Phytopath. Mediterr.* 41: 45-54.

Mayton, H. S., Olivier, C., Vaughan, S. F. and Loria, R. 1996. Correlation of fungicidal activity of *Brassica* species with allyl isothiocyanate production in macerated leaf tissue. *Phytopath.* 86: 267-271.

Meyer, W. A., Sinclair, J. B. and Khare, M. N. 1973. Biology of *Macrophomina phaseolina* in soil studied with selective media. *Phytopath.* 63: 613-620.

Mihail, J. D. and Alcorn, S. M. 1984. Effect of soil solarization on *Macrophomina phaseolina* and *Sclerotium rolfsii. Plant Dis.* 68: 156-159.

Mitchell, R. and Alexander, M. 1961. Chitin and the biological control of *Fusarium* diseases. *Plant Dis. Reptr.* 45: 487-490.

Papavizas, G. C. 1961. Evaluation of various media and antimicrobial agents for isolation of *Fusarium* from soil. *Phytopath.* 57: 848-852.

Patil, R. G., Deshmane, N. B. and Pandhare, T. M. 1982. Effect of plant density and row spacing on charcoal rot incidence in four cultivars of sorghum. *Sorghum Newsltr.* 25: 100.

Phae, C. G., Saski, M., Shoda, M. and Kubota, H. 1990. Characteristics of *Bacillus subtilis* isolated from composts suppressing phytopathogenic microorganisms. *Soil Sci. Plant Nutri.*36:575-586.

Ramirez Villapudua, J. and Munnecke, D. E. 1987. Control of cabbage yellow *(Fusarium oxysporum* f sp. *conglutinans)* by solar heating of field soil amended with dry cabbage residues. *Plant Dis.* 71:217-221.

Ramirez-Villapudua, J. and Munnecke, D. E. 1988. Effects of solar heating and soil amendments of cruciferous residues on *Fusarium oxysporum* f sp. *conglutinans* and other micro-organisms. *Phytopath.* 78: 289-295.

Sang, J. P., Minchinton, I. R., Johnstone, P. K. and Trustcott, R. J. W. 1984. Glucosinolate profile in the seed, root and leaf issue of cabbage, mustard and rapeseed. *Canadian J. Pl. Sci.* 64: 77-93.

Sharma, S. K., Aggarwal, R. K. and Lodha, S. 1995. Population changes of *Macrophomina phaseolina* and *Fusarium oxysporum* f sp. *cumini* in the oil cake and crop residue amended sandy soils. *App. Soil Ecol.* 2: 281-284.

Sheikh, A. H. and Ghaffar, A. 1979. Relation of sclerotial inoculum density and soil moisture to infection of field crops by *Macrophomina phaseolina. Pak. J. Bot.* 11: 185-189.

Sheikh, A. H. and Ghaffar, A. 1984. Reduction in viability of sclerotia of *Macrophomina phaseolina* with polyethylene mulching of soil. *Soil Bio. & Biochem.* 16:77-9

Singh, R. S. and Singh, N. 1970. Effect of oil-cake amendments of soil on populations of some wilt causing species of *Fusarium. Phytopath.* 69: 160-167.

Skujins, I. and Fulgham, P. T. 1978. Nitrification in great basin desert soils. (*In*:) West NE and Skujins I (Eds.) *Nitrogen in Desert Ecosystems*. pp 60- 76. Dowden, Hutchinson & Ross Inc. Stroudsburg, Pennsylavania.

Smith, W. H. 1969. Germination of *Macrophomina phaseolina* sclerotia as affected by *Pinus lambertiana* root exudate. *Can. J. Microb.* 15: 1387-1391.

Spring, D. E., Ellis, M. A., Spotts, R. A., Hoitink, H. A. I. and Schmitthenner, A. F. 1980. Suppression of the apple collar root pathogen in composted hardwood bark. *Phytopath.* 70: 1209-1212.

Walker, I. C., Morell, S. and Foster, H. H. 1937. Toxicity of mustard oil and related sulfur compounds to certain fungi. *Amer.J. Bot.* 24: 536-541.

Ylirnaki A, Toiviainen A, Kallio H and Tikanmaki E, 1983. Survival of some plant pathogens during industrial-scale composting of wastes from a food processing plant. *Ann. Agril. Fen.* 22:77-85.

•••

Advances in Soil Borne Plant Diseases, 2008
Editors: Naik and Devika Rani, pp. 311-340
New India Publishing Agency, New Delhi

13

Biological Control of Soil Borne Plant Pathogens using *Trichoderma* Species

R. D. Prasad and M. K. Naik*

INTRODUCTION

Fungi naturally present in soil and rhizosphere of various crops serve as potential biological control agents of various plant pathogens. Species of fungi occurring in the genus *Trichoderma* have gained significant importance for use as agents for control of several diseases of cultivated crops. The last decade has witnessed a tremendous break-through in the research efforts on biological control of plant diseases in India especially by using species of *Trichoderma* and *Gliocladium*. Several economically important plant diseases of a variety of crops were effectively managed by using *Trichoderma* based biofungicides.

The work on biocontrol of pigeonpea wilt by *Bacillus subtilis* by Dr. R S Vasudeva and his associates, published as a series of papers in *Annals of Applied Biology* during 1950 to 1965 can be considered as a starting point for Indian work on biological control of plant pathogens (Mukhopadhyay, 1994). Work on biological control in India really gained a momentum in the 1980s with efforts of Departments of Plant Pathology from G. B. Pant University of

Directorate of Oilseeds Research,(Indian Council of Agricutural Research) Rajendranagar, Hyderabad-500 030, Andhra Pradesh, India

* Department of Plant Pathology, University of Agricultural Sciences, College of Agriculture, Raichur-584 101, Karnataka, India

Agriculture and Technology, Pantnagar, Uttaranchal and Tamil Nadu Agricultural University, Coimbatore, Tamil Nadu. At present, there are strong centers like Project Directorate of Biological Control (ICAR), Bangalore, Karnataka engaged in plant disease biological control and in recent years, numerous publications have appeared in the literature in this field of research. Many private and government agencies started production of fungal bioagents especially from Tamil Nadu, Karnataka, Andhra Pradesh and Maharashtra at commercial level but because of recent stringent registration guidelines set by Govt. of India hindered the pace as small scale producers were not able to register their products because of high registration costs involved. Many commercial products of *Trichoderma* are now available worldwide and also in India (Table 13.1) for control of various plant diseases.

The future holds tremendous possibilities for efficient use and manipulation of *Trichoderma* and other biological agents for management of crop diseases under varied agro climatic conditions. Recently, CIB (Central Insecticide Board), Government of India has developed and fine-tuned the regulatory requirements for the mandatory registration of biological control agents. The Government needs to develop suitable mechanism by which the quality parameters of these products are monitored without imposing much restriction in their registration (like toxicological data) as it has happened in many developed countries. Various instances of successful control of plant pathogens at greenhouse and field level with special emphasis on soil borne plant pathogens by these fungi, improvement of *Trichoderma* species, and instances of systematic work leading to biofungicides development were highlighted. Finally future strategies for research and commercial development of fungal biological control agents were discussed.

TRICHODERMA SPECIES AS BIOLOGICAL CONTROL AGENTS OF VARIOUS CROP DISEASES

Majority of the work done on plant disease biological control in India relates to soil borne diseases using either fungal or bacterial antagonists. Some efforts were also made to select bioagents for foliar diseases in some places.

Biological Control of Soil Borne Diseases

Antagonistic fungi like *Trichoderma* and *Gliocladium* spp. have generated more interest among researchers as well as private entrepreneurs. One of the earlier successes was reported by Padmanabhan and Alexander (1983) - mass incorporation of *Trichoderma viride* in seedling plots protected approximately 3 lakhs sugarcane seedlings from root rot caused by *Pythium*. An isolate of *Trichoderma harzianum* effectively suppressed fusarial wilt of pigeonpea

(Dwivedi 1984). Maiti and Sen (1985) integrated *T. harzianum* with nitrogenous fertilizers for managing *Sclerotium rolfsii*. Sreenivasaprasad and Manibhushanrao (1990, 1993) achieved success in suppressing *Rhizoctonia solani, S. rolfsii* and *Pythium aphanidermatum* in cotton, groundnut and tomato through the soil and seed application of *Gliocladium virens* (=*Trichoderma virens*) and *T. longibrachiatum*. *T. harzianum* and *T. koningii* were effective in controlling *S. rolfsii* in soybean (Deb and Dutta 1991). Singh (1991) reported reduction in white mold (*Sclerotinia sclerotiorum)* of pea under the field conditions through the application of *T. harzianum*. A 30-75% reduction in the incidence of stem rot of groundnut was obtained by Asghari and Mayee (1991) through the application of *T. harzianum* and *T. longibrachiatum* under field conditions. Seed treatment with *T. viride, T. harzianum* and *Gliocladium roseum* reduced collar rot of gram (Sugha *et al.*, 1993). Baby and Manibhushanrao (1993) reported control of rice sheath blight through the integration of fungal antagonist and organic soil amendments. *G. virens* was effective in suppressing *Aspergillus flavus* in groundnut, *R. solani* and *Fusarium solani* in paddy and brinjal under greenhouse conditions (Mishra and Narain, 1994). Integration of carbendazim with *T. harzianum* and *T. viride* seed treatment resulted in reduced incidence of dry root rot and increased yield of soybean in farmer's field (Vyas, 1994). *T. harzianum, T. hamatum* and *G. virens* were found to be effective in suppressing yellows (*F. oxysporum* f. sp. *zingiberi*) of ginger under field conditions (Dohroo, 1995). Sawant and Sawant (1996) developed a simple method using coffee fruit skin composted with biogas slurry for multiplication of *T. harzianum* for controlling diseases of horticultural crops. The experiment on management of root rot (*R. solani*) and *M. phaseolina* of cotton (*Gossypium* spp.) by integrating chemical, cultural and biological control in varieties HS-6 (*G. hirsutum*), HD-107 (*G. arboreum*) and hybrid HH-81 (Intra *hirsutum* hybrid) was conducted. Seed treatment with *T. viride* reduced plant mortality due to cotton soilborne diseases by 56%. *T. viride* used first as a seed dressing, followed by soil application, reduced mortality by 86%. Integrated use of *T. viride* and carbendazim reduced plant mortality by up to 72% (Mathivanan *et al.*, 1997). Soil drenching with carbendazim followed by seed treatment with the same fungicide, intercropping with moth bean (*Vigna aconitifolia*) and soil treatment with *T. harzianum* gave highest germination and lowest mortality (Jakhar *et al.*, 1998). Farmyard manure culture of *T. harzianum* was found effective for control of *R. solani* in soyabeans in the field at Jorhat. Disease index was significantly reduced and germination, plant height, root and shoot dry weight and grain yield were significantly increased (Dutta and Das, 1999). *T. harzianum* was effective against *S. sclerotiorum* causing chickpea stem rot. Application of *T. harzianum,* a week before sowing reduced the disease considerably in the field as compared to the application at sowing time (Sharma, 1999). Biological control of *P. aphanidermatum-Meloidogyne incognita* disease complex in chilli with organic amendments (farmyard manure and neem cake) and antagonistic organisms *T. viride, T. harzianum* (antagonists against *P. aphanidermatum*) and *Paecilomyces lilacinus* (antagonist against *M. incognita*) was reported by Karthikeyan *et al.,* (1999).

Intensive research and development efforts were made on biocontrol of soilborne plant pathogens using *Trichoderma* and *Gliocladium* spp. at the G. B. Pant University of Agriculture and Technology, Department of Plant Pathology, TNAU, Coimbatore, Project Directorate of Biogical Control (PDBC), Bangalore and Indian Spices Research Institute, Calicut etc. Systamatic research efforts made at Pantnagar during eighties resulted in identification fungal bioagents like *Trichoderma* spp. and *G. virens* (Mukhopadhyay, 1987, 1996; Mukhopadhyay *et al.,* 1992) for a range of plant pathogens (*S. rolfsii, R. solani, S. sclerotiorum, Fusarium* spp. and *P. aphanidermatum*). This group was able to demonstrate their efficacy in farmers fields and now it has reached a stage when some of the isolates are in the process of commercialization. *T. harzianum* was highly effective in suppressing *S. rolfsii* in sugarbeet (Upadhyay and Mukhopadhyay 1986) and wilt and root rot pathogens in chickpea (Kaur and Mukhopadhyay, 1992) under field conditions. Very high degree of protection in many crops against a range of pathogens through seed treatment with spores of *G. virens* was achieved (Mukhopadhyay *et al.,* 1992; De and Mukhopadhyay, 1994). The ability of *G. virens* to parasitize *R. solani in situ* has been demonstrated (Mukherjee and Mukhopadhyay, 1993). *G. virens* has been tested extensively in the farmers' fields and is being developed into a commercial formulation product.

Research efforts made at Department of Plant Pathology, Tamil Nadu Agricultural University (TNAU), Coimbatore in late eighties and early nineties resulted in popularization of *Trichoderma* based biofungicides in southern states. Many commercial firms have started producing these bioagents in large scale in Tamil Nadu, Karnataka and Andhra Pradesh. A talc-based seed treatment formulation of *T. viride* developed at TNAU has been used extensively by farmers of Tamil Nadu for protecting several crops from *Macrophomina phaseolina* infection (Jeyarajan *et al.* 1993). Studies at same center revealed that field application of *T. viride* was effective in suppressing *Macrophomina* root rot of soybean (Raghuchander *et al.* 1993). Seed pelleting with *T. harzianum, Rhizobium* sp. and carbendazim was found to be highly effective in managing sclerotial root rot of groundnut (Muthamilan and Jeyarajan, 1996). Seed treatment with *Trichoderma* spp. controlled *M. phaseolina* in sesame under field conditions (Sankar and Jeyarajan 1996a, b).

At Indian Institute of Spices Research (IISR), Calicut, Kerala considerable work has been carried out to select and demonstrate the efficacy of fungal and bacterial biological control agents against wilt diseases in pepper and other spice crops diseases (Table 13.2).

Biological Control of Foliar Fungal Pathogens in India

The research on biological control of foliar plant pathogens in India is still in infancy. In countries like USA and Israel *Trichoderma* some commercial formulations were already available for control of some foliar diseases. *T. harzianum* strain T-22 available under trade name Top Shield, Bio-Trek 22G, T22 Hopper Box in USA was found to be effective against *Botrytis cinerea*, Powdery and Downey mildews of various crops, turf grass diseases like brown patch (*R. solani*) and damping off (*Pythium* spp.). An example of a novel approach to the biocontrol of *B. cinerea* is the development of the product GREYGOLD, which consists of a mixture of the fungi *T. hamatum* and *Rhodotorula glutinis* and the bacterium *Bacillus megaterium* in USA. In Israel a commercial product "Trichodex" based on *T. harzianum* Strain 39 is available for control of *B. cinerea* gray rots and other foliar diseases in various crops. Currently no commercial products were available for foliar disease control in India but few reports have shown possibility of using bioagents for foliar disease management. Mukherjee and Haware (1993) isolated a *T. viride* isolate from chickpea rhizosphere which was found to be highly effective against grey mold *Botrytis cinerea* of chickpea both under growth room and field conditions (Mukherjee *et al.* 1995a). Field trials were conducted to test the efficacy of integrated control methods for the control of grey mould of chickpea caused by *B. cinerea*. The plots with treatments *viz.*, chickpea detopped at 60 DAS + *T. viride* + vinclozolin sprays showed the lowest disease severity and highest grain yield. The combinations of antagonist and fungicide were superior to fungicide or antagonist alone (Abha *et al.*, 1999). Field experiments were conducted to study the control of early blight of tomatoes (cv. Pusa Ruby) caused by *Alternaria solani*. Of the *T. viride* treatments, seed treatment combined with soil application was the most effective, recording the lowest disease incidence (Sawant *et al.*, 1999)

Studies on efficacy of chemical and biological treatments for the control of covered smut of barley caused by *Ustilago hordei* (*U. segetum*) showed that infection of smut fungus can be eradicated completely in fungicidal treatments as well as the *T. viride* treatments (Singh, 1999). Loose smut of wheat caused by *Ustilago segetum* var. *tritici* has been significantly controlled by treatment with isolates of *Trichoderma viride*.

Stem rot of soyabean caused by *Rhizoctonia solani* was managed effectively by seed treatment with *T. harzianum* with a significant increase in dry weight of root and shoot and yield over pathogen check (Das and Dutta, 1999). *T. viride* was found to be equally effective as that of fungicides in controlling early and late leaf spots (*Cercospora arachidicola* (*Mycosphaerella arachidis*) and *Phaeoisariopsis personata* (*M. berkeleyi*), respectively) and rust (*Puccinia arachidis*) in groundnut, and Alternaria blight (*A. helianthi*) in sunflower (Mathivanan *et al.*, 2000).

Biological Control of Post Harvest Fungal Pathogens in India

In India, having a tropical climate and poor storage facilities, a huge amount of the agricultural produce is lost in storage and transit due to post-harvest pathogens. Post-harvest application of pesticides is restricted due to residue problems. Moreover, post-harvest systems represent a more controlled system compared to the field conditions. Despite these, very few serious attempts have been made in India to combat the menace created by post-harvest pathogens. Although a couple of commercial formulations of antagonists are already available in USA for post harvest applications, researches in India in this field have still not gained momentum. An isolate of *Trichoderma* sp. (Td-1), isolated from a parasitized sclerotium of *S. rolfsii* was highly effective in suppressing *S. rolfsii* on ginger rhizomes, beet, carrot, bitter gourd and elephant-food yam under storage conditions (Mukherjee *et al.*, 1995c).

Potential of *Trichoderma* spp. against plant parasitic nematodes

Potential of *Trichoderma* based biofungicides against plant parasitic nematodes was also reported (Reddy, 1996; Rangaswamy *et al.*, 2000).

Combination of *Trichoderma* and bacterial biological control agents

Pseudomonas sp. was evaluated for its efficacy against rhizome rot of ginger caused by either *Pythium* or *Fusarium* spp, or both (mainly *P. myriotylum* and *F. solani*) alone and in combination with *T. harzianum* and also with fungicidal rhizome treatment. Combination of both bioagents resulted in better germination and plant stand, reduced disease and increased yield. Soil application of the bioagents was more effective compared with seed treatments. Integration of soil application of bioagents with fungicidal rhizome treatment using bavistin (carbendazim) + ridomil MZ (metalaxyl + mancozeb) increased the efficiency of disease control as compared with their individual treatments. Soil application of *T. harzianum* and rhizome treatment with *Pseudomonas* spp. and fungicides was the most effective among all the tested treatments (Ram *et al.*, 1999).

IMPROVEMENT OF *TRICHODERMA*

On improvement of *Trichoderma* no significant achievements that resulted in commercial exploitation were made though reports on improvement by protoplast fusion and mutation were reported (Karpagam, 1995; Mrinalini, 1995; Rajappan *et al.*, 1996; Mukherjee *et al.* 1997; Haware *et al.*, 1997; Selvakumar *et al.*, 2000). Mukherjee *et al.*, (1995a) obtained vinclozolin-tolerant isolates of *T. viride* by selection on a fungicide-amended medium. These isolates were as effective as the original wild type isolate in controlling Botrytis mold of chickpea (*B. cinerea*). Mukherjee and Mukhopadhyay (1993) have obtained a stable mutant (M7) of *T. virens* (=*G. virens*) through gamma-irradiation that is non-conidiating, but produces profuse chlamydospores. Two strains obtained through protoplast fusion of *T. harzianum* and *T. longibrachiatum* exhibited an enhanced antagonistic potential against *R. solani, Fusarium oxysporum* f. sp. *lycopersici, Venturia inaequalis, Curvularia lunata* and *Bipolaris oryzae* and tolerant to copper sulphate and carbendazim.

An extra cellular alpha-glucosidase from *T. viride* capable of inactivating a host-specific phytotoxin was characterized and transfer of the gene to rice plants would lead to enhanced resistance against sheath blight pathogen (Shanmugam *et al.*, 2001).

The new technologies of protoplast fusion, transformation and gene cloning have resulted in the development of new strains of *Trichoderma* in their countries. Induced mutagenesis has also been used successfully to develop new biotypes of *T. harzianum* and *T. viride* (Papavizas and Lewis, 1983, Papavizas *et al.*, 1982). These induced biotypes can tolerate between 50 and 500 μg of benomyl/ml of medium. Several biotypes induced by ultraviolet radiation were consistently more effective than the wild strains in suppressing the damping-off (*Pythium ultimum*) of peas, damping-off (*R. solani*) of cotton and radish, damping-off and blight (*Sclerotium rolfsii*) of bean, and white rot (*Sclerotium cepivorum*) of onion. Exposing conidia to N-methyl-n3-nitro-N-nitrosoguanidine (NTC) produced mutants tolerant to benomyl and TBZ up to 100μg a.i.ml^{-1} produced mutants of *T. hamatum*. Some mutants of *T. harzianum* and *T. viride* that were resistant to benomyl showed better suppressing activity of *R. solani* in soil, damping-off (*P. ultimum*) of pea, damping-off (*R. solani*) of cotton and radish, white rot (*Sclerotium cepivorum*) of onion, and damping-off and blight of beans (*S. rolfsii*).

Genetic recombination is a more powerful method for developing superior biocontrol strains than either selection or mutation. A new strain of *T. harzianum* (A2) obtained through protoplast fusion was effective in controlling Rhizoctonia damping-off of cotton seedlings when compared to parental strain.

Migheli *et al.*, (1995) used protoplast fusion technique for the production of new antagonistic strains of *Trichoderma* spp. Two British and two Italian strains of *Trichoderma* with different biocontrol potential against *Botrytis cinerea, Sclerotinia sclerotiorum* and *Pythium ultimum* were obtained by protoplast fusion but no significant increase in the activity was accomplished after fusion, the hybrids were generally less active than the parental strains. Biolistic procedure in which conidia of bioagents were bombarded using a helium driven biolistic device to accelerated M5 tungsten particles coated with plasmid or genome DNA was technically simple with increased transformation frequency and genetic stability in the transformed progeny compared to protoplast fusion.

Transformations other than fusion of whole protoplasts, such as transfer of whole natural plasmids between microorganisms or transfer of DNA segments have also been achieved. Genes for biocontrol traits, which are directly related to antagonism against plant pathogens, or fitness traits could be cloned to various bacterial and yeast antagonists. For example lytic enzymes like chitinase, protease and cellulose are responsible for lysing fungal cell walls. The chitinase gene (*chiA*), which was isolated from *Serratia marcescens* and encodes for chitinase, which is capable of lysing fungal cell walls, could be introduced into various microorganisms, including antagonists .There is also a possibility of introducing a promoter of this gene to achieve a high level of the antagonists gene expression constitutively. Enhancement of biocontrol activity by *T. harzianum* was attempted by Haran *et al.*, (1993). A *Serratia marcescens* chitinase gene was introduced into the *T. harzianum* genome, under the control of a 35S constitutive promoter from cauliflower mosaic virus. The transformants expressed significantly higher chitinase activity than wild *T. harzianum*. Gabor *et al.*, (1998) were able to clone a 42-kDa wild type *T. hamatum* endochitinase gene into a plasmid construct and subsequently obtained a recombinant *T. hamatum* over expressing the gene and they are also trying to obtain a genomic construct having multiple copies of the endochitinase gene for high expression of the enzyme. Geremia *et al.*, (1993) cloned a gene encoding the basic proteinase *prb1* from *T. harzianum* strain IMI 206040, and suggested that it plays a key role in mycoparasitism. Flores *et al.*, (1997) introduced protease gene *prb 1* into *Trichoderma harzianum* by protoplast fusion. The copy number of the gene was increased for over-expression. The transformed *T. harzianum* significantly reduced the disease caused by *Rhizoctonia solani* in cotton plants. Limon *et al.*, (1995) isolated and characterized a chitinase gene (*chit* 33) from *T. harzianum*, coding for a chitinase of 33 kda by cDNA analysis. Garcia *et al.*, (1994) were able to clone the cDNA of *T. harzianum* (*chit* 42), coding for an endochitinase of 42 kda using synthetic oligonucleotides corresponding to amino acid sequences of purified chitinase. Very recently plants were transformed with *Trichoderma harzianum* chitinase gene for induction of resistance to invading pathogen. There are tremendous possibilities for desirable transformations in biocontrol fungi provided that the genes coding for certain desirable traits are identified and cloned.

MECHANISMS INVOLVED IN BIOLOGICAL CONTROL BY *TRICHODERMA* SPECIES- INDIAN WORK

Most of the studies carried out in India for selection of potential bioagents of plant pathogens were based on simple laboratory, greenhouse and field efficacy tests. Serious efforts were not made really to look into the mechanisms involved in biological control by the selected bioagents. Many researchers reported mechanisms like mycoparasitism and antibiosis based on laboratory tests from India. But more important activities of fungal bioagents like competition, induced resistance, inactivation of host enzymes and plant growth promotion were not given due importance in selection of bioagents. This information is highly essential for selection of potential isolates and for further improvement of the antagonists.

Biological control mechanism involving *Trichoderma* species operates by way of mycoparasitism, antibiosis, competition, induced resistance and inactivation of host enzymes. A brief description of these mechanisms was described hereunder.

Mycoparasitism

Mycoparasitism has long been considered an important mechanism of action of biocontrol by *Trichoderma* spp. This is a complex process that involves tropic growth of the biocontrol agent toward target fungi, lectin-mediated coiling of attachment of *Trichoderma* hyphae to the pathogen, penetration by haustoria and lysis. The *Trichoderma* recognizes and attaches to the pathogenic fungus and begins to release extra cellular lytic enzymes like β-1,3 glucanase, chitinase, protease and lipase. The recognition mechanism is the basis for the specificity of the antagonists. Lectins (glycoprotein's) produced by some soil borne pathogenic fungi, such as *R. solani* and *S. rolfsii* are known to involve in the recognition and specificity of the interaction between *Trichoderma* and its host fungi.

Antibiosis

Trichoderma species releases antibiotics or other chemicals like alkyl pyrones, isonitriles, polyketides, peptaibols, diketopiperazines, sesquiterpenes and steroids which are harmful to the pathogen and inhibits its growth (Antibiosis). *Trichoderma* spp. are known to produce antibiotics like Trichodermin, Gliovirin, Viridol and Gliotoxins.

Competition

Competition for space or nutrients has long been considered one of the "classical" mechanisms of biocontrol by *Trichoderma* spp. Competition is an indirect mechanism by which *Trichoderma* and other bioagents whereby pathogens are excluded by depletion of food bases or by physical occupation of sites. All these three mechanisms may act combinedly since a single simple action in any biological systems is unlikely.

Induced resistance

Elicitation of resistance in plants by *Trichoderma* spp. is becoming a more researched topic these days. Some *Trichoderma* strains clearly are potent inducers of SAR-like responses. Strain T-39 of *T. harzianum*, which is the active ingredient in Trichodex, when inoculated onto roots or onto leaves provided control of diseases caused by *Botrytis cinerea* on leaves spatially separated from the site of application of the biocontrol agent (De Meyer *et al.*, 1998) which shows the induction of defense response in plants. Chitinases, peroxidases and terpenoid phytoalexins are known to involve in disease resistance in plants.

Tolerance to biotic and abiotic stresses through enhanced root development

Tolerance to stress through enhanced plant or root development is also suggested as one of the mechanisms action of *Trichoderma* spp.

Solubilization and sequestration of inorganic plant nutrients (Plant growth promotion)

In soil, various plant nutrients undergo complex transitions from soluble to insoluble forms that strongly influence their accessibility and absorption by roots. Some *Trichoderma* strains are known to enhance nitrogen use efficiency in plants. *Trichoderma* spp. also solubilizes rock phosphate, Zn metal, Mn^{4+}, Fe^{3+} and Cu^{2+}.

Inactivation of the pathogen's enzymes

Many plant pathogens depend upon production of pectolytic, cutinolytic and cellulolytic enzymes to infect living plants. Some of the *Trichoderma* strains are known to produce serine protease that is capable of degrading the pathogen's plant cell wall degrading enzymes and thereby reducing the ability of the pathogen to infect the plant.

Fungicidal compatibility

Though few researchers from India demonstrated tolerance of *Trichoderma* species to some of the commonly used fungicides, it is not necessary to concentrate in this area as there is always a possibility of obtaining *Trichoderma* from nature which are much more effective than the fungicides for use against soil borne diseases.

Studies revealed that *Trichoderma* are highly sensitive to the most commonly used seed treatment chemical bavistin and compatible with thiram only up to 50 ppm conc. (Pandey and Upadhyay, 1998).

LABORATORY AND COMMERCIAL SCALE MASS PRODUCTION

Biomass of *Trichoderma* must be produced in a timely and cost-effective way and it should survive each processing step, such as harvesting, drying, formulation, storage, and delivery. Much work has been done on production of *Trichoderma* spp. by liquid and solid-state fermentation methods in India. Production of these antagonists can be done easily using cheaper substrates. In general, product formed from solid or semi solid-state fermentation do not require sophisticated formulation procedures prior to use. For example, grain or other types of organic matter upon which antagonists are grown are simply dried ground and added to the area to be treated. There are several problems with solid-state fermentation, which may make the system in appropriate for commercial product development. The preparations are bulky, they may be subject to a greater risk of contamination and they may require extensive space for processing, incubation and storage. The liquid state fermentation is devoid of such problems and large quantities of biomass can be produced with in few days. Some of the media tried for production are listed in the Table 13.3.

It appears feasible for industry to produce various beneficial microbes against plant pathogens through liquid fermentation at the present time. In industrialized countries deep tank fermentation in inexpensive media such as molasses and brewers yeast for large-scale production of *Trichoderma* is being followed.

Studies at PDBC, Bangalore revealed that maximum amount of biomass and viable propagules of *T. harzianum* can be obtained within 96 h of fermentation in a fermentor with aeration, agitation, temperature controls.

Maximum biomass of *Trichoderma* spp. can be realized in short-time by using appropriate medium in a fermentor with aeration, agitation, temperature, pH and antifoam controls than in shake-flask cultures and will be more suitable for industrial production of *Trichoderma* spp. Various media used in liquid fermentation are listed in Table 13.4.

FORMULATION AND SHELF LIFE OF *TRICHODERMA*

A formulated product with agricultural application should posses several desirable characteristics. These include adequate market potential, ease in preparation and application, stability during transportation and storage, abundant viable propagules and good shelf life, sustained efficacy and accepted cost. Biomass of the antagonistic fungus either can be separated from medium and concentrated or entire biomass with medium can be incorporated in to dusts, granules, pellets, wettable powders or emulsifiable liquids. *Trichoderma* spp. can be formulated as pellets, dusts and powders. Various carrier materials used in *Trichoderma* formulations are talc, kaolin, bentonite and lignite etc. The carrier material may be inert material or a food base or in combination of both.

The viability of different species of *Trichoderma* in talc based powder formulation was ranged from 224-297 X 10^6 per gram of fresh product in a study by Sankar and Jeyarajan (1996a). The population was reduced by 19, 21, 21 and 29.9% in *T. viride, T. harzianum, T. longibrachiatum* and *T. virens*, respectively, after 75 days of storage. After 120 days 50% of the population was retained. In treated seeds there was a negligible reduction in the population of antagonists after 45 days and then a slight decrease. After 120 days 55-68% of the initial population of various *Trichoderma* species was maintained. Packaging and storage conditions are very important for longer shelf life bioagents in various formulations. Longer shelf life of *T. viride* was recorded with vermiculite wheat bran acid fermentor biomass (VBA-FB) formulation stored in milky white bags (205 X 10^6 cfu/g) (Nakkeeran *et al.*, 1997). Studies by the same researchers on storage temperatures revealed that 20-30°C was optimum for storage of the VBA-FB formulation in milky white bags. Although, in talc based formulation there was a slow decline in the population of *Trichoderma* during storage, a population of 230-242 X 10^6 cfu/g was observed at 20-30°C after 75 days of storage. *Trichoderma* can be safely stored as talc based formulations in milky-white bags at room temperature up to 75 days. Some significant findings on shelf life *Trichoderma* at PDBC were highlighted earlier.

RESEARCH ON BIOLOGICAL CONTROL OF FUNGAL PATHOGENS USING FUNGAL ANTAGONISTS AT PDBC, BANGALORE

Repository of Fungal Antagonists

More than five hundred isolates of *Trichoderma* and *Gliocladium* spp. were isolated from rhizosphere samples of various crops collected from all over India and maintained in germplasm collection of the Project Directorate of Biological Control, Bangalore, India. The identified antagonistic fungi include 10 species of *Trichoderma,* 3 species of *Gliocladium,* two *Chaetomium globosum* isolates and one *Verticillium lecanii* (Table 13.5).

Standardization of media for mass production of *Trichoderma*

Molasses-soy medium (MSM) standardized for mass production of *Trichoderma* by liquid fermentation at PDBC gave maximum yield of biomass, viable propagules and spores compared to standard molasses-yeast medium. The MSM medium could serve as a better alternative to MYM for commercial production of *Trichoderma* species (Prasad *et al.,* 2000a). As molasses availability is limited to areas near to sugar factories, attempts were made to utilize Jaggery as an alternative to molasses. Jaggery-soy medium gave more biomass and viable propagules compared to MYM but it was not superior to MSM (Prasad *et al.,* 2002).

Formulations and Shelf Life of *Trichoderma*

In studies on effect of various formulations on shelf life and bioefficacy, kaolin and talc were found to be better carriers of *T. harzianum* (Prasad *et al.,* 2000b).

In greenhouse tests, modified granular formulations containing powdered wheat bran, kaolin, acacia powder and biomass of isolates of *Trichoderma harzianum* (PDBCTH 10 and PDBCTH 8), *T. virens* (PDBCTVs 3 and ITCC 4177) and *Gliocladium deliquescens* (ITCC 3450) were evaluated for their effect on the reduction of chickpea damping-off caused by *Rhizoctonia solani,* reduction of pathogen inoculum and proliferation of the bioagents in the soil. Granular formulation of different bioagents significantly reduced damping-off. PDBCTH 10 and PDBCTH 8 isolates treatments recorded better plant stands (63 and 53%) than fungicide (captan) treatment (43%). Substantial increase in the population of all bioagents was observed in amended soils. (Prasad and Rangeshwaran, 2000c).

Studies on comparative efficacy of *Trichoderma* propagules

Trichoderma harzianum conidia produced by solid state fermentation and chlamydospores produced in liquid fermentation are formulated as powder and tested for their comparative bioefficacy and shelf life. Conidia and chlamydopores did not differ significantly in reducing root rot incidence in chickpea caused by *Rhizoctonia solani*. Conidial formulation retained optimum amount of viable propagules (>10^6 cfu/g) even after 180 days of storage at room temperature but in chlamydospore formulation viable propagules reduced to less than 10^6 by 150 days.

Biological control of soil borne plant pathogens

Intensive efforts were made at PDBC to screen many isolates of *Trichoderma* and *Gliocladium* species against *Sclerotium rolfsii* of sunflower. Two isolates of *T. viride*, four isolates of *T. harzianum*, one each of *T. hamatum*, *T. koningii*, *T. polysporum*, *G. virens*, *G. deliquescens* and *G. roseum* inhibited mycelial growth of the pathogen significantly. Among *Trichoderma* species, *T. harzianum* isolate PDBCTH 2 gave 61.4 per cent inhibition of mycelial growth followed by PDBCTH 8 (55.2%) and PDBCTH 7 (54.9%). Suppression of sclerotial production by the antagonists ranged from 31.8 to 97.8 %. The three *T. harzianum* isolates were superior under greenhouse conditions with PDBCTH 8 showing maximum disease control (66.8%) followed by PDBCTH 7 (66.0%), PDBCTV 4 (65.4%), PDBCTH 2 (61.6%) and were even superior to the fungicide, Captan (Prasad *et al.*, 1999).

Efforts made to select suitable biological control agents against *Sclerotium rolfsii*, the causal organism of root/collar rot of chickpea in micro plots. Two *T. viride* isolates (PDBCTV 23 and PDBCTV 24) and *T. harzianum* (PDBCTH 10) gave more than 50% disease control. These bioagents were found superior to the fungicide, captan. All bioagents tested have shown good plant growth promoting ability and rhizosphere competence (Prasad and Rangeshwaran, 2001b). Further the two effective antagonistic fungi viz., *Trichoderma harzianum* (PDBCTH 10) and *T. viride* (PDBCTV 23) were evaluated against natural incidences of wilt (*Fusarium oxysporum* f.sp. *ciceri*) and wet root rot (*Rhizoctonia solani*) of chickpea in a field plot. Soil application of these bioagents one week prior to sowing was found to be more effective than seed treatment in reducing wilt and wet root rot of chickpea. The two bioagents had positive effect on plant growth and yield.

Studies on biological control of wilt of pigeonpea

The biocontrol efficacy of two *Trichoderma harzianum* isolates was tested against wilt of pigeonpea in a wilt sick plot. Soil amendment with two bioagents (PDBCTH 10 and 15) resulted in a disease incidence of 22.5 and 28.8% respectively whereas in seed treatments 35.3 and 40.3% disease incidence was recorded. In general, soil application of the bioagent was found to be more effective than seed treatment in disease suppression. The seedling emergence and vigor index were also significantly high in bioagent soil amendments. This study clearly indicates the positive effect of the *T. harzianum* on plant health. Soil application of *T. harzianum* prior to sowing was found to be more effective than seed treatment in controlling fusarial wilt of pigeonpea under field conditions.

Studies conducted to know the efficacy of various doses of *Trichoderma harzianum* at various *Fusarium udum* levels in field yielded very interesting results. Fusarial wilt sick plots with three levels of pathogen concentration *viz.*, log 3.04, 4.98 and 5.34 colony forming units (i.e. approx. 0.1, 1 and 2 lakh propagules) per gram of soil were identified. Bioagent was applied as seed coat (@10 and 20 g/kg seed) and soil amendment (@ 0.5 and 1 kg/ 9 m^2 area). At $10^{3.04}$ cfu level of pathogen, the soil amendment of bioagent @ 0.5 and 1 kg gave 42.9 and 61.5% disease control respectively compared to control. Seed treatments at two levels resulted in less than 30% disease control. At higher pathogen levels disease control achieved ranged between 22 to 35.3% with soil treatment. Even at highest pathogen density, soil amendment @ 1 kg/m^2 could give about 30% disease control. Disease control obtained was between 4.36 to 13.7% with seed treatment at two higher pathogen levels. In general, soil application of the bioagent was found to be more effective than seed treatment in controlling disease at higher pathogen levels, which may be due to higher population of bioagent in the rhizosphere (Prasad *et al.*, 2002).

Biological Control of Foliar Plant Pathogens

The work carried out at PDBC on biological control of botrytis grey rot of rose resulted in identification of potential *Trichoderma* isolates (Prasad *et al.*, 1998).

Fungal biological control agents were used either alone or in combination with fungicides to manage black spot of rose caused by *Diplocarpon rosae* under greenhouse conditions. Black spot incidence in biological control agent and fungicide treatments alone or in combination treatments was significantly low (Disease ratings from 0.33 to 3.33) compared to pathogen check at all observation dates. At 100 days after first spray lowest defoliation rating was recorded in Chlorothalonil, *T. harzianum*-Chlorothalonil and *T. harzianum*-Mancozeb treatments (1.00). Highest vigour index was recorded in *T. harzianum* treatment. Highest flower production was recorded in *C. globosum*-Chlorothalonil treatment (4.33) followed by *T. harzianum* alone and *T. harzianum*-Chlorothalonil treatments (4.00) (Prasad *et al.*, 2001 communicated).

Verticillium leccanii parasitizing coffee rust in natural field conditions was isolated and further large-scale field tests are necessary to develop it as biofungicide against rust fungi.

Development of a faster *in vivo* screening technique for selection of potential Trichoderma:

Since faster *in vivo* methods for the selection of potential bioagent is a limiting factor, a rapid *in vivo* bioassay method for screening and selection was developed using *R. solani* -chickpea system and *P. capsici* -bell pepper system. The results obtained through rapid *in vivo* method showed that chickpea seed germination; seedling growth and vigor index were significantly high in all bioagent treatments as compared to pathogen check. Maximum vigor index (2239.4) was recorded in *T. harzianum* (PDBCTH 10) followed by *T. viride* (PDBCTV 32) (2226.3). Disease incidence was also low in bioagent treatments. Similarly bell pepper seeds treated with *P. capsici* along with different bioagents showed significantly high germination. Highest germination (98.9%) along with highest root and shoot length and vigor index was observed in *T. viride* (PDBCTV 23) treatment. *T. viride* (PDBCTV 31) treated seeds showed minimum disease incidence (5.6%) and pathogen check showed high disease incidence (68.9%). *T. viride* (PDBCTV 32) and *T. harzianum* (PDBCTH 10) were rated as efficient bioagents against *R. solani* and four *T. viride* isolates (PBCTV 6, 31, 33, 34) were found efficient against *P. capsici* (Srivastava *et al.*, 2001 in press).

The results recorded through faster *in vivo* bioassay methods were validated under greenhouse against *R. solani* of chickpea and *P. capsici* of bell pepper. *T. viride* (PDBCTV 32) showed minimum (13.3%) pre-emergence mortality and no post emergence mortality followed by *T. harzianum* (PDBCTH 10). Bell pepper crown/root rot and fruit rot was significantly reduced by *T. viride* (PDBCTV 31) treatment under greenhouse/growth chambers confirming the potentiality of bioagent as observed in rapid bioassay method.

Induction of Defense Response in Plants by *Trichoderma*

Efforts made to study the defense response induced by *Trichoderma* in plant system yielded encouraging results. Spectrophotometric studies revealed that bioagent treatments have induced expression of more quantities of proteins in chickpea roots. SDS - PAGE studies revealed that roots obtained from chickpea seeds treated *T. viride* (PDBCTV 32) showed the presence of a unique protein (44KDa) out of three proteins detected . Further studies are needed to study the enzymatic nature of the protein and its role in plant defense.

REGISTRATION AND QUALITY CONTROL

Recently, Central Insecticide Board (CIB) under the Ministry of Agriculture and Co-operation, Govt. of India has made registration of biopesticides like chemical pesticides compulsory through its gazette notification dated 26th March 1999 to have control on quality of the products. Till recently commercial producers of bioagents used to have permission from Agriculture Departments of State Governments and sell their products. At present a manufacturer who wants to market *Trichoderma* should apply to CIB for registration by submitting information on bioefficacy and toxicology data by spending considerable amount of money. Central Insecticides Board should relax some of the requirements like toxicology information of bioagents like *Trichoderma viride, T. virens* and *T. harzianum* as these fungi were already registered in USA, Israel and European countries and found to be safe to all test organisms. It is unnecessary to spend money in toxicology data generation and it may discourage small entrepreneurs who wish to produce bioagents. CIB could have been more stringent on bioefficacy data of bioagents involving more field tests and also information on host range of these antagonists. In the face of continuing environmental pressure only rational approaches to registration can lead to marketing of an increased number of reliable antagonists.

Quality control parameters set by CIB are inadequate for knowing potentiality of a bioagent. It is time now for Ministry of Agriculture to identify a Central Agency for quality testing not only in terms of amount of live propagules in the formulation but also their bioefficacy against plant pathogens. Quality of the products in these lines should be periodically checked by the identified agency and that will ensure availability of better products to farmers.

PRIORITY AREAS FOR RESEARCH AND DEVELOPMENT

A very good beginning has been made in India in the field of biological control research and development. However, if we compare our status global progress, the picture is not that satisfactory. We have been able to isolate large number of highly effective antagonist isolates and their efficacy has been demonstrated mostly in laboratory and greenhouse conditions but not in field conditions. No efforts were made to screen and select *Trichoderma* effective against wide range of plant pathogens. Though there are several antagonist preparations already available in India, we still do not have quality controlled formulated products for large scale use. However, the initiative and interest taken by several commercial organizations in commercializing *Trichoderma, Gliocladium* and *Pseudomonas* preparations is a very good beginning.

At present most of the work from India on fungal biological control agents of plant pathogens was on applied aspects but not basic aspects. There is a strong need to understand these systems thoroughly through basic researches especially on the biology and ecology of the antagonist in relation to pathogen levels, growth promotion, and induction of defense response and inactivation of pathogen enzymes by these antagonists.

THRUST AREAS IN BIO-CONTROL FOR FUNDING

- There is a strong need to develop suitable screening strategy (the one like germination towel test developed at PDBC) which will save time, money and energy and also quality of commercial product can be monitored more efficiently.
- There is a need to develop a uniform system of field-testing of biological control agents. These bioagents need to be tested in fields with uniform and reasonable level of pathogen (at least 30% level of disease incidence) in the field. Information on plant growth promotion and bioagent levels in treated plots needs to be recorded during bioefficacy testing by any organization. The bioagent testing centres like ICAR institutes and SAU's should follow the guidelines mentioned earlier for more meaningful information on bioagents efficacy.
- A thorough understanding of the mechanisms by which an antagonist suppresses the pathogen *in vivo* is of utmost importance in biocontrol research. Most of the reports on the mechanisms of biocontrol from India are not aimed to study mechanisms like plant defence induction and plant growth promotion by fungal bioagents. In many places elsewhere, several attempts have been made to elucidate the mechanism of biocontrol using mutational analysis, *in situ* detection of metabolites and using molecular biological tools in identification of genes involved in higher lytic enzyme production by the antagonists, their identification, isolation, cloning and transformation of biocontrol genes to plants. Efforts should be made to immediately start work on these lines.
- As fungal and bacterial bioagents have shown promise in control of some foliar and post harvest diseases efforts should be made to initiate work on these lines and these areas need more attention from funding agencies.

- Ministry of Agriculture, Government of India should take steps to relax some of the registration formalities currently required.
- Molecular biological techniques have been used extensively in biological control research in other countries to characterize strains of antagonists using molecular markers like AFLP, RFLP and RAPD techniques and to understand the mechanisms of biocontrol at the molecular level, to transform the *Trichoderma* with green fluorescent gene markers for monitoring population, strains rhizosphere competence, and to clone chitinase, proteinase, glycosidase genes to a potential bioagent or plants. Though initial efforts by researchers in India lead to availability of few bioagents to farmers, no significant work has been done so far on the applications of above-mentioned novel techniques in plant disease biocontrol research. Time is ripe to invest more money and energy in this field to change the facet of biocontrol research and its application in the new millennium.

TABLE 13.1. Commercial products of *Trichoderma*

Product/antagonist	**Target pathogen**	**Source**
Global		
Antifungus (*Trichoderma* spp.)	*Rhizoctonia solani, Pythium spp./bedding plants*	Grondontsmettingham De Ceuster, Belgium
Binab-TW (*Trichoderma* spp.)	*Verticillium*/Mushrooms	Binab USA, Madison, WI, USA
Binab-T (*Trichoderma* spp.)	*Chondrostereum purpureum*	Bio-Innovations AB, Sweden
F-Stop, Root Shield, Top Shield, Planter Box (*Trichoderma harzianum* strain T-22)	*Pythium ultimum, Phytophthora, Rhizoctonia solani, Fusarium* sp., Botrytis, Powdery mildews and downey mildews/ row crop seed treatment, soil application, foliar sprays	Bio Works, Inc., Cornell Research Foundation, Ithaca, NY, USA
Gliogard (*T. virens=Gliocladium virens*)	*Pythium ultimum, Rhizoctonia solani*	W. R. Grace & Co., USA
Promot (*Trichoderma* spp.)	Various fungi/fruits, vegetables	J. H. Biotech, Inc., Ventura, CA, USA

Contd...

Table 13.1. Contd...

Soilguard 12G (*T. virens*)	*Pythium* spp.	ThermoTriology Corp., USA
Supresivit (*T. harzianum*)	Various fungi	Bonegaard and Reitzel ,Denmark
T-35 (*T .harzianum*)	*R. solani, Fusarium* spp. / cucumbers, tomatoes	Makhteshim, Israel
Trichodermin-3 (*T. lignorum*)	*Fusarium* sp./field crops, vegetables	Russia, Bulgaria
	Botrytis cinerea/ fruits, ornamentals	Makhteshim, Israel
Tricho dowels, Tricho mini dowels, Trichoject, Trichoseal, Trichopel (*Trichoderma* sp.)	Various plant pathogenic fungi	Agrimm Biologicals, New Zealand,
TY (*Trichoderma* spp.)	*R. solani, Pythium* sp. *S. rolfsii*/ field crops, vegetables	Mycontrol, Israel
MR35 (*Trichoderma* spp.)		Israel,
Trichodex (*T. harzianum* T 39)		Israel
India		
Basderma (*Trichoderma* sp.)		Basaras Biocon (India) Pvt. Ltd., Chennai
Sun-derma (*T. viride*)	*R. solani, Macrophomina phaseolina*	Sun Agro Biosystems Pvt. Ltd., Chennai
Ecofit (*T. viride*)	*R. solani, M. phaseolina*	Hoechst Schering Agr Evo Ltd., Mumbai
Trieco (*T. viride*)	Root rot and wilt	EcoSense Laboratories (I) P. Ltd, Mumbai
Antagon - TV (*T.viride*)	*R. solani, M. phaseolina*	Green Tech Agroproducts (P) Ltd., Coimbatore
Trico (*T. viride*)	-	Pralshar Biotech (P) Ltd., Goa

Contd...

Table 13.1. Contd...

Biohit (*Trichoderma* spp.)	-	Indore Biotech Inputs & Research (P) Ltd.,
Biofit (*T. viride*)	-	Ajay Biotech (I) Ltd., Pune
Niprot (*T. viride* and *T. harzianum*)	Root rots and wilts	Biocontrol Research Laboratories (PCI), Bangalore
Bio guard (*T. harzianum*)	*Rhizoctonia, Sclerotium* and *Fusarium* diseases	PKVK, Pondicherry
V. Trijaivica-H (*T. harzianum*)	Wilt diseases of black pepper	VAST Centre, Adyanadka, Katnatataka
Spices Care (*T. harzianum*)	Root rot and wilts	BacTo Agro Culture Care Pvt. Ltd, The Nilgiris, Kerala
Trichozen T (*T. viride*)	Root rot and wilt	SRI biotech, Hyderabad
T. viride	Root rot and wilt	Margo Biocontro Laboratory, Bangalore
Tricho X (*T. viride)*	Root rot and wilt	Excel Industries, Mumbai
Trichopower (*T. viride*)	Root rot and wilt	KN Bioscience, Hyderabad
Spiceguard (*T. viride*)	Root rot and wilt	Visvamitra Biotech, Guntur, AP
Tricoat (*T. viride*)	Root rot, wilts	K. N. Biotech Limited, Hyderabad
Trivir (*T. viride*)	Fusarial wilts	Directorate of Oilseeds Research, ICAR, Hyderabad

TABLE 13.2. Fungal biological control agents of plant diseases reported from India

Crop	Disease	Pathogen	Bioagent
Lentil	Collar rot	*Sclerotium rolfsii*	*T. harzianum*
Sugarbeet	Root rot	*S. rolfsii*	*T. viride* and *T.harzianum*
Sunflower	-	*S. rolfsii*	*T.harzianum*
Groundnut	Stem blight	*S. rolfsii*	*T. harzianum*
Sugar beet	Root rot	*S. rolfsii*	*T. harzianum*
Groundnut	-	*S. rolfsii*	*T. harzianum*
Groundnut	Root rot	*S. rolfsii*	*T. viride*
Ginger	-	*S. rolsii*	*Trichoderma* sp.
Ginger		*S. rolfsii*	*T. pseudokoningii*
Chickpea	Stem rot	*S. sclerotiorum*	*T. harzianum* and *Absidia cylindrospora*
Rice	Sheath blight	*Rhizoctonia solani*	*T. viride* and *Aspergillus niger*
Pigeon pea	Root rot	*R. solani*	*T. viride*
Tomato	-	*R. solani*	*T. pseudokoningii*
Tomato	Damping off	*R.solani*	*Trichoderma* spp.
Tomato	Damping off	*R. solani*	*T. pseudokoningii*
Rice	Sheath blight	*R. solani*	*T. harzianum, T. viride* and *Aspergillus terreus*
Rice	Sheath blight	*R. solani*	*T. harzianum*
Rice	Sheath blight	*R. solani*	*T. viride*
Rice	Rice sheath blight	*R. solani*	*T. harzianum* and *T. viride*
Eucalyptus	Damping off	*R. solani*	*T. viride*
Cauliflower	Stem rot	*R. solani*	*T. viride*
Ginger	Rhizome rot	*R. solan* , *S. rolfsii* and *P. aphanidermatum*	*G. virens* and *T. longibrachiatum*
Chickpea	Dry root rot	*Rhizoctonia bataticola*	*T. harzianum*
Mung	Dry root rot	*Macrophomina phaseolina*	*T. viride*
Groundnut	Root rot	*M. phaseolina*	*T. viride*
Cowpea	Root rot	*M. phaseolina*	*T. harzianum, T. viride*

Contd...

Table 13.2. Contd...

Sesamum	Root rot	*M. phaseolina*	*Trichoderma* spp.
Sesamum		*M. phaseolina*	*Trichoderma* spp. and *Gliocladium virens*
Jasmine	Root rot	*M. phaseolina*	*T. viride*
Pigeon pea	Wilt	*Fusarium udum*	*Trichoderma* spp.
Cotton	Wilt	*F. oxysporum* f. sp. *vasinfectum*	*T. harzianum*
Tomato	Wilt	*F. oxysporum* f. sp. *lycopersici*	*Trichoderma* spp.
Pigeonpea	Wilt	*F. udum*	*Trichoderma* sp.
Pigeon pea	Vascular wilt	*F. udum*	*T. viride*
Pigeon pea	Wilt	*F. udum*	*T. harzianum*
Ginger	Rhizome rot	*Pythium pleroticum, Fusarium equiseti*	*T.viride*
Sugarcane	Seedling rot	*P. graminicola*	*T. viride*
Ginger	Rhizome rot	*P. myriotylum Fusarium solani*	*T. viride*
Tomato	Damping off	*P. indicum*	*T. viride*
Ginger	Rhizome rot	*P. aphanidermatum*	*Trichoderma* spp. and *G. virens*
Cardamom	Seedling rot/ Damping off	*P. vexans*	*T.viride* and *T.harzianum*
Cardamom	Seedling rot	*Phytophthora meadi*	*T. harzianum*
Citrus	Root rot	*Phytophthora*	*T. harzianum, T. viride* and *G. virens*
Betelvine	Foot rot	*Phytophthora* sp.	*T. viride*
Vegetable crops	Foot rot	-	*T.harzianum*
Ginger	Rhizome rot	Fungal pathogens	*Trichoderma* spp.
Sugarcane	Red rot	*Colletotrichum falcatum*	*Trichoderma* sp. and *Aspergillus* sp.
Sugarcane	Red rot	*C. falcatum*	*T.harzianum* and *T.viride*
Red gram	-	Seed mycoflora	*T.viride*
Sunflower	Plant mortality	Root pathogens	*T.viride*
Wheat	-	*Drechslera sativum*	*Trichoderma* sp. and *Streptomyces* sp.
Wheat	Loose smut	*Ustilago segetum* var. *tritici*	*T. koningii*
Mulberry	Leaf spot	*Cercospora moricola*	*Trichoderma* spp.
Chickpea	Grey mould	*Botrytis cinerea*	*T. viride*

TABLE 13.3. Substrates successfully used to produce *Trichoderma* spp. by solid state Fermentation

Substrate
Sorghum grain
Wheat bran-saw dust modified medium
Sand and sorghum medium
Tapioca rind, Tapioca refuse, FYM, press mud
Peat- bran
FYM, wheat bran, rice bran, peat soil, rice straw
Groundnut shell medium
Spent tea leaf waste and coffee husk
Wheat bran and biogas manure
Coffee-berry husk and poultry manure
FYM and coconut coir pith
Pigeonpea husk, tapioca waste (after starch extraction) and press mud.
Coffee fruit skin, poultry manure and coffee fruit skin composted with cow dung slurry
Decomposed Coconut Coir pith
Spent malt

TABLE 13.4. Growth media used for production of *Trichoderma* in liquid state fermentation

Growth media
Molasses and brewers yeast
Potato dextrose broth, V-8 juice and Molasses-yeast medium
Molasses-soy medium
Jaggery-soy medium

TABLE 13.5. Antagonistic fungi maintained in germplasm collection of PDBC

S. No.	Fungus	No. of isolates
1.	*Trichoderma harzianum*	52
2.	*T. viride*	35
3.	*T. hamatum*	6
4.	*T. virens*	22
5.	*T. koningi*	14
6.	*T. pseudokoningi*	4
7.	*T. piluliferum*	8
8.	*T. citrinoviride*	3
9.	*T. longibrachiatum*	2
10.	*T. polysporum*	4
11.	*Gliocladium deliquescens*	4
12.	*G. roseum*	2
13.	*G. catenulatum*	1
14	*Verticillium lecanii*	1
15.	*Chaetomium globosum*	2

REFERENCES

Abha, A., Tripathi, H. S., Rathi, Y. P. S. and Agarwal, A., 1999. Integrated management of grey mould of chickpea. *J. Mycol. Pl. Pathol.* 29: 116-117.

Asghari, M. A. and Mayee, C. D. 1991. Comparative efficacy of management practices on stem and pod rots of groundnut. *Indian Phytopath.* 44: 328-332.

Baby, U. I. and Manibhushanrao, K., 1993. Control of rice sheath blight through the integration of fungal antagonists and organic amendments. *Tropic. Agric.* 70: 240-244.

Das, B. C., Roy, S. K. and Bora, L. C., 1997. Mass multiplication of *Trichoderma* species on different *media. J. Agric.Sci. Soc. North East India* 10: *95-100.*

Das, B. C. and Dutta, P. 1999. Biological management of stem rot of soybean caused by *Rhizoctonia solani* Kuhn. *J. Agric.Sci. Soc. North East India.* 12: 217-220.

Dasgupta, B., Dutta, P.K., Muthuswamy, S and Maiti, S. 2003. Biological control of foot rot of betelvine (*Piper betle* Linn.). *J. Biol. Control.* 17: 63-67.

De Meyer, G. Bigirimana, J., Elad, Y. and Hofte, M., 1998. Induced Systemic resistance in *Trichoderma harzianum* biocontrol of *Botrytis cineria. Eur.J.Plant Pathol.* 107:279-286

De, R. K. and Mukhopadhyay, A. N., 1994. Biological control of tomato damping-off by *Gliocladium virens. J. Biol. Control.* 8: 34-40.

Deb, P. R. and Dutta, B. K., 1991. Studies on biological control of foot rot disease of soybean caused by *Sclerotium rolfsii* Sacc. *J. Pl. Dis. Prot.* 98: 539-546.

Dohroo, N. P. 1995. Integrated management of yellows of ginger. *Indian Phytopath.*48: 90-92.

Dwivedi, R. 1984. Biocontrol of fusarium wilt by *Trichoderma harzianum* Rifai. *Indian J. Agric.Sci.* 54: 513-514.

Flores, A., Chet, I. and Herrera Estrella., 1997. Improved biocontrol activity of *Trichoderma harzianum* by over-expression of the proteinase encoding gene Prb 1. *Curr. Genet.* 31:30-37.

Gabor, G., Kerenyi, Z., Dallman, G. and Hornok, L., 1998. Homologous transformation of *Trichoderma harzianum* with an endochitinase encoding gene, resulting in increased levels of chitinase activity. *FEMS Microbiol. Lett.*165: 247-252.

Garcia, I., Lora, J. M., Cruz J de la., Benitez, T., Llobell, A., Pintor Toro, J. A. and De la cruz, J. 1994. Cloning and characterization of a chitinase (*CHIT 42*) cDNA from the mycoparasitic fungus *T. harzianum. Curr. Genet.* 27: 83-89.

Geremia, R., Goldman, G. H., Jacobs, D., Ardiles, W., Vila, S. B., Van Montagu, M. and Herreera-Estrella, A. 1993. Molecular characterization of the proteinase-encoding gene, *prb 1,* related to mycoparasitism by *Trichoderma harzianum. Mol Microbiol.* 8: 603-613.

Haran, S., Schichler, H., Oppenheim, A. and Chet, I. 1993. Increased constitutive chitinase activity in transformed *Trichoderma harzianum*. *Biol. Control.* 3: 101-108.

Haware, M. P., Lenne, J. M. and Mukherjee, P. K., 1996. Integrated management of *Botrytis* gray mold of chickpea. *Proc. XI Int. Botrytis Symp.* 23-27 June, Wageningen p. 98.

Jakhar, S. S., Chauhan, M. S. and Duhan, J. C., 1998. Management of root rot of cotton (*Gossypium* species) caused by *Rhizoctonia* species. *Plant Dis. Res.* 13: 115-118.

Jeyarajan, R., Ramakrishnan, G., Dinakaran, D. and Sridhar, R. 1993. Development of product of *Trichoderma viride* and *Bacillus subtilis* for biocontrol of root rot disease. *(In:)* Dwivedi R (Ed.), *Biotechnology in India*. Bioved Research Society, Allahabad.

Karpagam, S. 1995. Variation in cultural morphology and biocontrol efficacy among protoplast regenerated isolates of *Trichoderma harzianum*. *Indian J. Mycol. Pl. Pathol.* 25: 102.

Karthikeyan, G., Sabitha Duraisamy, Sivakumar, C. V. and Duraisamy, S., 1999. Biological control of *Pythium aphanidermatum, Meloidogyne incognita* disease complex in chilli with organic amendments. *Madras Agric. J.* 86: 320-323.

Kaur, N. P. and Mukhopadhyay, A. N., 1992. Integrated control of 'chickpea wilt complex' by *Trichoderma* and chemical methods in India. *Trop.Pest Manag.* 38: 372-375.

Maiti, D. and Sen, C., 1985. Integrated biocontrol of *Sclerotium rolfsii* with nitrogenous fertilizers and *Trichoderma harzianum*. *Indian J. Agric. Sci.* 55: 464-468.

Mathivanan, N., Srinivasan, K., Chelliah, S. 1997. Evaluation of *Trichoderma viride* and carbendazim for integrated management of root diseases in cotton. *Indian J. Microbiol.* 37: 107-108.

Mathivanan, N., Srinivasan, K., Chelliah, S. 2000. Field evaluation of *Trichoderma viride* Pers. ex. S. F. Gray and *Pseudomonas fluorescens* Migula against foliar diseases of groundnut and sunflower. *J. Biol. Control.* 14: 31-34.

Migheli, Q., Whipps, J. M., Budge, S. P. and Lynch, J. M. 1995. Production of inter- and intra-strain hybrids of *Trichoderma* spp. by protoplast fusion and evaluation of their biocontrol activity against soilborne and foliar pathogens. *J. Phytopath.* 143:91-97.

Mishra, D. K. and Narain A. 1994. *Gliocladium virens* and *Streptoverticillium* as sources of biocontrol of few phytopathogenic fungi. *Indian Phytopath.* 47: 236-240.

Mrinalini, C. 1995. Implementation of protoplast fusants of *Trichoderma* spp. with enhanced biocontrol efficacy to control sheath blight of rice under greenhouse conditions. *Indian J. Mycol. Plant Pathol.* 25 :102.

Mukherjee, P. K., and Haware, M. P. 1993. Biological control of *Botrytis* gray mold of chickpea. *Int. Chickpea Newsltr.* 28 :14-15.

Mukherjee, P. K., and Mukhopadhyay, A. N. 1993. Induction of stable mutants of *Gliocladium virens* by gamma irradiation. *Indian Phytopath.* 46: 393-397.

Mukherjee ,P.K., Haware , M.P and Raghu, P. 1997 . Induction and evalution of benomyl-tolerent mutant of *Trichoderma viride* for biological control of *Botrytis* grey mold of chickpea. *Indian Phytopath.* 50: 485-489,

Mukherjee, P.K., Haware, M. P. and Jayanthi, S. 1995a. Preliminary investigations in integrated biocontrol of *Botrytis* grey mold of chickpea. *Indian Phytopath.* 48: 141-149.

Mukherjee, P. K., Thomas, P. and Raghu, K. 1995c. Shelf-life enhancement of fresh ginger rhizomes at ambient temperatures by combination of gamma-irradiation biocontrol and closed polyethylene bag storage. *Ann.Appl. Biol.* 127: 375-384.

Mukhopadhyay A N. 1987. Biological control of soil-borne plant pathogens by *Trichoderma* spp. *Indian J. Mycol. Pl. Pathol.* 17: 1-10.

Mukhopadhyay, A. N. and Mukherjee, P. K. 1996. Fungi as fungicides. *International J. Trop. Plant Dis.* 14: 1-17.

Mukhopadhyay, A. N., Shrestha, S. M. and Mukherjee, P. K. 1992. Biological seed treatment for control of soil-borne plant pathogens. *FAO Plant Prot. Bull.* 40: 21-30.

Mukhopadhyay, A. N. 1994. Biocontrol of soil borne fungal plant pathogens - current status, future prospect and potential limitations. *Indian Phytopath.* 47: 119-126.

Muthamilan, M. and Jeyarajan, R. 1996. Integrated management of *Sclerotium* root rot of groundnut involving *Trichoderma harzianum* and carbendazim. *Indian J. Pl. Pathol.* 26: 204-209.

Nakkeeran, S. and Devi, P. R. 1997. Seed borne microflora of pigeon pea and their management. *Plant Dis. Res.*2: 197-200.

Pandey, K. K. and Upadhyay, J. P. 1998. Sensitivity of different fungicides to *Fusarium udum, Trichoderma harzianum* and *Trichoderma viride* for integrated approach of disease management. *Veg. Sci.* 25: 89-92.

Padmanabhan, P. and Alexander, K. C. 1983. Seedling rot of sugarcane. Extension publication no. 5, Sugarcane Breeding Institute, Coimbatore.

Papavizas, G. C., Lewis, J. A., Abd-El Moity, T. H. 1982. Evaluation of new biotypes of *Trichoderma harzianum* for tolerance to benomyl and enhanced biocontrol capabilities. *Phytopath.* 72: 126-132.

Papavizas, G.C. and Lewis, J. A. 1983. Physiological and biocontrol characteristics of stable mutants of *Trichoderma viride* resistant to MBC. *Phytopath.* 73:407-411.

Prasad, R. D., Sreerama kumar, P. and Narayanan, K. 1998. Biological control of botrytis grey mold of rose. *J. Mycol. Plant Pathol.* 28: 61-63.

Prasad, R. D. and Rangeshwaran, R. 2000a. An improved medium for mass production of the biocontrol fungus *Trichoderma harzianum*. *J. Mycol. Pl. Pathol.* 30: 233-235.

Prasad, R. D. and Rangeshwaran, R. 2000b. Shelf life and bioefficacy of *Trichoderma harzianum* formulated in various carrier materials *Plant Dis. Res.* 15: 38-42.

Prasad, R. D. and Rangeshwaran, R. 2000c. Effect of soil application of a granular formulation of *Trichoderma harzianum* on seed rot and damping-off of chickpea incited by *Rhizoctonia solani*, saprophytic growth of the pathogen and bioagent proliferation. *J. Mycol. Pl. Pathol.* 30: 216-220.

Prasad, R. D. and Rangeshwaran, R. 2001. Biological control of root and collar rot of chickpea caused by *Sclerotium rolfsii*. *Ann. Plant Prot. Sci.* 9:297-303.

Prasad, R. D., Rangeshwaran, R. and Sunanda, C. R. 2001. Jaggery-An easily available alternative to molasses for mass production of *Trichoderma harzianum*. *Plant. Dis. Res.*17: 363-365.

Prasad, R. D., Rangeshwaran, R., Anuroop, C. P. and Phanikumar, P. R. 2002. Bioefficacy and shelf life of conidial and chlamydospore formulation of *Trichoderma harzianum*. *J. Biol.Control.*16:145-148.

Prasad, R. D., Rangeshwaran, R., Sunanda, C. R. and Vinita, J. 2001. Biological control of black spot of rose caused by *Diplocarpon rosae*. *Ann. Plant Prot. Sci.* 10: 256-259.

Prasad, R. D., R. Rangeshwaran, S. V. Hegde and C. P. Anuroop. 2002. Effect of soil and seed application of *Trichoderma harzianum* on pigeonpea wilt caused by *Fusarium udum* under field conditions. *Crop Protection*. 21: 293-297.

Raghuchander, T, Samiappan, R and Arjunan, G. 1993. Biocontrol of *Macrophomina* root rot of mung bean. *Indian Phytopathol.*46: 379-382.

Rajappan, K. and Ramaraj, B. 1999. Evaluation of fungal and bacterial antagonists against *Fusarium moniliforme* causing wilt of cauliflower. *Ann. Plant Prot. Sci.* 7: 205-207.

Ram, P., Mathur, K. and Lodha, B. C. 1999. Integrated management of rhizome rot of ginger involving biocontrol agents and fungicides. *J. Mycol. Pl. Pathol.* 29: 416-417.

Rangaswamy, S. D., Reddy, P. P. and Nagesh, M. 2000. Evaluation of bio-control agents (*Pasteuria penetrans* and *Trichoderma viride*) and botanicals for the management of root-knot nematode, *Meloidogyne incognita* infecting tomato. *Pest Manag. Hortic. Ecosys.* 6: 135-138.

Reddy, P. P., Rao, M. S., and Nagesh, M. 1996. Management of the citrus nematode, *Tylenchulus semipenetrans*, by integration of *Trichoderma harzianum* with oil cakes. *Nematologia Medit.* 24:265-267.

Reddy, P. P., Nagesh, M., Devappa, V. and Reddy, P. P. Kumar, N. K. K. Verghese, A. 1998. Management of *Meloidogyne incognita* on tomato by Integration of *Trichoderma harzianum*, *Glomus fasciculatum* and neem cake. *Advances in IPM for horticultural crops. Proc.First National Symp. Pest Manag.Hortic. Crops: Environmental Implications and Thrusts, Bangalore, India,* 15-17 October 1997, 349-352.

Sankar P and Jeyarajan R. 1996a. Seed treatment formulation of *Trichoderma* and *Gliocladium* for biological control of *Macrophomina phaseolina* in sesamum. *Indian Phytopath.* 49: 148-151.

Sankar, P. and Jeyarajan, R. 1996 b. Biological control of sesamum root rot by seed treatment with *Trichoderma* spp. and *Bacillus subtilis. Indian J. Mycol. Pl. Pathol.* 26: 217-220.

Sawant, I. S. and Sawant, S. D. 1996. A simple method for achieving high CFU of *Trichoderma harzianum* on organic wastes for field applications. *Indian Phytopath.*49: 185-187.

Sawant, G. G., Desai, P. V. and Padhiar, R. G. 1999. Effect of different spacing, fertilizer dose, use of *Trichoderma viride* and different fungicidal formulations on the occurrence of early blight disease of tomatoes. *Indian J. Environ. Toxicol.*9: 84-87.

Selvakumar R, Srivastava K D, Rashmi Aggarwal, Singh D V, Prem Dureja, Aggarwal, R, Dureja, P.2000. Studies on development of *Trichoderma viride* mutants and their effect on *Ustilago segetum tritici. Indian Phytopath.*53: 185-189.

Sharma, S. K., Verma, B. R. and Sharma, B. K. 1999. Biocontrol of *Sclerotinia sclerotiorum* causing stem rot of chickpea. *Indian Phytopath.* 52: 44-46.

Sharma, B. K. and Basandrai, A. K. 2000. Effectiveness of some fungicides and biocontrol agents for the management of Karnal bunt of wheat. *J. Mycol. Pl. Pathol.* 30: 76-78.

Shanmugam, V., Sriram, S., Babu, S., Nandakumar, R., Raguchander, T., Balasubramanian, P. and Samiyappan, R. 2001. Purification and characterization of an extracellular alpha-glucosidase protein from *Trichoderma viride* which degrades a phytotoxin associated with sheath blight disease in rice. *J. Appl. Microbiol.* 90: 320-329.

Singh, D. 1991. Biocontrol of *Sclerotinia sclerotiorum* (Lib.) de Bary by *Trichoderma harzianum. Trop. Pest Manag.* 37: 374-378.

Singh, D. P. 1999. Chemical and biological control of covered smut of barley. *J. Mycol. Pl.Pathol.*29: 256-257.

Sreenivasaprasad, S. and Manibhushanrao, K. 1990. Biocontrol potential of fungal antagonists *Gliocladium virens* and *Trichoderma longibrachiatum. J. Plant Dis. Prot.* 97 : 570-579.

Sreenivasaprasad, S. and Manibhushanrao, K. 1993. Efficacy of *Gliocladium virens* and *Trichoderma longibrachiatum* as biological control agents of groundnut root and stem rot diseases. *Int.J. Pest Manag.* 39: 167-171.

Srivastava, R. K., Prasad, R. D., Rangeshwaran, R., Wasnikar, A. R., Singh, S. P. and Rao, N. S. 2001 A rapid *in vivo* bioassay method for testing and selection of fungal antagonists of plant pathogens. *J. Biol. Control.* 16:173-176.

Srivastava, R. K., Prasad, R. D., Rangeshwaran, R. and Kannan, P. 2001 Detection and quantification of proteins induced by *Trichoderma* in Chickpea-*Rhizoctonia* system. *Plant Dis. Res.*17:252-255.

Sugha, S. K., Sharma, B. K. and Tyagi, P. D. 1993. Factors affecting development of collar rot of gram (*Cicer arietinum* L.) caused by *Sclerotium rolfsii*. *Indian J. Agric. Sci.*63: 382-385.

Upadhyay, J. P. and Mukhopadhyay, A. N. 1986. Biological control of *Sclerotium rolfsii* by *Trichoderma harzianum* in sugarbeet. *Trop. Pest Manag.* 32: 215-220.

Vyas, S. C. 1994. Integrated biological and chemical control of dry root rot on soybean. *Indian J. Mycol. Pl. Pathol.* 24: 132-134.

Yogendra Singh and Singh, Y. 1998. Biological control of Sclerotinia rot of rapeseed and mustard caused by *Sclerotinia sclerotiorum*. *Plant. Dis. Res.* 13: 144-146.

• • •

Advances in Soil Borne Plant Diseases, 2008
Editors: Naik and Devika Rani, pp. 341-354
New India Publishing Agency, New Delhi

14

Lesion Nematode, *Pratylenchus* Species of Pulses, their Biology Responsible Factors and Management

S.P.Tiwari[1] and M.N.Khare[2]

INTRODUCTION

The genus *Pratylenchus* was established in 1936. It has a cosmopolitan distribution prevalent in temperate, sub-tropical and tropical regions. The species of *Pratylenchus* are popularly known as lesion nematode or meadow nematode. They are migratory endoparasites. Various pulse crops are damaged by them. *P. thornei* attacks chickpea, *P.*zeae on mungbean (Walia and Seshadri, 1986); *P. mulchandani* on urdbean (Nanad Kumar *et al.*, 1969) and *Pratylenchus* sp.on lentil (Mishra and Gaur, 1980) and cowpea (Aung and Prot, 1990).

Besides pulses species of *Pratylenchus* attack a number of other crops also. The infested plants exhibit various types of symptoms and results in heavy yield losses. Investigations have been made on various aspects of host-nematode relationship, distribution, host range, yield losses, biology, life cycle, disease cycle, interaction with other microorganism and management through cultural, physical, chemical, biological methods and host resistance.

Department of Plant Pathology, Jawaharlal Nehru Agricultural University, Jabalpur, (Madhaya Pradesh) 482004 India

SYMPTOMS

The symptoms due to the nematode vary with the host growth stage. At seedling stage infection is manifested by patchy appearance, stunted growth and pale green colour of the lower foliage. Stunting pronounced as the growth advanced. Root lesions vary with the cultivars but small lesions are characteristic of this nematode. The lesion mainly appears on young feeder roots but may be formed any where on roots. The lesions enlarge, coalesce and may girdle the entire root. The secondary attack is by bacteria and fungi resulting in rotting. The severely infected plants are easily pulled out from the soil due to destruction of the root system. Moderately diseased plants exhibit varied stages of root damage. Generally the individual roots are discolored, stubby and mutilated.

The outer root tissues are destroyed and lesions are formed. The number and size of lesion depend on the quantum of the nematode. The lesions are elongate, narrow and dark colored. The nematodes leave the older tissue and invade the fresh ones. In mungbean var. Pusa Baishakhi more than six nematodes reduced the seed germination and also adversely affected the seedling growth (Walia and Seshadri, 1986).

The damaged tissues are liable to get secondary invasion of soil fungi. The nematodes usually feed on root hairs, epiblema, cortex of rootlets and main root. Jenkins and Taylor (1967) have reported conspicuous destruction of parenchymatous cells of the cortex by the nematode and dark, thickened cell walls in the feeding areas.

All stages of *P. penetrans* were observed in the cortex but no nematode was seen in the stele. In case of chickpea *P. thornei* causes vascular and parenchyma disorders in roots and suppress Rhizobium activity (Gresco and Sharma 1990).

HOST RANGE

The species of *Pratylenchus* infect pulse crops chickpea, green gram, black gram, lentil, pigeon pea, cowpea, grass pea, french bean and faba bean besides wheat, rice, barley, maize, sorghum, pearl millet, cotton, lesser millets soybean, peanut, sunflower, potato, ginger, turmeric, coffee, cabbage, cauliflower, carrot, radish, turnip, banana, mandarin, tobacco, apple, plume, grape, poppy, avocado, lima bean, common bean and alfalfa etc.

DISTRIBUTION

Lesion nematodes are common and worldwide and their distribution does not seem to be influenced greatly by climate. Some species are adapted to cool regions and others to warm. According to Christie (1959) they are somewhat more numerous in warmer parts of the temperate zone. The nematode *P. thornei, P. penetrans, P. vulvus, P. coffeae, P. zeae, P.mulchandi, P.minyus P. mediterraneus, P. brachyurus* have been observed in Algeria, Morocco, Tunisia (Di Vito *et al.*, 1994), India (Starr, 1979), Italy and Mexico (Schultz *et al.*, 1976), Spain (Castillo *et al.*, 1995), Syria (Greco *et al.*, 1987), Turkey (Ferraz, 1995), U.K (Smartt, 1976) and Mozaimbigue, USA (Smolik and Everson, 1987). Several species of *Pratylenchus* have been reported from India.

YIELD LOSSES

In chickpea losses in grain yield up to 58 per cent have been reported by Di Vito *et al.* (1992) at 2-nematodes/cm^3 soils. The losses varied in different cultivars, 20 percent in JG 74, 16.43 per cent in JG 315, 22.2 per cent in Annigiri when nematode population ranged between 125 to 525/200 cm^3 soil. Mean yield losses from 1991 to 1993 were 23, 22.5 and 11.9 per cent in chickpea cultivars Annigiri, JG 74 and JG 315 respectively (Anon., 1992, 1993).

In India *Pratylenchus coffeae* causes a loss of nearly $ 3 million to coffee annually (Van Berkum and Seshadri, 1970). It also causes severe damage to ginger (Kaul, 1987) and turmeric (Ganguly and Gill, 1993) crops in Himanchal Pradesh and Kerela. Prasad and Rao (1978) reported reduction of ear heads up to 33%, grain yield up to 55% and 1000-grain weight up to 15.1% in upland rice. The necrotic depressions and protuberances formed on potato tubers lowers the tuber quality by 10 to 20% (Kirjanova and Krall, 1971).

BIOLOGY

Members of the genus *Pratylenchus* are typically vagrant parasites of plant root, although in rare instances they may inhabit above ground parts as well. Both young and adult nematodes enter roots by forcing their way through or between the cortical cells and then feed on the cell contents as they migrate through the tissue (Jenkins and Taylor, 1967).

Species of *Pratylenchus* are small vermiform with length of 0.3 mm to 0.9 mm. They are not extremely slender since most species have an "a" value of 20 to 30. The tail tapers gradually to bluntly rounded terminus and in about three anal body diameters long. Female gonad is single, outstretched and interiorly directed with oocyte in a single row and a conspicuous sperm atheca in bisexually reproducing species. The postvulval uterine branch usually about one body diameter in length but longer in some species may contain a small amount of vertisol ovarian tissue. The vulva is located at about 70 to 80% of the body length in most of the species (Thorne, 1961).

One hundred nematodes/kg soils significantly reduced the plant growth characters (Walia and Seshadri, 1985). A tolerance limit of chickpea to P. thornei was worked out by Di Vito *et al.* (1992) to be 0.031-nematodes/cm^3 soils in Syria.

After four to six days the nematode within the root begins to move about quite readily and about this time the cortical tissue begins to break and appears to be killed. The cortical cells surrounding the nematodes get killed and disrupted and it is assumed that the nematodes die of starvation. The epidermal cells also appear to be necrotic, but retain their original shape and do not disintegrate. The cells of the root appear alive.

In order to see the survival of the nematode in soil, *Pratylenchus thornei* were extracted from the soil after six months plant growth and three months later, after a slow dehydration of soil and storage in dry condition resulted in gradual decline in the population.

Eggs hatch of *P. thornei* took place at all the temperature within the range 10-25° C and was optimal at 20 C. Temperature affected penetration rate in chickpea in cultivar JG 62 and UC 27. No difference in percentage penetration among host genotype was observed at 20 or 25 C. All migratory stages of *P. thornei* penetrated roots of chickpea from first to 11th day after inoculation (Castillo *et al.,* 1996 a). Sixty four percent of root samples with nematode densities ranging from 6-70 /g fresh root (Castello *et al.,* 1996 b) were found when survey of 63 fields in chickpea growing areas of Guadalquivir Valley of Andaluci, Southern Spain was conducted. Number of *P. brachyurus* remained the same on triticale but increased in soybean (Johnson *et al.,* 1998). In soybean 1000 and 5000 nematodes per plant, per pot exhibited growth reduction in plants and decrease in fresh root and top dry weight (Ferraz, 1995). There was a negative relationship (r = -0.73) between plant survived and nematode inoculum densities in case of alfalfa due to *P. penetrans* (Griffin, 1993).

INTERACTION WITH OTHER MICROORGANISMS

Root infection with 10 nematodes/cm^3 soil and 250-2000 chlamydospores significantly increased the root infection by *Fusarium oxysoprum* f. sp. *ciceri.* Plant growth was reduced by *P. thornei* infection in wilt susceptible and wilt resistant chickpea. The severity of root necrosis was greater. Nematode reproduction on the wilt susceptible cultivars was significantly increased by *F. oxysporum* f. sp. *ciceri* infection (Castillo *et al.,* 1998a). The severity of *Fusarium* wilt was not modified by concentration of *P. thornei* and *Fusarium oxysporum* f sp. *ciceri* in wilt susceptible and resistant chickpea cultivars. Root infection of *P. thornei* resulted in higher number of propugules of

F.oxysporum f. sp. *ciceri* in wilt susceptible cultivars but not in the resistant cultivars (Castillo *et al.*, 1990).

Nematode inoculum 10^4/plant greatly reduced the number of *Rhizobium* nodules. Babu *et al.*, (1996) have reported that *Glomus fasciculatum* individually and in combination with bio fertilizers gave higher yields by reducing the multiplication rate of the nematodes in ragi.

SURVEY

Pratylenchus species inhabit soil environments even when host roots are present, hence for estimation of size of their population all possible habitat should be considered. During survey of nematodes associated with chickpea in Madhya Pradesh, *Pratylenchus* was found dominant (24% frequency of occurrence) over other nematodes (Fig.14.1). The nematode was also found attacking soybean and sunflower. Castillo *et al.* (1996 b) surveyed 63 chickpea fields in Spain for *P. thornei* and obtained the nematode from 92% of soil and 84% of root samples, the population densities ranged from 2-48/100 cm^3 and 6-70/g of fresh root. The reproduction of the nematode increased with increase of soil temperature (15-20-25° C).

RESPONSIBLE FACTORS

Several factors like soil type, soil structure, soil texture, soil pH,fertility, fertilizer application, macro and micronutrients, temperature, moisture, rains, tillage practice, sowing time, depth, cropping sequence, crop rotation, fallow period, application of pesticides etc., influence the population of nematodes and severity of diseases caused by them. Some factors have been analyzed concerning species of *Pratylenchus*.

Soil type is said to exert considerable influence on species of *Pratylenchus* towards their distribution and population levels. Several species have been found in higher numbers in light, sandy soils (Jenkins and Taylor, 1967).Anwar and Khan (1992) in Pakistan observed sand and sandy loam soil most favourable for infection and reproduction of *P. zeae* in maize. Clay soil was least suitable and sandy clay loam and silt loams were intermediate. Similar observations were made by Kanwar *et al.* (1996) in Haryana (India), they obtained higher populations of the nematode in light soils, sandy or sandy loam. *P. penetrans* in barley exhibited a strong tendency to increase in number with increasing root zone depth which provides increased host root substrate (Edwards and Kinprinski, 1997).

In maize *P. zeae, P. neglectus, P. penetrans, P. brachyurus* and *P. crenatus* have been observed. A positive correlation was noted between *P. brachyurus* with long-term average annual rainfall (Jordacan *et al.*, 1989).

Jeong and Kim (1989) observed higher reproduction rate in *P. penetrans* on alfalfa and potato callus. Plant growth regulator 2, 4-D significantly increased the production of *P. penetrans* and *P. vulvus* on alfalfa and Korean pine callus. The optimum temperature for the growth was 24° C and 27° C and pH 5 and 7 respectively for the two nematodes.

Pratylenchus pernoxins needs high soil temperature (32-36 C) and soil moisture (7-11% water content) for reproduction of which millet and sorghum are good hosts. The nematode is able to enter anhydrobiosis during dry season with 16-49% survival rates (Banjard and Martiny, 1994).

MANAGEMENT

Several methods have been used in controlling the species of *Pratylenchus* attacking various crops.

Physical (Solarization)

Solarization of soil for 6-8 weeks greatly reduced *P. thornei* population (D. Vito *et al.*, 1991). Sauerborn and Saxena, (1987) also reported more than 90% control of nematode by soil solarization.

Cultural

Crop rotation with hairy indigo or marigold consistently reduced *Pratylenchus brachyurus* (Reddy *et al.*, 1996). Thompson *et al.* (1995) reported control of *P. thornei* in wheat by crop rotation, improved tillage and growing tolerant and resistant varieties. *P.* penetrans was less numerous under congenial tillage in tomato than under strip tillage (McKeown *et al.*, 1998).In Germany celery is infected by *P. bukowinensis, P. projectus* and *P. penetrans* for the control of which crop rotation of crops with other than parsley and carrots is recommended (Knuth and Schrameyer, 1991). Trigoe (1994) reported control of *P. coffeae* on *Colocasia esculenta* by crop rotation with peanut. In fallow - green gram - chickpea rotation, the population of *P thornei* in soil showed a definite pattern. Increase in population corresponding to the presence of chickpea and gradual reduction in the following months was observed by Sebestian and Gupta (1996).

Mountain and Elliot (1962) have demonstrated the effective control of *P. penetrans* by ploughing rye straw into infested soil during summer; the control was effective as fumigation with Dichloropropane-dichloropropene (DD) mixture. Addition of decomposed timothy hay into soil (Morgan and Collins, 1964); chopped paper or white pine straw dust (Miller and Edington, 1962); addition of chitin along with hemicellulosic wastes (Miller *et al.*, 1973); soybean and corn meal, cotton seed oil, groundnut oil, soybean oil 1% (W/w) (Walker *et al.*, 1967) also checked nematode population.

Mechanical tillage was better than ordinary tillage in reducing the number of *P. thornei* when barley is planted after wheat. The nematodes were increased by N fertilizers of previous crops (Thompson, 1995). In apple *Pratylenchus jordanensis* was controlled by early removal of previous infected orchards, incorporation of animal manure or green manure with urea, choosing an appropriate rootstock and maintenance of a layer of organic mulch around trees (Stirling *et al.*, 1995). In walnut, *P. vulnus* population was significantly reduced by using chitin-urea soil amendment (Westerdahl *et al.*, 1992). Pyrite application decreased the population of *P. zeae* in groundnut (Kushwaha and Prasad, 1992). Green manure and crop meal amendments checked *P. neglectus* in potato (Hafiz *et al.*, 1997). *P. penetrans* attacking oriental lilies was controlled by applying nematicides (oxamyl, fenamiphos, ethoprophos) with drip irrigation. Foliar spray of oxamyl also reduced the nematodes in soil and roots (Westerdahl *et al.*, 1992).

Botanicals

Chickpea seeds were coated with latex extracted from *Calotropis gigantean, Euphorbia pulcherrima* and *Carica papaya* and were sown in micro plots in fields infested with *P. thornei*. The latex treatments significantly reduced soil population of *P. thornei* at 45, 90 and 120 days after sowing (Hillocks *et al.*, 1995). Aspyrone was isolated from a culture filtrate of *Aspergillus melleus*, and the structure was determined by a comparison of spectroscopic data with those of an authentic sample. Nematicidal activity of 80.8%, at a concentration of 300 mg/litre was shown by aspyrone towards *Pratylenchus penetrans* (Kimura *et al*, 1996). Mustard and neem cake @ 400 kg/ha increased shoot and root weight after 90 days, however groundnut and linseed cake did not show any negative influence on the nematode.

Hutchinson *et al.*, (1960) have reported that population of *P. penetrans* was lower in soil when pumpkin pieces were left to rot, compared with rest of the soil.

Chemical

The application of aldicarb, carbofuran, fensulfothion or phorate @ 2 percent a.i. seed weight to chickpea significantly reduced final population of *P. thornei* (Walia and Seshadri, 1985). The application of oxamyl reduced *Pratylenchus penetrans* in total root and soil system of *Rabi* cabbage. Soil application of Carbofuran 1, 2 and 3 kg a.i./ha reduced *P.thornei* population and enhanced the yield of chickpea variety Annigiri, JG 74 and JG 315 over untreated control (Gill *et al.*, 1991). The application of carbofuran @ 2 kg a.i./ ha reduced the *P. thornei* soil population (364/200 cm^3) and root (231/10 g) population and increased the yield by 32.35 per cent over control. Seed soaking with monocrotophos (0.1%) was also significantly superior over control in reducing the nematode population and increasing the yield (Anon., 2000).

Resistance

For long term control against *P. thornei,* it is essential to search for resistant source. Attempts have been made to screen the available germplasm lines and varieties for resistance.

Chickpea accessions ICC 6928 and IPC 96-69 showed resistance to *P. thornei*. Six accessions of *C.bijugum*, three of *C. cuneatum*, 11 *C. judaicum* and one *C. yamashitae* were resistant to *P. thornei* (D. Vito., 1995). The chickpea lines GNG 543, GF 88428 and PKG-24 were rated as highly resistant to *P. thornei* as a result of pot tests. Thirty-three lines were rated as resistant and 68 moderately resistant. The nematode was confined to the cortex of the plant and showed no adverse effect on seed germination (Tiwari *et al.*, 1992).

Chickpea cultivars (cv.), GNS 3097, K 959, BDG 63 and BBG 119 supported less population, whereas PDG 83-84, ICC 43, KPG 24, H 8360, BG 329 and ICCC 42 were resistant. This has been confirmed in All India Coordinated Research Project on Plant Parasitics Nematodes. Chickpea varieties were screened against *P. thornei* at Jabalpur, M.P., India. Chickpea cv. JG 1 had minimum reproduction factor (14.9 as against 41.3 in JG 62). Inoculation of three days old seedlings with 1000 nematodes/500 cm^3 soil against 20 genotypes showed resistant reaction, however ten genotypes exhibited moderately resistant reaction, whereas 15 were found with moderate resistance (MR) during, 1991. Out of 24 genotypes, 14 were found with moderate reaction against *P. thornei* (Anon., 1992), whereas chickpea lines ICC 12245 and ICC 12267 showed resistant reaction (Anon., 1987). Chickpea genotypes ICC 11313, ICC 11317, ICC 11319, ICC 11323,ICC 12241, ICC 12242, ICC 12244, ICC 12246, ICC 12247, ICC 12249, ICC 12259, ICC 12254, ICC 12255, ICC 12259, ICC 12270, ICC 12273 and ICC 12275 were found to be wilt resistant besides moderate resistant reaction against *P. thornei* (Anon., 1989). One hundred varieties/lines of chickpea were screened against *P. thornei* for their resistance during 1995-99 and JG 62 served as standard check. Out of the varieties/lines screened KW-80-2, H-91-80 GNG-1078 and BG-1035 showed resistance reaction against the nematode, whereas, BG-1054, BGD 117 and BGD- 93 were moderately resistant (Anon., 2000).

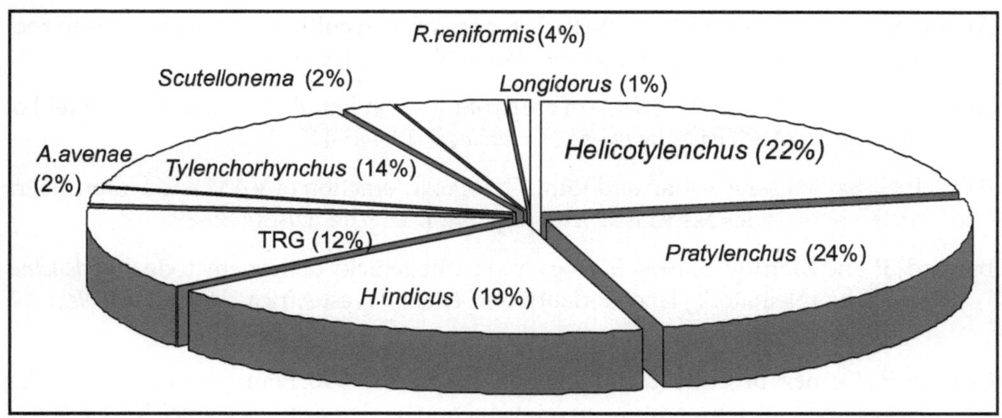

Fig.14.1. Plant parasitic nematodes of chickpea in Madhya Pradesh

FUTURE STRATEGY

Looking to the importance of pulses and the damage caused by lesion nematodes, it appears essential to streamline and strengthen the research. Factors influencing the nematode population, host- nematode interaction like temperature, pH, micro and macronutrients, drought, flood, tillage, planting date, soil type, soil structure, crop rotation, stubbles of previous crops, weeds, soil amendments, root structure etc., need detailed investigations. Bionomics of the nematode should be worked out. More information is needed on the host range and influence by biotic and abiotic factors in the agro-ecological on the nematode likes species of *Meloidogyne, Heterodera, Rotylenchulus reniformis, Hoplolaimus* etc., causing damage to pulse crops. It is necessary to make integrated approach against these nematodes leading to their control. Resistant varieties should be developed suiting to agro climatic conditions.

REFERENCES

Anonymous, 1987. All India Coordinated Project on Plant parasitic Nematodes with integrated approach for their control. *Proc. workshop*, College of Agriculture, Pune, Mahatma Phule Agric. Univ., Rahuri, Maharastra, June 9-12. pp: 128.

Anonymous. 1992. Annual report 1991-92. Directorate of Pulses Research. Indian Council of Agricultural Research. Kanpur pp:149.

Anonymous. 1993. All India Coordinated Project on Plant parasitic Nematodes with integrated approach for their control. *Consolidated Biennial Report* (1991-93), Project Coordinating Cell. Division of Nematology, IARI, New Delhi. pp:108.

Anonymous. 2000. All India Coordinated Project on Plant parasitic Nematodes with integrated approach for their control. *Quinquinnial Report* (1995 to 1999) Jabalpur Center, Department of Plant Pathology, JNKVV, Jabalpur (M.P.). pp:65.

Anwar, S. A. and Khan, M. S. A. 1992. Response of corn cultivars and soil types to root lesion nematode, *Pratylenchus zeae*. *Pak. J. Nematol.* 10:103-111.

Aung, T. and Prot, J. C. 1990. Effects of crop rotations on *Pratylenclus zeae* and on yield of rice cultivar UPL Ri-5. *Revue-de-Nematologie*. 13:445-447.

Babu, R. S., Sanakaranaryanan and Jothi, G. 1996. Interaction of VAM and biofertilizers on the nematodes associated with ragi. *Int. J. Trop. Pl. Dis.* 14:85-89.

Baujard , P. and Martiny, B. 1994. Ecology and pathogenicity of the nematode *Pratylenchus pernoxius* (Nemato: Tylenchulidae) from Senegal, West Africa. *Afro-Asian J. Nematol.* 4:7-10.

Castillo, P., Gomez. Bercina, A. and Jimenez. Diaz, R.M, 1990. Plant parasitic nematodes associated with temperature on reproduction by *Pratylenchus thornei. Nematologica.* 42:211-219.

Castillo, P., Jimenez Diaz, R. M., Gomez Barcina, A. and Vovlas,N.1995. Parasitism of the root lesion nematode *Pratylenchus thornei* on chickpea. *Plant Pathol.*44:728-733.

Castillo, P., Trapero Casa, J. L. and Jimenez Diaz, R.M.1996.The effect of temperature on hatching and penetration of chickpea roots by *Pratylenchus thornei*. *Plant Pathol.*.45: 310-315.

Castillo, P., Gomez Barcina, A. and Jimenez Diaz, R. M. 1996.Plant parasitic nematodes associated with chickpea in southern Spain and effect of soil temperature on reproduction by *Pratylenchus thornei*. *Nematologica*.42: 211-219.

Castillo, P., Mora Rodriguez, M. P., Navas Cortes, J.A. and Jimenez Diaz, R.M.1998. Interactions of *Pratylenchus thornei* and *Fusarium oxysporum* f.sp. *ciceris* on chickpea. *Phytopath.*88: 828-836.

Castillo, P. and Jimenez Diaz, R. M 1998.Interactions between *Pratylenchus thornei* and *Fusarium oxysporum* f. sp. *ciceris* in chickpea. 3rd European conference on grain legumes. *Opportunities for high quality, healthy and added-value crops to meet European demands*. Valladolid, Spain, 14-19 November 1998:259.

Castillo, P., Vovlas, N. and Jimenez Diaz, R. M. 1998.Pathogenicity and histopathology of *Pratylenchus thornei* populations on selected chickpea genotypes. *Plant Pathol.*47: 370-376.

Christie, J. R. 1959. Plant Nematodes. Their bionomics and control. Agricultural Experiment Station, University of Florida, Gainesville, Florida. pp:114-125

Culver, D. J., Ramming, D. W. and Mckenry, M. V. 1989. Procedures for field and green house screening of *Prunus* genotypes for resistance and tolerance to root lesion nematode. *J.American Soc. Hortic. Sci.* 114:98-99.

Di Vito, M., Greco, N., Saxena, M. C. and Di-Vito M. 1991.Effectiveness of soil solarization for control of *Heterodera ciceri* and *Pratylenchus thornei* on chickpea in Syria . *Nematologia Medit.*19: 109-111.

Di Vito, M., Greco, N., Saxena, M. C. and Di, Vito M. 1992 Pathogenicity of *Pratylenchus thornei* on chickpea in Syria. *Nematologia Medit.*20: 71-73.

Di Vito, M., Greco, N., Oreste, G., Saxena, M. C., Singh, K. B., Kusmenoglu, I. and Di, Vito M. 1994. Plant parasitic nematodes of legumes in Turkey. *Nematologia Medit.* 22: 245-251.

Di Vito, M., Greco, N., Halila, H. M., Mabsoute, L., Labdi, M., Beniwal, S. P. S., Saxena, M. C., Singh, K. B., Solh, M. B. and Di Vito-M. 1994. Nematodes of cool season food legumes in North Africa. *Nematologia Medit.*22: 3-10.

Di Vito, M.., Zaccheo, G., Catalano, F., and Di, Vito M.1995.Response of chickpea lines to *Meloidogyne artiellia* and *Pratylenchus thornei. Nematologia Medit.*23: 81-83.

Edwards, L. M. and Kimpinski, J. 1997. Relationships between soil penetration resistance in soil nematode burden in barley on Price Edward Island. *Bio. & Fert. Soil.* 24:13-17.

Ferraz, L. C. C. B. 1995. Pathogenicity of *Pratylenchus brachyurus* to three soybean cultivars. *Nematologia Brasileira.* 19:1-18.

Ganguly, S. and Gill, J. S. 1993. Nematode problems of crops in Himalayan Regions of India and their control. (*In*:) B.K. Dwivedi M. M. Alam and G. Pandey (Eds.) *Integrated Pest Management*. Bioved Research Society, Allahabad. pp:165-188.

Gill, J. S., Singh, R. V. and Sud, U. C. 1991. Achievements (1977-87). All India Coordinated Project on Plant parasitic Nematodes with integrated approach for their control (Complied). *Tech. Bull.*1: 1-27.

Greco, N., Saxena, M. C and Singh, K. B. 1987. Nematodes and their control in chickpea. *The Chickpea*. 271-281.

Greco, N., Vito, M-di, Saxena, M. C, Reddy, M. V. and Di-Vito-M. 1988. Investigation on the root lesion nematode *Pratylenchus thornei*, in Syria. *Nematologia-Medit.*16: 101-105.

Greco, N., Vito-M-di. and Di-Vito-M. Nematodes of food legumes in the Mediterranean Basin. *Bulletin-OEPP*

Greco, N., Vito, M, di, Saxena, M. C., Reddy, M. V. and Di Vito, M.1988. Investigation on the root lesion nematode *Pratylenchus thornei*, in Syria. *Nematologia-Medit.*16:101-105.

Greco, N. and Sharma, S. B. 1990. Progress and problems in the management of nematode diseases. In chickpea in the nine tees. *Proc. Second Int. Workshop on Chickpea Improvement.* ICRISAT, Patencheru, India. pp:135-137.

Greco, N., Vito, M. di, Saxena, M. C., Di Vito, M 1991.Soil solarization for control of *Pratylenchus thornei* on chickpea in Syria. FAO-Plant-Production and Protection Paper. (*In:*) *Proc. First Int. Conf. Soil Solarization,* Amman, Jordan, 19-25 Feb.1990 (Eds.) DeVay, J.E., Stapleton, J.J. and Elmore, C.L). 109: 182-188.

Griffin, G. D.1993. Comparative response of alfalfa to *Pratylenchus penetrans* populations. *J. Nematol.* 25:461-465.

Hafiz, S.L., Al-Rehiayani, S., Weiser, G. C. and Monser, E. M. 1997. New concepts in management systems for potato nematodes in Idaho. *Proc.First Int. Workshop of Afro-Asian Nematologists,* Menowfiya University. Egypt. pp:33-38.

Hillocks, R. J., Siddiqi, M. R and Khonga, E. B. 1995. Nematodes associated with subsistence crops in southern Malawi. *Afro-Asian J. Nematol.*5: 14-19.

Hutchinson, M. T., Reed , J.P. and Pramar, D. 1960. Observations on the effect of decaying vegetables matter on nematode populations. *Plant Dis. Reptr.* 44:400-401.

Jenkins, W. R. and Taylor, D. P. 1967. *Plant Nematology*. Rein Hold Publishing Corporation. New York. pp:68-77.

Jeong, M. G. and Kim, S. S. 1989. Effects of plant callus, temperature, pH, medium and growth regulating substances on the culture of *Pratylenchus penetrans* and *P. vulnus in vitro. Korean J. Pl. Path.* 5:359-363.

Johnson, A. W., Dowler, C. C., Baker, S. H. and Handoo, Z. A. 1998. Crop yields and nematode population densities in triticale- cotton and triticale – soybean rotations. *J. Nemato.* 30:353-361.

Jordocan, E.M., Wacle, D. de, Rooyen, P. J. Van, De Wacle, D. and Van Rooyen, P.J. 1989. Endoparasitic nematodes in maize roots in the western Transval as related to soil texture and rainfall. *J. Nemato.* 21:356-360.

Kanwar, R. S., Rana, B. P., Walia, K. K. and Walia , R. K. 1996. Studies on the pathogenic potential of *Heterodera zeae* and *Pratylenchus zeae* on maize in Haryana. *Ind. J. Nemato.* 26:7-14.

Kirjanova, E. S. and Krall, E. L. 1971. *Plant parasitc nematodes and their control.* Vo. II. Amerind Publishing Company. New Delhi.

Kimura, Y., Nakahara, S. and Fujioka, S. 1996.Aspyrone, a nematicidal compound isolated from the fungus, *Aspergillus melleus. Biosci. Biotec. Biochem.* 60: 1375-1376.

Knuth, P. and Schrameyer, K. 1991. Nematodes of Celery. *Gemuse Munchen.* 27:136-140.

Kushwaha, S. and Prasad, D. 1992. Effect of sources of sulphur and urea on the growth of groundnut and nematode population. *Curr. Nematol.* 3:7-10.

Kaul, K. J. 1987. Studies on nematodes associated with ginger in Himanchal Pradesh. Thesis, H. P. University, Shimla .pp. 198.

McKeown, A. W., Cerkauskas, R. F., Potter, J. W., Driel, L. Van and Van Driel, L. 1998. Long-term evaluation of cover crop and strip tillage on tomato yield, foliar diseases and nematode populations. *Can. J. Pl. Sci.* 78:341-348.

Miller, P. M. and Edington, L. V. 1962. Effects of paper and sawdust soil amendments on meadow nematodes and subsequent Verticillium wilt of tomato. *Plant Dis. Reptr.* 46: 745-746.

Miller, P. M., Sands, D. C. and Rich, R. 1973. Effect of industrial mycelial residue, wood fiber wastes and chitin on plant parasitic nematodes and soil-borne diseases. *Plant Dis. Reptr.* 57:438-442.

Mishra, S. D. and Gour, H. S. 1980. Reaction between nematization levels of the root knot nematode, *Meloidogyne incognita* and growth of lentil *Lens culinaris* Linn. *Ind. J. Entom.* 42: 262-263.

Morgan, G. T. and Collins, W. B. 1964. The effect of organic treatment and crop rortation on soil populations of *Pratylenchus penetrans* in strawberry culture. *Can. J. Pl. Sci.* 44:272.

Mountain, W. B. and Elliot, J. M. 1962. The effect of summer fallowing on the root lesion nematode *Pratylenchus penetrans*, and yield of flue-cured tolerance in Ontario. *Can. J. Pl. Sci.* 42:642-645.

Nandkumar, C., Khera, S. and Bhatnagar, G.C. 1969. *In vito* studies on the host range of *Pratylenchus mulchandani Indian Phytopath.* 22:521-522.

Prasad, J. S. and Rao, Y.S. 1978. Potential of *Pratylenchus indicus* Das. 1960, the root lesion nematode pest of up land rice. *Ann. Zool. Ecol. Anim.* 16:635-640.

Reddy, N. C., Soffee, A. R., Princ, C. M. and Dunn, R. A. Tropical legumes for green manure II. Nematode populations and their effects on succeeding crop yields. *Agron. J.* 78:5-10.

Sebastian, S. and Gupta P 1996. Evaluation of some oil cakes against *Pratylenchus thornei* on chickpea. *Int. Chickpea. Pigeonpea Newsltr.* 3: 39-40.

Sebastian, S. and Gupta P 1996. Evaluation of three plant lattices as seed treatment against *Pratylenchus thornei* on chickpea. *Int. Chickpea. Pigeonpea. Newsltr.* 3: 40-41.

Sebastian, S. and Gupta, P. 1997.Crop loss trial of chickpea infested with *Pratylenchus thornei*. *Ind. J. Nematol.*27: 142-143.

Schultz, O. F., Roth, M. and Jenkins, S. E.1976. Vegetable crop disease reports. *Fungicide and Nematicide Tests Results.* 31:75-111.

Smartt, J. 1976. *Tropical Pulses* . Southampton University.U.K. pp:348.

Smolik, J. D and Everson, P. D. 1987. Relationship of yield and *Pratylenchus* spp. population densities in dry land and irrigation corn. *Ann. App. Nematol.* 1:71-73.

Starr, J. L. 1979. Plant parasitic nematodes associated with sorghum pearlmillet, groundnut, pigeonpea, and chickpea at ICRISAT, Hyderabad, *Ind. J. Nematol.* 11:314-315.

Stirling, G. R., Dullahide, S. R. and Nikulin, A. 1995. Management of lesion nematode (*Pratylenchus jordanensis*) on replanted apple trees. *Aust. J. Exper. Agr.* 35:247-258.

Sauerborn, J. and Saxena, M. C. 1987. Effect of soil solarization on Orobanche spp. infestation and other pests in faba bean and lentil. *Proc. 4th Inter. symp. parasitic Flowering Plants.* ICARDA Syria 733-744.

Tiwari, S. P., Vadhera, I., Shukla, B. N. and Bhatt, J. 1992.Studies on the pathogenicity and relative reactions of chickpea lines to *Pratylenchus thornei* (Filipjev, 1936) Sher & Allen, 1953. *Ind. J. Mycol. Pl. Path.* 22:255-259.

Thompson, J. P., Mackenzie, J and Amos, R. 1995. Root lesion nematode (*Pratylenchus thornei*) limits response of wheat but not barely to stored soil moisture in the Hermitage long term tillage experiment. *Aust. J. Experi. Agr.* 35:1049-1055.

Trigoe, H. 1994. Control of the root lesion nematode (*Pratylenchus coffeae*) on taro (*Colocasia esculenta*) by crop rotation with peanut and/ or other crops. *Proc. Ass. Pl. Pro. of Kyushu.* 40:113-118.

Thorne, G. 1961. *Principles of Nematology.* McGraw-Hill Book Company, INC. New York. pp: 203-211.

Van Berkam, J. A. and Seshadri, A. R. 1970. Some important nematode problems in India. 10th International Nematological Symposium ESN, Peseara, Italy: 136-137.

Walia, R. K. and Seshadri, A. R.1985 Pathogenicity of the root-lesion nematode *Pratylenchus thornei* on chickpea. *Int. Chickpea Newsltr.* 12: 31.

Walia, R. K and Seshadri, A. R. 1985 (b). Chemical control of *Pratylenchus thornei* on chickpea through seed treatment. *Int.Chickpea Newsltr.* 13: 32-34.

Walia, R. K and Seshadri, A. R. 1986. Pathogenicity of root lesion nematode to greengram. *Ind. J. Agril. Sci.* 56:792-795.

Walker, J. T., Specht, C. H. and Mavrodinean, S. 1967. Reduction of lesion nematodes in soybean meal and oil amendment soil. *Plant Dis. Reptr.*51:1021-1024.

Westerdahl, B. B., Carlson, H. L., Grant, J., Radewald, J. D., Welch, N., Anderson, C. A., Darso, J., Kirby, D. and Shibua, F. 1992. Management of plant parasitic nematodes with a chitin- urea soil amendment and other materials. *J. Nematol.* 24: 669-680.

Westerdahl, B. B., Girand, D., Radewald, J. D., Anderson, C. A. and Darso, J. 1994. Management of *Pratylenchus penetrans* on oriental lilies with drip and foliar applied Nematicides. *J. Nematol.* 25:758-767.

• • •

Advances in Soil Borne Plant Diseases, 2008
Editors: Naik and Devika Rani, pp. 355-369
New India Publishing Agency, New Delhi

15

Soil Solarization by Polyethylene Mulching for the Management of Soil-Borne Diseases

Harender Raj

INTRODUCTION

Soil-borne pathogens cause heavy losses to most agricultural crops. Repeated planting of a crop in the same piece of land results in a high inoculum build-up, which forces the farmers either to change the crop or the land. Thus, the search for new, effective, inexpensive and non-hazardous methods for the control of soil-borne diseases is a continuous one. Over the years, efforts have been made to manage the soil-borne pathogens by employing fumigants, chemical pesticides, biological control agents, different type of soil amendments and also by causing modifications in different cultural practices. But, pathogen control is often hampered by the fact that both the inoculum of the pathogens and the lethal agents applied to the soil are affected by the soil's physical, chemical and biological environment. Keeping in view the bottlenecks with different methods of management of soil-borne pathogens; harnessing of solar energy i.e. soil solarization is an effective method to control soil-borne pathogens.

Department of Mycology and Plant Pathology,Dr. Y.S. Parmar University of Horticulture and Forestry Nauni, Solan - 173 230, Himachal Pradesh

Soil solarization is done by covering the moistened soil in summer months with thin and transparent polyethylene mulch for capturing solar energy for heating the soil by increasing soil temperature, which becomes lethal to the soil-borne pathogen. Polyethylene (popularly known as polythene), one of the most important plastic materials used in agriculture, was first introduced on a commercial scale in 1939 (Byrdson, 1970). However, Jones *et al.*, (1966) for the first time applied solar sterilization with polyethylene mulch in combination with fumigants and found it effective against southern blight of tomato caused by *Sclerotium rolfsii*. The use of clear polyethylene film to cover moistened soil and trap lethal amounts of heat from solar radiations was developed for the first time in Israel (Katan *et al.*, 1976). The pesticide activity of solarization was found to stem from a combination of physical, chemical and biological effects (Chen *et al.*, 1991; Katan 1987; Stapleton, 2000).

PRINCIPLES OF SOLAR HEATING

Soil solarization involves the use of thin transparent polyethylene mulch for covering the soil surface for capturing solar energy for heating the soil. The heat generated in the soil due to increase in soil temperature acts as a lethal agent in controlling or inactivating soil-borne pathogens in soil. On an average, one square centimeter area outside the earth's atmosphere and parallel to its surface receives 2 calories /cm^2/ minute of energy in the form of solar radiation but only half of it finally reaches the ground (Farrell and Rose, 1967). The energy is lost from the soil in the form of long wave radiations, through conduction, convection and water evaporation. Absorption of solar radiation varies according to the colour, moisture and texture of the soil. In general, the soil has a relatively high thermal capacity and is a poor heat conductor. Transparent, not black or any other colour, polyethylene mulch should be used since it transmits most of the solar radiations that heats the soil (Katan, 1981; Raj and Kapoor, 1993). Transparent polyethylene mulch reduces heat convection and water evaporation from the soil to the atmosphere and as a result water droplets are formed on the inner surface of the polyethylene sheet, which reduces the transmissivity of long wave radiation, thus resulting in better heating due to an increase in the greenhouse effect (Malik and Tran, 1973). Solar heating is carried out at relatively low temperature, as compared to artificial heating; thus, its effect on living and non-living soil components is less drastic.

In studies carried out in different countries, soil solarization with transparent polyethylene mulch is reported to cause increase in soil temperature to different levels. In Israel, during July-August in different years, the recorded soil temperatures of mulched soil at depths of 5 and 20 cm were

45-55°C, respectively (Katan, 1981). The average daily air temperature for those months was 25-26°C. In California, temperatures of tarped soils were reported to be reaching upto 60°C at the 5 cm depth (Grinstein *et al.*, 1979). Soil solarization with thin transparent polyethylene mulch is reported to cause substantial increase ranging from 7 to 14°C in 5 to 7.5 cm depths of solarized soils in India in different studies conducted at different places during summer months (Raj and Bhardwaj, 2000; Raj and Gupta, 1995; Raj *et al.*, 1995; Lodha *et al.*, 1997; Patel, 2001).

MECHANISM OF DISEASE CONTROL

Reduction in disease incidence occurring in solarized soils, results from the effects exerted on host, pathogen, surrounding microorganisms and soil environment as all these factors are involved in disease development (DeVay and Katan, 1991; Katan *et al.*, 1989). But the most pronounced effect of soil mulching with polyethylene is a physical one i.e. an increase in soil temperature, for several hours of the day. However, other accompanying processes such as shifts in microbial populations changes in chemical composition and physical structure of the soil (Chen *et al.*, 1991), high moisture levels maintained by the polyethylene mulch, and changes in gas composition of the soil, should also be considered when analyzing mechanism of disease control. Such processes may be especially useful where the cumulative effect of heat may be insufficient for disease control, i.e. at deeper soil layers or at marginal seasons.

The physical effect of soil solarization is exerted on the soil-borne pathogens by subjecting them to moist heat, at temperatures exceeding the maximum required for the growths and hence their viability is reduced. The thermal death rate of an organism depends on both the temperature level and exposure time, which are inversely related. The success of soil solarization is based on the fact that most plant pathogens and pests are mesophilic or unable to survive for long periods above 37°C. The heat sensitivity of these organisms is related to an upper limit in the flaccidity of cell membranes which lose their ability to function at high temperatures. Other causes of death of organisms at high temperatures involve the sustained inactivation of enzyme system, especially respiratory enzymes (Anonymous, 1991). In practice, populations of soil-borne fungal pathogens are greatly reduced at temperatures of 40-50°C, exposure time ranging from minutes to hours for the higher temperatures and upto days for the lower ones. This temperature range matches the one recorded at the upper soil layers. Wet heat in solarization is more lethal as in the presence of water, less energy is required to unfold the peptide chain of proteins, resulting in a decreased heat resistance (Precht *et al.*, 1973). The surviving propagules in treatment of soil solarization are also affected. Weakened propagules may possess lower inoculum potential and shorter longevity due to slower germination or growth reduces number and length of mycelial germinating from a multicellular propagule, reduced capacity to produce enzymes, ruptures in cell membranes and leakage of nutrients from cells (Katan, 1981). Thus, propagule counts in heated cultures under sterile conditions may underestimate the thermal effect on pathogen survival and pathogenic capacity under natural conditions.

In addition to physical effect of heat, microbial processes induced by solarization may contribute to disease control (Chen and Katan, 1980; Stapleton and DeVay, 1982). If induced by solarization, biological control may affect the pathogen by increasing its vulnerability to soil micro-organisms or increasing the activity of soil micro-organisms, which will finally lead to a reduction in disease incidence, pathogen survivability or both (Lifshitz *et al.*, 1983; Merriman, 1976). The sub-lethal heat results in either weakening of the walls of the pathogenic propagules by fluctuating environmental conditions (Smith, 1976) or a shift in the microbial population in favour of antagonists (Katan *et al.*, 1976; Merriman, 1976). Inoculum potential may be reduced due to antibiosis or competition enhanced by soil solarization. Reduced disease incidence was observed in cotton and tomato plants inoculated with sub-lethally heated *Fusarium* conidia (Katan, 1981). Solar heating was also found to reduce the inoculum potential of *Rhizoctonia solani* and its build up in the soil when beans were repeatedly planted in the infested field (Elad, 1980). The above examples demonstrate the potential of solar heating in inducing biological control.

SOLARIZATION TECHNIQUES

Generally, a layer of thin transparent polyethylene (25-50 µm) is applied to the soil surface prior to planting and the mulch is left in place for 4 to 6 weeks during the hot season in the appropriate climatic region (Pullman *et al.*, 1979). Optimal use, however, may require longer period of time and adjustments in scheduling for other production practices. Use transparent (not black or any other colour) and thin (25-30 µm) UV stabilized plastic (polyethylene or polyvinyl chloride) mulch is best in efficiency as it transmits better radiations (Macfayden, 1967; Katan *et al.*, 1976; Horowitz and Regev, 1980). However, combining solarization with soil amendments, biological control agents, reduced doses of pesticides and other methods of disease management may reduce the duration of mulching. Proper soil preparation is also essential to provide a smooth, even surface for the polyethylene mulch and allow water to penetrate evenly and deeply into the soil. Patch holes immediately, if caused during solarization, with duct tape to prevent heat loss. To maintain proper soil moisture, irrigation is given to the area to be solarized 1 to 4 days prior to applying the plastic tarp. Polyethylene mulch may be applied either in strips (usually 2 feet wide) over the planting beds or as a continuous sheeting glued, heat fused or held in place by burying the sides in soil. Because pathogens that survive heat treatment within or at the periphery of treated soils tend to multiply (a phenomenon known as the edge or boarder effect), continuous sheeting is though to be more effective than strip application. The edges of the sheets must be buried to a depth of 5 to 6 inches in the soil to prevent blowing or tearing of the tarp by the wind. The effects of solarization can extend through several growing season (Afek *et al.*, 1991). After solarization, the polyethylene mulch is removed and the crop can be raised in the solarized fields but the mulch can be retained and the crops like strawberry can be raised by making holes in the sheet.

DISEASE MANAGEMENT

Field experiments have been carried out, since 1974, in different parts of the world for evaluating the effectiveness of soil solarization in disease management. The method was experimented against different soil-borne pathogens, which are responsible for causing wilt, collar rot and root rot diseases in various crop plants. These pathogens multiply faster in soil due to repeated cultivation of a crop and are more difficult to control. In general, soil solarization was found effective against several soil-borne pathogens under different agro-climatic zones. The technique of soil solarization has now been tested and modified under local conditions in more than 50 countries, including the United States, Israel, Greece, Morocco, Japan, Italy, Jordan, United Kingdom, England and India (Hartz *et al.*, 1993; Katan, 1981; Ristaino *et al.*, 1991, 1996).

Soil solarization was found effective against the wilt pathogen *Verticillium dahliae* as the incidence in tomato was reduced by 65 per cent, while the reduction was more than 95 per cent in case of potatoes (Katan *et al.*, 1976; Grinstein *et al.*, 1979). Likewise, the incidence of pink root (*Pyrenochaeta terrestris*) in onion was reduced by 90 per cent (Katan *et al.*, 1980). These results therefore, indicate that the efficiency of soil solarization varies with soil-borne pathogens infecting different hosts. The effectiveness of solar heating with polyethylene mulch in controlling cotton wilt (*Fusarium oxysporum* f.sp. *vasinfectum*) could be seen even after three years of mulching (Katan *et al.*, 1980). Viability of propagules of *F.o.* f.sp. *vasinfectum* that had been buried at 30 cm depth, was reduced by 97.5, 58 and 0 per cent under double layer, single layer and non-covered soil, respectively after 31 days of mulching with transparent polyethylene (Ben-Yephet *et al.*, 1987). Subsequent work on soil solarization indicated that soil solarization not only effectively reduced pathogen population in the soil and decreased wilt incidence in cotton plants but improved plant growth, weed control and yield also. The work on soil solarization gained momentum in different parts of the world and a large number of workers studied the pattern of heating in mulched and non-mulched soils and its impact on soil-borne pathogens and disease incidence. In United States, soil solarization for either 30 or 60 days was effective in delaying wilt symptoms as well as reducing disease incidence of *Fusarium* wilt of watermelon and the temperature maxima in solarized soils were 60, 50, 42 and 37°C at depths 2, 10, 20 and 30 cm, respectively (Martyn and Hartz, 1986). Soil mulching with 50µm polyethylene mulch increased maximum temperatures upto 69°C in dry field and 58°C in wet field, 6 and 5°C more than in unmulched controls and resulted in reduction of sclerotia of *Macrophomina phaseolina* from the initial count of 350 to bare 7 sclerotia in 15 days of solarization at a depth of 5 cm (Lodha *et al.*, 1991). This resulted in reduction of disease incidence of dry root rot (*M.phaseolina*) in cluster bean with significant increase in grain yield. The various diseases which have been effectively controlled by soil solarization in different parts of the world are: *Verticilllium* diseases (tomato, brinjal, potato, cotton, safflower); *Rhizoctonia solani* (potato, onion); *Sclerotium rolfsiii* (peanuts, strawberry, bell pepper, beans); *Pyrenochaeta lycopersici* (tomato); *Pyrenochaeta terrestris* (onion); *Fusasrium* wilt diseases (cotton, melon, tomato, onion, mango, citrus), (Elad *et al.*, 1980; Grinstein *et al.*, 1979; Horowitz, 1980; Jocobsohn *et al.*, 1980; Katan *et al.*, 1976; Sando *et al.*, 1977; Raj and Gupta, 1995, 1996; Raj and Bhardwaj, 2002; Pullman *et al.*, 1981).

Soil solarization has been demonstrated to control *Phytophthora cinnamom* infested orchards in Spain (Lopez-Herrera *et al.*, 1997). Different workers from the United States also reported the effectiveness of soil solarization against different soil-borne pathogens of tomato, bell pepper, brinjal, strawberry and lettuce (Gamliel and Stapleton, 1993; Hatz *et al.*, 1993; Ristaino *et al.*, 1991). Diseases like belly rot (*Rhizoctonia solani*) of cucumber, wilt (*Fusarium oxysporum* f.sp. *ciceri*) of chickpea, damping-off (*Pythium aphanidermatum*) and black shank (*Phytophthora parasitica* var. *nicotianae*) of tobacco, and stem root rot (*Macrophomina phaseolina*) and red rot (*Thielaviopsis basicola*) of sesame are also controlled effectively by mulching the infested soil with transparent polyethylene mulch (Keinath, 1995; Rao and Krishnappa, 1995; Chattopadhyay and Sastry, 1996; Adams, 1971). Mulching with transparent polyethylene mulch (25 µm) was also found effective against pre- and post-emergence damping-off (*Pythium* spp., *Fusarium* spp.) of different vegetable crops and recorded higher seed germination, lower incidence of post-emergence damping-off and better seedling vigour in solarized nursery beds (Raj *et al.*, 1997). In a vegetable field with infestation of different soil-borne pathogens, soil solarization effectively reduced the incidence of wilt diseases of broccoli, cabbage, cauliflower, Chinese sarson, lettuce, onion and tomato from 10 to 21 in unsolarized plots to 0 to 6 per cent in solarized plots accompanied with 35-161 per cent increase in yield in different crops. The treatment of soil solarization can also be combined with soil amendments for enhancing its capability of controlling the soil-borne pathogens. Soil solarization when combined with crucifer residues in moistened soil resulted in increase of soil temperature to 57 and 50°C at 0-15 and 16-30 cm depths respectively and resulted in 93-99 per cent decrease in population of *Macrophomina phaseolina* at both soil depths within 15 days of mulching (Lodha *et al.*, 1997). Soil solarization is also effective against bacterial pathogens causing wilt and other diseases in plants. Population densities of *Agrobacterium* spp., fluorescent pseudomonads, $G^{+positive}$ bacteria and fungi were greatly reduced after solarization under English walnut (Stapleton and DeVay, 1982). Population of *Agrobacterium* causing crown gall disease of pome and stone fruits was decreased by 99 per cent. In sandy loam soil, the target bacteria was eliminated in 4 weeks, while in silty clay soil, the population was markedly reduced after 2 months (Raio *et al.*, 1997). Solarization increased the frequency of recovery showing antagonistic activity from the rhizosphere and roots of tomato plants (Gamliel and Katan, 1991). Mulching with transparent polyethylene for 32 to 49 days effectively checked the population density of *Pseuodmonas salanacearum* down to a depth of 15 cm (Chellemi *et al.*, 1994). Enhanced biological activity of beneficial biological control agents in deeper soil profile (15 to 30 cm) also contribute substantially in management of soil-borne plant pathogens (Chen *et al.*, 1991).

OTHER BENEFICIAL EFFECTS

Control of Nemic Diseases : Mulching of soil with transparent polyethylene affects every component of soil and so also plant parasitic nematodes. Significant reduction of 42 to 100 per cent was observed in solarized soil of *Meloidogyne, Heterodera, Pratylenchus, Paratrichodorus, Circonemella, Xiphinema* and *Pratylencus* spp. in comparison to non-solarized soil (Stapleton and DeVay, 1982). Soil solarization also caused 93 to 100 per cent reduction in populations of nematode species like *Heterodera cajani, Rotylenchulus reniformis, Helicotylenchus retusus, Pratylenchus* sp. and *Tylenchorhynchus* sp. parasitic to chickpea and pigeonpea (Sharma and Nene, 1990). Mulching of nursery beds with transparent polyethylene was also found effective in reducing the incidence of root knot disease in tomato and incidence of stunt, root knot and reniform nematodes in tobacco (Patel *et al.*, 1991; Patel and Markwana, 1992; Patel and Patel, 1997). The cyst nematode (*Heterodera avenae)* of wheat was completely controlled by solarization (Grinstein *et al.*, 1995).

Weed Control

Solarization results in an effective weed control lasting in some cases for a whole year or even longer (Horowitz, 1980; Katan *et al.*, 1980; Rubin and Benjamin, 1983; Elmore *et al.*, 1997). Transparent polyethylene was found more efficient than the black film in weed control (Horowitz, 1980). Most of the annual and many perennial weeds were effectively controlled, including species of the following genera: *Amaranthus, Anagallis, Avena, Ageratum, conyjoides, Azinga bractiosa, Artemesia* sp., *Capsella, Chenopodium, Cynodon, Centella, Cyperus rotundus, Digitaria, Elensine, Euphorbia helioscopea, Fumaria, Ipomea cornea, Lactuca, Lathyrus aphaca, Malvestrum tricuspidatum, Mercurialis, Montia, Notobasis, Oxalis corniculata, Phalaris, Poa, Portulaca, Parthenium hysterophorus, Polygonum* sp., *Renunculus muricatus, Sisymbrium, Solanum, Stellaria, Trifolium repens* and *Veronica biloba.* Reduction in weed population was more in wet soils under solarizaton than dry soil (Tiwari *et al.*, 1997). Soil solarization created a long term effect by causing a reduction in weed population extending for two to three season in tomato (Abdel-Rahim *et al.*, 1988). Mulching with polyethylene for 15 days resulted only reduction of population of *Orobanche* from 1493 per 40m^2 unsolarized plot to only 33 in wet mulched plots and also resulted in 27.1 per cent increase in yield of tomato fruits. Soil solarization can be effective for weed management even in cooler areas (Elmore *et al.*, 1993). The possible mechanism of weed control are direct killing of weed seeds by heat; indirect microbial killing of seeds weakened by sub-lethal heating; killing of seeds stimulated to germinate in the moistened mulched soil and killing of germinating seeds whose dormancy is broken in the heated soil (Katan, 1981; Tiwari *et al.*, 1997; Elmore *et al.*, 1997). Volatiles released and trapped in solarized soil may also play a role in weed control (Horowitz, 1980; Rubin and Benjamin, 1984).

Increased Growth Response : Plant growth in disinfested soil is, as expected, enhanced as compared to untreated, infested soil, as a result of pathogen control. Less expected is a phenomenon, the increased growth response of plants in a soil free of known pathogens and disinfested (Tietz, 1970). Different mechanisms have been suggested for explaining increased growth response in disinfested or solarized soil; increased micro and macro elements in the soil solution; elimination of minor and unknown pathogens or parasites; destruction of phytotoxic substances in the soil; release of growth regulator like substances; and stimulation of mycorrhizae or other beneficial microorganisms (Altman, 1970; Broadbent *et al.*, 1977; Chen and Katan, 1980; Millhouse *et al.*, 1979). Soil micro-organisms likely to be beneficial to plant growth are either less affected by soil solarization (e.g. *Bacillus* spp., actinomycetes) than are pathogenic organism or they are able to quickly recolonize treated soil (e.g. fluorescent pseudomonads) (Elad *et al.*, 1980; Grinstein *et al.*, 1979; Stapleton and DeVay, 1982; Ristaino *et al.*, 1991). Heating causes the release of soluble mineral nutrients from soil organic matter and heat killed soil biota and also induces the upward movement of mineral nutrients in the soil profile. Disease reduction in solarized soils was usually accompanied by an increase in yield and other growth parameters to various extents. In tomato, soil solarization resulted in increasing yield between 100 and 350 per cent in fields infected with *Pyrenochaeta terrestris* (Tamietti and Garibaldi, 1980; Tjamos and Faridis, 1980). Similarly, a field infested with tomato wilt pathogen (*Fusarium oxysporum* f.sp. *lycopersici*) recorded 10.8 times more number of fruits in plots mulched with transparent polyethylene (25 μm) for 40 days in comparison to non-mulched plots (Raj and Kapoor, 1993). Soil solarization also resulted in increase in yield of safflower and egg-plant by 113 and 215 per cent, respectively, in soil infested with *Verticillium dahliae* (Katan *et al.*, 1976; Pullman, 1979). Similarly, soil solarization of *Pyrenochaeta terretris* infested field of onion resulted in an increase of 60 to 125 per cent in yield (Katan *et al.*, 1980). In a field with high incidence of different soil-borne diseases of different crops, soil solarization resulted in an increase of 35 to 161 per cent in yield of different crops (Raj and Bhardwaj, 2000).

Soil solarization also results in improved growth in plants. Soil solarization not only reduced the incidence of different soil-borne diseases in cotton (*Fusarium oxysporum* f.sp. *vasinfectum*), onion (*Pyrenochaeta terrestris*), mango (*Fusarium solani*) and strawberry (*Sclerotium rolfsii*) but also resulted in an increase in shoot and root length and other growth parameters (Katan *et al.*, 1980, 1983; Raj and Gupta, 1996; Raj and Bhardwaj, 2002).

ADVANTAGES, LIMITATIONS AND FUTURE PROSPECTS

Soil solarization, as a disinfection method, has potential advantages. It is a non-chemical method which is not hazardous to the user and does not involve substances toxic to the consumer, to the host plant or to the other organisms. In comparison, most of the pesticides used for the management of soil-borne pathogens cause pollution of drinking water, toxicity to aquatic life, contamination of agricultural produce and severely affect the beneficial soil micro-organisms and soil-ecosystem. Soil solarization has a long term effect and the method has the characteristic to fit in the integrated disease pest management strategy.

Solarization also has some limitations and some potential negative side effects. It can not be used effectively and inclusively in very cooler areas. The soil should be free for about one month or more at the time of mulching. The method may be expensive for some crops and ineffective in the control of certain diseases. Early detection and characterization of undesired effects can be checked by combining or alternating solarization with other methods of disease management.

In future, development of economic, biodegradable and novel plastic, or other materials, more efficient than polyethylene in trapping solar energy will be more useful and thus reducing our dependence on climate and making this method available to cooler regions. Furthermore, soil solarization can be combined with other chemical, physical and biological methods (e.g. fertilizers, soil amendments, reduced dosages of pesticides and biological control agents) for enhanced management of soil and root pests and diseases. Among combinations of different methods with soil solarization; combining solarization with readily available residues of cabbage or other *Brassica* spp. as well as composted animal manures provides improved disease management via enhanced release of biotoxic volatile compounds from the amendments in heated soil and amendments of such materials and their effectiveness can be further explored (Ramirez-Villapudna and Munnecke, 1987; Gamliel and Stapleton, 1993a, 1993b).

CONCLUSION

The principle of soil solarization has been now well established and has been tested in more than 50 countries of the world. The ultimate goal is to make this method practically applicable in different agro-climatic zones. As soil solarization affects the soil-ecosystem; disciplines of Soil and Soil Microbiology are also involved in the process and thus interdisciplinary efforts are needed to answer the intricacies involved in the process. This method should not be regarded as a universal method but rather as an additional method which can be used efficiently in integrated disease/pest management strategy. In disease management, we never aim complete control and in such a scenario, soil solarization is quite effective to reduce the soil-borne pathogens of different crops to that economic threshold where they are not able to cause much loss. Thus, soil solarization is a new, additional option, which can be used efficiently in integrated disease management strategies in different crops. The adoption, applicapability and dissemination of this technique in the future will depend on further research on the limitations of the method and refinement of the technology to make it more cheap and efficient. But, the growing concern about residues of pesticides in agricultural produce, drinking water and water based products and on the other hand growing demand for agricultural produce, produced through organic farming techniques, certainly highlight the importance of this technique in diseases.

REFERENCES

Abdel Rahim, M. F., Satour, M. M., Mickail, K. Y. and El Eraki, S. A. 1988. Effectiveness of soil solarization in furrow irrigated Egyptian soils. *Plant Dis.* 72:143–146.

Adams, P. B. 1971. Effect of soil temperature and soil amendments on *Thielaviopsis* root-rot of sesame. *Phytopath.* 61:93-97.

Afek, U. Menge, J. A. and Johnson, E. L. V. 1991. Interaction among mycorrhizae, soil solarization, metalaxyl and plants in the field. *Plant Dis.* 75:665-672.

Altman, J. 1970. Increased and decreased plant growth responses resulting from soil fumigation. *(In:)* Touson, T. A., Bega, R. V. and Nelson, R. E. (Eds.). *Root Diseases and Soil-Borne Pathogens.* pp: 216-221. Univ. Calif. Press, Berkley.

Anonymous, 1991. Soil Solarization *(In:)* DeVay, J. E., Stapleton, J. J. and Elmore, C. L. (Eds.) F.A.O. Plant Production and Protection Paper 109.pp:396.

Ben Yephet, Y., Stapleton, J. J., Wakeman, R. J. and DeVay, J. E., 1987. Comparative effects of soil solarization with single and double layers of polyethylene film on survival of *Fusarium oxysporum* f.sp. *vasinfectum. Phytoparasitica.*15:181-185.

Byrdson, J.A. 1970. Plastic Materials (2nd Edition). Iliffe Books, London. pp:597.

Broadbent, P., Baker, K. F. Franks, N., Holland, J. 1997. Effect of *Bacillus* spp. on increased growth of seedlings in steamed and non-treated soil. *Phytopath.* 67:1027–1034.

Chattopadhyay, C. and Sastry , R. K. 1996. Management of sesame stem root–rot through soil sterilization –An Eco friendly strategy. *Indian J. Mycol. Pl. Pathol.* 26:123.

Chelemi, D. O., Olson, S. M. and Mitchell, D. J. 1994. Effect of soil solarization and fumigation on survival of soil-borne pathogens of tomato in Northern Florida. *Plant Dis.* 78: 1167–1172.

Chen, Y. and Katan, J. 1980. Effect of solar heating of soils by transparent polyethylene mulching on their chemical properties. *Soil Sci.* 130:271-277.

Chen, Y., Gamliel, A. Stapleton, J. J. and Avaid, J. 1991. Chemical physical and microbial changes related to plant growth in disinfected soils. *(In:)* Katan, J. and DeVay, J.E. (Eds.). *Soil Solarization* pp: 103-129. CRC Press, Boca Raton, USA.

DeVay, J. E. and Katan, J. 1991. Mechanism of pathogen control in solarized soils. (*In:*) *Soil Solarization.* CRC Press, Boca Raton, Florida, USA. pp:87-101.

Elad, Y. Katan, J. and Chat, I. 1980. Physical, biological and chemical control integrated for soil-borne diseases in potatoes. *Phytopath.* 60: 1291

Elmore, C. L., Roncaroni, J. A. and Giraud D. D. 1993. Perennial needs respond to control by soil-solarization. *Califorina Agri.* 47:19-22.

Elmore, C. L. Stapleton, J. J. Bell, C. E. DeVay, J. E. 1997. Soil solarization: A non- pesticidal method for controlling diseases, nematodes and weeds. UC DANR Pub. 21377, Oakland, C.A. pp:14.

Farell, J. and Rose, A. H. 1967. Temperature effects on microorganisms. *(In:)* Rose, A.H (Ed.) *Thermobiology* pp:147-218. Academic Press, London. pp:653.

Gamliel, A. and Katan, J. 1991. Involvement of fluorescent *Pseudomonas* and other micro organisms in increased growth response of plants in solarized soils. *Phytopath.* 81: 494-502.

Gamliel, A. and Stapleton, J. J. 1993 a. Characterization of antifungal volatile compounds evolved from solarized soil amended with cabbage residues. *Phytopath.* 83:899-905

Gamliel, A. and Stapleton, J. J. 1993 b. Effect of soil amendment with chicken compost or ammonium phosphate and solarization on pathogen control, rhizosphere microorganisms and lettuce growth. *Plant Dis.* 77: 886–891.

Grinstein, A. Katan, J. Abdul Razik, A. Zeydan, O. and Elad, Y. 1979. Control of *Sclerotium rolfsii* and weeds in Peanuts by Solar heating of soil. *Plant Dis Reptr.* 63: 1056-1060

Grinstein, A, Kritzman, G. , Hetzroni, A., Gamliel, A., Mor, M. and Katan, J. 1995. The border effect of soil solarization. *Crop Protection.* 14: 315-320.

Hartz, T., DeVay, J. and Elmore, C. 1993. Solarization is an effective soil disinfection technique for strawberry production. *Hort. Sci.* 28: 104-106.

Horowitz, M. 1980. Weed research in Israel. *Weed Sci.* 28: 457-460.

Horowitz, M. and Regev, Y. 1980. Mulching with plastic sheets as a method of weed control. *Hassadeh.* 60:395-399.

Jacobson, R., Greenberger, A., Katan, J., Lev, M. and Alon, H. 1980. Control of Egyptian broomrape (*Orobanche aegyptiaca)* and other weeds by means of solar heating of the soil by polyethylene mulching. *Weed Sci.* 28: 312-316.

Jones, J. P., Overman, A. J. and Geraldson, C. M. 1996. Effects of fumigants and plastic films on the control of several soil-borne pathogens of tomato. *Phytopath.* 56: 929–932.

Katan, J. Greenberger, A. Alon, H. and Grinstein, A. 1976. Solar heating by polyethylene mulching for the control of diseases caused by soil-borne pathogens. *Phytopath.* 66: 683–688.

Katan, J. Rotem, I. Frinkel, Y. and Daniel, J. 1980. Solar heating of the soil for the control of pink root and other soil-borne diseases in onion. *Phytoparasitica* 8:39–50.

Katan J. 1981. Solar heating (solarization) of soil for control of soil-borne pests. *Ann. Rev. Phytopath.* 19:211-236.

Katan, J., Fisher, G., and Grinstein, A. 1983. Short and long terms effect of soil solarization and crops sequence on *Fusarium* wilt and yield of cotton in Israel. *Phytopath.* 73: 1215–1219.

Katan, J. 1987. Soil solarization. *(In:)* Chet, I. (Ed.) *Innovative Approaches to Plant Disease Control* pp: 77–105. John Wiley and Sons, New York.

Katan, J., DeVay, J. E. and Greenberger, A. 1989. The biological control induced by soil solarization. *(In:)* Tjamos, E.C. and Beckman, C.H. (Eds.) *Vascular Wilt Diseases of Plants.* pp :293-299. Springer Verlag, Berlin.

Keinath, A.P. 1995. Reductions in inoculum density of *Rhizoctonia solani* and control of belly rot of pickling cucumber and solarization. *Plants Dis.* 79:1213-1219.

Lifshitz, R., Tabachnik, M., Katan, J. and Chet, I. 1983. The effect of sub-lethal heating on sclerotia of *Sclerotium rolfsii. Can J. Microbiol.* 29:1607-1610.

Lodha, S., Singh, M. and Sharma, B.M. 1991. Solarization brings down soil–borne pathogens in arid lands. *Indian Farmg.* 40:12-13.

Lodha, S., Sharma, S.K. and Aggarwal, R.K. 1997. Solarization and natural heating of irrigated soil amended with cruciferous residues for improved control of *Macrophomina phaseolina. Plant Pathol.* 46:186-190.

Lopez–Herrera, C.J., Perez-Jimenze, R.M., Basallote–Ureba, M.J., Zeabonilla T. and Melero-Vara, J.M. 1997. Effect of soil solarization on the control of *Phytophthora* root- rot in Avacodo. *Plant Pathol.* 46: 329–340.

Macfayden, A. 1967. Thermal Energy as a factor in the biology of soils. *(In:)* Rose, A.H (Ed.). *Thermobiology* pp:535–553, Academic Press, London & New York. pp:653.

Malik, M. A. S. and Tran, U. V. 1973. A simplified mathematical model for predicting the nocturnal output of solar stile. *Solar Energy.* 14: 371-385.

Martyn, R. B. and Hartz, J. K., 1986. Use of solarization to control *Fusarium* wilt of watermelon. *Plant Dis.* 70:762-766.

Merriman, P. R. 1976. Survival of sclerotia of *Sclerotinia sclerotiorum. Soil Biol. Biochem.* 8:385–389.

Millhouse, D. E. and Munnecke, D. E. 1979. Increase in growth of *Nicotiana glutinosa* as partially related to accumulation of ammonium–nitrogen in soil fumigated with methyl bromide. *Phytopath.* 69: 793- 797.

Patel, D.J. Patel, H.V. and Patel, S.K. 1991. Comparative efficacy of LDPE films covers for soil solarization in control of root–knot nematodes in tomatoes nurseries. *Plasticulture.* 3: 33–36.

Patel, H. R. and Makwana, M. G. 1992. Integrated management of phytonematodes in bidi tobacco nursery through soil solarization, rabbing and carbofuran. *Tobacco Res.* 18:33-37.

Patel, B. K. and Patel, H. R. 1997. Effect of nematicides, soil solarization rabbing and green mannuring on some physical and chemical properties of soil nematode infested bidi tobacco nursery. *Tobacco Res.* 23:19–26.

Patel, D.J. 2001. Soil solarization for management of soil-borne plant diseases *J. Mycol. Pl. Pathol.* 31: 1-8.

Pullman, G. S., DeVay, J. E., Garber, R. H. and Weinhold, A. R. 1979. Control of soil-borne pathogens by plastic tarping of soil. (*In:*) Schippers, B. and Grams, W. (Eds.). *Soil-Borne Plant Pathogens*, pp 439-444, London, New York, San Francisco, Academic Press. pp:686.

Pullman, G. S., DeVay, J. E., Garber, R. H. and Weinhold, A. R. 1981. Soil solarization for the control of *Verticillium* wilt of cotton and the reduction of soil-borne populations of *Verticillium dahliae, Pythium* spp., *Rhizoctonia solani* and *Thielaliopsis basicola. Phytopath.* 71: 954–959.

Precht, H., Christophersen, J. Hensel, H. and Larcher, W. 1973. *Temperature and Life.* Springer, Berlin. pp:779.

Raj Harender and Kapoor, I. J. 1993. Soil solarization for the control of tomato wilt pathogen (*Fusarium oxysporum* Schl.) Z. *Pflkrankh Pflschutz.*100:652–661.

Raj Harender and Gupta V. K.1995. Soil solarization for the control of citrus wilt (*Fusarium solani*) in the nursery. *Indian J. Mycol Pl. Pathol.* 25: 111.

Raj Harender, Sharma, N. K. and Gupta, V. K. 1995, Effect of soil solarization on population of *Fusarium solani. Biol. Cult. Tes. Cont. Pl. Dis.* 10:55.

Raj Harender and Gupta, V. K. 1996. Soil solarization for the control of mango wilt (*Fusarium solani*) in nursery. *Indian J. Agricul. Sci.* 66: 258–262.

Raj Harender, Bhardwaj, M. L. and Sharma, N. K. 1997. Soil solarization for the control of the damping-off of different vegetable crops in the nursery. *Indian Phytopath.* 50 : 524- 528.

Raj Harender and Bhardwaj, M.L. 2000. Soil solarization for controlling soil-borne pathogens in vegetable crops. *Indian J. Agricult. Sci.* 70:305–307.

Raj Harender and Bhardwaj, Umang 2002. Use of soil solarization in integrated disease management of root- rot (*Sclerotium rolfsii*) of strawberry. *Indian J. Mycol. Pl. Pathol.* 32: 416.

Raio, A., Zoina, A. and Moore, L.W. 1997. The effect of solar heating of soil on natural and inoculated agro-bacteria. *Plant Pathol.* 46: 320–328.

Ramirez Villapudua, J., Munnecke, D.M. 1987 control of cabbage yellows (*Fusarium oxysporum* f.sp. *conglutinans)* by solar heating of fields amended with dry cabbage residues. *Plant Dis.* 71: 217–221.

Rao, V. K. and Krishnappa, K. 1995. Soil solarization for the control of soil – borne pathogen complexes with special reference to *Meloidogyne incognita* and *Fusarium oxysporum f.sp. ciceri. Indian Phytopath.* 48 :300–303.

Ristaino, J. B., Perry, K. B. and Lumsden, R. D. 1991. Effect of soil solarization and *Gliocladium virens* on sclerotia of *Sclerotium rolfsii,* soil microbiota and the incidence of southern blight in tomato. *Phytopath.* 81: 1117–1124.

Ristaino, J.B., Perry, K.B. and Lumsden, R.D. 1996. Soil solarization and *Gliocladium virens* reduce the incidence of southern blight (*Sclerotium rolfsii*) in bell pepper in the field. *Biocontrol Sci. and Tech.* 6/4: 583–593.

Rubin, B. and Benjamin, A. 1983. Solar heating of the soil; Effect on weed control and on soil incorporated herbicides. *Weed Sci.* 31: 819–825.

Rubin, B. and Benjamin, A. 1984. Solar heating of the soil; Involvement of environmental factors in the weed control process. *Weed Sci.* 32: 138–142.

Sando, Z., Even, S. and Hoffman, A. 1977. Control of *Orobanche* by methyl bromide and by thermal disinfection. Hassadeh 57: 2321–2323.

Sharma, S. B. and Nene, Y. L. 1990. Effect of soil solarization on nematode parasitic to chickpea and pigeonpea. *Supple. J. Nematol.* 22:658-664.

Smith, A. M. 1976. Ethylene in soil biology. *Ann Rev. Phytopath.* 14: 53-73.

Stapleton, J. J. and DeVay, J. E. 1982. Effect of soil solarization on population of selected soil – borne microorganisms on growth of deciduous fruit tree seedlings. *Phytopath.* 72: 323-326.

Stapleton, J. J. 2000. Soil solarization in various agricultural production systems. *Crop Protection.* 19:837-841.

Tamietti, G. and Garibaldi, A. 1980. Control of corky root in tomato by solar heating of the soil in greenhouse in Riviera Ligure. *La Difesa delle Piente* 3:143-150.

Tietz, H. 1970. One centenium of soil fumigation; Its first years. *(In:)* Tousson, T.A., Bega, R.V. and Nelson, R.E. (Eds.). *Root Diseases and Soil- borne Pathogens* pp: 203-207, Univ. Calif. Press, Berkeley.

Tjamos, E.C. and Faridis, A. 1980. Control of soil-borne pathogens by solar heating in plastic houses. In: *Proc. 5th Congr. Mediterr. Phytopath. Union.* pp: 2-84. Hellenic Phytopathol. Soc., Athens.

Tiwari, R.K.S., Parihar, S.S. and Chaure, N.K. 1997. Soil solarization for the control of *Sclerotium rolfsii* causing sclerotial root-rot of chickpea. *Plant Protection.* 25:142-145.

•••

Advances in Soil Borne Plant Diseases, 2008
Editors: Naik and Devika Rani, pp. 371-413
New India Publishing Agency, New Delhi

16

Integrated Management of Phytophthora Diseases

M. Gunashekaran

INTRODUCTION

Many economically important diseases are caused by *Phytophthora* spp. There are as many as 60 described *Phytophthora* species, most of which are primary invaders of plant tissues with limited saprophytic ability. Some important diseases with respect to symptoms, etiology and management have been described in this chapter as given below.

1. Blight of pepper
2. Quick wilt of black pepper
3. Wilt of betel vine
4. Blight of cardamom
5. Black shank of tobacco
6. Bud rot of coconut
7. Gummosis of citrus
8. Diseases of soybean
9. Leaf blight and rot of onion

Central Plantation Crops Research Institute, Kasaragod - 671 124, Kerala

10. Late blight of potato
11. Late blight of tomato
12. Root rot of avocado
13. Sudden oak death
14. Root and crown rot of tomato
15. Blight cucurbits
16. Root rot of papaya
17. Heart rot of pineapple.
18. Blight of taro
19. Black stripe of rubber
20. Abnormal leaf fall of rubber
21. Gummosis of durian
22. Crown, collar and root rot of apple and pear
23. Stem blight of pigeon pea
24. Seedling dieback of cocoa
25. Black pod disease of cocoa
26. Stem canker of cocoa
27. Chupon blight and twig dieback of cocoa.

BLIGHT OF PEPPER (*Capsicum annuum* L.)

Phytophthora blight, root and crown rot can be serious diseases affecting pepper or Chilli (*Capsicum annuum* L.) as well as eggplants, black pepper, cacao, tomatoes, and the entire cucurbit family. Because it affects such a wide range of vegetables, growers are challenged to develop adequate rotational strategies. Consequently, control must depend upon cultural, chemical and selection of resistant varieties when available.

Symptoms

Phytophthora blight affects both seedlings and mature plants. Infected seedlings show typical damping-off symptoms. Infection of older plants usually begins at or below the soil line, water-soaked, dark brown lesions on the lower stems (collar rot phase) usually extend upward for an inch or more above the soil line and may expand to girdle the stems, preventing upward movement of water and nutrients. This often results in a sudden wilting of foliage. Root infections may also occur, which kill roots and cause wilting of the plant without the appearance of stem cankers. The foliar phase of this disease commonly occurs at forks in the branches, resulting in dark, girdling cankers and wilting fruits, Infected leaves develop circular or irregular, dark green, water-soaked lesions which dry and appear light tan. The disease on the pods first appear as a small, water-soaked, dull green spot that rapidly elongates under favorable conditions to cover more than one quarter of the pod, which then shrivel and darken, A mass of white fungal growth may develop inside the fruit, and seeds usually turn dark brown or black. A fine, grayish-white to tan mold may also become evident over the lesion on the fruit surface. Under humid conditions, fungal growth develops extensively over the entire fruit, and the infected fruit often collapse or "melt down" in the field or in storage.

Causal Organism

Phytophthora blight is principally caused by *P. capsici,* although a similar fungus. *P. parasitica* has also been reported to cause fruit rot on chilli and cucurbits. Both of these fungi are dependent on free water in soil for infection. Because of this, initial infection by *Phytophthora* usually occurs on plants growing in poorly drained areas of fields.

The fungi survive as thick-walled zoospores in soil and as mycelium in infected plant tissues. Once introduced, these fungi can survive up to 15 months in moist soil in the absence of host plants. They can also be carried on seed or transplants. Once stem infection occurs, the fungi produce spores on infected stem tissues, which are then carried by splashed rain onto nearby plants. Lower branches of adjacent plants can also be infected by rain-splashed soil contaminated by run-off water. Fruits in contact with the soil are especially prone to infection. Spores are produced on newly infected fruit and stems, and new infections can develop quickly in a short time. Warm, wet weather is most favorable for disease development and spread of the pathogen. Disease progress declines when dry weather prevails.

Management

Selection of cultivars that are resistant and planting in soils that is well drained is the most effective methods of control. Since zoospores can develop in roots and stems of diseased plants, removal of the plant debris from infected fields should reduce the over-wintering inoculum. The collar rot phase of the disease can be controlled with repeated applications of Metalaxyl (Ridomil)

(Papavizas and Bovers, 1991). Crop rotation can be practiced so that the susceptible (chilli, tomato, eggplant and cucurbits) are grown only every 3 to 4 years to reduce the amount of *Phytophthora* present, in the soil. Use of non-susceptible crops such as corn, small grains, beans, crucifers, or potatoes is desirable in rotation. Severity of the disease can be reduced by irrigating the field less frequently (Ristaino, 1991). Application of fungicides may reduce disease development and spread, if used in combination with cultural practices such as crop rotation and raised beds. Application of fungicides at the time of planting and two or more (depending on the fungicide used) applications after planting are required.

QUICK WILT OF BLACK PEPPER (*Piper nigrum* L.)

The 'quick wilt' disease has caused major losses to black pepper (*Piper nigrum* L.) in India, Vietnam and Indonesia and has also been reported in Thailand.

Symptoms

The leaves become pale green, the vines droop and the leaves curl inward along the edges, become yellow and then drop prematurely. After leaf drop, the fruit dries. Within 7 to 14 days, most of the leaves are lost, resulting from the rotting of the main root system and girdling of the stem at the crown region. The root infection is usually unnoticed until the leaves begin to yellow. Deep brown lesions appear on leaves near the ground. The fimbriate edges of the lesions are a diagnostic symptom. In dry weather, the lesions are not active, and the fimbriations become indistinct. With periods of recurrent rainfall, lesions may become concentrically zoned. Leaf infection results in defoliation. Leaf rot phase does not occur in some areas in which foot rot is prevalent. Immature leaves of *P. belle* are more susceptible to this pathogen. Stems can be infected at any place in the vine. The infected portion of the vine then becomes water soaked, the foliage turns yellow, and leaves fall. The infection gradually progresses both up and down the vine, resulting in die-back of the affected stem.

Occasionally, only a single vine may be affected. The aboveground wilt symptoms result from the complete destruction of the main lateral roots and the girdling of the stem at the crown region. The infection of the roots is usually undetected until after the appearance of the foliar yellowing. In advanced stages, the affected tissues emit a foul odour. All parts of the plant-roots, unsuberized stems, leaves flowers and fruit are susceptible. Vines more than 3 years old are more susceptible.

Causal organism

The cause of quick wilt disease is *P. capsici,* which is pathogenic to black pepper *(Pipernigrum)* and betle vine *(Piper belle* L.).

Epidemiology

The pathogen survives in infected plant debris, and infection occurs throughout the main roots. Leaf infection can occur after rain splash has dispersed inoculum from the soil. The disease is more severe in the wet season when temperatures vary between 19 and 23 °C, and symptoms are usually delayed if the temperature remains above 28°C (Nambiar and Sarma, 1982). Soils low in organic matter has been shown to increase disease incidence.

Management

- Planting on well drained soils that have been free of black pepper for at least one year is desirable.
- Soil may be treated with copper based fungicide such as Bordeaux mixture, after removing diseased plants and debris.
- Metalaxyl and to a lesser extent, Fosetyl-Al as foliar spray were found to be effective (Ramachandran and Sarma, 1985).
- Addition of 2 kg organic matter (farmyard manure or neemcake) fortified with bio-control agents like *Trichoderma harzianum* or *T. viride* to the soil is recommended.

WILT OF BETEL VINE *(Piper betle L.)*

Betel vine or pan *(Piper betle L.)* is widely cultivated in moist tropical, sub-tropical regions of India, Bangladesh, parts of south-east Asia and Papua New Guinea for its leaves. It is utilized mainly for the masticatory purposes.

Symptoms

Initially the leaves develop a slight droop, which affects the whole canopy. As droop becomes more pronounced, the leaves become flaccid and the condition at this stage may be accompanied by a paling of colour to a yellow green. Finally the leaves and shoots turn yellow and wither, eventually drying out to a pale brown colour. Under humid conditions, circular, brown, water-soaked spots rapidly enlarge to cover the entire leaf and eventually turn black. Fimbriate edges of the lesions are the diagnostic symptom. A white fuzz of mycelium develops on the under side of the leaf if wet conditions prevail. When wet periods are interspersed with warm or dry periods, lesions develop concentrically, on the leaves. The disease may also extend into the petioles and occasionally into the stem. Leaf rot may result in 10 to 30% loss of the crop (Saksena, 1977).

More rapid development of the disease occurs at high humidity and temperatures above 20° C. The waterlogged soils and overhead shade provide ideal conditions for the development of foot rot. Foot rot results in sudden yellowing of otherwise healthy-looking vines followed by wilting of the tender shoots and leaves. When such plants are uprooted, the roots appear dark brown or black. The disease usually begins in the feeder roots and gradually progresses through the main roots to the collar region of the plant.

Causal organism

Wilt of betel vine is caused by *Phytophthora nicotianae.* The disease in the betel vine orchards develop suddenly during prolonged rainy and cloudy periods which are favorable for the development of zoospores on the infected soil. With the rain splashes, the inoculum was carried over to the lower leaves first and then elsewhere on the aerial parts. It was observed that the irrigation water from the ponds adjoining the orchards can also act as an important source for infection of leaves as the pathogen was isolated from these irrigation ponds. There is a regular practice of spraying water on the leaves, the source of water is usually from the contaminated ponds. The growers usually clean the leaves of betel vine in these ponds and very often throw away the rotten leaves containing *Phytophthora* in and around the ponds. This leads to the contamination of the pond water with *Phytophthora* during the rains.

Management

Irrigation water that is contaminated by debris infested with *P. parasitica* should not be used. Weeds should be removed from the immediate vicinity of planted area, since the pathogen over-winters in several weed hosts (Saksena and Mehrotra, 1970). Cuttings dipped in a streptomycin solution followed by sprays with 1 % Bordeaux mixture at 15 days intervals were protected from the disease at planting time.. Phenyl amide (e.g., Metalaxyl) or phosphonate (e.g., Fosetyl-Al) fungicides are also found to be effective. Amending the soil with corn straw supplemented with ammonium nitrate or with *Trichoderma viride* was found effective (Mehrotra and Tiwari, 1976).

BLIGHT OF CARDAMOM *(Elettaria cardomum* (L.) Maton)

Cardamom *(Elettaria cardomum* L.) is often referred to as the "queen of spices". In India the production is drastically reduced by blight caused by *P. parasitica* (Nair and Menon, 1982).

Symptoms

Blight is indicated by dark green or water-soaked lesion on mature leaves; the lesions gradually enlarge and turn necrotic. Under wet and warm conditions, the whole leaf becomes infected, shrivels, or shreds but remains attached to the plant. The leaf sheaths covering pseudo stem show brownish discoloration with water-soaked margins at the soil levels. A characteristic reddish pink colour can also be seen on the inner sheath even before rotting. The disease progresses under favorable conditions into the panicles. The panicles and capsules become water-soaked. When young, immature capsules are infected, they usually shed within 3 to 5 days. Infection of mature capsules results either in shriveling or reduction of the quality of the crop; the infected areas of the panicle usually turn yellow and eventually shrivel.

The tissue on infected rhizomes becomes dark brown, soft, and watery. Shoots arising from the rhizome either are infected early or may survive for a season before succumbing to the disease, infected roots are discolored at the tips. The disease then spreads towards the base of the root, in advanced stages, the epidermis and cortex may disintegrate, exposing the vascular tissues. Blight is usually observed at the beginning of the monsoon season and becomes severe 2 to 3 months later. The pathogen survives in infected rhizomes for long periods. Affected rhizomes and seeds are sources of inoculum for spread of the pathogen (Nair and Menon, 1982).

Causal organism

The pathogen is *P. parasitica* survives in infected rhizomes for long periods. The pathogen survives up to 32 weeks in dry soil.

Management

Blight can be effectively controlled by spraying 1 % Bordeaux mixture or 0.2% copper oxychloride (Nambiar and Sarma, 1974).

BLACK SHANK OF TOBACCO (*Nicotiana tabacum* L.)

Black shank disease has gradually spread throughout the tobacco growing regions of the world and is now considered to be a serious problem nearly every place where tobacco is grown.

Causal organism

The causal agent of black shank disease of tobacco has been known as *P. parasitica* var. *nicotianae* Tucker.

Symptoms

Plants of all stages of growth, whether in the seedbed or in the field, are susceptible to attack. Young seedlings (35- 40days) in particular are highly susceptible to damping-off in the nursery and cause heavy loss of transplants under favorable conditions. However, if cool weather prevails, the disease may develop so slowly that symptoms are missed. When infected seedlings are transplanted, root rot usually occurs soon after. Leaves of infected plants wilt in warm weather. In advanced stages, dry dark brown or black lesions develop in the cortical tissue on the stem near the soil line. These lesions rapidly expand upward and may cover as much as half the length of the stem (Shew and Lucas, 1991).

Leaves of infected plants nearing maturity suddenly wilt irreversibly, turn yellow, and droop. As the disease progresses, the entire root system and the base of the stem decay. When the stem of a diseased plant is cut longitudinally, the pith is characteristically separated into plate like disks. On some resistant varieties, disease occurs only on the roots with little visible damage above ground other than stunting of growth.

Temperatures above 20° C, high humidity and rainfall favour disease development. High temperatures are rarely a limiting factor because the pathogen grows at temperatures up to 35 to 36° C.

Management

- Control by short crop rotation is relatively ineffective because the pathogen is known to survive in the soil for 5 years or longer in the absence of tobacco.
- Spot application of neemcake @ 100 g / plant hole in combination with *Trichoderma harzianum* (2 g / plant) (multiplied in neemcake/ FYM) and Ridomil MZ 72 WP significantly reduced the disease and increased the yield (Narayanaswmy *et al*,. 2004).

BUD ROT OF COCONUT *(Cocos nucifera* L.)

Bud rot is a fatal disease of palms, characterized by rotting of the terminal bud and surrounding tissues. Even though it affects palms of all ages, young palms in low and moist situations are more susceptible to the disease. Incidence of the disease has been reported from almost tropical countries of the world wherever palms are grown for their economic or aesthetic value.

Butler (1906) first reported bud rot disease of coconut from India. In India it is quite common along the West and East Coast tracts. Radha and Joseph

(1374) reported a disease incidence of 1.2 to 10.9 per cent in Kerala, and 35 to 40 per cent in gardens having large number of palms in certain locations.

Symptoms

The first externally visible symptom is the withering of the spear leaf which subsequently turns brown and bends. In course of time, the younger leaves, closer to the spindle also show similar symptoms. Internal tissues show discoloration assuming a pale pink coloration delimited with a brown border (Quillec and Renard, 1984). The base of the spindle rots and can be detached with a gentle pull. It emits a foul smell. One by one, the inner leaves also fall away, is leaving only mature leaves in the lower whorl at the trunk apex. The palm ultimately succumbs to the disease if timely control measures are not taken.

Causal organism

In 1906, Butler reported that the bud rot disease was caused by *Pythium palmivorum* which was later renamed as *Phytophthora palmivora* Butl. The occurrence of *P. katsurae* causing bud rot has been reported from Hawaii and also from Ivory Coast.

Management

Control measures are effective only when they are adopted in the initial stage of the disease. If the disease is detected when the central shoot begins to wither, removal of affected portions followed by the application of 10% Bordeaux paste on the affected portion can check the disease. Bordeaux paste has to be applied after thorough cleaning and removal of the infected material. The treated portion should be given a protective covering to prevent washing off of the paste during rain. As a prophylactic measure, adjacent healthy palms should also be sprayed with 1 % Bordeaux mixture. Though application of 1 % Bordeaux mixture as prophylactic is very effective, there are difficulties in bringing this into practice. Since the disease occurs during the rainy season, the chemical application in the crown is difficult, as it involves climbing trees during inclement weather and also there is only short span of time available to undertake the spraying operations in extensive area. Intercrops like black pepper are grown in coconut garden and the fungicide application may result in contamination of these products. Moreover, spurious copper sulphate sold in the market affects the efficacy of the treatment. Bordeaux mixture toxicity causes blisters on nuts and other tissues of dwarf palms. In copper sensitive palms, like Chowghat Orange Dwarf (COD), Malayan Yellow Dwarf (MYD) etc. Keeping perforated sachets containing Mancozeb (3g of Indofil M-45 or Dithane M-45) in the leaf axils during rainy season is found to be useful.

Rohni Iyer (2000) reported that stem injection / root feeding with systemic fungicides like Aureofungin - sol (46.5%) at 36.4g / palm, Calixin (Tridemoph 80%) 21 ml / palm, Aliette (Phosetyl-Al 80% wp) 21g/ palm and Akomin (Phosphonic acid) 16.8 ml/palm can protect the crown from *Phytophthora* attack for a period of 8 weeks.

Antagonistic fungi like *Trichoderma* sp., *Myrothecium roridum* and *M. verrucaria* are reported to suppress the growth of *P. palmivora* and *P. katsurae.* It has been reported that antibodies produced by these fungi may have a role in suppression of the pathogen in the soil (Tuset *et al.*, 1992).

GUMMOSIS OF CITRUS *(Citrus* spp.)

Foot rot or gummosis is caused by two species of *Phytophthora. P. citrophthora* prefers climates below 30°C, while *P. parasitica* (= *P. nicotianae)* is more active in warm and humid regions above 35°C (Whiteside, 1970). The incidence is especially high in trees established with the graft union at or below the soil surface, exposing susceptible scion tissue to the two pathogens. In Punjab more than 67.5 % of the citrus trees in the existing plantations have been found infected with the disease (Thind *et al.* 1999).

Symptoms

The most diagnostic symptoms of *Phytophthora* foot rot are found at or below the soil level. Longitudinal cracking of bark, accompanied by profuse gumming, usually is positive evidence of infection. Soil removal around affected trees reveals bark which is water soaked, slimy, reddish-brown, and black. The bark surrounding the diseased area turns yellow and soft.

The infected tree may bear a heavy crop of poor quality fruit because of the girdling effect on the phloem. Fruit infections, although not common, may occur during or shortly after warm rainy periods during summer. These infections are caused by motile spores being splashed onto fruit. This stage of the disease is called brown rot, since the diseased areas on the fruit are brown in color. Active growth of trunk lesions occurs during May and June which starts declining considerably during the hot summer months, then resumes during September and October. The leaves become, yellow and drop and the tree may die, especially on highly susceptible rootstocks. The top branches of the tree may be the first to die.

Lemons and grapefruit are the most susceptible of the citrus trees, while mandarins and trifoliate oranges are highly resistant.

The host range of *P. citrophthora* includes citrus, papaya, carambola, capsicum, chickpea, cocoa, rubber, tomato and avocado. The host range of *P.*

nicotianae includes pineapple, tobacco, cocoa, cashew, jackfruit, papaya, capsicum, citrus, rubber, sweet potato, tomato, apple, cassava, black pepper, guava, sesame, eggplant, corn and many other species.

Epidemiology

The fungus can survive in the soil for many years and needs moisture to maintain its activity. The disease was most likely spread to each of the citrus growing regions on infected rootstocks. Rain splash may disperse the soil borne pathogen to the *lower* parts of the tree. The disease is favored by poor aeration beneath the tree, wet conditions and poor orchard drainage.

Management

Management of Phytophthora gummosis focuses on preventing conditions favorable for infection and disease development. All scion cultivars are susceptible to infection under the right environmental conditions.

- Selection of resistant rootstocks, such as trifoliate rootstocks.
- Pruning of low hanging branches so that they are at least 60cm from the ground. Removal of weeds from around the trunk to improve aeration and promote faster drying.
- Not to allow soil to cover the bud union when planting or during cultivation.
- Nursery stock should be propagated in glasshouses, on branches 35-40 cm from the ground to avoid the spread of the disease into nearby stock.
- When planting or replanting in soil infested with *Phytophthora* or when a susceptible rootstock has to be used, fumigation may be helpful.
- Affected tissue should be removed and the wound painted with fungicide. In optimal growing conditions, the wound will heal. Light pruning, maintenance of adequate moisture and application of recommended dosage of fertilizer will encourage the trees to grow.

ROOT AND STEM ROT OF SOYBEAN (*Glycine max* (L.) Merril)

Phytophthora root rot is a major disease of soybean, especially in areas where soybeans have been cultivated for many years. Yield losses can be substantial; entire fields have been destroyed.

Symptoms

Symptoms associated with *Phytophthora sojae* infections include seed rots, pre- and post- emergence damping off of seedlings and stem rot of plants at various growth stages. On older plants, leaves may become yellow and plants will wilt with wilted leaves remaining on the plant. The lateral and tap roots are rotted and destroyed. A dark brown discoloration that can turn into a girdling lesion often appears on the lower portion of the stem. Disease is usually patchy in the field, often occurring in low or flooded areas. Disease development is most rapid at soil temperatures above 60°F with high soil moisture. This disease is most common in low-lying areas of field, in poorly drained or compacted soils, and in soils with high clay content. It also occurs on well-drained hillsides during growing seasons with the availability of excessive moisture (Sinclair and Backman, 1983; Mc Gee, 1993).

Causal organism

Phytophthora diseases (root rot and damping oft) of soybean are caused by *Phytophthora sojae.* The pathogen survives primarily as "resting" spores in the soil or in association with infested crop debris. When there is high soil moisture the spores germinate and infect the roots. Infection and disease development can occur at any stage of plant development, but are most commonly observed and damaging at the seedling or young plant stage. Disease is most common in heavy, compacted clay soils and fields subject to flooding. Flooding rains, especially near planting, favor disease development: Reduced tillage, especially no-till, is reported to increase damage. The pathogen does not naturally infect other crops.

Management

Once *P. sojae* is established in a field, it cannot be eradicated. The disease must be "managed" with the use-of seed treatment fungicides and resistant varieties. Knowledge of the races present and how varieties with different resistance genes have performed in a grower's field is critical to selection of variety for maximum profitability.

- Crop rotation is not an effective method to reduce disease because the zoospores are very long lived in soil.
- Seed treatment with fungicides (eg. Metalaxyl) can be used for management of early season seed rot and damping off caused by *Phytophthora sojae.*
- Use of resistant varieties is the most effective means of managing *Phytophthora* diseases in soybean.

LEAF BLIGHT AND ROT OF ONION (*Allium cepa* L.)

In 1989, a detailed comparison was made on the species of *Phytophthora parasitica* and *Phytophthora nicotianae.* Researchers concluded from these studies that both names were being applied to a single species. Since *Phytophthora nicotianae* was described first, in 1906, it has nomenclatural priority and all the isolates of *P. parasitica* are now placed in *Phytophthora nicotianae*. Many growers are familiar with *Phytophthora parasitica*. The corrected name is now *Phytophthora nicotianae*.

Phytophthora nicotianae has a very wide host range and occurs on many species of plants. In has been isolated from field crops such as eggplant, bean, parsley, watermelon, papaya and pineapple and it is a major pathogen of ornamentals, tobacco, and tomato.

Symptoms

On onion leaves, spots begin as small, irregularly shaped, water soaked lesions on the young and mature leaves. In a few days these spots expand, girdling the leaf and causing the tissue above the infection point to wither. As the infection progresses, healthy tissue is invaded, eventually killing the entire leaf. Infected plants have a mix of healthy and withered leaves with some leaves showing a characteristic half infected-half healthy symptom.

Causal organism

Phytophthora nicotianae is the cause of the disease. In addition to leaf and stem infection in onions, *P. nicotianae* also causes root rots on poinsettia, tomato, pineapple, and anthurium; crown and collar rots of watermelon; and fruit rots of tomato, papaya, and eggplant. This pathogen is especially serious because studies have shown that it is not host specific. Disease can spread to adjacent plants of another species. The fungus produces sporangia that can either germinate directly, or can release up to 20 smaller zoospores. Both types, sporangia and zoospores, will send out fungal hyphae that penetrate onion leaves and cause disease.

Free moisture is necessary for the spread of the fungus. Sporangia need to be in water in order to germinate or release zoospores. Splashing water from raindrops or irrigation helps to move spores from infected plants to nearby healthy plants. The motile zoospores can swim in pools of standing water, drainage ditches or irrigation systems to infect plants far away from the original disease site.

Management

Effective disease control begins with prevention and proper water management to minimize excess moisture on the plants. Avoid handling of diseased and healthy plants at the same time since spores can be unknowingly transferred by workers during handling. Immediately removal and destruction of suspected diseased or plants is recommended. Monitor plants regularly in order to detect disease outbreaks early. Ridomil can be applied to green onion up to 14 days prior to harvest.

LATE BLIGHT OF POTATO (*Solanum tuberosum* L.)

Potato late blight *(Phytophthora infestans)* is a serious disease of the potato plant. *P. infestans* is perhaps the best known of all the species of *Phytophthora* because of the tremendous destruction it wreaked on the potato crops in Ireland during 1840s. The economic chaos from the loss of this vital food crop directly led to about million deaths and an equal number the surviving migrated to Europe and America (Hooker, 1981; Rich, 1983).

Causal organism

Phytophthora infestans (Mont.) de Bary is the causal agent of late blight of potato. Although, the pathogen has limited host range, affects mainly solanaceous hosts.

Until the early 1980s, A1 was the predominant mating type detected in all regions of the world except Central Mexico, where both A 1 and A2 mating types coexisted at a 50:50 ratio. A2 isolates are now found in nearly all potato growing regions of the world.

Symptoms : The early stages of blight are easily missed, and not all plants are affected at once. Symptoms include the appearance of dark blotches on leaf tips and plant stems. White mould will appear under the leaves in humid conditions and the whole plant may quickly collapse. Infected tubers develop grey or dark patches that are reddish brown beneath the skin, and quickly decay to a foul smelling mush caused by the infestation of secondary soft bacterial rots. Seemingly healthy tubers may rot later when in store.

The spores over winter on infected tubers, particularly those that are left in the ground after the previous year's harvest, and are spread rapidly in warm wet conditions when blight can have devastating effects, destroying entire crops. Reports indicate on formation survival and germination of zoospores (Chang and Ko, 1991; Drenth *et al.,* 1995).

The sporangia develop on the leaves, spreading through the crop when temperatures are above 10°C and humidity is over 75% for 2 days or more. Rain can wash sporangia into the soil where they infect young tubers, or else they may be blown in from distances of up to miles by the wind. (Plate 16.1)

Following conditions precede late blight epidemics:

1. Night temperatures below the dew point for at least 4h.
2. Low night temperature that does not drop below 10 ^{0}C.
3. The mean period of cloudiness not less than 0.8 of the day
4. Rainfall of at least 0.1 mm on the following day.

Management

- No potato cultivar is immune to all strains of *P. infestans*, but some cultivars with general resistance are available commercially in different regions of the world (Rich, 1983). If the climate in which the potatoes are grown is relatively dry, this resistance may significantly reduce disease severity and hence may require less fungicide application (Fry *et al.*, 1983).
- Good drainage and good air movement will help reduce moisture levels in the canopy. Fields bordered by trees and dense vegetation should be avoided. The shape of the field may affect the ease of fungicide application.
- Rotations for 2 to 3 years with non host crops are recommended. If zoospore production becomes widespread, rotation plans may be modified to accommodate this new source of inoculum. The fungus survives in infected tubers which decay relatively fast, but zoospores may survive in soil for many years.
- In the absence of zoospores, tubers infected in the previous season are the most important source of initial inoculum. Surviving tubers may be found in cull piles and tubers left in the field at harvest. Culled potatoes should be left on the soil surface to freeze, trucked to a landfill, or buried at least 3 feet deep. They may be fed to livestock. Volunteer plants in the spring should be destroyed to minimize initial inoculum.
- Only certified seed tubers should be planted. However, currently even certified seed tubers are allowed up to 1 % incidence of late blight.
- Soil can be made deeper around the base of the plants after emergence of the young potato plants. Hilling helps in early weed control and minimizes tuber infections from sporangia that wash off the leaves of infected plants into the soil.
- It is important to minimize the time that leaves are wet to help prevent foliar infection. Irrigation should be timed so that length of the night dew period is not extended i.e., no late afternoon, early evening, or morning irrigation just as the plants are drying. Excessive irrigation can wash some of the "hilled" soil away from the base of the plants, exposing tubers to greater potential infection.

- Excessive nitrogen fertilization delays maturity and may reduce yield. Delayed maturity results in foliage exposed to potential infection for a longer time, increasing the risk of late blight.
- Scouting alerts growers that the potential for a severe disease outbreak exists. Scout where moisture is likely to persist (low areas, near hedges and trees) and where fungicide applications may be difficult because of obstructions (corners, trees, utility poles). Infected plants should be burned or ploughed wherever isolated "hot spot" is noticed. Current season temperature, relative humidity and rainfall data are used to predict disease outbreaks, based on historical patterns. Growers apply fungicides when the forecast signals the likely disease development, which may be more risky than routine preventive fungicide applications (Forecasting systems for late blight include the Hyre system, the Wallin system, and BLITECAST which incorporates both systems) some commercial microcomputer units are available, at least in United States and European countries, to provide forecasts.
- Fungicide applications are an important means of late blight management, particularly in humid areas. Combined application of protectant and systemic fungicides are effective and has not resulted in pathogen resistance after many' years of use. Systemic fungicides (eg., Metalaxyl) are absorbed into plant tissue and can offer some post-infection control.
- Some resistant strains of *P. infestans* have been detected. The rapid spread of the newly introduced strains is of special concern because many are resistant to Metalaxyl. In the early 1990s, some growers lost entire potato crops. Some new fungicides, such as Acrobat and Curzate, are being used against late blight. Farmers are encouraged to concentrate on the prophylactic use of protectant fungicides.
- At the time of harvest, it is possible that some late blight infections are present in the foliage. To prevent inoculation of the tubers during harvesting, foliage should be killed (Rich, 1983). In the past, vines were killed with flames. Herbicides are currently used to ensure that plants are killed completely before harvest. Infected tubers should be removed before storage and disposed of properly. Blighted tubers do not generally store well under typical storage conditions.

LATE BLIGHT OF TOMATO (*Lycopersicon esculentum* Mill.)

Late blight of tomato and potato are the most important diseases caused by *P. infestans*. Both hosts are members of plant family Solanaceae.

Causal organism

Late blight of tomato is caused by *P. infestans* (Mont.) de Bary.

Symptoms

Leaf lesions are water soaked with definite margins and a gray green colour. The infected tissues become brown and necrotic and the leaf dies. Stems and petioles are also affected; eventually the plants will be killed if the disease is not controlled (Jones *et al.*, 1991). The foliage symptoms are similar to those on potato. On tomato fruits, gray green, water soaked spots enlarge rapidly. The lesions then turn dark brown, and although the disease areas become wrinkled, the rotted tissues remain firm. The unaffected parts of the fruit ripen to a normal red colour. Occasionally the lesion may have rings in a zonate pattern that superficially resemble those of buckeye rot caused by *P. parasitica;* even, in fruit rot caused by *P. infestans.* The rings are much closer together. Seeds from infected fruit often give rise to infected tomato seedlings (Vartanian and Endo, 1985). Disease development is similar to that of late blight on potato. Some new biotypes of *P. infestans* can be aggressive on both tomato and potato (Legard *et al.*, 1995).

Management

- Bordeaux mixture used as foliar spray before infection is effective but toxic to the foliage of tomato, so other, more insoluble copper based compounds, dithiocarbamates or phenylamide fungicides are more often used.

- Although Metalaxyl (phenylamine) fungicide is effective, resistance to Metalaxyl within populations of *P. infestans* on potato has become a problem in Europe and America. The addition of broader spectrum fungicide Mancozeb to Metalaxyl or other phenylamide fungicides may at least partially alleviate this buildup of biotypes with resistance to *P. infestans.*

ROOT ROT OF AVOCADO *(Persea indica* and *P. americana* Mills)

Root rot is now recognized to be the most important disease wherever avocado trees are grown. The disease is an economically limiting factor to production in many areas of California, Australia, South Africa and many other countries. In California, about 60 and 70% of the orchards are affected by *Phytophthora* root rot (Coffey, 1992).

Causal organism

Phytophthora root rot on avocado is caused by *P. cinnamomi*. It is the most important pathogen of avocado in the United States. Some economically important plants affected by this pathogen include quinine, eucalyptus, cherry, plum, peach, pineapple, cinnamon and almond.

Symptoms

Occasionally, *P. cinnamomi* infects large roots and some times causes a canker at the base of the tree that extends upwards approximately 2 m; however the more common cause of trunk canker of avocado is *P. citricola.* A majority of the feeder roots of infected trees are blackened, brittle or necrotic. After *Phytophthora* root rot occurs, the leaves on infected trees become smaller than normal and pale green to yellow colour. Wilting occurs as the disease progress followed by heavy leaf fall and die back of branches. In older, more mature trees, dieback of branches and sparse foliage give the tree very characteristic "stags head" appearance. Diseased trees quite often set an abnormally heavy crop of small fruit, probably because of the girdling effect. Trees eventually are reduced to leafless skeletons and are killed (Coffey, 1987).

Disease development

The disease is more severe and develops most rapidly under poor drainage conditions. Disease progress is optimum in wet soils with a pH of 6.5 and at a temperature of about 24° C; however, disease occurs within a range of about 13 to 32° C. The pathogen easily spreads in moist soil. *P. cinnamomi* can survive in diseased or dead roots up to 6 years if the soil is moist. Trees downhill from an infected area often infected by surface or subsurface drainage of irrigation water or runoff rain water.

Management

Control of *P. cinnamomi* is difficult because of its wide host range and its ability to survive in symptomless plants as chlamydospores in roots at soil depths at which fumigation is not always effective.

The most effective control measure is to exclude *P. cinnamomi* by planting disease free nursery stock on clean land. Seeds from fruits that have fallen to the ground might be infected and should not be used in nurseries. If necessary, *P. cinnamomi* in infected seeds can be eradicated by treatment with hot water at 49 to 50° C for 30 min.

Metalaxyl can be used as a soil drench to suppress the development of root rot on trees growing in infested field soil. The phosphonates (Fosetyl-Al and phosphorous acid) are highly effective because they are translocated upward and downward, a unique quality of the fungicide for control of root rots. Fosetyl -Al at a dosage of 1 lit per tree at 3 g (a.i) per lit controlled root rot effectively (Coffey, 1987).

SUDDEN OAK DEATH (*Gravilia robusta*)

It was first discovered in California in 1995 when large numbers of Tanoaks *(Lithocarpus densiflorus)* died mysteriously, and was described as a new species of *Phytophthora* in 2000. It has subsequently been found in many other areas including Britain, Germany, and some other US states, either accidentally introduced on nursery stock, or already present undetected. Heavy loss of oaks, or of related susceptible genera, due to sudden oak death (SOD). *P. ramorum* infection could result in significant ecological effects, including changes in forest composition, loss of wildlife food and habitat, increased soil erosion and a significant increase in fuel loads in heavily populated urban- forest interfaces.

Causal organism

Sudden oak death (SOD), caused by *Phytophthora ramorum,* is a serious disease, of tanoak and two species of oaks, causing crown dieback, stem bark lesions, and basal cankers typical of other *Phytophthora* species (Erwin and Ribeiro 1996). Other species of *Phytophthora, P. citricola* and *P. cactorum* are associated with oaks in urban landscape situations in California (Mircetich *et al.,* 1977).

Symptoms

Diseased tanoaks and oaks exhibit symptoms that include crown dieback or wilting, stem bark lesions or cankers, and tarry, reddish to dark brown exudations characteristic of *Phytophthora* root and collar rot. The exudations do not always develop on tanoak, especially on smaller diameter branches. The cankers are typically found on the lower portion of the trunk, but may occur to 12 feet or higher. Branch cankers, especially on tanoak are known to occur. Infected hosts often appear to die suddenly, with crowns turning from apparently healthy green to yellow-green to brown within a few weeks (Swiecki, 2000). Infected trees die relatively fast once crown symptoms develop. Disease severity varies considerably from location to location.

Biology

Phytophthora sp. associated with SOD is a previously undescribed species in the genus. DNA sequencing indicates that the closest relative is *Phytophthora lateralis* Tucker & Milbrath.

Numerous sporangia are produced in culture. The sporangia are deciduous, suggesting air-borne dispersal. Production of thick-walled chlamydospores is prolific in culture. Sexual spores have not been observed. Infected phloem tissue is discolored while infected xylem tissues typically exhibit a dark, often streaky discoloration. The current known distribution of the pathogen is in coastal environments of northern California. Nursery stock (living plant material), and infested firewood with attached bark are the most likely means of long-distance transport of the pathogen. Not much control measures are available for the disease.

ROOT AND CROWN ROT OF TOMATO (*Lycopersicon esculentum* L.)

Symptoms

Plants of all ages are susceptible. Typical damping-off symptoms occur on seedlings. Symptoms may vary from cankers limited to the crown and upper part of the root to cankers forming on majority of the roots. In young plants, infection of roots and crowns result in rapid wilt followed by complete collapse of the plant. On older plants, cankers that are originally moist but later become soft, dry, crumbly develop on the roots. Occasionally, the vascular tissues become discolored immediately above and below the cankers. Leaves become bronze and later necrotic. Necrosis begins at the tip and progresses toward the petiole, after which the entire leaf dies. The pathogen is not host specific.

Causal organism

Root and crown rot of tomato is caused by *Phytophthora capsici*. Root rot can also be caused by *P. parasitica*.

Management

Planting tomatoes in well drained soil on a raised beds helps to prevent severe expression of the disease. Saturation of the soil for 5 hr or longer favours infection (Jones *et al.*, 1991). Rotation to nonsusceptible crops will decrease the inoculum in the soil. Application of potash helped to retard symptoms (Erwin and Ribeiro, 1996).

BLIGHT OF CUCURBITS (*Cucumis sativa*)

Seedlings show typical symptoms of damping-off. Mature plants may wilt suddenly as a result of root infection, in the absence of obvious stem or vine lesions. Stem and leaf petiole lesions are light to dark brown, water soaked and irregular, eventually becoming dried and papery. Fruit symptoms begin as small, water-soaked circular lesions, which expand to become large, soft, sunken areas covered with white fungal growth. Infected fruits often collapse or "melt down" in the field or in storage.

Causal Organisms

Phytophthora blight of cucurbits and pepper is caused principally by the soil-borne fungus, *Phytophthora capsici,* although a similar fungus, *P. parasitica,* has also been reported to cause fruit rot on peppers and cucurbits. Some cucurbit isolates of *P. capsici* were virulent on pepper while others were not (Ristaino, 1990). These fungi are dependent on free water in soil for infection. Because of this, initial infection by *Phytophthora* usually occurs on plants growing in poorly drained areas of fields.

The fungi survive as thick-walled, resistant zoospores in the soil and as mycelium in infected plant tissues. Once introduced, these fungi can survive up to 15 months in moist soil in the absence of host plants. They can also be carried on seed or transplants. Once stem infection occurs, the fungi produce spores on infected stem tissues, which are then carried by splashed rain onto nearby plants. Lower branches of adjacent plants can also be infected by rain-splashed soil contaminated by run-off water. Fruits in contact with the soil are especially prone to infection. Spores are produced on newly infected fruit and stems, and new infections can develop quickly in a short time, warm, wet weather is most favourable for disease development and spread of the pathogen. Disease progress declines when dry weather returns.

Management

- Use only certified disease-free planting material.
- Produce cucurbits plants on raised beds to retard initial stem infection.
- Avoid planting cucurbits in poorly drained fields.
- Practice crop rotation so that peppers or cucurbits are grown only every 3 to 4 years to reduce the amount of *Phytophthora* present in the soil. Use non-susceptible crops such as corn, small grains, beans, or crucifers in rotation.
- Application of fungicides, Metalaxyl or Fosetyl-Al, may reduce disease development and spread if used in combination with cultural practices such as crop rotation and raised beds.

ROOT ROT OF PAPAYA (*Carica papaya* L.)

Phytophthora fruit, stem, and root rots of papaya was first reported in the Philippines in 1916 and then in Ceylon in 1924. The disease also occurs in Malaysia, Hawaii, Australia, Brazil, Spain and Taiwan.

Heavy fruit losses caused by *Phytophthora* sp. frequently occur during rainy periods. Rains also can result in severe decline and death of papaya trees because of root rot in poorly drained areas. Losses due to *Phytophthora* root rot in southeastern Queensland; Australia exceeded 8000 plants in 1955 and 1956, and in 1975 more than 20% of the plants in one papaya plantation in central Taiwan was destroyed by the disease.

Symptoms

Infected young fruit on trees usually show water-soaked lesions exuding milky latex. Frequently, the disease continues to develop and causes the infected fruit to shrivel and become mummified before falling to the ground. When mature fruits on trees are infected, lesions are covered with whitish mycelial and sporangial masses.

The top portion of the fruit-bearing region of the stem is very susceptible to infection during rainy periods. Stem cankers in this area causes many young fruit and leaves to fall prematurely and renders the tree top susceptible to wind damage. Older portions of stems also become infected after an extended rainy period and develop horizontal water-soaked lesions along old leaf scars. The infected area may enlarge and weaken the stem, causing the plant to break off in a strong wind.

In poorly drained areas, *P. palmivora* initially attacks lateral roots of papaya. The disease later extends to the tap root and the whole root system becomes brown, soft, and shredded. Trees become stunted; leaves turn yellow and hang limply around the stem, leaving only a few small leaves at the apex of the tree. The infected tree eventually dies.

Papaya roots are especially susceptible to *P. palmivora* in the first 3 months after planting of seeds. During this period, root infection results in yellowing of leaves, premature defoliation and eventual death of the seedlings. Occasionally, the fungus destroys only a portion of the roots before the plant becomes somewhat resistant with age. Under dry conditions, the disease may cease to develop and the plant resumes normal growth. Often, however, with a heavy load of fruit, the tree may be easily blown over by wind because of the damage to the root system.

Causal organism

P. palmivora (syn.: *Phytophthora faveri* Maubl.), previously misidentified as *P. parasitica* Dastur is the most important cause of this disease on papaya.

Biology

P. palmivora produces abundant sporangia on V-8 agar under continuous fluorescent light. However, light is not required for sporangia production on infected papaya fruit. Sporangia are usually produced in clusters sympodially. Sporangia are papillate and ovoid with the widest part close to the base. They are easily washed off and each detached sporangium contains a short pedicel. The average size of the sporangia is 50 x 33 Hm with a length of about 1.6 times longer than it is wide. Sporangia germinate directly in a nutrient medium by producing germ tubes that develop into mycelial masses. In water, however, zoospores are released from germinating sporangia. Zoospores aggregate and form distinct patterns at 16° C in water.

Chlamydospores produced in infected papaya fruit and 100% papaya juice is thick-walled. However, chlamydosporcs produced in papaya juice at lower concentrations or in other kinds of fruit juice are mostly thin-walled. In the presence of nutrients, chlamydospores germinate by producing germ tubes that continue to grow and form mycelial masses. In water, chlamydospores germinate by producing short germ tubes, each with a sporangium at the tip.

Sexual reproduction in *P. palmivora* requires the presence of opposite mating types known as A_1 and A_2. Both A_1 and A_2 isolates can produce oospores by selfing when stimulated by sex hormones produced by A_2 and A_1, respectively. Light is inhibitory to oospore formation but stimulatory to oospore germination. Mature oospores can be induced to germinate by treatment with 0.25% $KMnO_4$ for 20 min and incubation under light during germination.

Although sporangia and zoospores may survive in soil for short periods, chlamydospores are the main survival structure for *P. palmivora* in nature. Oospores are capable of long-term survival but do not play significant role in the disease- cycle because sexual reproduction in *P. palmivora* requires the presence of opposite mating types, and the chance for this to occur in nature is very low.

During rainy periods, chlamydospores in soil may germinate in water to produce sporangia and release zoospores. The impact of falling rain drops may splash zoospores into air in droplets. The zoospore-containing droplets may be further dispersed by wind and become the inoculum for infecting fruit and occasionally stems of papaya in the fields. The pathogen produces abundant sporangia on the surface of infected fruit that are further dispersed by wind-blown rain and cause outbreaks of Phytophthora fruit rot in the same and nearby orchards. Chlamydospores formed in fallen fruit survive in soil and serve as the main source of inoculum for infection of roots of papaya seedling in subsequent plantings.

Phytophthora root rot of papaya seedlings is most serious during rainy periods. Under waterlogged conditions, *P. palmivora* may attack roots of papaya older than three-months of age, the time at which they become resistant to the pathogen under normal conditions. Therefore, Phytophthora root rot may occur on papaya at any age in poorly drained areas. Waterlogged conditions appear to weaken the defense mechanism of papaya roots against invasion by the pathogen. Mobility of zoospores of *P. palmivora* under such conditions also may contribute to the severity of the disease due to their attraction by papaya roots.

Epidemiology

Rain and wind are the two major factors in the epidemiology of *Phytophthora* fruit rot of papaya. Rain splash is needed for liberation of sporangia of *P. palmivora* from the surface of infected fruit into the atmosphere and for projection of the soil inoculum into air. Wind is required for dispersal of the inoculum once it reaches the air. Therefore, wind-blown rain is essential for initiation of the primary infection and the development of epidemics in papaya orchards.

Favorable temperature is also a contributing factor to the severity of *Phytophthora* diseases because of its effect on growth and sporulation of the pathogen. *Phytophthora palmivora* has an optimum temperature for growth at 30° C, a maximum temperature of 36° C and a minimum temperature of 12° C. The pathogen produces the most sporangia at 25° C but no sporangia are produced at temperatures higher than 35° C or lower than 15° C.

Management

Root rot of papaya seedlings, caused by *P. palmivora* in replant fields can be controlled with the virgin soil technique. Virgin soil (soil in which papaya has never been grown in before) is placed in planting holes about 30 cm in diameter and 10 cm deep with a mound of about 4 cm height. Roots of papaya plants are protected by the virgin soil during the susceptible stage, and become resistant to the pathogen when they extend to the infested soil.

Cultural practice is also important in the management of *Phytophthora* diseases of papaya. Incidence of Phytophthora root rot can be greatly reduced-by improving drainage in the orchards. Infected fruit on the trees and those that have fallen to the ground should be removed to reduce the inoculum for aerial infection of fruit and stems, and infection of seedling roots in subsequent plantings.

Phytophthora fruit rot of papaya can be controlled by fungicides. Applications of mancozeb or basic copper sulfate to the root column as prophylactic spray are effective in protecting papaya fruit from infection by *P. palmivora.*

HEART ROT OF PINEAPPLE *(Ananas cosmosus)*

Symptoms

Heart rot disease of pineapples is characterized by yellow-red colored leaves that are easily pulled from the plant and white leaf bases that bear a distinct water-soaked grey-brown transverse lesion. The lesion progresses slowly from the base of the leaf to the tip. Rotting pineapple fruit tissue may also smell badly. If the pineapple plants develop root rot, above ground symptoms include stunting and low yield.

Causal organism

Heart rot of pineapple is caused by *P. nicotianae* (= *P. parasitica).* A disease with similar symptoms, caused by *P. cinnamomi* occurs in temperate (temperatures between 20 and 27 ^{0}C) countries. The host range of *P. nicotianae* includes pineapple, tobacco, cocoa, cashew, jackfruit, papaya, capsicum, citrus, rubber, sweet potato, tomato, apple, cassava, black pepper, pistachio, guava, sesame, eggplant, corn and many other species. In the wet season, the disease has a tendency to be more serious. High soil pH is said to promote disease development.

The pathogen can be transmitted through suckers, and if pineapples are grown for two seasons, the disease incidence is likely to be very high. The pathogen can survive in the soil or in plant debris as chlamydospores for many years. The chlamydospores germinate and hyphae infect roots of young leaf and stem tissue. The infection progresses up the root to the stem apex causing the heart rot symptoms. The pathogen can proliferate through the production of zoospores from sporangia if conditions are conducive for their production. In these situations the disease may spread through irrigation patterns.

Management

- Ensure adequate soil drainage, especially in the lower parts of the field.
- Add organic manure.
- Do not use suckers from diseased crops as planting material.
- Clear pineapple residues from the soil that are acting as mulch.
- Rotate with non-susceptible crops such as sugarcane, corn, soybean and peanuts.
- Dip the suckers in Bordeaux mixture or Captafol or any of the systemic fungicides (Aliette or Ridomil) prior to planting.

BLIGHT OF TARO (*Colocasia esculenta* L.)

Symptoms

The leaves will develop water-soaked purple to brown circular lesions from 2 mm to 2 cm on the lobes and the sides of the leaves. The lesions gradually enlarge and become dark brown with yellow margins. White growths may be seen around the lesion and a yellow to red exudates is released from the centre of the spots. This darkens and hardens when dry. In the advanced stages of the disease, irregularly shaped lesions partially or completely cover the leaf, and the leaves remain attached to the plant. A post-harvest rot of infected rhizomes can also occur, where the pale brown, firm and dry lesions rot under poor conditions.

Causal organism

Phytophthora colocasiae is primarily a foliar pathogen with a limited host range, distributed throughout the taro growing regions of Asia on vegetatively propagated planting material and in soil. *P. colocasiae* causes blight of taro (*Colocasia esculenta*) and yam (*Alocasia spp.*), rhizome crops that are staple foods for many developing countries.

Epidemiology

The sporangia are produced in wet conditions and are disseminated by rain splash. Infection usually begin where there is water resting on the leaf, in day temperatures between 25 and 28°C, night temperatures 20-22°C and periods of rainy, overcast and humid weather. The pathogen survives in infected plant debris and rhizomes, producing zoospores and chlamydospores to maintain the inoculum levels in the soil indefinitely.

Management

Burn infected leaves and remove corms from the soil after harvest to prevent the survival of the pathogen in the infected debris. To reduce the humidity and disease spread within the crop, avoid planting in shaded areas, intercrop with non-susceptible plants and increase the spacing between plants. Copper oxychloride sprays or the application of Bordeaux mixture to the foliage may be beneficial. Dipping the harvested corms in 1% sodium hypochlorite may limit post- harvest rot, so long as material is dried prior to storage. Metalaxyl, Captafol, and chloroneb used as sprays at 15-days interval gave good control (Agarwal and Mehrotra, 1987).

BLACK STRIPE OF RUBBER *(Hevea brasiliensis* Muell.)

Distribution

Black stripe is a serious problem in Brazil, Cameroon, China, Cote d'Ivoire, India, Indonesia, Malaysia, Thailand, and Vietnam. An earlier study had indicated that China, Cote d'Ivoire and Sri Lanka have serious problems with black stripe. The severity of the disease was then lower in India, Malaysia and Thailand although it was an endemic problem in all these countries. In Brazil, Nigeria and Philippines the disease is generally sporadic and controllable, with occasional local, very severe outbreaks in Brazil.

Causal organisms

Black stripe is the most severe of *Phytophthora* diseases of rubber. Several species of *Phytophthora* have been reported to be responsible for black stripe. The common species are: *Phytophthora palmivora* (Butl.) Butl., *P. meadii* Mc Rae, *P.botryosa* Chee.

In 1984, several other species of *Phytophthora* were identified as causes of black stripe infection in China: viz, *P. citrophthora* (Smith & Smith) Leonian, *P. cactorum* (Lebert & Cohn) Schroeter, *P. capsici* Leonian.

Symptoms

Superficially, the symptoms can be confused with tapping injury a series of sunken and slightly discolored areas just above the cut. Later, vertical fissures appear in the renewing bark; when these are removed, dark vertical lines are visible. As the infection progresses, the stripes coalesce forming broad lesions, finally spreading the full width of the panel. When the disease is severe, it extends vertically in the wood as far as below the tapping cut and 2-5cm upwards on the regenerating bark. Pads of coagulated latex sometimes form beneath the bark causing extensive bark splitting and bleeding.

Occasionally, infection occurs on untapped bark resulting in a wound, called "canker", this may arise on bark previously affected by black stripe or on wounds caused by sprouts or wires. The early symptoms of canker are not obvious but in the more advanced stage, the bark bursts and latex oozes out. Pads of coagulated latex form under the bark causing it to bulge and split open.

Spread

Black stripe infections are associated with wet weather, being favored by prolonged cool and rainy periods. The fungus is commonly present in the soil and spores are spread by water droplets. In areas where abnormal leaf fall occurs the spores are washed down from the canopy. Canker development is more common in leaf fall areas via heavy inoculum from the tree canopy; infected tapping knives are also an important means of dissemination.

Management

Dipping the tapping knives in a bleach or fungicide solution prior to cutting will limit the spread. Infected trees should not be tapped until the weather is dry. Black stripe can be kept under control during continuation of tapping by regular applications of an effective fungicide provided the disease is detected before infection becomes severe. Metalaxyl (Ridomil) give control as protectant and therapeutic.

ABNORMAL LEAF FALL OF RUBBER *(Hevea brasiliensis* Muell.)

Distribution

Abnormal leaf fall and leaf wither have been reported in most rubber growing countries. In China, India and Sri Lanka the disease is a significant annual problem, whereas it is sporadic in Cote d'Ivoire, Malaysia, Nigeria and Thailand. Active, regular control measures were regarded as essential.

Causal organisms

Phytophthora palmivora (Butl.) Butl; *P. meadil* McRae and *P. botlyosa* Chee are documented as causal agents of disease. Opinions among countries differ about which species of *Phytophthora* are responsible. In India abnormal leaf is attributed to *P. meadii.* While in Malaysia it is considered to be caused by *P. palmivora.*

Symptoms

Abnormal leaf fall affects mature leaves. The fungus attacks the petioles causing the leaves to shed while they are still green. The fallen leaves exhibit one or more dark brown lesions with one or two white spots of coagulated latex in the middle of each, usually towards the base of the petiole. Sometimes the leaf blades are affected with initially translucent water-soaked lesions, which, in course of time, develop various colours between brown and black. When lesions are present, they are concentrated more towards tile apex and margin of leaflets. Without petiole infection, the affected leaflet changes colour from green to shades of yellow and red before being shed.

The fungus, produces a downy white growth on the surface also attacks green fruit pods in all stages of development. The affected pods do not abscess, but remain on the tree blackened, mummified and undehisced. Terminal shoots may be attacked, resulting in dieback of green shoots.

Another symptom of foliage attack by *Phytophthora* observed in Brazil is leaf wither. The disease differs from abnormal leaf fall in that the fungus attacks the green shoots rather than the leaves. Young leaves wither and remain attached to the dead or dying shoot, although the leaves themselves are not attacked.

Management

The disease can be prevented by one spray of an oil-based copper fungicide before the onset of the rainy season. In India, the use of annual pre-monsoon spraying to control abnormal leaf fall has increased yields of rubber by up to 50 per cent (Pillai, 1982).

GUMMOSIS OF DURIAN (*Durio zibethinus* Murr.)

This disease occurs especially during the wet season. The causal pathogen of this disease is *Phytophthora palmivora.*

Symptoms

The first and most obvious symptom is defoliation, caused by coalescing cankers that girdle the stem or branches. Stem cankers are not always readily visible on the outer bark. If the bark is peeled away from the collar region, the wood appears watery grey, often with red streaks that intensify in colour upon exposure to air. In the advanced stages of the disease reddish fluid exudes from cankers that dry into a rusty deposit. The pink or light brown cortical tissue darkens to brown and blackish streaks (Holliday, 1980).

CROWN, COLLAR AND ROOT ROT OF APPLE (*Malus pumila* Mill.) AND PEAR (*Pyrus communis* L.)

Collar rot on apple trees affects the scion cultivar if trees were planted with the graft union close to or in contact with the soil. Crown rot, on the other hand, usually affects the rootstock. The factors that favour disease development and control are similar for diseases of apple and pear and hence are described together.

Causal organism

P. cactorum is the causal agent of crown, collar and root rot of apple and pear. Other species of *Phytophthora* that cause crown and root rot of apple include *P. syringae* (Harris, 1991), *P. megasperma* (Jeffers *et al.*. 1982), *P. citricola, P. cambivora* and *P. cryptogea* (Jeffers and Aldwinckle, 1988).

Symptoms

Apple trees initially infected by *Phytophthora* are difficult to distinguish from healthy trees unless one looks carefully for symptoms on the crown (rootstock), collar (the scion), or root. By the time visible symptoms are observed the foliage of the tree, considerable development of root, collar and or crown rot symptoms has already developed, and at this stage it is difficult to save the tree by applying therapeutic control measures. Crown and collar rot usually cause an abnormal light green colour of the foliage. Later terminal growth is stunted and the small chlorotic leaves droop. As the disease progresses, the leaves drop and fruits are exposed to sun burn, which results in a loss of quality. Crown or collar rot eventually leads to girdling of the trunk and death of the infected trees.

To diagnose collar or crown rot, the soil from the crown region of the plant should be removed so that characteristic lesions that occur on the crown, collar, or roots can be seen. Scrapping or cutting away the outer bark from infected crown reveals dead bark (often with a reddish tinge). A sharp demarcation line can be observed between diseased, reddish brown bark and healthy, white bark. In some cultivars, a moist exudate covers the cankered areas. Rapid extension of the cankered areas leads to girdling and death of the tree in a single season, although in the majority of cases, the cankers are confined to one side of the trunk during first season of infection.

Root rot phase of the disease can be detected only by removal of sufficient soil to expose substantial part of root system. Failure of some rootstocks to remain upright during windstorms is often the result of the rotting of feeder roots.

Fruit rot caused by *P. cactorum* is generally minor economic importance; however, fruit rot caused by *P. syringae* can lead to economic losses. Fruit infection usually follows contact with soil particles that splash upward during heavy rain. The infected tissue becomes spongy, mealy, or granular because of the dissolution of the middle lamellae between the cells. After the infected fruits become mummified, it can be a source of inoculum in the form of zoospores that form in the decayed tissue.

Disease management

Control is difficult, since *P. cactorum* survives for several years in orchard soils either as mycelium or zoospores (Harris and Bielenin, 1986). The use of sanitation in the production of disease-free nursery tree is important. *Phytophthora* contamination of nursery tree stock can be reduced or eliminated by soaking roots in Metalaxyl, Fosetyl-Al, or copper hydroxide for in at a concentration of 1,000 mg (active ingredient) per liter (Jeffers, 1992; Tidball and Linderman, 1990).

Therapeutic treatments with Metalaxyl and Fosetyl-Al are effective if applied to trees during the early stages of infection. Because Metalaxyl is systemic and highly effective, it is presently one of the chemicals of choice of many apple growing regions. Fosetyl-Al moves upward and downward, has curative properties when used as foliar, trunk, or trunk injection application on infected trees (De Boer *et al.*, 1990). Copper formulations painted on the trunk at the base of the tree are effective for prevention of infection from the splashing of contaminated soil but are not effective for control of internal infection that progress from the roots into the crown and trunk. Biological control of *P. cactorum* by use of the bacterium, *Enterobacter aerogenes* (Utkhede, 1987) and the fungus, *Trichoderma* was also reported.

STEM BLIGHT OF PIGEON PEA (*Cajanus cajani* (L.) Mill.)

This disease may be confused with *Fusarium* wilt, caused by *Fusarium udum* Butl, another serious disease of pigeon pea (red gram). The wilting symptom of *Fusarium* closely resembles that of *Phytophthora* blight. Disease incidence is high in water stagnating areas.

Causal organism

P. cajani (=P. drechsleri f. sp. *cajani)* is the causal agent of stem rot of pigeon pea.

Symptoms

P. cajani causes stem rot and does not affect roots. The first observable symptoms on foliage are desiccation and upward wilting of leaflets, usually without chlorosis, followed by withering of the petioles and small stems. Dark brown to black lesions encircle the stem at or near the base and may extend

15 -30 cm up the stem (Pal *et al.*, 1970; William *et al.*, 1975; Kannaiyan *et al.*. 1980). Lesions with well defined margins may also develop on the branches, petioles or leaves. These lesions initially have a flat surface but later become slightly depressed in newly formed lesions, the bark and the cambium, but not the older xylem tissue, may become discoloured. Roots are not attacked, even in advanced stage of disease development. Occasionally large galls may develop on the stem at the soil line. Stem breaks easily at lesion sites on the main stems during strong winds. Plants are susceptible to infection from the seedling stage to maturity.

The disease can be managed by growing disease resistant varieties.

SEEDLING DIEBACK OF COCOA (*Theobroma cacoa*)

Seedling dieback also known as seedling blight is very common in the cocoa nurseries of Kerala, Karnataka and Tamil Nadu during rainy season. Intensity of the disease varies from locality to locality as well as with the age of seedlings. Younger seedlings are more susceptible to the disease.

Symptoms

Defoliation and dieback of seedlings are the characteristic symptoms of the disease. Generally the infection starts from the tip of the stem and proceeds downwards as dark brown to black water-soaked linear lesions. The lesions also extend to the leaves through the petioles resulting in wilting and subsequent defoliation of the seedlings. The infection also initiates from the collar region, cotyledonary stalk or leaves as dark brown to black discoloration. In all the cases, the infection spreads to the entire stem causing wilting, defoliation and ultimate death of the seedlings.

Causal organism

Phytophthora palmivora Butl has been found to be the causal organism of this disease (ChandraMohanan, 1979). Severe infection of grafted and budded seedlings caused by *P. palmivora* has also been observed in India. In such cases, infection mainly starts from the grafted or budded region and proceeds upwards and downwards. Infection continues to spread internally after rainy season leading to high mortality.

Pre-emergence damping off of the seedlings was also noticed in several nurseries especially when sowing of beans was carried out during rainy season. Pre-emergence death of seedlings is mainly due to the inoculum present in the soil. Primary source of inoculum for the seedling blight is through soil splashes. Later, infected seedlings serve as the source of inoculum for the spread of the disease in the nursery. Removal and destruction of infected seedlings from the nursery are very important management practices to check the secondary spread of the disease.

Management

The disease incidence can be considerably reduced by improving the drainage facilities in the nursery and by providing proper shade. Drenching the seedlings with Bordeaux mixture or Copper oxychloride just before the onset of monsoon and thereafter at frequent intervals is essential in effective management of the disease in nurseries with high disease intensity. A combination of seed dressing and soil drench with Kocide at a concentration of 0.91 kg in 45 liters of water has been found to be very effective in controlling pre and post-emergence seedling deaths caused by *P. palmivora* (Asare- Nyako *et al.,* 1972).

BLACK POD DISEASE OF COCOA (*Theobroma cacoa*)

Black pod disease was first noticed in Guyana and West Indies and referred as black cocoa. At present, it is prevalent in all the cocoa growing countries (Zentmyer, 1988). ChandraMohanan and Kaveriappa (1981) reported black pod as the major disease of cocoa in India based on the district wise distribution of cocoa diseases and the percentage of gardens showing disease incidence. Losses from black pod vary from country to country; Nigeria (30%), Brazil (30.8%), Ghana (25-30%), Cameroon (95%), Dominican Republic (10-20%) and Togo (10-80%) (Lass, 1985). In India, the incidence of this has been found to vary from 12.93 to 29.78%, depending upon locality and garden (ChandraMohanan, 1985).

Symptoms

Pods of all ages are susceptible to the disease. The infection appears as one or more small, chocolate brown circular lesion(s) anywhere on the pod surface. Within four to seven days, the lesion enlarges assuming an elliptical shape. As the lesion advances, a whitish growth of the fungus consisting of mycelia and sporangia is produced over the dark brown pod surface. The lesion increases rapidly and covers the whole pod surface. After about 15 days of infection, the fungus invades the whole pod and beans and the pod turns black in colour. By this time, several saprophytic microorganisms colonize the rotten pod. The beans in ripe pod may escape partly or wholly from infection as the beans get separated from the pod husk on ripening (Gregory, 1974; Thorold, 1975).

Causal organism

Phytophthora palmivora (Butl.) Butl., *P. capsici* Leonian, *P. megakarya* Brasier and Griffin and *P. citrophthora* (Smith and Smith) Leonian are the major *Phytophthora* species causing this disease in various cocoa growing countries (Zentmyer, 1988). In addition to these four major species, several other species of *Phytophthora* viz., *P. heveae* Thompson, *P. botryosa* Chee, *P. meadii* Mc Rae, *P. drechsleri* Tucker, *P. nicotianae* Breda de Hann var. *nicotianae* Waterhouse, *P. nicotianae* var. *parasitica* (Dast.) Waterhouse, *P. megasperma* Drechsler and *P.katsurae* Ko and Chang have been described as minor pathogens of cocoa (Liyanage and Wheeler, 1989; Waterhouse, 1974a; Waterhouse, 1974b and Zentmyer, 1988).

The species of *Phytophthora* causing cocoa diseases in India was first identified by ChandraMohanan *et al.*, (1979) as *P. palmivora* based on detailed studies on Phytophthora diseases of cocoa.

Sastry and Hegde (1989) found *P. meadii* as the species causing black pod disease in certain localities of Uttara Kannada district of Karnataka state. *P. capsici* and *P. citrophthora* were isolated from black pod affected cocoa pods in very few localities in Kerala state. But, they were not recorded as the major pathogens of cocoa. Thus, *P. palmivora* has been found to be the predominant species causing black pod disease of cocoa in India. Though, A1 and A2 mating types occur in India, A2 is the predominant mating type in *P. palmivora* population in this country.

Disease development

Black pod disease occurs during rainy season when humidity is high and the temperature is constantly optimum. In India, this disease is a serious problem during south-west monsoon period (June-September). Soil phase of *Phytophthora* is very important. Under favorable climatic conditions, spores are liberated to the soil surface and rain splashes carry sporangia to the pods, which are at the lower part of the trunk causing infection. Pods, which are touching the soil surface, get infected directly. Sporangia are produced abundantly when there is high humidity, rainfall and constant optimum temperature. The pods thus infected act as the source of inoculum for the secondary spread of the disease. The sporangia are spread by rain splashes, insects and rodents (Thorold, 1975). Tent building ants and scolytid beetles have also been reported as agents carrying sporangia from diseased pods to healthy pods (Mc Gregor and Moxon, 1985). Thus, the disease spreads, from pods at the lower part of the trunk to that at the upper part and branches in a climbing up pattern.

Management

Phytosanitation, fungicidal application and host resistance are the three major aspects in the management of black pod disease. Since infected pods form the main source of secondary infection, all the diseased pods should be removed at weekly intervals or during each harvest and buried in the soil. Proper spacing between plants and pruning are also important to regulate the shade. Periodic removal and destruction of infected pods alone will help to reduce the disease incidence to the extent of 50%. Spraying of Bordeaux mixture (1%) at 15 days interval starting from the onset of south- west monsoon along with periodic removal of infected pods is effective in controlling the disease in severely affected gardens. Experiments conducted during last several years revealed that the disease could be effectively controlled by the use of copper based fungicides.

In Costa Rica, of the nine cultivars some cultivars (EET 59, EET 376, UF 713, UF 715, SCA 6, SCA 12, Pound 7, Catongo and Diamantes 800) were found resistant to pod rot (Lawrence, 1978). In India, the cocoa Accession C 78 has been found to be comparatively less susceptible to wound inoculation by *P. palmivora* (ChandraMohanan, 1982).

STEM CANKER OF COCOA (*Theobroma cacoa*)

The term canker is generally used for a disease symptom in which there is sharply limited necrosis of the cortical tissue (Ainsworth and Bisby, 1961). But cocoa canker is not a sharply limited necrosis. In India, stem canker was first reported in 1978 in cocoa plants grown as mixed crop in arecanut garden in Karnataka state (ChandraMohanan, 1978).

Symptoms

Stem canker appears at different parts of the tree including jorquette and fan branches. This disease is difficult to detect in the early stages of development. The symptoms on the surface of bark can be detected only by close examination. The size and shape of external lesions as well as symptoms on the external bark vary. The external symptom appears as grayish brown water soaked lesion with a broad dark brown to black margin. A reddish brown liquid oozing out from such lesions dries up and forms a rusty deposit.

Based on the studies on the symptoms of the disease occurring in West-Coast of India different kinds of external symptoms, have been reported. The external symptoms appear from December after the rainy season. Dark brown, round to oval discoloration of the bark formed as a result of exudation of reddish brown liquid from the point of infection is the usual symptom.

Sometimes the lesions are water soaked and grayish brown. In severe cases, canker lesions coalesce to form larger lesions. Cankers at the collar region are bigger and spread faster. The collar infection appears as dark brown irregular water soaked lesion with reddish brown liquid oozing out. The collar infection then spread to the taproot and main stem. Ultimately, the portions above the infected portion show wilting and finally leading to death. Cankers also develop without any external symptom mainly on seedlings of two or three years and on branches of trees. Such infected seedlings appear weak. Such cankers can be detected only by examining the internal tissue.

When the outer bark of canker infected portion of the stem is removed, the tissues beneath always show a characteristic reddish brown discoloration. Lesions in the internal tissues coalesce leading to extensive rotting. The infection spreads from the cortical tissues to the vascular tissues and reaches the wood. Wood infection appears as grayish brown to black discoloration with black streaks. When canker girdles the stem, dieback occurs. Leaves wilt, turn yellow and fall off. Pods also show wilting. Finally the whole tree dies. Spread of infection in the internal bark is faster than the spread on the surface of bark (Rao and ChandraMohanan, 1993).

Causal organism

P. palmivora (A_2 mating type) has been reported as the only species causing stem canker of cocoa in India.

Management

The canker often develops from the pods infected by *P. palmivora.* Infection from the pods spreads to the peduncle and then to the cushion and bark causing canker. Hence, such infected pods should be removed and destroyed. Stem canker can be controlled in the initial stages by the excision of diseased bark followed by wound dressing with Bordeaux paste. High yielding cocoa trees and rare germplasm collections with advanced stages of canker disease can be rejuvenated by cutting the whole tree well below the canker lesion and allowing a fresh chupon to develop from the basal portion of the stem.

CHUPON BLIGHT AND TWIG DIEBACK OF COCOA (*Theobroma cacoa*)

Chupon blight and twig dieback are caused by *P. palmivora.* The infection usually initiates in the axils of leaves at the tip of twigs or chupons. It appears as water soaked lesions. Infection also starts anywhere on the leaf blade or petiole and extends backwards into the stem. In any case, the chief characteristic symptom is the appearance of water soaked lesion, which soon turns dark brown to black. The lesions coalesce to form bigger lesions. The lesion on stem spreads longitudinally in all directions and turns dark brown to black and is shrunken. When the lesions girdle the stem, the portion above the point of infection wilts showing twig dieback or chupon blight.

Lesions on leaves generally start from the apex or margin of the leaves, more at the apical portion and usually enlarge and coalesce forming large blighted areas. This leads to much defoliation and subsequent dieback (ChandraMohanan, 1979). The disease occurs during south-west monsoon season and becomes severe under conditions of high humidity, improper pruning and heavy overhead shade. This phase of *Phytophthora* infection of cocoa is very important as it plays a major role in the secondary spread of the disease as well as for the high incidence of black pod disease and canker.

Removal and destruction of infected twigs and chupons are very important in the effective management of all *Phytophthora* diseases of cocoa. In cases of severe incidence, the plants may be sprayed with one per cent Bordeaux mixture or any other copper fungicide after removing the infected twigs and chupons (Chandra Mohanan, 1994).

PHYTOPHTHORA PROBLEMS -A Summary

Phytophthora spp. occur worldwide and cause serious damage to numerous cultivated crops, forestry and natural vegetation. Although more than 50 species have been formally identified, species identification is a specialized, time consuming and difficult task. Thus in many cases taxa are only identified as *Phytophthora* sp. Yet adequate identification is fundamental to the implementation of effective disease control strategies.

Symptoms of *Phytophthora* damage

Phytophthora usually infect the root, crown, and lower stem tissue of plants, causing deterioration of root, crown, and stem rots. Infected plants may be stunted and chlorotic and may wilt and die. Infected roots will usually be mushy and discoloured and will not be able to function in absorption of water and nutrients, resulting in symptoms of stunting and plant death.

Unfortunately, certain species of *Phytophthora* can also cause serious aerial blights. Aerial *Phytophthora* blight has become a particularly serious problem over the past several years, and it generally is more damaging to woody plants. Brown, sunken stem lesions and brown, desiccated terminal growth are characteristic of aerial *Phytophthora* blight. Rainy weather and overhead irrigation encourage plant infection as well as splash inoculum from plant to plant.

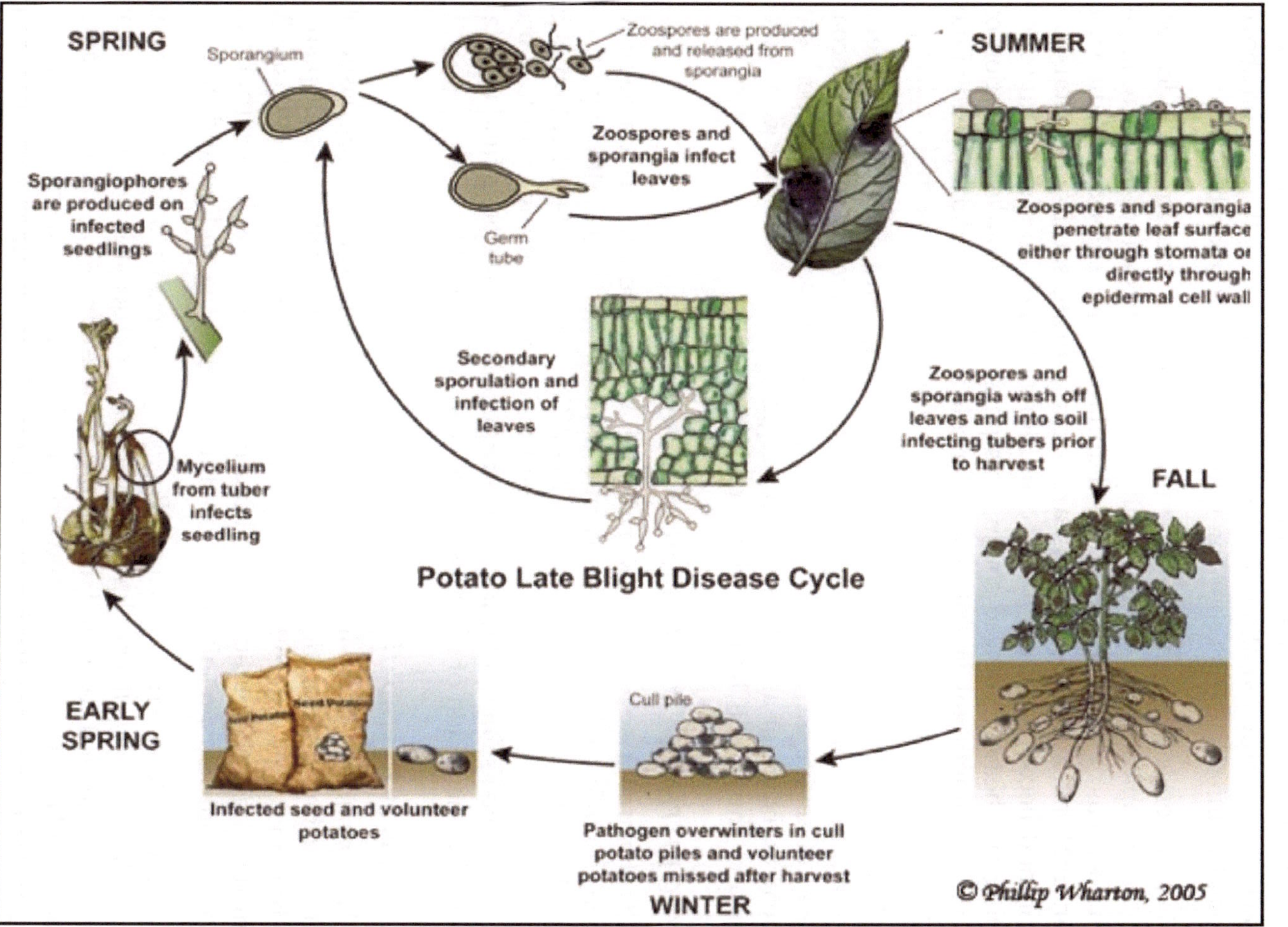

PLATE 16.1: Disease cycle of *Phytophthora infestans* causing late blight of potato.

Conditions that favor *Phytophthora* diseases

Phytophthora are favored by wet conditions; *Phytophthora* produce zoospores that can move in water. Rainy weather and overhead irrigation splash spores from plant to plant. Soil saturation from over watering and/or soil compaction encourages diseases. Cultural stresses that weaken or damage plant roots allow sugars and amino acids to "leak" from plant roots, thereby attracting the zoospores.

Management strategies to minimize *Phytophthora* diseases

a. Practice strict sanitation grow seedlings/plants on raised beds if possible. Remove infected plants from the growing area.

b. Avoid excessive soil compaction when transplanting. Soil compaction reduces soil aeration and can significantly reduce drainage.

c. Avoid excessive drought as well as excessive irrigation. Stress from either factor can weaken plant roots and predispose them to infection.

d. Schedule irrigations as needed to avoid prolonged soil saturation. This is also important in greenhouse and nursery crops.

e. Prune infected twig/branch/shoot tissue from infected plants as soon as symptoms develop.

f. *Phytophthora* infection is enhanced by wounding from pruning and tender, young growth is more susceptible to infection.

g. Schedule fungicide spray applications following pruning and during flushes of new shoot growth.

REFERENCES

Agarwal, A. and Mehrotra, R. S. 1987. Control of *Phytophthora* leaf blight of taro *(Colocasia esculenta)* by fungicides and roguing. *Phytoparasitica.* 15: 299 -305.

Agrios, G. N. 1988. *Plant Pathology,* 3rd edition. Academic Press, Inc.: San Diego. 803 pp.

Ainsworth, G.C. and Bisby, G.R. 1961. *Dictionary of the fungi.* CMI, Kew 547 p.

Asare-Nyako, A., Amponash, J. D. and Nuamah, G. E. A. 1972. Nursery diseases of cocoa in Ghana I. Pre and post emergence deaths of seedlings caused by *Phytophthora palmivora. Ghana J. Agril Sci.* 5: 127-133.

Butler, E. J. 1906. Some diseases of palms. *Agric. J. India* 1: 299 -310.

ChandraMohanan, R. 1978. Cacao canker caused by *Phytophthora palmivora. Plant Dis. Reptr.* 62: 1080- 1082.

Chandra Mohanan, R.1979.Cocoa seedling dieback. *Indian Cocoa. Arecanut & Spices Journal* 3: 5-6.

ChandraMohanan, R. 1982. Studies on the reaction of pods of different cocoa accessions to *Phytophthora palmivora. Planter (Kuala Lumpur)* 58:99-103.

ChandraMohanan, R. 1994. Diseases of Cocoa. (*In*:) K.L. Chadha and P. Rethinam (Eds.) *Advances in Horticulture* Vol. 10- *Plantation and Spice Crops Part* 2. Malhothra Publishing House, New Delhi 1001 -1014 pp.

ChandraMohanan, R., Anandaraj, M. and Joshi, Y. 1979. Studies on *Phytophthora* diseases of cocoa occurring in India. *Proc. PLACROSYM II.* 33, 5-342 pp.

ChandraMohanan, R. and Kaveriappa, K. M. 19(J) Occurrence and distribution of cocoa diseases in South India. Proc. *8th Int. Cocoa Res. Conf.* 18-23, Oct., 1981. Cartagena, Columbia 445-449 pp.

Chang T. T. and Ko, W. H. 1991. Factors affecting the germination of oospores of *Phytophthora infestans. J. Phytopath.* 133: 29- 35.

Coffey, M. D. 1987. Phytophthora root rot of avocado. An integrated approach to control in California. *Plant Dis.* 71: 1046-1052.

Coffey, M. D. 1992. *Phytophthora* root rot of. avocado. (*In*:) J. Kumar *et al.*, (Eds.) *Plant Diseases of International Importance.* P 423-444. Vol. III. *Diseases of Fruit Crops.* Printice Hall, Englewood Cliffs, N. J. 456 pp.

De Boer, R. F., Greenhalgh, F. C., Pegg, K. G., Mayers, P. E., Lim, T. M. and Flett, S. 1990. Phosphorus acid treatments control *Phytophthora* diseases in Australia. *EPPO Bull.* 20:193 -197.

Drenth, A. Janssen, E. M. and Govers, F. 1995. Formation and survival of zoospores of *Phytophthora infestans* under natural conditions. *Plant Pathol,* 44: 86 -94.

Erwin, D, C, and Rebairo, O. K, (Eds.) 1996, *Phytophthora Diseases Worldwide.* American Phytopathological Society, St, Paul, Minnesota, USA, 562 pp.

Fry, W. E., Apple, A. E. and Bruhn, J. H. 1983. Evaluation of potato late blight forecasts modified to incorporate host resistance and fungicide weathering. *Phytopath.* 73: 1054-1059.

Gemewat, P. D. and Prasad, N. 1964. Further studies on *Phytophthora* blight of *Sesamum. Indian Phytopath.* 17: 273 -283.

Gregory, P. H. (ed.) 1974. *Phytophthora Diseases of Cocoa.* Longman groups Ltd., London 348 p.

Harris, D. C. 1991. The Phytophthora diseases of apple. *J. Hortic. Sci.* 66: 513 -544.

Harris, d. C. and Bielenin, A. 1986. Evaluation of selective media and bait methods for estimating *Phytophthora cactorum* in apple orchard soils. *Plant Pathol.* 35: 565 - 574.

Holliday, P. 1980. *Fungus diseases of field crops.* Cambridge University Press, Cambridge, U. K. 602 pp.

Hooker, W. J. (ed.) 1981. *Compendium of potato diseases.* American Phytopathological Society, St. Paul, Minnesota, USA. 141 pp.

Huang, T. H., Chen, D. W. and Leu, L. S. 1976. *Phytophthora* fruit and root rot of papaya in Taiwan. *Plant Proto Bull.* 18: 293-308.

Hunter, J. E., and Buddenhagen, I. W. 1969. Field biology and control of *Phytophthora parasitica* on papaya (*Carica papaya*) in Hawaii. *Ann. Appl. Bioi.* 63: 55-60.

Hunter, J. E., and Kunimoto, R. K. 1974. Dispersal of *Phytophthora palmivora* sporangia by wind-blown rain. *Phytopath.* 64: 202-206.

Jeffers, S. N. 1992. Pre-plant root treatments to reduce the incidence of *Phytophthora* species on dormant apple rootstocks. *Plant Dis.* 76: 12 -19.

Jeffers, S. N.. Aldwinckle, H. S. Burr, T. J. and Arneson, P. A. 1982. *Phytophthora* and *Pythium* associated with crown rot in New York apple orchards. *Phytopathology.*72: 533-538.

Jeffers, S. N. and Aldwinckle, H. S. 1988. *Phytophthora* crown rot of apple trees: Sources of *Phytophthora cactorum* and *P. cambivora* as primary inoculum. *Phytopath.* 78: 328 - 335.

Jones, J. B. Jones, J. P. Stall, R. E. and Zitter, T. A. (Eds.). 1991. *Compendium of Tomato Disease.* American Phytopathological Society, St. Paul, Minnesota, USA. 100pp.

Kannaiyan, J., Ribeiro, O. K., Erwin, 1980. Phytophthora blight of pigeon. pea in India. *Mycologia.* 72: 169 -181.

Ko, W. H. 1971. Biological control of seedling root rot of papaya caused by *Phytophthora palmivora. Phytopath.* 61: 780-782.

Ko, W. H. 1987. Biological control of *Phytophthora* root rot of papaya with virgin soil. *Plant Dis.* 66: 446-448.

Liyanage, N. I. S. and Wheeler, B. E. J. 1989. *Phytophthora katsurae* from cocoa. *Plant Pathol.* 38: 627-629.

Legard, D. E., Lee, T. Y. and Fry, W. E. 1995. Pathogenic specialization in *Phytophthora infestans:* Aggressiveness on tomato. *Phytophpath.* 85: 1356 -1361.

Lass, R.A. 1985. *Diseases.* (*In:*) G.A.R. Wood and R.A. Lass (*Eds.*). *Cocoa,* Longman, London 265-305 pp.

Liyanage, N. I. S. and Wheeler, B. E. J. 1989. *Phytophthora katsurae* from cocoa. *Plant Pathol.* 38: 627-629.

Mehrotra, R. S. and Tiwari, D. P. 1976. Organic amendments and control of foot rot of *Piper betle* caused by *Phytophthora parasitica* var. *piperina. Ann. Microbiol.* 127 A: 415 -421.

Mircetich, S. M., Campbell, R. N. and Mathero, M. E. 1977. *Phytophthora* trunk canker of coast live oak and cork oak trees in California. *Plant Dis. Reptr.* 61: 66 -70.

Nair, R. R. and Menon, M. R. 1982. 'Azhukal' disease of cardamom. (*In:*) K. K. N. Nambiar (Ed.). *Proc. Workshop on Phytophthora disease of tropical cultivated plants.* Central Plantations Crops Research Institute, Kasaragod, Kerala, India. 284pp.

Nambiar, K. K. N. and Saffila, Y. R. 1974. Chemical control) of capsule rot of cardamom. *J. Plantn. Crops.* 2: 30 -31.

Narayanaswamy, H.., Pradeep, S., Chandrashekhar, S. Y. and Basavaraja, M. K. 2004. Integrated management of black shank disease of tobacco in field crop. *Plant Dis. Res.* 19: 66 -67.

Nambiar, K. K. N. and Sarma, Y. R. 1982. Some aspects of epidemiology of foot rot of black pepper. (*In:*) K. K. N. Nambiar (Ed.). *Workshop on Phytophthora diseases of tropical cultivated plants.* Central Plantations Crops Research Institute, Kasaragod, Kerala, India. 284pp.

Pal, M., Grewal, J. S. and Sarbhoy, A. K. 1970. A new stem rot of arhar caused by *Phytophthora. Indian Phytopath.* 23:583 -587.

Papavizas, G. C. and Bovers, J.H. 1991. Comparative fungitoxicity of captafol and Metalaxyl to *Phytophthora capsici. Phytopath.* 71: 123 -127.

Parris, G. K. 1942. *Phytophthora parasitica* on papaya *(Carica papaya)* in Hawaii. *Phytopath.* 32: 314-320.

Pillai, P. N. R. 1982. Abnormal leaf falls of rubber caused by *Phytophthora* spp. In: workshop on Phytophthora diseases of tropical cultivated plants (Ed. K. K. N. Nambiar). Central Plantations Crops Research Institute, Kasaragod, Kerala, India. 284pp.

Radha, K. and Joseph, T. 1974. Investigations on the bud rot disease (*Phytophthora palmivora* Butl.) of coconut. Final Report, PL 480 Scheme. 32 pp.

Ramachandran, N. and Sarma, Y. R. 1985. Efficacy of three systemic fungicides in controlling *Phytophthora* infections on black pepper. *Indian Phytopath.* 38: 160 -162.

Rao, G. K. R. and ChandraMohanan, R. 1993. Symptomatology of stem canker of cocoa in India. *The Planter (Kuala Lumpur)* 69: 153-158.

Rich, A. E. 1983. *Potato diseases.* Academic Press, New York. 238pp.

Ristaino, J. B. 1990. Intraspecific variation among isolates of *Phytophthora capsici* from pepper and cucurbrt fields in North Carolina. *Phytopath.* 80: 1253 -1259.

Ristaino, J. B. 1991. Influence of rainfall, drip irrigation, and inoculum density on the development of *Phytophthora* root rot and crown rot epidemics and yield in bell pepper. *Phytopath.* 81: 922 -929.

Rohni Iyer. 2000. Effect of systemic fungicides on control of bud rot of coconut. (*In:*) *Int. Conf. on Integrated Plant Disease Management for Sustainable Agriculture.* Indian Phytopathological Society, IARI, New Delhi.

Saksena, S. B. 1977. *Phytophthora parasitica* the scourge of pan *(Piper betle* L.). *Indian Phytopath.* 30: I -6.

Saksena, S. B. and Mehrotra, R. S. 1970. Studies on foot and leaf rot of *Piper betle* L. Host: range and role of cuttings in the spread of the disease. *J. of Indian Bot. Soc.* 49: 24 -29.

Sastry, M. L. N. and Hegde, R. K. 1989. Variability of *Phytophthora* species obtained from plantations of Karnataka. *Indian Phytopath.* 42: 421-425.

Shew, H. D. and Lucas, G. B. (eds).1991. *Compendium of tobacco diseases.* APS Press, American Phytopathological Society, St. Paul, Minnesota, USA.

Sinclair, J. B. and Backman, P. A. 1.983. *Compendium of soybean diseases,* 3rd ed. American Phytopathological Society, St. Paul, Minnesota, USA. 116 pp.

Teakle, D. S. 1957. Papaw root rot caused by *Phytophthora palmivora* Butl. *Queens-land J. Agric. Sci.*14: 81-91.

Tidball, C. J. and Linderman, R. G. 1990. *Phytophthora* root and stem rot of apple rootstocks from stool beds. *Plant Dis.* 74: 141 -146.

Thind, S. K., Sharma, J. N. and Josan, J. S. 1999. Comparative efficacy of fungicides and their methods of application for the control of foot rot of citrus. *Plant Dis. Res.* 14: 155-160.

Thorold, C. A. 1975. *Diseases of cocoa.* Claredon Press, Oxford 423 p.Trujillo, E. E., and Hine, R. B. 1965. The role of papaya residues in papaya root rot caused by *Pythuim aphanidermatum* and *Phytophthora parasitica. Phytopath.* 55: 1293-1298.

Turller, P. D. 1965. Behavior of *Phytophthora palmivora* in soil. *Plant Dis. Reptt.* 49: 135-137.

Utkhede, R. S. 1987. Chemical and biological control of crown and root rot of apple caused by *Phytophthora cactorum. Can. J. Plant Pathol.* 9: 295 -300.

Vartanian, V. G. and Endo, R. M. I 985.Survival *of Phytophthora infestans* in seeds extracted from infected tomato fruits. *Phytopath.* 75: 375 -378.

Waterhouse, G. M. 1974a. *Phytophthora palmivora* and some related species. (*In:*) P.H. Gregory (Ed.). *Phytophthora diseases of cocoa* Longman Group Ltd., London 51-70 pp.

Waterhouse, G. M. 1974b. Other *Phytophthora* species recorded on cocoa. (*In:*) P.H. Gregory (Ed.). *Phytophthora diseases of cocoa*. Longman Group Ltd., London 71-79 pp.

Whiteside, J. O. 1970. Factors contributing to the restricted occurrence of citrus brown rot in Florida. *Plant Dis. Reptr.* 54: 608 -612.

William, F. J., Amin, K. S. and Baldev, B. 1975. Phytophthora stem blight *of Cajanus cajani. Phytopath.* 65: 1029 -1030.

Zentmyer, G. A. 1988. Taxonomic relationships and distribution of *Phytophthora* causing black pod disease of cocoa. (*In:*) *Proc.10th Inter.Cocoa Research Conf,* Douala, Cameroon 291-295 pp.

• • •

Advances in Soil Borne Plant Diseases, 2008
Editors: Naik and Devika Rani, pp. 415-427
New India Publishing Agency, New Delhi

17

Integrated Management of Soil Borne Diseases

V.K. Gupta and R.C. Sharma[1]

INTRODUCTION

Plants are susceptible to different types of pathogens, which are responsible for production of varied symptoms and causing losses. In the absence of the crops pathogens perpetuates in different manners and survival in the soil is one of the most important amongst these. Because of the constant presence of the pathogen in the soil, these become more destructive. These may reduce the plant growth, results in rotting of different plant parts, reduce yield or may cause the death of the plant. The destructiveness of soil borne pathogen increases because of the ability to survive in the soil as dormant structures for several years. Efficient control of soil borne diseases is thus difficult but can be achieved by employing various disease management practices in an integrated manner (Verma and Varma, 1992; Gupta, 1996, Gupta and Sharma, 1999).

The philosophy of integrated disease management aims at integrating all the available methods of plant disease control in ecologically sound manner which takes into consideration the economic, social and legislative parameters (Kommedahl and Williams, 1983). An appropriate combination of disease management methods could thus result in improved pathogen control with

Post Graduate Studies, Himachal Pradesh Agricultural University, Palampur-176062 (Himachal Pradesh)

1 Department of Mycology and Plant Pathology, University of Horticulture and Forestry, Solan-173230 (Himachal Pradesh)

long-term efficacy. Combination of disease management practices may have additive or synergistic effects and such an approach is especially desirable in the case of soil borne diseases which are entirely different epidemiologically. Hopefully, the present situation, which emphasizes the use of integrated disease management practices, will stimulate the development/refinement of non-chemical methods of disease management, which have been neglected in the past due to convention of application of chemicals (Verma and Sharma, 1999).

DISEASE MANAGEMENT PRACTICES

The use of disease management practices depends upon various factors like the crop in question and its economic value, method of crop production, and nature of soil borne pathogens. Relatively inexpensive disease management strategies can be used in case of low value field crops, while expensive methods are suitable for high value horticultural crops. Root rot pathogens that reduce the number of feeder roots during the growing season require different management practices than structural root rots, damping off and vascular wilt pathogens. Some of the important practices, which components of integrated management of soil borne diseases are discussed below.

Cultural Practices

Variation in cultural practices such as fertigation, irrigation, sanitation, time of sowing, increasing soil temperature through soil solarization, intercropping, tillage practices and crop rotation, grafting etc. have been shown to be effective in reducing the incidence of soil borne diseases.

Fertigation

Application of nutrients to the plants influence the severity of soil borne diseases by affecting the pathogen population, overall plant health and growth of biocontrol agents. In the rhizosphere the nutrients affect the saprophytic growth and survival of the pathogens. Punja *et al.*, (1986) showed that application of ammonium bicarbonate (84 kg N/ha) reduce the viability of *Sclerotium rolfsii* sclerotia in the soil due to increased leakage of components from sclerotia resulting in enhanced antagonism. Higher C:N ratio has been found to be more effective in reducing the survival of *Macrophomina phaseolina* in India (Sharma and Bhowmik, 1989).

Application of phosphatic fertilizers is also known to enhance host resistance by increasing production of phytoalexins (Gottstein and Kuc, 1989). Copper and ammonium fertilizers also elicit phytoalexin production in cotton. Management of *Pythium* and *Phytophthora* pathogens by phosphoric acid is mediated through increased host resistance (Cohen and Coffey, 1986). Use of gypsum in groundnut crop reduces the incidence of *Macrophomina* rot (Bhowmik *et al.*, 1985). Form of the element used also influence diseases in different ways. Nitrate and ammonical forms of nitrogen have different effects on disease development as the soil population of fluorescent pseudomonas increase in the later and are four times higher compared to nitrate nitrogen. In case of take all disease of wheat, this results in lower disease incidence.

Deep sowing of seed in different crops generally increases the damping off incidence. In most of the nurseries, shallow sowing up to a depth of about 0.7 cm gives satisfactory control of various diseases (Sharma, 1989).

Crowded crop stands increase the chances of the roots of adjoining plants to come in contact in the soil resulting in dissemination of the pathogen from infected to healthy plants. Spread of *Armillaria mellea* and *Verticillium albo-atrum* is fast in closely spaced stands of plants. Dense sowing of low germination capacity seed provides the attacking fungi with greater amount of substrate for inoculum build up.

Stratification of hard coat seeds before sowing helps in quicker germination. Sowing of stratified seed reduces the damping off incidence. Sowing with the help of V- shaped drills increases the incidence of damping off due to over crowding of seeds. By sowing with the help of flat drills of 2.5 cm or broader the disadvantage of V-shaped drill can be avoided.

Irrigation and Water Management

Irrigation is one of the most important cultural practices and no other cultural practice probably has a greater influence on the plant- microbe-environment (Bruehl, 1975). Its over as well as under-use affects disease intensity differently in different crops depending upon the requirement of the pathogen. Irrigation affects diseases by crop intensification, change in growing season and sowing time. Heavy and frequent irrigation favours pathogens like *Pythium* and *Phytophthora. Phytophthora* rot of chillies *(P. capsici)* is less in plants receiving alternate row irrigation than in plants having irrigation in every row (Biles *et al.*, 1992). Decreasing the number of irrigations also reduces *Sclerotinia sclerotiorum* in lettuce and *Pythium* rot of groundnut. Irrigation can also be useful for reducing diseases favoured by water stress such as charcoal root rot caused by *Macrophomina phaseolina.* Maintaining soil moisture near field capacity during tuber formation reduces potato scab. Management of Panama wilt disease of banana through flooding is one such classical example. Many soil borne pathogens, such as *Sclerotium rolfsii,* will not affect crops in arid areas unless the crop is irrigated. Irrigation technique also influences the extent of disease. Triple irrigation by drip method results in disappearance of *Sclerotium rolfsii* from deciduous fruit orchards because trickling lowers soil temperatures and increase the activity of microorganisms antagonistic to the pathogen (Palti, 1981), but in watermelon, drip irrigation results in small root system and such plants are more susceptible to Monosporascus wilt (Cohen *et al.,* 2000).

Irrigation water distributed through canals and channels also distribute a number of soil borne fungal, bacterial and nematode pathogens from one field to another. Pathogen infested water is also a problem for greenhouse crops.

Site selection

Infection by *Sclerotium rolfsii* can be minimized by avoiding heavily infested fields. Problem fields may be identified by cropping history and sampling soil for soil borne pathogens. Correlations between inoculum density and disease development have been developed for a number of plant pathosystems. For example carrots should not be planted if the initial inoculum density of *Sclerotium rolfsii* is above one sclerotia for 300 cm^3 of soil unless other stringent control measures are adopted (Punja *et al.,* 1984).

Grafting

Grafting on root stock resistant to soil borne pathogens is an effective method to combat soil borne diseases (Sharma *et al.,* 1999). This practice is of great importance with regard to *Phytophthroa cactorum* infection on pome and stone fruits where resistant root stocks are available. Apple root stocks M_2, MM_{105}, MM_{113} and MM_{114} have been found to be resistant to *Phytophthora cactorum* under Indian conditions (Gupta and Mir, 1983). Grafting melons and watermelons onto *Cucurbita* and melon root stocks to manage Fusarium wilt is common in the Mediterranean basin and Southeast Asia (Lee, 1994).

Tillage practices

Various tillage practices like minimum tillage, zero tillage, reduced cultivation and sub-soiling influence the soil borne diseases (Gupta and Sharma, 1999). Tillage practices also affect diseases indirectly by controlling weeds and volunteer plants (Cook, 2000).

Deep ploughing reduces the contact between roots and pathogen or expose the inoculum to natural heating and desiccation. Exposure of top layers of soil to strong radiation and thus heating and drying it thoroughly, has long been practiced in the warmer areas as a means of disinfestation, chiefly for the management of nematodes.

Dry fallowing for disease management is restricted to extensive farming, where other methods are not economical. Wet fallowing, viz. leaving land free from crops while irrigating it intermittently which is practiced for weeks rather than months is considered useful for pathogens whose propagules can germinate in absence of root exudates from their hosts, such as *Sclerotium rolfsii* and *Verticillium dahliae* (Palti, 1981; Bockus and Shroyer, 1998),

In conservation tillage at least 30 per cent residue cover on the soil surface is retained after planting. This system includes no-till, slot planting, ridge till, strip till, mulch till and other reduced or minimum tillage practices. All these tillage practices influence plant diseases. Many diseases important in conventionally planted crops disappear with no-till while minor disease problems such as Rhizoctonia root rot may become major problem under these conditions (Rovira, 1986).

Verticillium albo-atrum causing hop wilt can also be eradicated from infected soil by keeping it under grass for 3 to 5 years provided that weeds are strictly controlled.

Soil solarization

Soil solarization is trapping of solar irradiation by tightly covering the soil with transparent polyethylene sheets (Katan, 1981). It is an environment friendly technique which results in killing of pathogens in upper soil layers due to high temperature, weakening of pathogen propagules by sub-lethal heating in lower soil layers and increase in the biological activity of heat tolerant biocontrol agents. It is mainly a pre-sowing treatment and also effective as post-planting treatment in case of *Verticillium* affected olives and prunes.

Verticillium wilt and Fusarial wilt of large number of crop plants are controlled by soil solarization. Besides these, *Pythium, Phytophthora, Sclerotinia, Scierotium, Rhizoctonia solani* and *Thielaviopsis basicola* are also managed by this technology (Tjamos, 1996). This technique is also useful in reducing bacterial canker of tomato caused by *Clavibacter michiganense* sub sp, *michiganense* (Tjamos *et al.,* 1992). Heat tolerant pathogens like *Monosporascus eutypodies* are however, not controlled by this method, *Macrophomina phaseolina* require longer treatment period and high solar radiation. Sub-lethal doses of raised temperature by soil solarization in lower soil layers also make the pathogen propagules more susceptible to the attack of biocontrol agents.

Thermo-sanitation

The technique of managing diseases through fire and flaming is known as thermo-sanitation. Burning of infected crop residue in the fields help in eradicating pathogen propagules. Burning of rice stubble and straw is a common practice to control stem rot *(Sclerotium oryzae)* and sheath blight *(Corticium sasaki)*. Take-all of wheat *(Gaeumanomyces graminis)* and potato and pepper mint wilt *(Verticillium dahliae)* are other diseases which are primarily controlled by fire and flame (Palti, 1981).

Crop rotation

Growing of crop plants in recurring succession and in a definite sequence on the same land is termed as crop rotation. The soil borne root infecting pathogens likely to affect a crop are in the first place those left over from the preceding crop. Crop destruction by *Pythium, Phytophthora, Verticillium, Fusarium, Rhizoctonia* and *Sclerotinia* is the result of faulty crop rotation. Generally the crop rotation is recommended with botanically unrelated species.

Soil borne pathogens that infect plants of one or few species or even families of plants can be reduced by planting for 3 or 4 years, crops belonging to species or families not attacked by the particular pathogen. Crop rotation is effective against soil invaders and less effective against soil inhabitant pathogens.

Chemicals in control of soil borne diseases

Chemicals are being used to manage soil borne diseases throughout the world. These are applied in the soil as pre- or post-planting application. Chemical soil treatments are being given in high value cash crops as well as in nurseries. Chemicals can be applied by different methods as below:

Soil fumigation

Fumigation of soil with broad spectrum chemicals such as formaldehyde, methyl bromide, chloropicrin and metham sodium is being done for reducing the population of soil borne pathogens.

These chemicals which are highly toxic are applied in the soil either as drench or in pre-plant furrows. At low dosages they weaken the pathogen propagules which are easily attacked by the antagonistic and parasitic organism whereas at high doses these kill them. Soil conditions such as its texture, presence of moisture, aeration and organic matter also affect the efficacy of these chemicals (Gupta and Sharma, 1999).

Soil Drenching

Certain fungicides are applied in the soil as dusts, granule suspensions and drench to reduce the pathogen population. Fungicides used for soil treatment include carbendazim, metalaxyl, diazoben, quintozene, captan and chloroneb, although the last two are used primarily as seed treatments.

Soil drenching with systemic fungicides is recommended both for the control of soil borne and foliar diseases. Carbendazim, thiabendazole and benomyl have been extensively used against soil borne diseases (Sharma, 1996). Kitazin is commonly used in paddy soils through irrigation water to control rice blast. Fungicides like prothiocarb, propamocarb and metalaxyl are particularly useful to control soil borne oomycetous pathogens (Cohen and Coffey, 1986).

Fosetyl-Al (Aliette) fungicide is one of the rare products which is used to suppress soil borne pathogens after application as foliar spray. The chemical or its breakdown products move systemically from the foliage to the roots where it can control pythiaceous fungi (Utkhede, 1988). Challenging problem in the use of chemicals for the management of soil borne diseases by systemic fungicides is the development of resistance in the pathogens to site specific acting chemicals (Brent, 1995; Sharma, 1996; Sharma *et al.*, 1998).

Biocontrol Agents

Use of biocontrol agents for the management of soil borne diseases involves the application of these agents to soil and planting material. Some biocontrol agents are more effective than others (Schneider, 1982). Another phenomenon called development of suppressive soils against certain diseases has also been noticed. Transfer of small amount of suppressive soil many a times converts conducive soils to suppressive ones. In such suppressive soils agents responsible for disease suppression have been identified and added back to normal fields to make them suppressive. Natural suppressiveness of soils to soil borne diseases can be increased by cultural practices (Mukerji and Garg, 1988). Biological antagonisms although subject to numerous ecological limitations, are expected to become an important part of the integrated disease management of any soil borne disease. Some of the biocontrol agents of soil borne diseases are discussed briefly in the following paragraphs:

Fungi

Several soil borne diseases such as vascular wilts of different plants *(Fusraium oxysporum),* take all of wheat *(Gaeumnnomyces graminis),* root rot of fruit and forest trees *(Phytophthora cinnamomi)* and damping off *(Pythium* spp.) develop well in some soils, now called conducive soils, whereas they develop much less in other soils, known as suppressive soils (Hornby, 1983). Suppression of these pathogens in suppressive soils is due to the presence of a number of microorganisms and important among these are different fungi like *Trichoderma, Gliocladium, Penicillium* and *Sporidesmium* (Mukerji and Garg, 1988). Much of the work on fungal biocontrol agents is centered on the genus *Trichoderma. Phytophthora* root rot of papaya in *P. palmivora* infested soils was controlled by raising plants in holes filled with suppressive soils. However, in some diseases continuous cultivation of the same crop in a conducive soil, after some years of severe disease occurrence, eventually leads to reduction in disease through increased population of microorganisms antagonistic to the pathogen. In take all disease continuous cultivation of wheat in a field has led to reduction of disease and this is considered as a classical example of suppressive soil (Weller, 1988; Agrios, 1997).

The mycelium and resting spores of soil borne pathogens such as *Pythium, Phytophthroa, Rhizoctonia, Sclerotinia* and *Sclerotium* are parasitized and lysed by several mycoparasitic fungi. Among the most common mycoparasitic fungi, *Trichoderma* sp. and *Laetisaria arvalis* are known to be parasites of *Rhizoctonia* and *Pythium. Sclerotinia* diseases are effectively controlled by antagonists *Sporidesmium sclerotiovorum, Gliocladium virens* and *Coniothyrium minitants.* Verticillium wilt of egg plant is known to be controlled by *Talaromyces flavus.* Hypovirulent strains of *Rhizoctonia solani* reduces the damping off in sugarbeet seedlings caused by virulent strains of the same pathogen (Utkhede and Gupta, 1996; Gupta and Sharma, 1999, Bertrand, 2000).

Bacteria

In addition to fungi, bacteria belonging to genera *Bacillus, Enterobacter* and *Pseudomonas* have been shown to parasitize and/or inhibit pathogenic fungi. Spore forming bacteria such as *Bacillus* spp. have the additional advantage of longer shelf life and survival ability under dry conditions. Suppressive soils also contain these antagonistic bacteria, which reduce number of soil borne diseases. Normal soil amended with soil containing a strain of *Streptomyces* species antagonistic to *Streptomyces scabies,* the cause of potato scab, resulted in reduction in the disease levels and production of potato tubers significantly free from scab (Agrios, 1997). Strain K 1026 of *Agrobacterium radiobacter* modified from strain K 84 has been registered as 'Nogall' for the control of crown gall. *Streptomyces griseoviridis* has been developed as 'Mycostop' product for the control of damping-off diseases (Lahdenpera *et al.,* 1991). Siderophore producing fluorescent pseudomonads such as *Pseudomonas fluorescens* and *P. putida* are quite effective for the control of *Pythium* species.

Plant growth promoting rhizobacteria increase plant growth by the control of soil borne pathogens through production of antibiotics and siderophores. Plant growth promoting rhizobacteria also inhibit deleterious rhizobacteria which otherwise produce phytotoxins such as hydrocyanic acid. These organisms may enhance phosphorus uptake by plants and secrete plant growth regulating hormones (Schippers *et al.* 1987).

Amoeba

Amoeba, *Arachnula impatiens* perforates the spores of number of fungi in the soil. Since these amoebae are widely distributed in the soil, they might be playing an important role in the management of soil borne pathogens (Old and Patrick, 1979).

Mycorrhizae

Mycorrhizal fungi play a significant role in promoting plant health. These were known to be present in forest trees since long but have now been found to be normally associated with feeder roots of most plants including cereals, vegetables, ornamentals and trees. Among the two groups of mycorrhizae, ectomycorrhizae are formed primarily on the forest trees. These fungi enter roots and grow around the cortical cells only, replacing part of the middle lamella between the cells, and forming the so, called Hartig net. The other group called endomycorrhizae is the most common. The roots infected with these fungi appear normal and are just like non-mycorrhizal roots in shape and size. The hyphae of endomycorrhizal fungi grow into cortical cells of the feeder roots either by forming specialized feeding hyphae (haustoria), called arbuscules or by forming large, swollen, food storing hyphal swellings, called vesicles. Most endomycorrhizae contain both vesicles and arbuscules and are, therefore, also called vesicular-arbuscular (VA) mycorrhizae (Agrios, 1997).

The association of Ectomycorrhizal fungi namely *Suillus granulatus, Pisolethus tinctorius* and *Cenococcum graniforme* with pine and spruce roots make them resistant to soil fungi like *Rhizoctonia, Phytophthora* and *Pythium spp* causing various root symptoms (Gupta and Sharma, 1999). Application of fungicides around pecan trees resulted in increased activity of ectomycorrhizal fungus, *Scleroderma bovista* which protect these plants from the infection of *Pythium* sp. Important VA mycorrhizae which suppress various soil borne diseases are *Glomus mosseae,* G. *intraradices* and G. *fasciculatum.* Application of G. *mosseae* in apple replant problem sites resulted in decreased disease incidence and increased plant growth.

Immunization and host resistance

Plants lack an antibody producing system, but through genetic engineering scientists have developed plant varieties, which produce specific antibodies against specific pathogens called plantibodies. It is not yet known whether such plantibodies will effectively protect plants from pathogens or not. Inoculation of plants with mild strain of a virus, which protects them from causing infection by more severe strain of the same virus, is called cross protection. The phenomenon when different types of pathogens are involved, is termed as induced or systemic acquired resistance. Both these phenomena have not been explored to the desired extent in the management of soil borne diseases.

Incorporation of resistance by conventional breeding techniques in susceptible cultivars has been achieved against many vascular wilt pathogens such as *Fusarium* and *Verticillium* spp. in a few crops. This approach has however, not been found useful in many crops and other root pathogens such as *Pythium, Rhizoctonia* and *Gaeumannomyces* (Nelson and Maloney, 1996). Crops like cotton, tobacco, wheat and banana are polyploidy and this makes incorporation of resistance against wilt pathogens in these crops difficult. The resistance is controlled either by single dominant gene or it is inherited polygenically. It is easier to incorporate single gene dominant resistance than the other one.

CONCLUSION

Management of soil borne diseases is most successful and economical when all the required information pertaining to the crop, diseases affecting it, history of these in the previous years, resistance level of the host and environmental conditions expected to prevail is available. Integrated disease management practices are useful in reducing the quantity and effectiveness of primary inoculum. The best available host can be used which may delay the onset of the disease and slow down the secondary cycles of pathogens. Sowing disease free seed or propagative material is of paramount importance. Seed or propagative material should be treated with fungicides/bactericides and/or biocontrol agents. Proper cultural practices, which reduce the disease incidence, should be followed. The drainage in the fields should be well drained. Even after adopting all these strategies if disease appears, soil drenching with fungicides may be given in high value cash crops. In a nutshell, several strategies have to be employed which include regulatory inspection for healthy seed or nursery crop production, manipulation in cultural practices, use of biocontrol agents, use of chemicals and host resistance. Each one of these measures must be used and combined for reaching a target of no disease in the crop.

REFERENCES

Agrios, G. N. 1997. Plant Pathology. 4th Ed. Acad. Press, San Diego, p. 635.

Bertrand, H. 2000. Role of mitochondrial DNA in the senescence and hypovirulence of fungi and the potential for plant disease control. *Ann. Rev. Phytopath.* 38: 397-422.

Bhowmik, T. P., Sharma, R. C. and Singh, A. 1985. Effect of gypsum, row spacing and groundnut varieties on the incidence of root rot disease caused by *Macrophomina phaseolina. Int. J. Trop. Pl. Dis.* 3: 69-72.

Biles, D. L., Lindsey, D. L. and Liddell, C. M. 1992. Control of Phytophthora root rot of Chilli peppers by irrigation practices and fungicides. *Crop Protection.* 11: 225-228.

Bockus, W. W. and Shroyer, J. P. 1998. The impact of reduced tillage on soil borne plant pathogens. *Ann. Rev. Phytopath.* 36: 485-500.

Brent, K. J. 1995. Fungicide Resistance in Crop Pathogens: How can it be managed? *Int. Group Nat. Assoc. Manuf. Agrochem. Prod., Brussels.* 48 p.

Bruehl, G. W. 1975. *Biology and Control of Soil borne Plant Pathogens. American Phytopathological Society St. Paul.* p.216.

Cohen, R., Pivonia, S., Burger, Y., Edelstein, M., Gambiel, A. and Katan, J. 2000. Towards integrated management of Monosporascus wilt of melons in Israel. *Plant Dis.* 84: 496-505.

Cohen, T. and Coffey, M. D. 1986. Systemic fungicides and the control of oomycetes. *Ann. Rev. Phytopath.* 24: 311-38.

Cook, R. J. 2000. Advances in plant health management in the twentieth century. *Ann. Rev. Phytopath.* 38: 95-116.

Gottstein, H. D. and Kuc, J. A. 1989. Induction of systemic resistance to anthracnose in cucumber by phosphates. *Phytopath.* 79:176-179.

Gupta, V. K. 1996. Integrated management of soil borne diseases. pp: 40-64. *(In:)* R.S. Utkhede and V.K. Gupta (Eds.). *Management of Soil Borne Disease,* Kalyani Publishers, Ludhiana.

Gupta, V. K. and Mir, N. M. 1983. Efficacy of fungicides on the survival of *Phytophthora cactorum* propagules in soil and control of disease. 10th *Int. Congr. Plant Proto Brighton,* UK, Nov., 1983, 1006 (Abstr.).

Gupta, V. K. and Sharma R. C. 1999. Concepts and practices in the management of soil borne diseases. PP: 333-334. *(In:)* D.J. Bagyaraj, A. Verma, K.K. Khanna and H.K. Kehri. (Eds.) *Modem Approaches and Innovations in Soil Management,* Rastogi Publications; Meerut.

Hornby, D. 1983. Suppressive soils. *Ann. Rev. Phytopath.* 21: 65-85.

Katan, J. 1981: Solar heating (solarization) of soil for the control of soil borne pests. *Ann. Rev. Phytopath.*19: 211-236.

Kommedahl, T. and Williams, P .H. 1983. *Challenging Problems in Plant Health.* Ame. Phytopath. Soc., St. Paul. p:538.

Lahdenpera, M. L., Sismon, E. and Uoti, J. 1991. Mycostop a novel biofungicide based on Streptomyces bacteria. pp. 258-263. *(In:)* A.B.R. Beemster, G.J. Bollen, M. Gerlagh, M.A. Ruissen, B. Schippersn and A. Tempel (Eds.) *Biotic Interactions and Soil borne Diseases,* Elsevier, Amsterdam.

Lee, J. M. 1994. Cultivation of grafted vegetables 1. Current status, grafting methods and benefits. *Hort. Science.* 29: 229-235.

Mukerji, K. G. and Garg, K. L. Eds. 1988. *Biocontrol of Plant Diseases,* CRC Press, Boca Raton. p:198.

Nelson, E. B. and Maloney, A. P. 1996. Application of recombinant DNA technology for the management of soil borne diseases. pp: 65-99. *(In:)* R.S. Utkhede and V.K. Gupta, (Eds.) *Management of Soil Borne Diseases,* Kalyani Publishers, Ludhiana.

Old, K. M. and Patrick, Z. A. 1979. Giant soil amoebae, potential biocontrol agents. *(In:)* B. Schippers and W. Gams, (Eds.). *Soil borne pathogens.* pp:617-628. Academic Press. London.

Palti, J. 1981. *Cultural Practices and Infections Crop Diseases,* Springer Verlag, Berlin. p.243.

Punja, Z. K., Smith, V. L. and Jenkins, S. F. 1984. Relationship of disease incidence to inoculum density in *Sclerotium rolfsii* root rot of processing Carrot. *Phytopath.* 79: 813 (Abstr.).

Punja, Z. K., Carter, J. D. Campbelly, G. M. and Rossell, E. L. 1986. Effects of calcium and nitrogen fertilizers, fungicides and tillage practices on incidence of *Sclerotium rolfsii* on processing carrots. *Plant. Dis.*70:819-824.

Rovira, A. D. 1986. Influence of crop rotation and tillage on Rhizoctonia base patch of wheat. *Phytopath.* 76: 669-673.

Schippers, B., Baker, A. W. and Bakker, P. A. H. M. 1987. Interactions of deleterious and beneficial rhizosphere microorganisms and the effect of cropping practices. *Ann. Rev. Phytopath.* 25: 339-358.

Schneider, R. W. 1982. *Suppressive Soils and Plant Disease.* American Phytopath. Soc., St. Paul, p:88.

Sharma, R. C. 1989. Forest Nursery Diseases and their Management. *Tech. Bull.* SAF, 2, UHF, Solan, p:24.

Sharma, R. C. 1996. Fungicide use in soil and development of resistance. pp. 282 - 300. *(In:)* R.S. Utkhede and V.K. Gupta, (Eds.). *Management of Soil Borne Diseases,* Kalyani Publishers, Ludhiana.

Sharma, R. C. and Bhowmik, T. P. 1989. Effect of soil moisture and C : N ratios on survival of *Macrophomina phaseolina* in groundnut shells in soil. *Int. J. Trop. Plant Dis.* 7: 61-64.

Sharma, R. C., Sharma, S. and Sharma, J. N. 1999. Present concepts and strategies of disease management in pome and stone fruits. *Int. J. Trop. Pl. Dis.* 17:109-120.

Sharma, S., Sharma, R. C. and Dohroo, N. P. 1998. Fungicide resistance: Development and its management. *Int. J. Trop. Pl. Dis.* 16: 103-117.

Tjamos, E,.C. 1996. Chemical treatment and soil solarization for management of soil borne diseases. *(In:)* R.S. Utkhede and V.K. Gupta, (Eds.). *Management of Soil Borne Diseases,* pp: 261-281. Kalyani Publishers, Ludhiana.

Tjamos, E. C., Antoniou, P. and Panagopoylos, G. C. 1992. Control of bacterial canker of tomato by application of soil solarization. *Phytopath.* 82: 1076 (Abstr.).

Utkhede, R. S. 1988. Chemical and biological control crown and root rot of apple caused by *Phytophthora cactorum. Can. J. Pl. Pathol.* 9:295-200.

Utkhede, R. S. and Gupta, V. K. 1996. *Management of Soil Borne Diseases,* Kalyani Publishers, Ludhiana, pp:354.

Verma, J. P. and Varma, A. 1992. *Farming Systems and Integrated Pest Management.* Malhotra Publishing House, New Delhi. pp:332.

Verma, L. R. and Sharma, R. C. 1999. *Diseases of Horticultural Crops: Fruits,* Indus Publishing Co., New Delhi, p.724.

Weller, D. M. 1988. Biological control of soil borne plant pathogens in the rhizosphere with bacteria. *Ann. Rev. Phytopath.* 26: 379-407.

• • •